Science at the Borders

PUBLISHING FOR THE WORLD
125 Years
THE JOHNS HOPKINS UNIVERSITY PRESS

SCIENCE AT THE BORDERS

*Immigrant Medical Inspection and
the Shaping of the Modern Industrial Labor Force*

Amy L. Fairchild

The Johns Hopkins University Press
BALTIMORE AND LONDON

The Johns Hopkins University Press
2715 North Charles Street
Baltimore, Maryland 21218-4363
www.press.jhu.edu

Library of Congress Cataloging-in-Publication Data

Fairchild, Amy L.
Science at the borders : immigrant medical inspection and the
shaping of the modern industrial labor force / Amy L. Fairchild.
p. cm.
Includes bibliographical references and index.
ISBN 0-8018-7080-1
1. Immigrants—Medical examinations—United States.   2. United
States. Public Health Service.   3. United States—Emigration and
immigration.   4. Labor policy—United States.   I. Title.
RA4448.5.I44 F35 2003
362.1'086'91—dc21   2002006242

A catalog record for this book is available from the British Library.

*To my teachers
(including Max, who may not know to
count himself among them)*

CONTENTS

ACKNOWLEDGMENTS

WORK ON THIS PROJECT was supported by a dissertation improvement grant from the National Science Foundation and a dissertation fellowship from the National Endowment for the Humanities. Many individual archivists and scholars also supported me and provided more leads than I was ever able to pursue. At the National Archives in College Park, Maryland, Washington, D.C., and San Bruno, California, my thanks go to Marjorie Ciarlante, Aloha South, and Neil Thompson (and to Rosemary for the back massages generously provided to all those who spent hours bent over allergy-aggravating papers). I thank the archivists at the Louisiana State Museum; Joel Wurl at the Immigration History Research Center in Minnesota; and John Parascandola and Marian Smith, the historians at the Public Health Service and Immigration and Naturalization Service. At Ellis Island, I am indebted to Paul Sigrist and Jeffrey Dosik. Ed Morman, in addition to providing research support on this and so many other projects while he was at the New York Academy of Medicine, also offered a critical reading of selected chapters. Thanks, too, go to Erika Lee for directing me to a wealth of materials regarding Chinese immigration in San Francisco. Finally, I thank the docent who took me on an exhilarating, illicit tour of the deteriorating immigrant hospital at Angel Island.

I am grateful to all those individuals who read early chapters, helped to wade through materials, helped to reconcile quantitative and qualitative evidence, and offered unwavering support and good advice. Foremost among these is Gerard Carrino, my economist husband who was my chief quantitative guide, copy editor, and fan. Ronald Bayer, who was singularly responsible for my decision to do doctoral work at Columbia University, is an editor, scholar, teacher, colleague, and, most important, friend beyond compare. It was he who first very gently told me that I was missing the forest for all of the statistical trees. Both Ron and Nancy Leys Stepan fostered my interest in history and public health in unprecedented ways. Both have altered my thinking and helped me shape my expectations and career plans. Nancy gave me her unique, priceless scholarly and emotional support from start to finish. The combination of Ron and Nancy made it possible for me to merge my interests in history and contemporary policy.

A wider community of scholars also deserves thanks. Julia Rodriguez, Charles Forcey, Michael Sappol, and Barron Lerner gave invaluable material advice on chapters, and great insight and encouragement throughout the process of writing and defending the dissertation. Judy Whang both pushed me and tolerated my presence on her couch for a considerable portion of the many months I spent at the National Archives. David Rothman first mentored me in my graduate work in history; his critical eye was essential in pushing me to ask hard questions of the primary material. Members of the dissertation reading group in the History Department at Columbia read drafts of various chapters. In particular, I am indebted to the comments and critique of Mae Ngai, Rebecca McLennan, Jeff Sklansky, Adam Rothman, Michael Berkowitz, Elizabeth Blackmar, and Anders Stephanson. Ira Rezak arranged for me to present my work on this topic at a seminar at the State University of New York, Stony Brook, where I received excellent comments. Charles Greifenstein arranged for me to present my work at the College of Physicians in Philadelphia, where the observations of Charles Rosenberg proved invaluable. Jay Dobkin offered the insights of a physician with a broad vision of public health. Jackie Wehmueller at the Johns Hopkins University Press helped me in the final round of editing and boosted me up with her marvelous enthusiasm.

While these scholars have served as the teachers and colleagues most directly touching this work, others also had a profound impact. We get so little opportunity to thank our early educators that it seems the least I can do to thank all of them generally and to single out the most remarkable. Whether they remember me or not, those rare teachers in my life whom I remember with special fondness and deep respect include Audrey Whitworth, Geraldine Kidwell, David McCutchan, Mary McFarland, Gary Biggers, Myron Gutmann, Betty Sue Flowers, and Kenneth Foote.

This book, and my particular interests, would never have taken the shape they did had I not spent several years working at the New York State Department of Health, AIDS Institute. There, Eileen Tynan always expressed a keen interest in my project and helped to facilitate it. Sonja Noring not only showed great enthusiasm about the project from the start, but, with her sharp editorial skills, gave an edge to early versions.

I thank Cheryl Healton, my first department chair, for making possible my years at the AIDS Institute and also for creating my dream job in which I could work closely with David Rosner and Ronald Bayer in an innovative new Center for the History and Ethics of Public Health in the Department of Sociomedical Sciences at the Joseph L. Mailman School of Public Health. I thank Richard Parker, our new chair, for recognizing the importance of history to public health and helping this center to flourish. Although I had taken courses with David Rosner while he was at Baruch College and CUNY, and knew that it would be tremendously fun to work with him, I could never have imagined and certainly would never have dared ask for the kind of backing he has provided me as a junior colleague. He has been a steadfast supporter and unparalleled promoter. With great care and considerable time and effort he helped me to bring out the argument in this book. The intellectual debt I owe to him in this book is tremendous, the personal debt greater still.

My job at Columbia also brought Martina Lynch into my life. As a colleague, Martina worked to help free my time. As a dear friend, she offered bottomless interest and enthusiasm. As a tireless helper, she carted almost as many books back and forth from the library as our doctoral students Elizabeth Robilotti and James Colgrove, and, with them, endured the suspension of library privileges until, very tardily, I paid the inevitable fines. Sheena Morrison, a new doctoral student, missed out on the library fun but was tremendously helpful checking the final tables for this work. Valeri Kiesig, another new arrival, helped with the final task of providing answers to Lys Ann Shore, the superb copy editor. Other scholars, some from the Columbia community, who deserve thanks for their comments on this work in its various stages include Josh Freeman, Matthew Frye Jacobson, Priscilla Wald, Gerald Markowitz, and Gerald Oppenheimer.

Projects like this are never started and never completed without accumulating an enormous debt to family members. John Higham, in writing *Strangers in the Land,* saw himself "addressing the immediate world in which I had grown up."[1] Similarly, Alan Kraut was drawn to the study of immigrants because, "the son of a factory worker and grandson of a tailor, I was two generations removed from . . . [the] streets of that great incubator of new Americans, Manhattan's Lower East Side."[2] While not all historians have brought the

[1] John Higham, *Strangers in the Land: Patterns of American Nativism, 1850–1925* (New York: Atheneum, 1967), 332.
[2] Alan M. Kraut, *Silent Travelers: Germs, Genes, and the Immigrant Menace* (New York: Basic Books, 1994), ix.

same sense of their immigrant heritage to their studies, even lost heritage figures prominently in the stories they tell. For Howard Markel, "a personal dividend of these studies has been the opportunity to learn more about my cultural heritage as a second-generation East European Jewish American. The world of my grandparents seems much clearer and far less distant as a result."[3]

Unlike many historians, I do not bring a strongly felt immigrant heritage to this endeavor. (I will always think of myself as a Texan.) My particular debt, then, is of the emotional and material rather than the lineal sort. My grandparents instilled in me the importance of hard work. My parents, in addition to providing decades of love, assistance, and financial generosity (which seems to have no end), taught me to pursue those truths that were true for me. Although she complained about the archival chill, my sister Mia graciously helped with tedious work in the Texas State Archives. Susan, my other sister, has always offered praise to which I can never live up. My parents-in-law contributed constant encouragement, in addition to the material gifts of a car, a microfilm reader, and a printer.

Sometimes one thank-you is not enough. My husband generously helped me, to his own detriment, and managed to love me throughout. Thank you, Jerry. Thanks, finally, go to Max, who proved to be the best kind of impediment to completing the book in a timelier manner.

---

[3] Howard Markel, *Quarantine! East European Jewish Immigrants and the New York City Epidemics of 1892* (Baltimore: Johns Hopkins University Press, 1997), xiv.

Abbreviations

BSI	Board of Special Inquiry, IS
CGAR	Commissioner General of Immigration, Annual Report
EIOHP	Ellis Island Oral History Project
ICD	International Classification of Diseases
IHRC	Immigration History Research Center, University of Minnesota, St. Paul
IS	Immigration Service
LPC	"likely to become a public charge"
LSA	Louisiana State Archives, New Orleans
NARA	National Archives and Records Administration
NSDMR	number of standard deviations from mean rate
PHS	Public Health Service
RG 85	Record Group 85, Records of the Immigration Service, NARA, Washington, D.C.
RG 90	Record Group 90, Records of the Public Health and Marine Hospital Service, NARA, College Park, MD
SGAR	Surgeon General of the Public Health Service, Annual Report

SCIENCE AT THE BORDERS

IMMIGRATION BY THE NUMBERS

Rethinking the Immigrant Medical Experience

ELLIS ISLAND retains a strong grip on our collective memory as a nation. Indeed, Ellis Island serves as shorthand for the process of entry into the nation, whether it took place on this coast or on another. We think of Galveston as the "Ellis Island of the South" or San Francisco's Angel Island as the "Ellis Island of the West." We think of the early immigrant experience in the United States as unfolding "after Ellis Island." Ellis Island is more pristine now than it ever was in its heyday. Even those of us whose family members did not pass through its doors, under the watchful eyes of medical inspectors and immigration officers, feel the lingering tension upon setting foot on the island made from the soil of the city's subway tunnels, from the very sweat of immigrants. Although less frequented, other ports of immigration evoke the same collective memory. At Angel Island, the echoes of foreign voices seem to reverberate throughout the hospital and dormitory rooms, and the fear and frustration of detainment still settle in the air, stirred in the dust and falling tiles. In both places the past presses heavily against the present. The inexpressible tension and anxiety that the immigrants experienced reaches across time, stretching from mind to mind, filling in the void that even words of remembrance can never quite fill.

Among the immigrants' many apprehensions, the fear of rejection loomed foremost in the minds of most as they undertook passage from abroad. The immigrant medical examination, in particular, forged memories that would last a lifetime. Immigrants were forewarned of the experience overseas through immigrant aid guides, steamship brochures, and the initial steamship company medical and quarantine examinations needed to secure passage to America. The healthy and unhealthy alike anticipated with enormous trepidation the final U.S. Public Health Service (PHS) medical exam that would take place on the threshold to this nation.[1] When a PHS officer formally diagnosed an immigrant with a disease or defect, throwing his or her admissibility into question, that individual was considered "medically certified" (for an example of a typical medical certificate, see figure I-1).

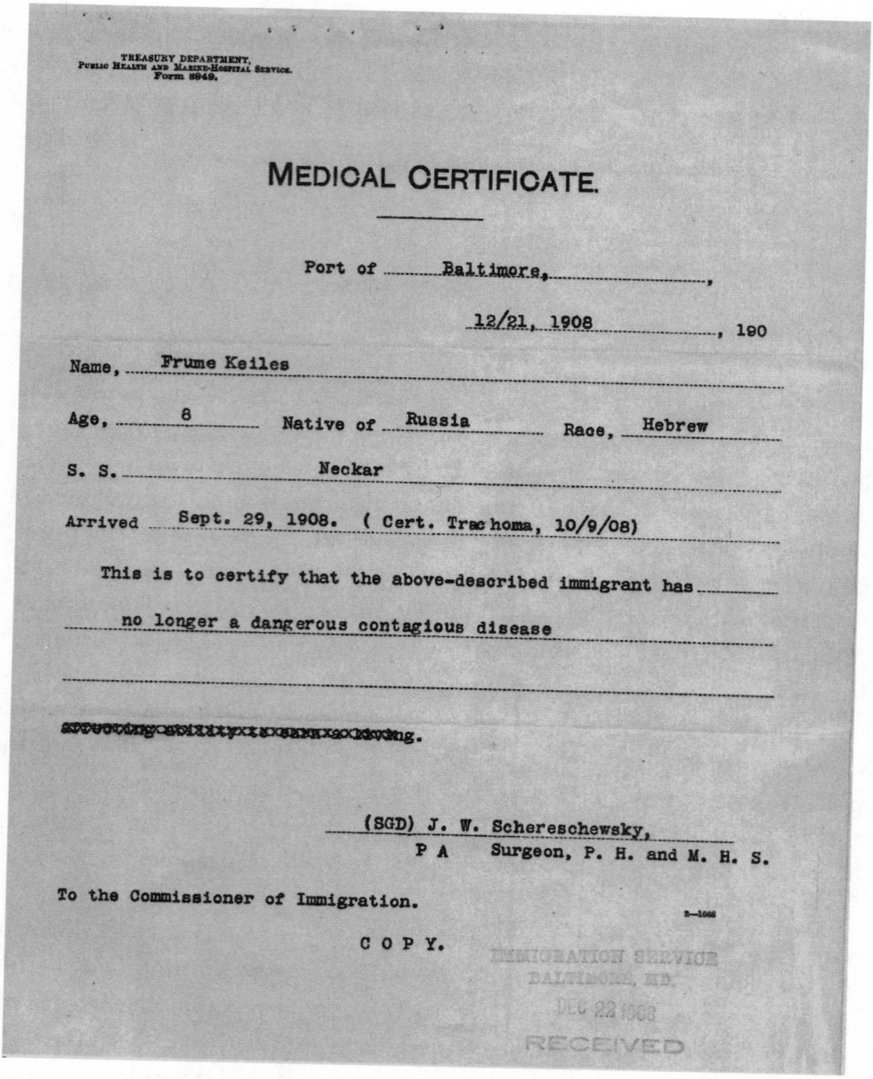

TREASURY DEPARTMENT,
PUBLIC HEALTH AND MARINE-HOSPITAL SERVICE.
Form 1949.

MEDICAL CERTIFICATE.

Port of Baltimore,

12/21, 1908, 190

Name, Frume Keiles

Age, 8 Native of Russia Race, Hebrew

S. S. Neckar

Arrived Sept. 29, 1908. (Cert. Trachoma, 10/9/08)

This is to certify that the above-described immigrant has no longer a dangerous contagious disease

[illegible]ng.

(SGD) J. W. Schereschewsky,
P A Surgeon, P. H. and M. H. S.

To the Commissioner of Immigration.

2—1066

C O P Y.

IMMIGRATION SERVICE
BALTIMORE, MD.
DEC 22 1908
RECEIVED

FIGURE I-1. Typical immigrant medical certificate. (RG 90, NARA)

Emma Greiner, who entered the United States at Ellis Island in 1925 as a young girl, remembered developing a troubling eye infection on the ocean voyage. "Here was a tremendous psychological experience," she noted, recalling her escalating anxiety regarding the worsening infection and the impending eye exam awaiting her at Ellis Island. "I still had this terrible, terrible worry about my eyes, that we were going to Ellis Island and that there they would make the decision whether I would be accepted or whether I would be rejected, which is something [she laughs] for an eleven-year-old girl."[2] The daughter of an immigrant, Marie Jastrow explained that all immigrants dreaded the medical and immigration exams: "My father crossed the Atlantic and landed on Ellis Island at the beginning of October 1905, well aware of the ordeal that awaited him. He had heard stories. The threat of deportation haunted every immigrant; at least every steerage immigrant. . . . It was the towering spectre in every immigrant's mind. Complete, convulsive fear of being judged unfit for America created waves of terror in the people who sat waiting for inspection."[3]

All those old enough to comprehend the immigrant medical exam feared it, but for those caught in the medical net the experience was nightmarish. Doukenie Papandreos, a fifteen-year-old Greek immigrant, anxiously watched those in line before her as they underwent the medical exam on Ellis Island in 1919. She imagined herself and family as "lambs to the slaughter" as they slowly approached the somber doctor in his tall military boots and uniform, who turned aside one or two out of every ten immigrants for more intensive medical examination. Doukenie found herself among those "turned off the line." During the subsequent days while she was held at Ellis Island, she recalled, "I couldn't enjoy nothing. I was afraid they were going to send me back. And I was dreaming that if they try to send me back, I'm going to fall into the river and die. I couldn't go back."[4]

Although the accounts of Greiner and Papandreos represent the medical fears and experiences of many third-class or steerage passengers arriving at ports in the eastern United States, immigrants faced more considerable medical obstacles to entry at the nation's Pacific coast and Mexican border immigration stations. At Texas border stations such as Brownsville and El Paso, PHS medical inspectors stripped, showered, and disinfected large groups of immigrants, searched them for lice, and performed physical examinations in a regime eerily akin to those used later and for a much different purpose at the concentration camps of Nazi Germany. The PHS suspected that many of the European and Asian immigrants crossing at these remote land stations—who seemed to correspond to the second- and third-class steamship "type" of immigrant—had failed entry at one of the nation's seaports.

All second- and third-class Asian immigrants arriving in San Francisco endured a physical exam similar to that conducted along the Mexican border, in addition to laboratory testing for parasitic infection, which required detention at Angel Island for one or more days, which could easily turn into weeks or months, sometimes years. Those infected faced a choice: submit to deportation or pay for lengthy treatment. Those with the cash, and perhaps some without it, typically chose treatment. On the walls of San Francisco's Angel Island detention barracks, Chinese immigrants etched out their hopes, despairs, resentments, and fears.[5]

> When I began reflecting, I became sad and composed a poem.
> It was because my family was poor that I left for the country of the Flowery Flag.
> I only hoped that when I arrived it would be easy to go ashore.
> Who was to know the barbarians would change the regulations?
> They stab the ear to test the blood and in addition they examine the excrement.
> If there is even a shadow of hookworms, one must be transferred to undergo a cure.
> They took several dozen foreign dollars.
> Imprisoned in the hospital, I was miserable with grief and sorrow.
> I do not know when I will be cured.[6]

We cannot know whether the unknown Chinese immigrant who lamented on the walls of Angel Island found his cure and entered this country of the "Flowery Flag." Asian immigrants were more likely than other groups to arrive in the United States, only to be turned back, but in sharp contrast to the fears immigrants expressed in their narrative accounts of passage to America, statistics testify that the immigrant medical exam excluded relatively few: most immigrants duly entered the nation—even those arriving along the Pacific coast and Mexican border.

The PHS inspected more than 25 million arriving immigrants from 1891 to 1930, but it issued only some 700,000 medical certificates signaling disease or defect. The Immigration Service (IS), which made final decisions regarding admission, denied entry to roughly 79,000 of those whom the PHS medically certified. On average, 4.4 percent of all immigrants were certified annually from 1909 to 1930, peaking at more than 8.0 percent in 1918 and 1919 but only about 11 percent were deported. Thus, the deportation rate for medical causes never exceeded 1 percent (fig. I-2). Disease, moreover, was never the most important cause of immigrant rejections from the United States (figs. I-3

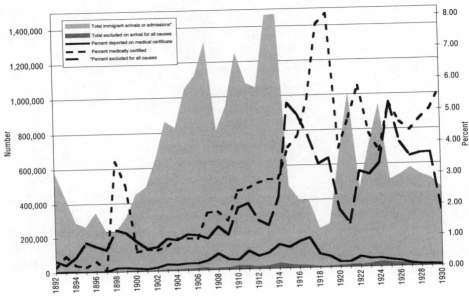

FIGURE I-2. Total immigration to, and total exclusion from, the United States, 1892–1930. (SGAR and CGAR, 1892–1930)

and I-4). Immigrants were rejected mainly for causes related to economic "dependency," independent of any medical factor, most often as "likely to become a public charge" (LPC).[7]

FRAMEWORKS OF ANALYSIS

Despite the statistics, Dr. Alfred Reed, an outspoken PHS officer strongly associated with Ellis Island, boldly declared that the immigrant exam conducted on the waters off Manhattan was "the most important feature of the medical sieve spread to sift out the physically and mentally defective."[8] Although touted—and, indeed, initially intended—as such, the immigrant medical exam was a far more flexible tool, for immigration represented a point of tension among a variety of interests in the United States in the Progressive era: it operated at the intersection of contradictory needs, intentions, impulses, and interests in a nation in flux. Science had to find its place on a spectrum of needs and demands. The exam, which literally touched millions of immigrants, served a much more complex social function than simple exclusion on the basis of disease, class, or race.

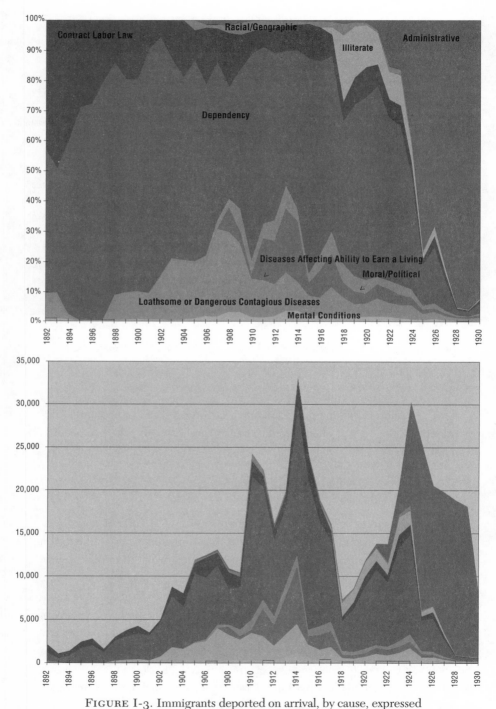

FIGURE I-3. Immigrants deported on arrival, by cause, expressed
as a percentage of total deported on arrival, 1892–1930.
FIGURE I-4. Number of immigrants deported on arrival by cause, 1892–1930
(for categories indicated by degrees of shading, see fig. I-3). (CGAR, 1892–1930)

In the period of mass immigration to the United States, there were various ways to protect the nation. Industrialists, plagued by a constricted labor market and rebellious workers, sought to create a disciplined work force; the social reform movement strove to inculcate immigrants with American values, ideas, and behaviors; the nativists and the labor movement, to a greater or lesser extent at different moments, strove to exclude immigrants, particularly the new immigrants, altogether. These various impulses could potentially push and pull the immigrant medical exam in different directions.

The immigrant medical exam, rather than losing coherence or meaning in the face of so many competing pressures, became aligned with the forces demanding fit industrial citizens, usually to the benefit of industry but sometimes to the benefit of labor. The PHS was originally founded in 1798 as the U.S. Marine Hospital Service to care for naval officers, seamen, and marines, based on the belief that sick sailors placed a heavy and unmanageable burden on the nation's public hospitals.[9] The original mission of the medical officers was not merely to treat a particular group of individuals, but to protect the economic health of the nation they served.

First and foremost, then, the immigrant medical inspection became an important part of a subtle yet pervasive nationwide endeavor to discipline the labor force. In sharp contrast to the common image of immigrant medical inspection, the procedure that immigrants endured at Ellis Island and many other stations around the nation was, in general, part of a process of *inclusion*. Between 1891 and 1930, while federal immigration law sent home only half a million immigrants on the grounds of diseases or defects, it brought *all* the arriving immigrants—25 million of them—under the scrutiny of the PHS. These immigrants represented the nation's industrial work force, and it was imperative that they work efficiently, obediently, unflaggingly. The assembly line of flesh and bone developed to defend the nation from diseased immigrants served as the inaugural event in the life of the new working class—one that would impress upon each immigrant the national hierarchy and his or her low place in it. Only when groups of immigrants failed to conform to societal expectations about the fit industrial worker did the immigrant medical exam serve to exclude those groups at the nation's borders.

This book describes how science answered the nation's competing claims regarding immigrants at different moments and in different regions of the country. There was no single regime for immigrant medical inspection. Yet there was a consistent logic to the ways that the system cut across race and

class. There was sense to the manner in which it responded to competing social, cultural, economic, and industrial pressures that created two distinct imperatives: to discipline the industrial working class, and to exclude immigrant
groups that did not make a decidedly positive contribution to the industrial
work force.

RACE AND CLASS

The prospect of employment in a rapidly expanding industrial economy
brought millions of immigrants from Europe, Asia, and Latin America to the
brink of America between 1891 and 1930—a period of intense social, economic,
and political anxiety in the United States. By 1900 immigrants from southern
and eastern Europe—Italians, Poles, Russians, Czechs, and Magyars—had surpassed that from the more familiar northern and western European countries. These new groups of immigrants differed from "native-born" Americans
in appearance, cuisine, language, and worship; to many, they seemed dirty, il-
literate, and poverty-stricken. Growing social and economic pressures posed
by industrialization, sprawling urban cities, violent labor uprisings, economic
depression, fears of middle-class "race suicide," the changing structure of
American authority, and a fractured sense of American unity all fueled growing nativist sentiment in the United States.[10]

The emphasis on nativism in the literature of immigration has enabled historians to throw into relief the meaning of race and ethnicity in American society. For example, historians have explored how efforts to prevent the spread
of disease at the nation's borders served as a vehicle for attempting to restrict
immigration along ethnic or racial lines.[11] Questions of social class and labor
have, of course, always been part of the background in discussing immigrants,
disease, and restriction, for historians have understood that the national response to disease and immigrants was driven by both social and economic factors. The new immigrants, after all, were poor. These immigrants were largely
unskilled and unpropertied, and they took a place at the bottom of the nation's
social hierarchy. To talk about race necessarily involved the intimately related
background issues of class and labor. Social historians have been so taken by
the exclusionary languages and logics of Progressive-era culture, particularly
in the context of immigration, that race has become the primary framework
for understanding immigration policy in general and the immigrant medical
exam in particular.

I was reluctant to discuss my treatment of race and class explicitly at the outset of this book lest it be read as prioritizing one over the other as the category that helps to explain the immigrant medical exam and its meaning. Worse, I feared reifying either category and unintentionally taking away some of the murkiness of these concepts. The historian Ira Berlin maintains that the notion that race is a social construct "has won few practical battles," in large measure due to a failure "to demonstrate how race is continually redefined, who does the defining, and why."[12] Race, in the hands of the PHS and IS, had distinct meanings, which did not entirely overlap with each other, and which also differed from other understandings of race, such as that used by the eugenicists, which has been widely used to explain the American reaction to immigration in the Progressive era. Different peoples were marked as racially distinct in different ways, using different criteria.

Understandings of race varied from region to region as well as over time, shifting with new laws, migration patterns, economic trends, and relations with foreign governments. Race and class could align in unexpected fashion. The language of race could infuse the language of class. For example, class could imply fixed, biological characteristics without explicitly overlapping with any immigrant group that was otherwise defined in terms of race, language, or nationality.[13] What is clear is that both race and class, in all their complexity, affected how the PHS and other stakeholders viewed particular diseases. Setting aside difficult terms such as race or class in quotation marks has become a standard and well-warranted means of acknowledging these complexities. I felt, however, that I could not adopt a consistent or coherent precedent regarding the use of quotation marks for problematic social and historical constructions. The term *race*, like the term *class*, remains fluid in this book.

CITIZENSHIP

Questions of racial exclusion were intimately linked to a broad national effort to define citizenship after the Civil War.[14] The notion of citizenship, however, remained in flux, taking different forms and emphasizing different elements at particular moments. Race was always implicated in notions of citizenship, regardless of how it was formulated. Understanding how race was constructed in America is critical because racial categories implied an ordering of society—that is, determinations about who belonged where in the social hierarchy and who was both capable and deserving of what democratic privileges and rewards.[15] The notion not just of race but of *whiteness* represented the foundation on which legal definitions of citizenship were built in the Constitution and then reevaluated following the Civil War.[16]

It would be easy to read the history of immigration restriction as a story of racial exclusion that turned on some concept of whiteness. In the arena of immigration, after all, racial exclusion figured early on and quite dramatically in the Chinese Exclusion Act of 1882. The nation would continue to limit Asian immigration to the United States throughout the Progressive era and into the 1920s: The 1907 Gentlemen's Agreement negotiated the voluntary restriction of Japanese and Korean immigration to the United States; the 1917 Immigration Restriction Act created the Asiatic Barred Zone, which denied admission to immigrant laborers from several Asian countries;[17] and the 1924 Immigration Restriction Act excluded all those ineligible for citizenship, which achieved restriction of Japanese immigrants and broadened the basis for restriction of those from China.[18]

It is important to understand, however, that the term *whiteness* had little place within the language of the PHS or the IS in this period, and only with great care can we make inferences about how their ranking and ordering of immigrant groups may have spoken to an underlying idea of whiteness or some such broad racial division.[19] While it is clear that both medical and lay immigration officials saw the world in terms of a multitude of unequal races, they placed little emphasis on drawing fine, racial distinctions between peoples with the goal of exclusion during the period covered in this study. Rather, the overwhelming impulse was to absorb immigrants into the laboring body.

The roots for this emphasis are found in the post–Civil War period, which was initially characterized by a profoundly *inclusionary* impulse. The Fourteenth Amendment, ratified in 1868, took decisions about citizenship out of the hands of the states, and made citizenship and equal protection under the law a constitutional birthright. In effect, it rejected what the historian Eric Foner has termed a "racialized definition of freedom."[20] The Naturalization Law of 1870, accordingly, affirmed the right of persons of African descent to vote. However, signaling the beginnings of the splintering of this broad and egalitarian notion of citizenship—typically read as inclusion in political or civic life—the Naturalization Law, which explicitly limited citizenship to free "white" men, also denied citizenship to first-generation Asian immigrants. The fissure in an unrestricted, democratic vision of citizenship deepened in the 1880s as the rights of citizens to participate in American public life and to enjoy equal protection under the law began to fracture along racial lines. For example, the Supreme Court eviscerated the Fourteenth and Fifteenth Amendments, affirming the doctrine of separate but equal and ultimately granting leeway not only to private organizations but also to the states themselves to abridge the rights of citizens based on race.

Europeans were never excluded from citizenship on the basis of race nor caught in the snare of Jim Crow. But, of course, the restriction of European immigration was the subject of considerable public debate. Significantly, within the arena of immigration, a tension between inclusion and exclusion would remain, for even if certain segments of society wanted to limit immigration, the nation needed cheap labor.[21] Immigration to the United States during the Progressive era represented not only a host of potential citizens, but also a vast influx of laborers who—regardless of whether they became American citizens or eventually returned to their own countries—would participate in and, indeed, help to shape American economic life. Although the overwhelming impulse was to absorb immigrants into the laboring body, the tension between inclusion and exclusion—between democracy and capitalism—would contribute to the emergence of two interrelated conceptions of citizenship that differentially weighed questions of race and labor—one political and one industrial.[22]

The impulse to exclude Asian immigrants amplified the two distinct varieties of citizenship. Protecting a racialized notion of political citizenship, Congress deemed the Chinese unfit for democratic self-rule. The 1882 exclusion act not only barred Chinese laborers from entry into the United States, but also barred those Chinese who did enter from achieving citizenship. In 1884 the Democrats would insist in their campaign handbook that "[the Chinese] come to this country not to partake in the responsibilities of citizenship; they come here with no love of our institutions; they do not hold intercourse with the people of the United States except for gain." The U.S. government, the handbook insisted, is "designed for a higher class of intelligence and virtue than that which belongs to the people who are migrating to this country from Asia in such swarms."[23]

But Congress was also responding to the threat that the Chinese posed to industrial civilization. The 1882 exclusion act did not bar the entry of all Chinese immigrants. It targeted laborers only. The movement to exclude the Chinese, then, was not rooted exclusively in questions of fitness to participate in American public life, but had firm roots in American industrial life. For example, in 1881 San Francisco's Assembly of Trades and Labor Unions studied the presence of the Chinese in California industries in response to the perception that Chinese laborers posed serious competition for white workers' jobs and depressed their wages. Samuel Gompers of the American Federation of Labor charged in 1893 that the Chinese degraded American workers by undercutting their wages and standard of living.[24] How could the American worker compete with the Chinese worker, who would work for less, eat offal and refuse, and live in filth and squalor?[25]

The exclusion of the Chinese in some sense set the terms for the larger debates over immigration in the Progressive era, establishing two models interconnected by questions of race and labor: one of fitness for civic participation, and one of fitness for industrial participation. The movement to restrict immigration from southern and eastern Europe prioritized questions of fitness for self-government and emphasized the racial inferiority—inherent genetic and intellectual inferiority—of the new immigrant streams.[26] In the 1890s Senator Henry Cabot Lodge and his Immigration Restriction League associated the literacy test with protection of American character and citizenship.[27] The literacy test was specifically seen as a means of restricting the entry of southern and eastern Europeans after Francis Amasa Walker, an economist at the Massachusetts Institute of Technology who headed the U.S. Census of 1870, sounded an alarm over the declining birth rate among native-born Americans and shifting sources of immigration. The literacy test promised to restrict the entry of "beaten men from beaten races" who possessed "none of the ideas and aptitudes" necessary for democratic self-rule.[28]

Critically, however, expansive notions of racial restriction stemming from civic concerns did not find their way into actual immigration legislation until well after the turn of the twentieth century.[29] Few groups other than the Chinese found themselves singled out as unsuitable for both civic and industrial citizenship.

Federal law did not forcefully take up the issue of the immigrant as a civic participant until 1917, when Congress enacted the literacy test over President Woodrow Wilson's veto. Earlier, only the immigration law of 1903 showed the first glimmers of concern for civic citizenship; carried on a small wave of anti-radical sentiment following the assassination of President William McKinley by Leon Czolgosz, the law mandated the exclusion of anarchists as clearly unsuited for participation in a democratic society. Although the new immigration law was associated with undesirable civic orientations such as anarchism, the exclusion of anarchists was not intended as an overt means of racial restriction. Rather, with its entry into the control of immigration in 1882, Congress remained legislatively focused on the immigrant as industrial participant, at least through immigration restriction in the 1920s.[30]

Until the 1880s the states exercised exclusive authority over immigration.[31] Coastal states, such as New York, often imposed a fee on shipping lines to finance the support of destitute immigrants. In 1876, when the Supreme Court declared such fees an unconstitutional infringement on foreign commerce, the onus fell on the individual states to support needy immigrants. Led by New York, the eastern states immediately pressed Congress for federal control of immigration. The movement for passage of the law sought to stifle the boss system that was foiling efforts at municipal reform. It was also driven by the philanthropic impulses of eastern relief agencies, which sought to prevent corporate exploitation of poor immigrants.[32] Congress responded in 1882 by imposing a head tax on each arriving immigrant and prohibiting the entrance of convicts, lunatics, idiots, and persons likely to become a public charge, leaving to the states the enforcement of exclusion.[33] Thus, while the law did stem from growing concerns about immigrants' civic responsibility, it tempered this concern with recognition that immigrants were members of an industrial society. But only fears about dependency were ultimately expressed within the law, for the perceived threat that dependent immigrants would become the responsibility of the individual states garnered support for some measure of immigration control. As the immigration historian John Higham explains, the measure was not intended to *restrict* immigration, but rather to *control* it by leveling a head tax and preventing the entry of those who could not support themselves.[34]

Concerns about dependency were consistent with notions of industrial citizenship within an economic model of freedom in a newly emerging bureaucratic nation-state in which men of science were believed to run the government in an efficient, expert fashion, preferable to the power of a political machine. Beginning in the 1880s, corporations gained recognition as "persons" sharing in constitutional rights in a context where the courts increasingly reduced rights of citizenship to "unfettered liberty of contract."[35] The problem of citizenship, conceived as a problem of race and slavery in the wake of the Civil War, was gradually winnowed down to "little more than the denial of the laborer's right to choose his livelihood and bargain for compensation."

Eric Foner thus views the 1880s as the beginning of a long era in which "the courts consistently viewed regulation of business enterprise—especially interventions in contractual labor relations such as laws establishing maximum hours of work and safe working conditions—as a paternalistic insult to free labor, a throwback to the thinking characteristic of slavery." Even labor and its allies tended to view citizenship within an industrial framework, though they saw it in terms of rights and protections. Industrial citizenship meant economic freedom and self-government in the workplace for all wage earners. Failure to achieve such economic freedom amounted to the inverse of industrial citizenship—"industrial slavery." The judicial branch of the federal government viewed industrial citizenship from the corporate side of the coin. According to Foner, "between 1880 and 1931, by one count, nearly two thousand injunctions were issued prohibiting strikes and labor boycotts."[36] Like industry and the courts, the bureaucratic state tended to view industrial citizenship in terms of the obligations of, and expectations for, the worker.[37] And the worker was the immigrant.

DISCIPLINE, EXCLUSION, AND POWER

With the immigration law of 1891 the federal government took complete authority over immigration and created the machinery for federal officials to inspect and exclude immigrants. The law required medical officers of the PHS to inspect and issue a medical certificate to all immigrants suffering from a "loathsome or a dangerous contagious disease." By 1903 the PHS began to formally classify diseases in accordance with the law—and with national industrial expectations. In addition to "loathsome and dangerous" contagious diseases, which it labeled Class A conditions mandating exclusion, the PHS created a new category of Class B diseases or conditions—those rendering the immigrant "likely to become a public charge" (LPC).[38] The unfortunate immigrant suffering from hernia, heart disease, or deformities that rendered him unsuitable for industrial living was now vulnerable to medical certification and exclusion. Both Class A and Class B designations took on economic meaning in the hands of the PHS. This new way of categorizing laborers, of categorizing both inability and ability to perform in the work force, elaborated on the formulation of what made a good industrial citizen: one who would remain healthy, be a useful worker, and not become dependent on the charity of the nation.

Although the PHS expressed an interest in excluding immigrants unfit for laboring in an industrial society, the exam did not function in this fashion except on the West Coast and the Mexican border. This was because adapting the industrial worker to the line was of such paramount social and economic importance. The PHS gained authority over immigrant medical inspection at a time when industry—which had been ambivalent regarding the merits of immigration restriction versus control—became a staunch and lasting opponent of immigration restriction. During the Progressive era, according to Higham, industry "perfected a system for mastering" the cheap, unskilled immigrant labor force.[39] The immigration law, through the mechanism of the medical inspection, ultimately provided another tool for such mastery.

Fears of dependency were connected intimately to fears about bodies—specifically, about unhealthy bodies—and bodies were needed to fuel industry. Not all of these bodies would become naturalized citizens. Many immigrants came to the United States only to work for a while before returning to their native countries. Like industry's scientific managers, federal immigration officials would concern themselves primarily with the so-called "dumb ox" variety of immigrant favored by industry: those with strong backs but lacking the capacity to manage their own labor.[40] Within a new context of assembly-line production and industrial management, the laborer became an expendable cog in a vast machine.

The immigrant medical inspection represented a new technique for discipline in the new social and economic order, signaling a transformation in the nature of discipline and power, from corrective to preventive, from violent to normative. When historians have focused on the disciplinary function of public health, emphasis has traditionally been on efforts to correct *deviant* behavior.[41] The aim of discipline in the industrial era, however, was neither correction nor amelioration; it was a normative expression of power intended not simply to prevent deviant behavior but to promote adoption of core industrial values, to create a cadre of good industrial citizens.[42]

It is this notion of power as a productive process that Michel Foucault illuminates. Foucault asserts that whether we are talking about a punitive or a normative political economy of power, "it is always the body that is at issue—the body and its forces, their utility and their docility, their distribution and their submission." He insists that "power relations have an immediate hold upon [the body]." "They invest it, mark it, train it, torture it, force it to carry out tasks, to perform ceremonies, to emit signs." The body's economic use contributes to its political investment:

It is largely as a force of production that the body is invested with relations of power and domination; but, on the other hand, its constitution as labour power is possible only if it is caught up in a system of subjection . . . ; the body becomes a useful force only if it is both a productive body and a subjected body. This subjection is not only obtained by the instruments of violence or ideology; . . . it may be calculated, organized, technically thought out; it may be subtle, make use neither of weapons nor of terror and yet remain of a physical order.[43]

Power in the context of instilling social norms and values retains both productive and coercive elements, at least in the context of the United States in the Progressive era. It is dependent both on a physical ordering of bodies, creating hierarchies of belonging and worth, and on a very practical, real-world need to control bodies, to make them perform as demanded and expected. Finally, it is dependent on excluding bodies—that is, on containing the diseased immigrant by sending him or her home. The immigrant body "was entering a machinery of power that explores it, breaks it down and rearranges it."[44] Power at America's gates, then, involved conveying a set of expectations regarding what it meant to be, not necessarily a good citizen, but a good *industrial* citizen. On the nation's factory floors, power required removing "the manager's brain" from "under the workman's cap."[45] This amounted to eliminating the worker's self-discipline.

COMPLEMENTARY STORIES

In this book, then, I tell a story of science and power—of the power of an industrial mindset to penetrate science and make the immigrant medical exam into a tool for defining and shaping the nation's laboring classes. I approach this story from two angles. In Part I, I tell a story of immigration and power that is national in scope. It is a story of language, authority, and large numbers, focusing on the broad meaning of medical categorization and the intent behind the medical examination of millions of arriving immigrants. It is a story not of exclusion, but of inclusion.

Chapters 1 and 2 elaborate on the main thesis of this book: that the immigrant medical examination was shaped by an industrial imperative to discipline the laboring force in accordance with industrial expectations. These chapters lay out an argument about the expression of power inherent in the different forms of examination, both medical and nonmedical, that the immigrant underwent upon arrival. Accordingly, the emphasis falls on those immigrants inspected, instructed, and admitted in a series of public examination spectacles, rather than on those excluded. Chapter 3 focuses on the intent and form of the immigrant medical examination, exploring how the ideology of industry and the imperative to discipline the laboring body penetrated it, giving it a socializing rather than an exclusionary function. Medical authority, within a disciplinary framework, rested largely on the absence of scientific technology. The PHS was able to prize broad professional judgment over scientific technology. As we shall see, the microscope and the laboratory became increasingly important to the immigrant medical exam in the years preceding World War I, but only to the extent that the nation became more interested in excluding rather than receiving immigrants.

Part II shifts the emphasis from large numbers to small numbers, from discipline to exclusion, looking closely at those factors that enabled the PHS and, to some extent, the IS to attempt to use the immigrant medical exam for broader restrictive purposes. Chapter 4 discusses regional variations in the medical exam. Because each region in the United States expressed unique concerns about immigration and the industrial economy, there were sometimes notable regional differences in the practice and outcome of immigrant medical inspection. A variety of factors springing from deep regional concerns about labor, race, class, and even civilization figured into the immigrant medical examination, which became the appropriate medium to address key regional needs and prejudices in social and scientific terms. As these factors came together in each region, a shift took place in the urgency of exclusion.

Chapters 5 and 6 explore in more depth the ways in which race, class, labor, and disease—the determinants of exclusion—combined to create variations in the practice of immigrant medical inspection that resulted in different regional patterns of medical certification and exclusion. Chapter 5 shows that although the value of a new set of scientific technologies, innovations, or understandings was inevitably at issue whenever questions of exclusion were at stake, a convergence of disease, race, and industrial citizenship was required for the PHS to adhere to a mission of exclusion.

In chapter 6 I turn more directly to questions of exclusion by focusing on the fraction of immigrants actually turned away by the PHS and IS. That fraction offers testimony to the physical expression of science and power, and its relation to class and race. The previously neglected immigration statistics—perhaps impersonal but not necessarily insensitive to identity—tell stories of exclusion turning not on raw intellect or civic capacity but on the place an immigrant was likely to fill in the work force. Defining and acting on racial distinctions when inspecting and excluding immigrants underscore how different constituents—particularly science and the state—conceived of race in the modern industrial era.

The PHS, as an ostensibly neutral agent of bacteriology or science, stood in a unique position to make and act on powerful statements about racial differences in the human population, framing different races either as agents of specific diseases or as susceptible to specific diseases. Nevertheless, the PHS confounds historiographical expectations of the role of science in the social construction of race. Despite the opportunity for the Public Health Service to reconceptualize both disease and the diseased in bacteriological terms during a period of intense social, economic, and political anxiety in the United States, it was the Immigration Service, an organization of civil servants, political appointees, and bureaucrats, that took the lead in defining race. Although it incorporated notions of health and fitness, the construction of race used by the IS was a complex mix of cultural, social, and economic understandings and expectations of different immigrant groups.

In chapter 6, then, we find that while there were racial differences in medical certification and medical deportation rates, these differences were limited to a very few groups. There is evidence that the PHS began to draw a line through Europe, certifying some southern and eastern Europeans at higher rates. The IS, however, erased this line by excluding, and at extraordinary rates, only several groups of Asians and Latin Americans—a phenomenon rarely discussed in broad histories of immigration before 1930. Thus, rather than demonstrating clear differences between southern and eastern Europeans and northern and western Europeans, the evidence shows that the racial fault lines in the United States worked primarily to distinguish all Europeans from "coolies" (Chinese, Japanese, Koreans, and "Hindus"), Mexican "peons," and other Latin American immigrants.

The epilogue recounts the saga of decline, looking overseas and exploring immigrant medical inspection in the period after 1924. The Immigration Restriction Act limited the flow of immigration from any one nation to 2 percent of the foreign-born population of that nationality according to the 1890 U.S. Census, marking the end of open immigration and a new national policy of exclusion. The medical inspection was moved to U.S. consular offices in European ports of departure.[46] The year 1924 thus marked the end of the Progressive-era tradition of immigrant medical inspection and triggered a dramatic refiguring of power. Certification and exclusion rates rose tremendously, reflecting a new rigor and philosophy of medical examination. PHS officers no longer examined immigrants in a public display of power or exercised social and medical authority on the line. The PHS relinquished its moral and social authority to define and manage the industrial work force, settling into a role of technical adviser in an industrial context in which corporations adopted new strategies to control labor.

Yet it was not so much the law as industrial, ideological, and cultural forces that combined to spell the demise of the earlier tradition of immigrant medical inspection. Labor turmoil accompanying World War I forced corporate America to regard the worker's thinking and organizational capabilities as a potential threat. The "dumb ox" of the scientific manager's day became a thinking man to be reckoned with. Large corporations experimented with welfare capitalism and, more important, labor segmentation that resulted in new combinations and rotations of races and ethnicities in the workplace. At the same time mass consumer culture began to erode ethnic culture. The new culture deemphasized the need to discipline the immigrant labor force at the borders as the PHS and IS had done in earlier decades. In this new environment, the cultural imperative for the federal officer to categorize, define, and discipline the laboring class came to a rather abrupt end.

This book may rely more heavily on numbers than most, but it is not a quantitative history of immigration. Different views of immigration as a quantifiable phenomenon give the book its structure, but immigration statistics differentiate it in terms not of style or method but rather of perspective. While they do not represent the neutral voice of science, the carefully accumulated immigration statistics persistently, stubbornly suggest that we ask a new set of questions—not about rates and proportions of immigrants excluded or about the stated intent of the immigrant medical exam, but about the complex function of medical examination and the meaning of immigrant diseases in a modern industrial nation, about the nature and determinants of scientific and social power, and finally about class, labor, and industrial citizenship.

PART I

NUMBERS LARGE

*Immigrant Medical Inspection
as an Inclusionary Tool*

CHAPTER ONE

IMMIGRANTS AND THE NEW INDUSTRIAL ECONOMY

Loathsome contagious disease.—A loathsome disease is a disease which excites abhorrence in others by reason of the knowledge of its existence. The term contagious as used in the law shall be regarded as synonymous with communicable. By loathsome contagious disease is meant a loathsome communicable disease, essentially chronic in character.

Dangerous contagious disease.—By a dangerous contagious disease is meant a communicable disease, essentially chronic in character, which may result in the destruction of one of the most important senses or loss of life.

—*Book of Instructions for the Medical Inspection of Aliens,* 1910

ONE OF THE MOST enduring icons of the American immigration experience is the Statue of Liberty, the towering figure who, we imagine, welcomed and embraced the arriving immigrant. Just as moving as the statue itself are the bronzed words of Emma Lazarus adorning the base:

> . . . cries she
> With silent lips. 'Give me your tired, your poor,
> Your huddled masses yearning to breathe free,
> The wretched refuse of your teeming shore.'[1]

Viewed sentimentally, the Lazarus poem suggests a fundamental sympathy for the "tempest-tost" immigrant seeking refuge in a land of opportunity. John Higham explains that Lazarus, moved by the plight of Jews fleeing Russian pogroms in 1881, intended the poem as a "message of succor." He notes that immigrants themselves believed the Statue of Liberty extended a warm welcome. He argues, however, that the statue did not become a figure widely perceived as beckoning immigrants and forging a sense of common heritage until the 1940s. During the peak period of immigration over which it presided, this symbol of "republican stability," Higham asserts, "remained for most citizens an aloof, impersonal symbol, conveying a warning."[2]

But the Lazarus poem, when read most simply, reflects the national decision to admit millions of immigrant laborers. Once we begin to think about immigration legislation in terms of admitting and processing an unprecedented "flood" of bodies to join the industrial working class, the form the immigrant medical examination took and the function it served begin to make sense. The decision to engage in the mass processing of individuals necessarily shaped the organization of inspection procedures, favoring the expression of power as a productive process for disciplining the body for industrial work, rather than as a negative process of exclusion. But because of ambivalence regarding immigrants within the labor force and the nation, the poem reflects the unfolding, complex drama of language, bodies, and authority in industrial-era America.

Immigrants evoked a mix of responses from different sectors of society concerned with the industrial economy. In this chapter I focus on the responses of labor, business, and the federal government. While the forces in favor of exclusion were strong, and did leave their mark on the effort to control immigration at the borders, the imperative to discipline the labor force exerted a stronger influence on the immigrant medical experience.

For organized labor, which by the 1890s was firmly in favor of immigration restriction, immigrants represented not only competition, but also the changing structure of American production.[3] Between the Civil War and World War I, the United States was transformed from a society of artisans to one of the world's industrial powers. Before the 1880s, although the United States was in the early throes of the industrial revolution and work was becoming increasingly mechanized, individual workers and their unions retained control over production. Manufacturers subcontracted work out to unions, which negotiated the pay rate, hours, and daily production quotas. As David Montgomery explains of the iron and steel industry, "both the management of the production process and the craft union of workers rested on the same social basis." Workers, therefore, exerted collective control and self-discipline in the workplace.[4] Although industry sought to wrest control from workers, both sides agreed that the new industrial equipment that increased the speed and efficiency of production lacked one critical component, controlled exclusively by the skilled labor force—"brains."[5]

While skilled workers retained control over production until the late nine-teenth century, common laborers were largely excluded from unions and col-lective bargaining because they were not central to the production process. They might move and haul and clean, thus facilitating production, but they did not produce. As mechanization allowed smaller and smaller crews to produce efficiently, the proportion of laborers in the workforce steadily increased.[6] More and more, the common laborer was drawn from a new peasant work force arriving from southern and eastern Europe. Unions saw the new laborer as living outside the craftsman's ethic of collective behavior: "There were no men invited such as Slavs and 'Tally Annes,' / Hungarians and Chinamen with pigtail cues and fans."[7] This "dangerous class" of unskilled labor was perceived as "inadequately fed, clothed, and housed"; accordingly, it "threatened the health of the community."[8] A new means of production, combined with a new method of industrial management and supervision that increasingly allocated workers to discrete tasks representing only a step in production, opened the door for this "unworthy" labor force to become the engine of America's in-dustrial might after the 1890s.

Labor leaders within the federal government and the unions attacked a newly arriving immigrant work force in physical terms because they perceived them as making an assault on the American laboring body. For example, Ter-rence Vincent Powderly, the nation's first commissioner general of immigra-tion and a former labor leader, identified a space in which trachoma, favus, itch, heart disease, feeblemindedness, blindness, and other communicable and chronic diseases were understood as "contagious." Wrote Powderly, who oversaw the creation of a vast machinery for regulating immigration beginning in 1891:

Just take a day off, if you're willing and able;
Please sit for a time at one side of our table
And listen to the stories, some false and some true,
That are told every day to the Board of Review.

. . . .

One comes as a student (he shows his diploma),
But we can't let him in 'cause he has trachoma;
A well-meaning lady, kind hearted, would have us
Admit those afflicted with itch and with favus.

. . . .

The feeble in body, the feeble in mind,
Come to us, jostling the deaf, dumb, and blind;
Imbecility, lunacy, the criminal, too,—
They'll all come before you at the Board of Review.
They come with hearts valvular, chronic cardiac,
Tubercular lungs, and with curvature back,
Ankylosis, psychosis, and hernia, too,—
They all come before you at the Board of Review.
. . . .
Led to believe that they don't have to please us
They come with all kinds of contagious diseases . . .[9]

The physical danger that immigrants posed was seen as far reaching, threatening not only labor but also American institutions and the industrial economy. Asked Powderly:

Is it right that our country, its asylums, its jails,
Should in years become crowded because this Board fails
To bar out the tainted in body, in mind?
Think of this when you ask us to be overkind!
. . . .
America's claim to the world's leadership
May lose in its strength, it may weaken and slip,
If we give all our time to those we reject
And pay little heed to the kind we select.

In sharp contrast to labor, those with economic power—the nation's industrialists—had little interest in excluding immigrants. Discipline, however, was in the interests of industry. Daniel Nelson argues that new machine processes came to stand at "the heart of the factory, and mechanical innovation." Thus, the use of electrical power to drive machinery "played a major role in the transformation of the manufacturing system." Specifically, technology enhanced "the managers' ability to control the manufacturing process."[10] The influx of unskilled immigrant laborers represented not only new and cheaper labor, but also an opportunity to reassert control over the laboring force and to rethink management techniques. A new cadre of scientific and industrial managers redefined the ability to work in terms of performing segmented tasks.

This put management on a more rational footing. By 1911, Frederick W.

Taylor's "Principles of Scientific Management"—first articulated in the 1890s—had profoundly shaped ideas about how to organize work efficiently. Taylor saw scientific management as a rational alternative to unionization and the consequent division between the worker and manager. It represented a means of undercutting unions and workers' autonomy, for it sought control over the "brains" of production.[11] Taylor stated that his "philosophy of scientific management" takes problems, initiative, and forethought out of "the hands of each individual workman" and "places their solution in the hands of the management."[12]

The new industrial organizers saw the laboring force as "men of mighty thews and sinews under poor control, lacking the brain development, experience or training which would fit them for anything but routine muscular effort."[13] Labor poet Alter Abelson imagined that the shop overseer viewed his workers as brainless, and pictured the worker thinking, "I wish I were a baster, plain, / Without a brain-storm in my brain."[14] A task-oriented approach did not demand intelligence on the part of the worker. Wrote Taylor of pig-iron handling, "This work is so crude and elementary in its nature that the writer firmly believes that it would be possible to train an intelligent gorilla so as to become a more efficient pig-iron handler than any man can be." Managers' knowledge exceeded the mental capacities of the common laborer: "the science of handling pig iron is so great and amounts to so much that it is impossible for the man who is best suited to this type of work to understand the principles of this science, or even to work in accordance with these principles without the aid of a man better educated than he is." Even "the more intelligent mechanics, that is, . . . men who are more capable of generalization" were deemed incapable "either through lack of education or through sufficient mental capacity" of understanding the principles of scientific management.[15] In "The Song of the Boss," Abelson shows us industry's ideal worker:

> The best of hands I have is one
> All deaf and dumb; for naught will stun
> His nerves, nor will he swear or curse
> When I offend, or grudges nurse.[16]

While there is debate about the actual influence of Taylorism and the extent to which the principles of scientific management were applied,[17] critical to my point is the understanding that Taylorism was a response to the technological transformation of work. "More than most prominent men," Nelson explains, "Taylor was a product of his environment." Taylor articulated a core set of industrial values that resonated with widely shared conceptions of the proper place and role of the worker as contrasted to the manager. His rise was part and parcel of a trend toward systematic organization and technical or scientific expertise in economic and political life—what Robert Wiebe classically described as the "search for order." Accordingly, after 1910, Frederick W. Taylor was not merely an engineer who gained wide respect in his field and in industry. Rather, Taylor "became (with Henry Ford and Herbert Hoover) one of a trinity of early-twentieth-century technician-philosophers."[18]

Although not all American manufacturers adopted Taylorism, the early twentieth century saw the "irreversible transformation" of work in the nation's leading growth industries. Montgomery argues, "Skilled workers in large enterprises did not disappear, but most of them ceased to be production workers. Their tasks became ancillary . . . while the actual production was increasingly carried out by specialized operatives." At Bethlehem Steel, for example, while craftsmen served as foremen, 95 percent of the production positions were unskilled, retrained laborers.[19] Mechanization, in particular, helped to undercut the control, autonomy, and decision-making capacity of the worker. In mining, machines and not men cut half of the nation's coal by 1913. In the 1880s one Kentucky miner, who intuitively grasped the implications of Taylorism and mechanization, explained, "Anyone with a weak head and strong back can load machine coal. But a man has to think and study every day like you was studying a book if he is going to get the best of the coal when he uses only a pick."[20]

Within this new industrial ideology, defective, worn-out cogs could be discarded and easily replaced. Indeed, easy replacement was essential. Paul Miceli penned:

> Here in the land of far famed liberty
> Men are treated as part of a machine;
> Hired and fired without necessity
> According to set rule, and set routine.
> By younger men the old were soon replaced,
> Because they had outlived their usefulness;
> Cast off like some worn part in discard placed,
> Without regard to those it brought distress.[21]

Easy replacement was not only essential but also inevitable. In his poem "In the Sweat-Shop," Morris Rosenfeld, who was born in Poland in 1862 and worked in New York City sweatshops, found no answer to the question, "Pray, how long will the weak one drive the bloody wheel?" But he was certain of one thing: "when the work will have killed him another will be sitting in his place and sewing."[22]

The worker himself, accordingly, began to fill a role analogous to machines. One manager bragged of a prized employee, "She is a sure machine." A shirt worker in 1903 described the nature of the "1000 souls" engaged in factory work as "purely mechanical," producing "results as nearly as possible identical to one another, and all to the machine itself."[23] The poets, however, captured the grim reality of factory life most hauntingly:

> The machines in the shop roar so wildly
> that often I forget in the roar that I am;
> I am lost in the terrible tumult,
> My ego disappears, I am a machine.
> I work, and work, and work without end;
> I am busy, and busy, and busy at all times.
> For what? And for whom? I know not, I ask not!
> How should a machine ever come to think?[24]

Even for workers in factories whose structure continued to bear little resemblance to the scientifically managed organization, new modes of production shaped the nature of work and the ethos of scientific management affected the milieu.[25] Thus, although Rose Cohen's father worked as a tailor doing piecework, and not on a modern assembly line, the ethic was the same. After only a few weeks in America watching her father's routine, Ruth asked, "Father, does everybody in America live like this? Go to work early, come home late, eat and go to sleep? And the next day again work, eat, and sleep? Will I have to do that too? Always?" She soon discovered the grinding, mechanical expectations of piecework for herself when, at the age of twelve, she "climbed the dark, narrow stairs of a tenement house on Monroe Street" to begin work as a feller, sewing the lining of men's coat sleeves. Some sewed buttons, some pressed, but each worked at the same monotonous task day in and day out. Wrote Rose, "You with your eyes close to the coat on your lap are sitting and sweating the livelong day. The black cloth dust eats into your very pores. You are breathing the air that all the other bent and sweating bodies in the shop are throwing off, and the air that comes in from the yard heavy and disgusting with filth and the odour of the open toilets."[26] Indeed, even office workers were touched by the imperative for speed and efficiency, and learned the monotony of repetitive work.[27] In "A Typist Plaint," a stenographer laments,

> I stenograph, type and flame
>   Each weary, sunny hour;
> Numbed fingers, palms grown lame. . . .
> I type and am a prey
>   To commerce's letters, dull,
> All day, in office gray. . . .
>   Which makes my living null. . . .[28]

For all workers, skilled or unskilled, it was the clock that heralded the new importance of speed in production work: "my boss cries: 'Speed-' . . . Ah, speed, the cry of greed. . . . The clock-ticks never miss."[29] In Rosenfeld's "The Sweat-Shop," the clock similarly drives the worker on:

> The clock in the workshop does not rest;
> It keeps on pointing, and ticking, and waking in succession.
> . . . .
> In its sound I hear only the angry words of the boss;
> In the two hands I see his gloomy look.
> The clock, I shudder,—it seems to me it drives me
> And calls me "Machine," and cries out to me: "Sew!"[30]

The federal government's interests in exclusion versus discipline were more complex and consequently are harder to tease apart. Both the Public Health Service and the Immigration Service voiced an interest in exclusion, and the inspection regimes set up at the nation's immigration stations were intended to accomplish just that. Within the leadership of the IS, which made final decisions regarding admission or exclusion for all immigrants, the rhetoric of exclusion resounded. Some shared his roots in organized labor, and all of Powderly's successors in the IS during this period favored strict exclusionary practices.[31] For example, Frank Sargent, who followed Powderly as director of the IS, equated diseased immigrants with criminals and moral degenerates, arguing that their "mere presence is a menace to society." He insisted, "Further restrictive legislation is needed if the United States is to maintain its present industrial prosperity and to protect itself from pauperism and disease." The United States must not become the "'dumping ground' for the diseased and pauperized peoples of Europe."[32] As Powderly had put it, the United

States was not to be "the hospital of the nations on earth."[33] Herman J. Schulteis evoked similar imagery in warning against admitting socialist labor leaders, such as New York City's Joseph Barondess, to the United States: "We should guard against an invasion of such hordes as we would against an armed host or a pestilence."[34]

When he started out with the PHS, Dr. Victor Heiser was assigned to examine immigrants in Boston, where he "worked so well . . . that [he] was promoted to the chief center of immigration at New York." Heiser explained, "I believed that health should be regarded from the economic as well as from the humanitarian viewpoint. To be without it was to be without earning power."[35] Another official, Dr. Victor Safford, interpreted his job in these words: "It seems safe to say that it was the intention of those who have framed the provisions relating to a medical examination in our immigration laws . . . to provide a means of stopping the entry of the classes of aliens who were showing themselves to be a direct economic burden on this country after arrival."[36]

The PHS, representing both the state and science, was an ideal agent for determining the meaning of disease and shaping the consequences of being diseased for the marginal immigrant. Bacteriology offered new possibilities for protecting individuals and nations from infection, but it was also a social and cultural phenomenon: Our "traditional" public health policies, born of bacteriology, encompass ways of looking at the world and at relationships between groups that are not necessarily "scientific" or value-free. The categorization of immigrant diseases and laboring bodies, therefore, was open to the influence of ethnic, social, and economic perceptions of immigrants. Judith Walzer Leavitt, Naomi Rogers, and Nancy Tomes have shown that in the American system bacteriology, like other branches of medicine, was informed by class, race, and gender norms, and worked to "pathologize" private behavior.[37] In the very different setting of Africa, Maynard Swanson, John Cell, and Randall Packard argue that bacteriology was used to justify new forms of social segregation.[38] Thus, bacteriology needs to be understood not only as a science that failed to break the popular links between race, class, gender, and disease, but also as a science that could help to forge those links. In a nation in which different groups had very different interests in the immigrant laborer, bacteriology offered a means to objectify social fears using the ostensibly neutral language of science.

With the immigration law of 1891, the federal government took control over immigration and created the machinery for federal officials to inspect and exclude immigrants. In 1892, four PHS officers examined immigrants at seven ports across the nation—some serving more than one port. By 1910, seventeen commissioned and sixty noncommissioned PHS officers inspected immigrants at seventy-five American ports or immigration stations. The number of ports of entry continued to mount, with more than one hundred officers inspecting arrivals at one hundred twenty-six immigration stations around the nation by 1930.

Federal law required medical officers of the PHS to inspect and issue a medical certificate to all immigrants suffering from a "loathsome or a dangerous contagious disease." Loathsome and dangerous contagious diseases—also known as Class A conditions—included trachoma (also known as granular conjunctivitis), an infectious eye condition that could lead to blindness; favus, a fungal infection of the scalp and nails; venereal diseases; parasitic infections; and tuberculosis, perhaps the paradigmatic disease associated with immigrants and economic devastation.[39] A subset of Class A conditions included mental conditions such as insanity, feeblemindedness, imbecility, idiocy, and epilepsy (table 1-1).

By 1903, in addition to loathsome and dangerous contagious diseases (labeled Class A conditions mandating exclusion), the PHS had created a new category of Class B diseases or conditions: those rendering the immigrant "likely to become a public charge."[40] The PHS gave its officers little guidance in determining what constituted a Class B condition. Immigrants who should receive Class B certificates were defined only as "those who present some disease or defect, physical or mental, which may be regarded as conclusive or contributory evidence to justify the exclusion . . . of the person in question as an alien 'likely to become a public charge.'"[41] Officers were merely informed that "the certificate in each case should be sufficiently explicit to enable the inspectors whose duty it is to pass final judgment on these cases to form an opinion as to what degree the disease or deformity will affect the immigrant's ability to earn a living."[42] Conditions rendering the immigrant likely to become a public charge included hernia, valvular heart disease, pregnancy, poor physique, chronic rheumatism, nervous affections, malignant diseases, deformities, senility and debility, varicose veins, and poor eyesight (table 1-1).[43]

Congressional exclusion of people with diseases affecting ability to earn a living led both the PHS and IS to believe that officers were legally required to state on the medical certificate whether a disease affected ability to earn a living. This imperative created lasting problems and tension between the two services. While the PHS instructed its officers that this determination was to be based on medical opinion, IS officers had ultimate responsibility for deciding whom to exclude, so that the IS was the final arbiter of whether a disease or condition did materially affect an immigrant's ability to earn a living.[44] Nevertheless, the IS pushed the PHS to put the full weight of its medical authority into Class B certifications. The IS complained about the PHS practice, adopted after passage of the 1907 legislation, of certifying Class B conditions with the annotation, "which *may affect* ability to earn a living" (emphasis added). The IS directed PHS officers to state explicitly that a particular disease "does (or does not) affect ability to earn a living." In making such a determination, the IS also insisted that "the occupation of the immigrant must be the deciding point."[45]

Dr. George Stoner at Ellis Island anticipated this requirement and gave examples of how assessments of ability to earn a living might be incorporated into the medical certificate: "This is to certify that the above named alien has Hernia which affects his ability to earn a living—as a laborer. . . . This is to certify that the above named alien has Loss of Left Foot, which does not affect her ability to earn a living—as a housewife. It is corrected by an artificial foot."[46] Nevertheless, Stoner and other PHS officers were uncomfortable making decisions about whether an immigrant's disease would affect his or her ability to earn a living.[47] This seemed a task more properly assigned to the IS because of the occupational and financial information needed to make such decisions. As an example Stoner cited the problems stemming from a diagnosis of "old age": "That an advanced age of eighty or eighty-five years, does affect ability to earn a living, goes almost without saying. In cases of this kind, however, the alien may be surrounded and supported by able bodied or well-to-do children. Or later, before a Board of Special Inquiry, may produce evidence to show that he is not only a well-to-do, but actually a rich man."[48]

Ultimately, Stoner advocated dropping any formal medical annotation regarding an immigrant's ability to earn a living. He was already of the opinion that "too many excluding certificates are being rendered," many of which would be more accurate if they read "does not affect ability to earn a living." Moreover, it was unclear whether such diseases would "incapacitate" the alien from earning a living or merely "affect" that ability. But even the question of whether a disease "affected" ability to earn a living was tricky. As an example, Stoner noted that "certain forms of psoriasis . . . might seem worse than more serious affections of different character" to a lay IS Board of Special Inquiry

TABLE 1-1. U.S. Public Health Service Classification of Disease (the "Immigrant Nomenclature"), 1903–1930

Class A. Loathsome and Dangerous Contagious (Exclusion Mandatory)		

1903	*Dangerous Contagious Diseases (Subdivision I)*	
	Trachoma	
	Pulmonary TB♦[1]	
	Loathsome Diseases (Subdivision II)	
	Favus ♦	
	Syphilis	
	Gonorrhea	
	Leprosy	
	Insane Persons (Subdivision III)	
	Requires certificate of two physicians	
	Idiots (Subdivision IV)	

Class A		

1910	Idiots	**Blastomycosis**
	Imbeciles	**Frambesia (yaws)**
	Feeble-minded persons	**Mycetoma (Madura foot)** ♦
	Epileptics	Leprosy ♦
	Insane persons	Syphilis
	Tuberculosis ♦[2]	Gonorrhea ♦
	of respiratory tract	**Soft chancre** ♦
	of intestinal tract	**Endemic haematuria** ♦
	of genitourinary tract	*Dangerous Contagious Diseases*
	Loathsome Contagious Diseases	Trachoma
	Favus ♦	Filariasis (Filaria sanguinis
	Ringworm of scalp♦	hominis) ♦[4]
	Sycosis barbae ♦[3]	**Uncinariasis (hookworm)** ♦
	Actinomycosis	**Amoebic infection (Amoeba coli)** ♦

Class A		

	Idiots	Frambesia (Yaws)
1917	Imbeciles	Mycetoma (Madura foot) ♦
	Feeble minded	Leprosy ♦
	Epileptics	**Oriental sore (cutaneous leishmaniasis)** ♦
	Insane	Syphilis[6]
	Constitutional psychopathic inferiority	Gonorrhea ♦
	Chronic alcoholism	Soft chancre ♦
	Mentally defective	*Dangerous Contagious Diseases*
	Tuberculosis[5]	Trachoma
	◄ **Expanded to TB in any form**	Filariasis ♦
	Loathsome Contagious Diseases	Amoebiasis ♦
	Favus ♦	**Schistosomiasis** [7]
	Ringworm of scalp and nails	Leishmaniasis ♦
	Sycosis barbae ♦[3]	**Trypanosomiasis (sleeping sickness)** ♦
	Actinomycosis♦	**Paragonomiasis** ♦
	Blastomycosis♦	**Clonorchiasis**

Class B	Affecting Ability to Earn a Living	(Exclusion Discretionary)

Hernia
Valvular heart disease
Pregnancy
Poor physique ("chickenbreast," symptoms of pulmonary TB without evidence of bacillus)
Chronic rheumatism
Nervous affections (locomotor ataxia, spastic paraplegia)
Malignant diseases (carcinoma, sarcoma)
Deformities (kyphosis, lordisis, scoliosis, mutation of extremities, etc.)
Senility and debility
Varicose veins
Eyesight (refractive errors, optic atrophy, choroiditis, retinitis pigmentosa, etc.)
General considerations (disease/deformity that cannot be placed in other classes)

Class B		Class C Less Serious
Hernia	Eyesight	▶Pregnancy (may be classified as B if appropriate)
Heart disease (no longer limited to valvular disease)	**Cutaneous affections**	
Permanently defective nutrition and marked defective skeletal and muscular development	**Eruptive fevers**	
	Anaemia	
	Tuberculous affections of the skin, glands, bones, and joints	
Chronic arthritis and myositis	General considerations	
Nervous affections	Poor physique	
Malignant new growths	Chronic rheumatism	
Deformities (see above)	Debility	
Senility		
Varicose veins		

Class B	Class C
	Pregnancy
Hernia	
Heart disease	
Permanently defective nutrition and marked defective skeletal and muscular development	
Chronic arthritis and myositis	
Nervous affections	
Malignant new growths	
Deformities (see above)	
Senility	
Varicose veins	
Eyesight	
Chronic malaria ◆	
▶ Uncinariasis (hookworm) ◆	
Pellagra	
Beriberi	
Cutaneous affections	
Eruptive fevers	
Anaemia	

continued

TABLE 1-1. Continued

Class A °

Loathsome or Dangerous Contagious Diseases	Soft chancre
1930 Favus	Trachoma
Ringworm of scalp, nails, **or beard** ◀	Amoebiasis ◀
Actinomycosis ◀	Leishmaniasis ◀
Blastomycosis ◀	Trypanosomiasis ◀
Mycetoma ◀	Filariasis ◀
Leprosy ◀	Schistosomiasis ◀
Yaws ◀	Paragonomiasis ◀
Syphilis	~~Clonorchiasis~~
Gonorrhea ◀	~~Sycosis barbae~~

°Categories are the same as in 1917 except that "Loathsome Contagious" and "Dangerous Contagious" diseases
are collapsed into a single category.
◀ Moved up to a higher category from previous regulations.
▶ Moved down to a lower category from previous regulations.
Bold indicates that a disease was added to a category in that set of regulations.
~~Strikeouts~~ indicate that a disease included in previous regulations was removed.
◆ Microscopic confirmation required for certification.
[1]Sputum
[2]Sputum or intestinal or urinary discharges
[3]Microscopic confirmation recommended if possible.
[4] Examination of freash drop of blood taken at night optional.
[5] Pronounced clinical symptoms sufficient for certification.
[6] Wasserman recommended if feasible; results to be taken with caution.
[7] Microscopic confirmation discussed but not explicitly required.
SOURCES: *Book of Instructions for the Medical Inspection of Immigrants*, 1903; *Book of Instructions for the
Medical Inspection of Aliens*, 1910; *Regulations Governing the Medical Inspection of Aliens*, 1917; *Regulations
Governing the Medical Examination of Aliens*, 1930.

(BSI), but that they should not necessarily be considered as affecting ability to
earn a living.[49] In 1907, though the IS also recommended omitting such language from medical certificates, the form of the medical certificate remained
unchanged.[50]

The relative distribution of immigrant diseases according to the broad categories of immigrant certifications, from 1909 to 1930, is shown in figures 1-1
and 1-2.[51] The PHS subdivided Class A for reporting purposes. Class A1, the
mental conditions, consistently represented less than 5 percent of total certifications. The loathsome and dangerous contagious diseases, Class A2, represented slightly more than 10 percent of certifications from 1910 to 1921,
thereafter only about 5 percent. Diseases "affecting ability to earn a living"
represented the vast majority of certifications, reflecting the relative importance of the economic threat that immigrant disease presented, and the PHS's
willingness to render such certificates despite the protests of Stoner.[52]

Not only Class B conditions affecting ability to earn a living, but also the
loathsome and dangerous contagious diseases, like tuberculosis, took on eco-

Class B	Class C
No specific diseases listed for either category.	Same criteria as in 1917.

nomic meaning in the hands of the PHS, which defined contagious immigrant diseases as "essentially chronic." The PHS also determined that a key element in the definition of a "dangerous contagious disease" was the stipulation that it "may result in the destruction of one of the most important senses." Chronic, debilitating disease represented the permanent inability of an immigrant to function in society; it represented dependency.

There is no better illustration of this point than the rationale used to classify trachoma as a Class A condition. In 1903, when it published the first set of regulations governing immigrant medical inspection, the PHS neatly summed up the dual rationale for excluding immigrants with trachoma: "The object is not only to prevent the introduction into this country of a communicable disease, but also to keep a class of persons from whom so large a proportion of the inmates of institutions for the blind and recipients of public dispensary charity are recruited."[53] Trachoma leads to blindness, and blindness to dependency: "Sight is not only a most valuable asset in earning a living, but it is the medium of some of the greatest joys of life."[54] Dr. Victor Heiser urged that officers always remain vigilant: "Trachoma, a contagious inflammation of the eyelids, had always to be watched for with special care. It was estimated fifteen percent of the blindness in the United States institutions at that time was due to this disease. Clinics in our large cities were overrun with cases which proved stubborn to treat and often impossible to cure."[55]

In 1897, trachoma became the first disease officially declared by the PHS to be "a loathsome or a dangerous contagious disease." The decision met rapid challenge. Nazaret Saropian, a twenty-two-year-old Armenian immigrant with trachoma, argued that because of the lack of medical consensus regarding its

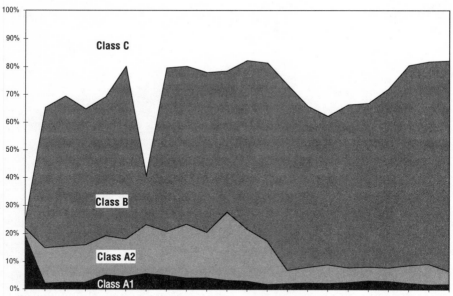

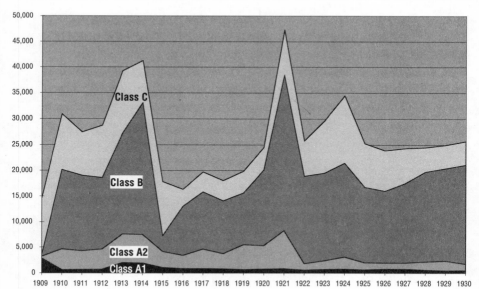

FIGURE 1-1. Immigrants certified by medical category, expressed
as percentage of total, 1909–1930. (SGAR, 1909–30)
FIGURE 1-2. Number of immigrants certified by medical category, 1909–1930.

contagiousness and responsiveness to treatment, trachoma was neither a loath-some nor a dangerous contagious disease.[56] The question, "being one purely of medical science," was referred to Surgeon General Walter Wyman. Wyman responded by citing a leading medical authority on trachoma, who described a disease that "will spread slowly through an orphan asylum, tenement house, or other place where the poor are crowded together, unless special means are taken to prevent this result, leaving its victims handicapped for life and often nearly blind." Wyman noted that the disease was "seldom seen except among recent immigrants from the eastern end of the Mediterranean, Polish and Rus-sian Jews, Armenians and others from that locality." He emphasized that "the presence of acute trachoma of the conjunctiva of immigrants should be a good and sufficient reason for turning them back whence they came. A large por-tion of these cases within a few months after their arrival become incapaci-tated and are public charges." Based on this opinion, in which trachoma was directly associated with immigration and dependency, Wyman—ignoring the question of loathsomeness—confirmed previous PHS conclusions that tra-choma was "both dangerous and contagious." The IS dismissed Nazaret Saropian's appeal and promptly deported him.[57]

"Contagion," then, was not understood in purely bacteriological terms. Or, rather, bacteriology had both social and medical implications. Just as pathogens multiplied in, spread throughout, and depleted the bodies of in-fected individuals, immigrants with disease infected the economic body. Thus, Marcus Braun, the IS special investigator sent to investigate immigration con-ditions along the Mexican border shortly after the turn of the century, con-cluded that contagious conditions "not only present the dangerous feature of spreading . . . , but can also be looked upon from the stand point of public charges, since most of these people infess [sic] the Dispensaries, Hospitals and other public Institutions."[58] By 1907 the PHS had come to associate trachoma not only with contagion and financial ruin, but also with racial economic de-generation: "we must concede to trachoma a high place among the factors that go to decrease materially not only the economic efficiency of the individual sufferer, but that of the race or people as a whole among whom it is preva-lent. . . . [T]he resulting visual impairment [of the sufferer] cannot fail greatly to reduce his efficiency and consequently his value to society at large. He is therefore restricted in his activities and may therefore become a public charge."[59]

The difficulty many immigrants had in paying for the medical care they received at Ellis Island or other immigration stations convinced the IS and PHS that disease, in general, was the major cause of dependency. Neither service was eager to grant diseased immigrants medical treatment.[60] Both were convinced that many immigrants simply could not afford treatment, and perhaps did not even deserve it. In the case of Mrs. Emilie Beller, for example, the commissioner general of the IS felt that the situation "seems to be going from bad to worse as time elapses. Certainly the Department's generous exercise of clemency in this matter is not appreciated by the alien nor apparently by [her Chicago lawyer and physician]." Claiming that the "feeling of disgust which must arise from the lack of gratitude in places where it ought to be expected to exist" did not influence his objectivity, the commissioner general concluded that not only was hospital treatment an "abuse and a nuisance," but also it afforded diseased aliens an opportunity to escape—as did Mrs. Beller—or spread contagion, almost always leaving the government to foot expensive bills that immigrants could not pay.[61]

Clearly smarting from episodes like that of Emilie Beller, in 1913 the IS reviewed its most troublesome cases up to that time. The evidence seemed overwhelming: again and again, immigrant families defaulted on their hospital bills. The PHS and IS regarded as a harsh lesson experiences like that with Josef Abdallah, a twelve-year-old Turkish child certified with favus, in December 1907. Josef's father, a naturalized citizen working in Wisconsin, successfully petitioned for his son's treatment at Ellis Island. The PHS physicians were reluctant to comply, describing his as a "marked case" requiring "treatment for *at least* twelve months, if not longer, and may prove practically *incurable.*" The father, however, "exhibited when here a roll of bills containing $3000 and bank book with substantial balance" and declared himself "willing to pay $5000 for the cure and landing of son." The IS authorized his treatment. More than a year later, in March 1909, though they extracted the medical expenses from the Immigrant Fund, the IS deported Josef after his father failed to pay a $235 hospital bill, "pleaded indigence," and "wrote several untrue letters."[62] The Abdallah case was one of many that the IS and PHS used to underscore the inability of immigrants to afford steep hospital bills.

As in the Abdallah case, immigrants granted hospital treatment at Ellis Island and other ports were often deported for inability to pay hospital expenses associated with Class A conditions. The IS deported roughly 30 percent of all immigrants certified for Class A conditions from 1891 to 1930. Nonetheless, the perception persisted that an elaborate system of regulatory, administrative,

and legal exceptions to immigrant medical exclusion built up after 1906 presented a tremendous barrier to the exclusion of diseased immigrants at domestic ports: "It is a matter of common knowledge that only a part of the immigrants certified for mandatorily excludable diseases in ports of the United
States are actually deported. Lack of funds, political influence, and a host of
other factors operate to make deportation difficult or impossible."[63]

Over time the IS granted medical treatment to more and more immigrants,
often justifying it on humanitarian grounds. Section 37 of the Immigration Act
of 1903 specified that the secretary of commerce and labor, at his discretion,
could allow the wife and minor children of an immigrant who had declared his
intention to become a citizen to receive medical treatment until cured, provided the disease had been contracted on board ship while the individual was
en route to the United States. In 1907, however, Section 19 of the Immigration Act merely stated that diseased aliens could land only with the "express
permission of the Secretary of (Commerce and) Labor." The law opened the
door for any diseased immigrant to appeal for treatment upon arrival in the
United States.[64]

Only 13 percent of those who applied for hospitalization after 1907—which
included not only those immigrants medically certified but also those in need
of treatment for a condition not covered under the immigration law, such as
diarrhea—were denied treatment. Most immigrants, however, did not apply
for treatment of Class A conditions because, if the request was granted, the
immigrant was required to pay all medical expenses.[65] In 1919 the Ellis Island
hospital charged the following per diem rates for hospital care: $2.75 for adults
and children, $1.50 for nursing infants and children under five accompanying
a sick parent, $3.25 for communicable diseases, $4.00 for cases of insanity, and
$2.25 for the care of seamen.[66] Ellis Island's hospital records show that the
majority of immigrants were treated for short-term acute infections, such as
measles, chickenpox, dysentery, minor eye infections, pregnancy, and child-
birth.[67] Although immigrants suspected of having excludable conditions like
tuberculosis and syphilis were admitted to the hospital and "treated" while
they were under "observation," the hospital did not keep these immigrants
with the intent to cure them unless they were granted permission to receive
treatment. Chronic conditions, such as trachoma and favus, could exact a considerable toll with no guarantee of success. In Baltimore, the IS reported that
in 1914 eleven immigrants certified with Class A conditions (two cases of favus,
five cases of ringworm of the scalp, and four cases of trachoma) were granted

treatment. At the end of the year, two immigrants were still receiving treatment for trachoma and five for ringworm, "very slow progress toward a cure having been effected." The expense was great, totaling $4,055 at the time of the report. At Ellis Island, a far greater number of immigrants were granted hospital treatment. By 1921 the PHS claimed that 90 percent of immigrants certified with Class A conditions received treatment.[68] The IS concluded that "another year's experience but emphasizes the inadvisability of granting hospital treatment except in cases of exceeding merit, where the assurances for payment are beyond question."[69]

PHS constructions of disease and disability resonated powerfully in American popular culture, playing particularly well off nativist sentiments and fueling the call for exclusion. Although immigrants had been associated with disease long before the rise of bacteriology, contagious disease provided a powerful, provocative, popular means of representing the immigrant "menace" to Americans.[70] Beginning in the 1870s, the American public was introduced to and rapidly became familiar with "germs," "microbes," "bacteria," and "microscopic parasites."[71] There was, nonetheless, little widespread popular or medical appreciation of the precise implications of contagion; the popular conception of germs was vague. Germ theory easily coexisted and overlapped with older miasmatic and filth theories of disease causation, all of which often linked immorality to disease.

Vague understanding of germs exaggerated their power in the popular imagination.[72] Dust, gases, "effluvia" exuded by the human skin, fabric, public soap, drinking cups, soda fountain glasses, pencils, playgrounds, books, money, ice, and flies were implicated in the popular press as carrying the germs of disease.[73] Such was the popular sense of constant assault by germs that the futility of avoiding them became the subject of jest. One soloist in a popular operetta of 1915 sang:

> In these days of indigestion
> It is often times a question
>   As to what to eat and what to leave alone;
> For each microbe and bacillus
> Has a different way to kill us,
>   And in time they always claim us for their own.
> There are germs of every kind
> In any food that you can find

In the market or upon the bill of fare.

Drinking water's just as risky

As the so-called deadly whiskey.

And it's often a mistake to breathe the air.

[Chorus]

> *Some little bug is going to find you some day,*
> *Some little bug will creep behind you some day,*
> *With a nervous little quiver*
> *He'll give cirrhosis of the liver;*
> *Some little bug is going to find you some day.*[74]

Immigrants were often quite literally equated with germs, threatening to establish "colonies" or "little centres of foreign inoculation" within the national body.[75] To writers in the popular press, the immigrant represented societal "contamination with criminality, contagious diseases, mental delinquency and hereditary handicaps."[76] As one commentator wrote in 1905, "An emigration tide unless thoroughly policed carries with it the germs of anarchy, crime, disease, and degeneracy."[77] The real "danger of allowing Europe to *drain* her social system into the United States" was "a wide contamination of society."[78]

Yet in this era in which germ theory easily coexisted with older sanitarian or miasmatic notions of disease transmission, the immigrant was also powerfully linked to disease through more subtle metaphors of contagion. Water metaphors, for example, described immigrants in terms of "swamps," "tides," "streams," "floods," "cesspools," and "drains."[79] Immigrants "poured" into the nation and "saturated" tenement districts.[80] Garbage metaphors evoked images of "human refuse," "dregs," "waste," "riff-raff," "flotsam," or "pollution" being "dumped" on the shores of America.[81] In the insect metaphor, immigrants were perceived as arriving in "swarms" or "hordes."[82] Often, these metaphors were extended. In 1896, for example, Francis Walker, the superintendent of the U.S. Census, declared that the "foul and stagnant pool of population in Europe, which no breath of intellectual or industrial life has stirred for ages, should not be decanted upon our soil."[83] Many of the refuse and water metaphors found expression in political cartoons, which depicted immigrants as garbage dumped into America's backyard and as water or sewage drained from Europe (figs. 1-3 and 1-4).

Just as imagery such as water or dirt could convey the filth of the immigrant, seemingly neutral representations of the immigrant could, in turn, suggest wa-

Knott in Dallas News

Do we want any more trash dumped into our back yard?

FIGURE 1-3. "No Dumping Here." (*The Independent* 105 [7 May 1921]: 485; reprinted with permission of the *Dallas Morningstar News*)

REGULATE THE FLOW AT THE SOURCE
—Page in Louisville *Courier-Journal*

FIGURE 1-4. "Guarding the Gates against Undesirables." (*Current Opinion* 76 [April 1924]: 401; reprinted with permission of the Louisville *Courier-Journal*)

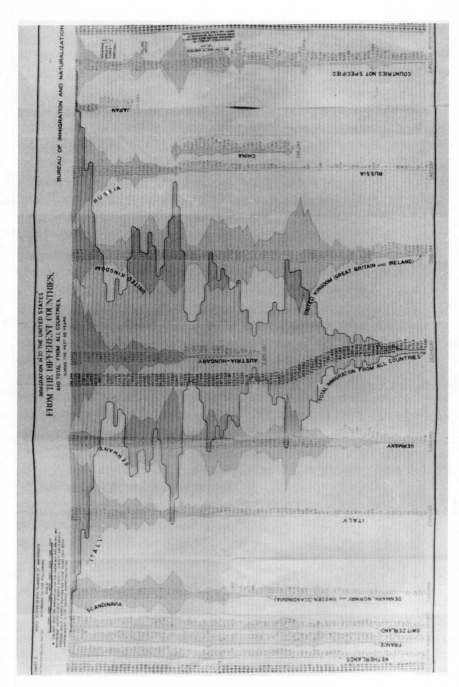

FIGURE 1-5. A 1908 graphic showing the shifting streams of immigration. (CGAR, 1908)

ter and its powerful connotations. The ostensibly objective graphic images that the IS used to represent the numbers and nationalities of immigrants arriving in the United States suggest the rising tide and shifting streams of immigration (fig. 1-5). So uncontrolled were these incoming waves of immigrants that they cascaded one upon the other, creating near chaos.

Metaphors, theorists argue, pervade and structure our normal conceptual and discursive systems.[84] They are not just colorful or decorative language. George Lakoff and Mark Johnson argue that "the essence of metaphor is understanding and experiencing one kind of thing in terms of another." Shared understanding must be rooted in experience. Although we still use many of the same metaphors to describe immigration today, water, insects, dirt, and germs all carry different meanings for us.[85] The metaphors in use at the turn of the twentieth century constituted a systematic discursive expression of a bacteriological or, perhaps, a "bacterio-sanitary" worldview.[86]

Although each metaphor was directly associated with germs and disease, each also expressed culturally relevant aspects of the nation's experience with immigrants. The complex network of coherent and consistent metaphors all reinforced not simply the notion that the immigrant was pathogenic and in need of exclusion, but also the notion that the immigrant must be controlled and disciplined.[87] Water was a force that, in the form of a flood or deluge or simply an open sewer, could not be contained. Insects transgressed class boundaries. Flies, for example, were commonly referred to as "germs with legs."[88] Moreover, immigrants—particularly southern and eastern European immigrants—were widely credited with creating the conditions in which the fly could breed; born of the immigrant neighborhood, the fly could transgress social and geographic boundaries to threaten the health of middle-class and even wealthy families.[89] Garbage and refuse represented pollution of the public space with private filth. In the specific context of immigration control, the metaphors of disease and immigration not only expressed nativist fears of contamination but also and more critically the need to order and manage a defined set of problems that emerged at the intersection of immigration, industrial production, and social dependency.

The industrial economy created a tension between the imperative to exclude and the imperative to discipline. Dramatic changes in industrial production brought about the unprecedented expansion of American industry, but at the price of great economic fragility. Thus, at the dawn of the modern industrial era, the world's emerging power began to grapple with the problem of unemployment, introducing a new dimension into defining and managing a necessarily fluid industrial work force while providing a compelling rationale for excluding those deemed to be destined to dependency. Government agencies and social reformers thus began to redefine the concept of "inability to work."

Pre-industrial means of production had been characterized by workers' control over production. Workers lived in communities that allowed them to integrate agricultural and manufacturing pursuits, substituting one type of work or production for another with changes in the season or economic cycles. The collapse of that system left workers vulnerable to "forced" unemployment beginning in the 1870s. According to Alexander Keyssar, as workers increasingly located in urban areas and the labor supply swelled to accommodate the demands of a rapidly growing industrial power, tens of thousands of industrial workers became "utterly dependent upon their industrial earnings in order to survive."[90] The survival of the working class, noted the Massachusetts Bureau of Labor Statistics, was "contingent upon continuous health and continuous work. But with the exception of some few in-door employments, continuous work is the exception and not the rule."[91] Boston's South End House estimated that 12 percent of the population of the city's North End and 9 percent of the West End received charitable assistance in 1901–2. This figure included all of the two districts' casual and intermittent workers.[92]

America may have been the world's leading industrial power, but, as Keyssar observes, "at least once in the course of each decade, there occurred a wave of bank and business failures, the engines of progress coughed and sputtered, wages dropped and some men and women were 'thrown out of work.'" Many "minor" recessions and depressions accompanied the six "major" economic downturns that the nation experienced from 1870 to 1921. "In all," according to Keyssar, "the business cycle rose and fell thirteen times between 1870 and 1921; roughly two out of every five years contained periods of recession or depression."[93] Pauline Newman, who worked for a time in the Triangle Shirtwaist Factory as a young girl, noted that unemployment struck abruptly: "If the season was over, we were told, 'You're laid off. Shift for yourself.' How did you live? After all, you didn't earn enough to save any money."[94] Abe Koosis, who came to New York City from Russia as a sixteen-year-old boy in the early 1920s, described a similar experience: "I worked in about a dozen places, some jobs lasting a week, some a year. I worked in a tin can factory, in a belt factory, in a luggage factory, as a pleater of women's skirts, as a shipping clerk, and for a publisher. There was no long-term employment. When things got busy, people would be employed; when it slowed up, people were discharged, without even a day's notice."[95] In Massachusetts, industry laid off one in three workers for up to four months in the ten years following 1885. Such a portrait of unemployment characterized periods of economic expansion as well as those of depression and recession. Eastern industrial states like New York and Massachusetts sought to maintain an adequate supply of reserve labor to meet the unpredictable yet urgent demands of expanding enterprise.[96]

Sickness and ill health during bouts of unemployment could mean the difference between survival and destitution. Commented a Massachusetts furniture polisher in 1879, "For a family of three to save one hundred dollars a year, the head of the family must earn twelve dollars each week the whole year round. . . . Suppose a man is idle a couple of months, or sick: what is going to become of his family?"[97] The sickness or death of working family members often precipitated or attended the descent into poverty. A case of rheumatism, for example, launched the Jenkins family in Philadelphia into poverty in the late 1920s. Mrs. Jenkins reported, "Last year Frank had rheumatism; so when they didn't have enough work to keep every one busy, he was one of the first to be laid off. He was out four months. Things got so bad then we went to live with his mother." Work thereafter was only sporadic for the thirty-year-old roofer. The Jenkinses' youngest daughter died of measles and pneumonia. "I couldn't help feeling that if he'd been working all winter she'd have been stronger and maybe not died," mused Mrs. Jenkins. "But of course you never know about them things."[98]

Settlement workers in Omaha, Nebraska, and Detroit, Michigan, observed that even when the families of "broken work" escaped illness, it was still the factor that always threatened ultimate ruin.[99] The U.S. Department of Labor's study of unemployment in Springfield, Massachusetts, and Racine, Wisconsin, for example, described the decline of an Italian laborer unemployed for fourteen months. Soon after he lost his job in 1920, his wife fell ill and required an operation, while his brother died, reducing family income and resulting in further accumulation of debt from funeral expenses. Within a year, the family had accumulated some twelve hundred dollars in debt.[100]

Thus, even in the "prosperous" 1920s, settlement house workers continued to find equally high rates of sickness accompanying extended unemployment. Of 150 cases observed by settlement house workers in major cities around the nation, 77 percent of families endured the effects of malnutrition, starvation, and the more severe consequences of illness with colds leading to pneumonia or tuberculosis, or death. While children and women bore the primary burden of unemployment—representing 50 percent and 44 percent of cases, respectively, of illness or death in homes suffering sickness—primary wage earners suffered from illness, disability, or death in 34 percent of families. Thus, Mrs. Domico, a thirty-year-old single mother of three working as a children's dress operator in Philadelphia, confided to a worker in the House of Industry her fear "that she will become ill and lose her position, and she has no money for doctor bills."[101]

Most of the laboring class relied primarily on the resources of family and friends during lean times, and less often on union unemployment benefits, savings, and simply the accumulation of debt.[102] It was unusual, at least in the Northeast from the 1870s through the 1920s, for the unemployed worker to rely on public or private charity or relief organizations unless he became "unemployable."[103] Nonetheless, the prevailing perception was that the link between unemployment and reliance on charity was strong. Assessments of the immigrant contribution to the burden of dependency in the United States thus received a new attention and specificity.[104] Organizations such as the Immigration Service, the Immigration Commission of the U.S. House of Representatives, the Children's Bureau, the American Council for Nationalities Service, and individual reform, penal, and charitable institutions began documenting the link between immigrants' health and dependency.

In 1909 the U.S. Immigration Commission surveyed the causes of need among 31,374 cases given charity assistance from 1850 to 1908. Nearly 50 percent of cases received charity due to the death or disability of a breadwinner or other member of the family. In another 18 percent, need resulted from the "neglect or bad habits of the breadwinner," which included spousal desertion, incarceration, intemperance, and neglect.[105] The Children's Bureau drew a more definitive conclusion. Based on a 1918 study in Gary, Indiana, the agency affirmed that "illness of self or of some other member of the family was the major cause of nonemployment."[106] U.S. agencies, commissions, and charities reached two main conclusions: (1) immigrants were over-represented in institutions serving the dependent; (2) physical disability was the "outstanding problem which leads to dependency."[107]

Although industry became interested in the problems of illness, employers were not concerned with dependency, as was the federal government; their main concern was labor turnover and the resulting inefficiency.[108] In an era in which the worn-out machinery of human bodies—discarded from the assembly line—found little place in industrial society, health was at a premium. But understanding illness, rather than unemployment, as the factor leading to dependency placed the onus of dependency on the immigrant rather than on the American industrial economy that thrived on a large, highly mobile, and responsive unskilled labor force. Among Jewish garment workers, for example, the imperative was for the worker to maintain his own body: "Stronger immigrant bodies meant more productive workers better able to withstand the alternating frenetic and slack seasons and the poor workplace conditions of garment work." Thus the director of the Educational Alliance in New York City, an organization providing a gymnasium for workers, proclaimed, "Let a young man develop his body, and he will neither shrink from . . . danger nor shirk manual work which falls [to] his lot."[109]

So while much of public and federal discourse and even immigration law favored exclusion, the economy made exclusion impractical. Industry relied on a constant supply of labor to draw on at a moment's notice and then discard when the need for immediate production diminished, and Congress accordingly formulated immigration law "along conservative lines [in order to] avoid measures so drastic as to cripple American industry."[110] Thus, a new, industrial-era conception of assembly-line production, interchangeable tasks, and chronic, cyclical unemployment undercut the impulse to exclude and reinforced the necessity of disciplining the laboring body. It was not simply the case that the worker bound for dependency had to be barred at the nation's threshold; rather, at the nation's threshold, *all* workers had to learn the rules and expectations of industrial society. Immigrant laborers—more vulnerable than native-born workers to unemployment during the many periods of depression[111]—had to learn that they were expected to remain fit throughout the inevitable spells of unemployment that they would be required to weather. The worker had to be taught quickly the need to endure the cyclical rise and fall of the nation's economy. This meant that immigrants had to be adaptable to work in a variety of industrial settings at whatever task was demanded of them. They had to survive hard times without falling into poor health and without relying on the kindness of anyone but family and friends. Bridget Fitzgerald, who came to the United States from Ireland in 1921 at age eighteen, understood perfectly: "You know what you needed then mostly? I'll tell you. Strong and healthy, that you won't become a public charge, because then, I mean, you go right back."[112] Just as immigrants were taught these lessons on the factory floor, they were introduced to them at the nation's gates.

While the dual fears of dependency and social contamination drove the impulse to exclude immigrants and shaped the classification of immigrant diseases, the needs of the industrial economy shaped the kind of examination that the PHS would conduct at the nation's borders and gave it a surprising meaning and purpose. Although touted as an exclusionary tool, the exam served a normative function. It conveyed a system of classification, of social ordering and industrial expectations to immigrants who feared and, ultimately, remembered it. Although she laughed at all she had forgotten about her early years in America, Sadie Guttman Kaplan at one hundred years of age remembered her arrival in New York as a child of twelve, insisting, "Ask me about Ellis Island. I can tell you about it."[113]

CHAPTER TWO

THE FUNCTION OF
MEDICAL INSPECTION

Restriction, Instruction, and Discipline
of the Laboring Body

Which is the first main requisite of the immigrant?
Health. Without health your life is ruined. Better for you never to have
been born than to find yourself in a far off foreign country—without
health. Therefore, your first requisite is to preserve your health. . . .
Every cent spent for the preservation of health is worth to you—a dol-
lar, one hundred dollars! Of what avail is your country to you, politics,
party, religion, systems—if you have no health?
      —Yugoslav newspaper, 1919, quoted in a Foreign Language
  Information Service report for submission to the Commonwealth Fund

THE IMMIGRATION SERVICE and the Public Health Service frequently experienced tremendous public pressure to admit or treat individual
immigrants rejected for one reason or another. How or why particular cases
came into the limelight is unknown, but when they did, the press consistently
cast the agencies as villains unfairly picking on innocent immigrants.[1] Even
Senator Henry Cabot Lodge, an outspoken advocate for immigration restriction and co-founder of the Immigration Restriction League, appealed to the
surgeon general for the release of one Marian Zatarian, the daughter of a naturalized citizen. By 1907, Miss Zatarian had been detained at the port of
Boston for more than two years. During this time she received treatment from
private physicians for trachoma. Lodge accepted the word of her "friends" that
she had been cured. The PHS, however, continued to rule that she suffered
from trachoma and fully expected her to be deported as incurable. The U.S.
Supreme Court ruled that the girl was deportable as an alien under the immigration law if the PHS confirmed her disease. Lodge urged the surgeon gen-

TABLE 2-1. Immigrant Appeals of Excluding Decisions to Secretary of Labor

| | Outcome All Appeals | | | Appeals by Major Nonmedical Categories | | | | | | Appeals by Disease Category | | | | | | | | | | | | All Medical Appeals | |
| | | Denied | | LPC/Paupers and Beggars | | Denied | Contract Labor | | Denied | Mental | | Denied | LDCD | | Denied | Class B | | Denied | Appeals for Treatment | | Denied | | |
Year	Total*	No.*	%	No.	Denied No.	%	No.	Denied No.	%	No.	Denied No.	%	No.	Denied No.	%	No.	Denied No.	%	No.	Denied No.	%	Total Certified	%Appealing Certification†
1906	3,821	1,968	51.50	3,227	1,606	49.77	430	284	66.05	13	12	92.31	4	3	75.00							11,188	0.15
1909	1,839	928	50.46	1,237	641	51.82	350	144	41.14	12	6	50.00	30	5	16.67							14,558	0.29
1910	5,146	3,063	59.91	4,222	2,527	59.85	430	308	71.63	25	13	52.00	54	13	24.07							30,909	0.26
1912	6,137	3,178	51.78	4,670	2,381	50.99	706	476	67.42	35	12	34.29	67	21	31.34							28,563	0.36
1913	6,947	4,139	59.58	4,641	1,848	39.82	988	655	66.30	45	20	44.44	80	30	37.50							41,520	0.30
1915	5,975	4,101	68.64	4,574	3,079	67.32	772	702	90.93	49	20	40.82	19	18	48.65				177	61	34.46	17,720	0.49
1916	6,987	4,723	67.60	3,919	2,538	64.76	1,038	930	89.60	23	22	95.65	21	8	42.11				312	82	26.28	2,071	2.03
1917	5,241	2,833	54.05	3,050	1,490	48.85	437	361	82.61	72	31	43.06	21	17	80.95				184	44	23.91	20,417	0.46
1918	3,618	2,555	70.62	1,719	1,030	59.92	432	396	91.67	7	1	14.29	31	22	70.97				122	15	12.30	16,971	0.22
1919	4,121	3,109	75.44	1,866	1,460	78.24	624	462	74.04	26	18	69.23	14	11	78.57	424	336	79.25	76	33	43.42	19,848	2.34
1920	4,812	2,950	61.31	2,195	1,187	54.08	618	497	80.42	25	13	52.00	15	15	62.50	265	184	69.43	59	7	11.86	24,152	1.30
1921	7,422	3,541	47.71	3,680	1,449	39.38	581	473	81.41	64	30	46.88	80	24	42.50	646	255	39.47	180	33	18.33	47,275	1.67
1922	12,828	5,244	40.88	3,714	1,627	43.81	577	443	76.78	41	29	70.73	39	18	46.15	1,020	412	40.39	1133	73	6.44	25,724	4.28
1923	14,506	6,247	43.06	4,485	1,873	41.76	977	662	67.76	53	37	69.81	13	13	59.09	1,293	456	35.27	555	94	16.94	29,725	4.60
1924	15,070	8,150	54.08	3,167	1,807	57.06	712	470	66.01	76	59	77.63	110	76	69.09	993	445	44.81	378	10	2.65	33,816	3.49
1925	8,493	4,480	52.75	1,762	930	52.78	426	326	76.53	33	27	81.82	37	30	81.08	817	277	33.90	606	110	18.15	25,226	3.52
1926	5,699	3,371	59.15	1,348	840	62.31	456	350	76.75	37	26	70.27	20	13	65.00	417	95	22.78	213	18	8.45	23,876	1.99
1927	5,565	3,706	66.59	1,103	13	1.18	398	330	82.91	29	24	82.76	26	13	50.00	256	78	30.47	448	14	3.13	24,292	1.28
1928	4,844	2,996	61.85	973	680	68.89	248	193	77.82	28	22	78.57	23	19	82.61	257	94	36.58	378	15	3.97	24,473	1.26
1929	4,972	3,113	62.61	1,058	780	73.72	296	234	79.05	39	26	66.67	34	26	76.47	234	116	49.57	443	9	2.03	24,943	1.23
1930	4,343	2,427	55.58	1,065	752	70.61	245	131	53.47	32	24	75.00	37	33	89.19	157	82	52.23	353	9	2.55	25,659	0.88

*Represents appeals and denials made during the fiscal year. Includes appeals and denials for categories such as polygamists not shown on this table.
†Represents percent appealing medical certifications only, not total percentage appealing any excluding decision.
KEY: LPC, likely to become a public charge; LDCD, loathsome and dangerous contagious diseases.
SOURCE: CGAR for each year.

eral to certify that she did not have a contagious disease. The surgeon general refused, and the IS deported Zatarian on the basis of the medical certificate.[2]

The Zatarian case was a dramatic one in which the PHS pitted its medical authority against the power of the press and the formidable Senator Lodge. In practice, however, relatively few immigrants appealed excluding decisions for medical causes or made formal applications for medical treatment. They likewise failed to appeal in instances in which they were deported for becoming dependant on charity after admission to the United States. For example, Gotlieb Herdenreder, "alias Harcheurader," became a public charge at the Tuberculosis Dispensary of Western Reserve University in Cleveland, Ohio. During his deportation hearing, his examiners—apparently concerned that he did not understand what was happening after he openly admitted he had no money, no friends, and no family—stopped midway through the questioning to explain to him the purpose of the examination and the implications for him, his wife, and child. When asked if he wished counsel or wanted to notify any friends so that they might seek counsel for him, he said no. The only argument he gave against his deportation was that "I am feeling better now and it is better here than in Russia and when the weather becomes warm I will be able to go to work."[3]

Table 2-1 shows that only a handful of immigrants had either the political influence or financial resources to launch an appeal of an excluding decision. From approximately 1906 to 1930, only 1.6 percent of immigrants medically certified ever appealed an excluding decision. The odds of success for those daring to launch an appeal were, at best, even. Overall, 62 percent of immigrants appealing exclusion for a mental condition and 54 percent of immigrants appealing exclusion for a loathsome or dangerous contagious disease were denied admission; 42 percent of immigrants appealing exclusion for a Class B condition were denied admission. On the nonmedical front, 53 percent of immigrants rejected as likely to become a public charge and 75 percent of immigrants rejected as contract laborers who appealed were denied admission.

No doubt, lack of political and financial resources on the part of immigrants contributed to their failure to struggle more actively to remain in America. Most did not even request hospital treatment—a step that could potentially have forestalled deportation and made an appeal unnecessary. The nature of the entrance ordeal worked to discourage resistance on the part of immigrants, regardless of whether they were marked for deportation or not. The process communicated something to the masses of immigrants who arrived, not simply to the few deemed unacceptable. This chapter refigures the story of the medical inspection as a story of large numbers, of those admitted. Subsequent chapters retell the story of small numbers, of those excluded. Both stories suggest that the immigrant medical exam, and the entire enterprise to regulate immigration in the Progressive era and beyond, though overtly intended to alter the flow of immigration to the United States, controlled immigration in a profound manner.

RESTRICTING BODIES: THE IMMIGRANT MEDICAL EXAM ABROAD

The law squarely placed the burden of preventing the passage of undesirable immigration at its source by requiring steamship companies to ensure that the immigrants they transported were eligible to enter America under U.S. law.[4] Accordingly, a whole network of European inspection stations was established along the routes that immigrants traveled to major ports of departure. The medical inspection abroad not only served a disciplinary function for those who finally secured passage to the United States, it also represented the most effective means of excluding diseased immigrants from the nation.

Mary Antin has provided one of the most widely read descriptions of the inspection procedures immigrants endured—procedures that foreshadowed the events to follow on American shores. Mary, her mother, and two siblings traveled from the Pale of Settlement across the German frontier to the port of Hamburg in 1894. Medical inspection procedures began even before they reached Hamburg, in Versbolovo, which Antin describes as "the last station on the Russian side [where] we met the first of our troubles." They knew that such inspections awaited them, but in this case "the blow [came] from where we little expected it, being, as we believed, safe in that quarter." In Berlin, the medical procedures intensified when they were fresh off the train (though not so fresh after hours in a compartment so "crowded by passengers or their luggage" that it became a "hot and close and altogether uncomfortable ... prison").[5]

> Our things were taken away, our friends separated from us; a man came to inspect us, as if to ascertain our full value; strange-looking people driving us about like dumb animals, helpless and unresisting; children we could not see crying in

a way that suggested terrible things; ourselves driven into a little room where a great kettle was boiling on a little stove; our clothes taken off, our bodies rubbed with a slippery substance that might be any bad thing; a shower of warm water let down on us without warning; again driven to another little room where we sit, wrapped in woolen blankets till large, coarse bags are brought in, their contents turned and we . . . hear the women's orders to dress ourselves,—"Quick! Quick!"—or else we'll miss [the train]. We are forced to pick out our clothes from among all the others, with the steam blinding us; we choke, cough, entreat the women to give us time.[6]

Antin observed, "None of us were sick now, yet hear how we were treated! Those gendarmes and nurses always shouted their commands at us from a distance, as fearful of our touch as if we had been lepers." Yet this was not the last inspection. After traveling for a day and a night in crowded, horse-drawn cars, her group arrived in Hamburg where, for the third time in their journey, they were "once more lined-up, cross-questioned, disinfected, labeled, and pigeonholed."[7] Antin writes as though the procedure were now routine.[8] She notes that her narrative is filled with the phrases "'we were told to do this' and 'told to do that,'" suggesting that by this point passengers were not only "herded at the stations, packed in the cars, and driven from place to place like cattle," but began to obey like cattle.[9] Immigrants would continue to receive such orders. Describing in his diary his entry at Ellis Island in May 1906, an immigrant known only as Totonno summed up his day: "Pushed here, pushed there. Get in this line; no, that line; get in that line; no, over there, rush over there, and wait."[10]

Antin vividly describes the quarantine period, comparing it to imprisonment: "[our] last place of detention turned out to be a prison, 'Quarantine' they called it, and there was a great deal of it—two weeks of it. Two weeks within high brick walls, several hundred of us herded in half a dozen compartments— numbered compartments,—sleeping in rows, like sick people in a hospital; with roll-call mornings and night . . . with never a sign of the free world beyond our barred windows; with anxiety and longing and home sickness in our heart." To her, "the fortnight in quarantine was not an episode; it was an epoch, divisible into eras, periods, events."[11]

There are no steamship company records available to help quantify the impact of the steamship company examination on the number of immigrants prevented from leaving for America. The prospect of rejection undoubtedly de-

terred some individuals with serious disability or illness or even precarious
health from expending scarce resources in an uncertain attempt to immigrate
to the United States. Both steamship company brochures and immigrant aid
pamphlets distributed abroad warned the immigrant of the nature and consequences of the medical exams abroad and at U.S. borders. Isaac Bashevis
Singer, the novelist who immigrated to the United States from Poland in 1935
at age thirty-one, shared boyhood memories of immigrants' fears of the medical exam awaiting them in the United States: "So many immigrants, I remember, before they went to America, went to doctors to cure their eyes and
all kinds of sicknesses which they suspected might hinder them of entering the
United States. In my case, when I came to this country, they only asked me if
I was a communist, and I said, 'God forbid!'"[12]

PHS involvement in and discussion of the inspection abroad suggests that
officers considered it a significant means of controlling undesirable immigration to the United States. The federal Quarantine Law of 1893 authorized the
United States to station PHS officers in European ports to monitor the prevalence of epidemic diseases and to inspect and pass vessels wishing to embark
for U.S. ports.[13] Government officials in Naples, Italy, granted PHS quarantine officers permission to inspect departing immigrants under U.S. immigration law. In 1903, the United States made a similar arrangement with China
and Japan.[14]

Although the PHS officers in Italy, China, and Japan lacked the authority
to refuse passage to immigrants, officers examined each immigrant appearing
on the ship's passenger manifest, focusing on trachoma, favus, and tuberculosis. PHS officers then forwarded a "descriptive list" of the diseased immigrants
to the steamship companies for rejection.[15] They also forwarded this list to the
PHS officers stationed in domestic ports. Thus, steamship companies, which
risked a one-hundred-dollar fine plus return passage for each rejected immigrant, had every incentive to follow the advice of the PHS and deny passage
to diseased immigrants. Victor Heiser, the PHS quarantine officer who initi-
ated the immigrant exams in Naples, firmly believed that the steamship companies honored and even welcomed their medical recommendations. He reported that official records did not reflect "the great number who seek advice
privately before sailing. A great many intending emigrants present themselves
at the consulate in order to get an opinion as to whether they would probably
land or not. When advised to the contrary, they do not present themselves at
the inspection; hence there is no record of them." Although the PHS gave only

advice and made no official rejections, steamship companies were only too "glad to avail themselves of the advice of one who has had practical experience in these matters in the United States."[16] Michael La Sorte observed that the steamship agents "were as anxious as their charges that everyone be able to pass through American customs without incident. Each rejected individual reflected on the agent's reputation and affected his right to claim his fee."[17]

Although the available data are limited and represent only the efforts of the PHS abroad, two kinds of data hint that the inspections abroad may have had a tremendous impact on the flow of immigration to the United States. The first consists of the recollections of immigrants who recalled that prospective immigrants "were inspected thoroughly."[18] Marjorie Kellhorn, an eighteen-year-old who left Ireland in 1906, "spent three days in Queenstown being examined by doctors." She recalled that "there were many sent back," and she feared that she would be included among them: "Several people had been turned away for heart murmur. One fellow had something wrong with a finger and he was turned down. I had some marks on my side from a sickness called 'shingles,' but there they called it 'wildfire,' . . . and I was worried I would be turned back because I was also asthmatic. I had spasms. When I got a cold it stayed longer than the average person, and I'd wheeze a lot, but I passed."[19]

The second kind of data concerns the percentage of immigrants rejected by PHS inspectors working abroad. While PHS recommendations did not necessarily reflect on the actions taken by steamship companies operating without the help or supervision of the PHS, it is telling that PHS inspectors working abroad sometimes rejected staggering percentages of immigrants. La Sorte, using the records of Italian immigration officials, confirmed that in sharp contrast to the exam given to immigrants bound for South America, the PHS and steamship exam for Italians proceeding to the United States was rigorous. Italian reports indicated that "in 1906, of the more than 25,000 intending emigrants turned away for medical reasons by Italian examiners, 8,000 had trachoma and another 7,000 were suspected carriers. In 1907, those not admitted on board ship at Italian ports for disease and other reasons numbered 35,196."[20]

Across the Pacific, rejections were also high. Between 1904 and 1910 the PHS typically recommended for rejection a substantial percentage of intending Japanese and Chinese immigrants—anywhere from 5 percent to 65 percent examined, compared to 4 to 6 percent examined in Naples. For example, during that period the PHS recommended rejection for more than 30 percent

(45,768 out of 149,291) of the Asian immigrants examined for trachoma alone.[21] Such high certification rates prompted the United States in 1904 to attempt to extend this strategy to European ports. The IS regarded the matter of preventing the passage of diseased immigrants "as one of such vital importance . . . that it feels that it would be derelict in its duty if it were to omit to resort to any legal means at its command to put a stop to the evil."[22]

But the imperative to exclude prevailed over an imperative to discipline only in the case of Asian immigrants, who were never deemed suitable industrial citizens. Outside of Italy, success with European governments was limited. Lack of precedent presented the major obstacle. Arrangements for immigrant medical inspections typically rested on the initiative and energy of local U.S. consulates. In his autobiography, Fiorello La Guardia described his three years as the consular agent in Fiume, Hungary (now Rijeka, Croatia). La Guardia, who later became mayor of New York City, served in Fiume from 1903 to 1906, though he could legally serve only as the acting consular agent until he reached age twenty-one in 1904. On his arrival in 1903, the Cunard Line had just started carrying passengers from Fiume to New York. La Guardia, with little official guidance, found that he was required to "certify to the health of all passengers and crews and give the ship a certificate that it had cleared from a port free from contagious diseases or illnesses subject to quarantine regulations." To his dismay, La Guardia also found that "the rules and regulations made no mention of specific duties of consular officers stationed at ports of embarkation and placed on them no definite responsibilities in connection with immigrants and the process of immigration."[23]

Surprised that there was no procedure for examining immigrants before departure, La Guardia retained local physicians to inspect immigrants for diseases covered under both the quarantine and immigration laws. Although local Hungarian officials had no objections—La Guardia claimed they deemed it a humane measure to prevent the needless exclusion of immigrants upon arrival in the United States—the Cunard line adamantly opposed La Guardia's plan. When the day for implementation came, Cunard officials refused to allow any inspections.[24] Calling Cunard's bluff, La Guardia refused to sign the bill of health. Without La Guardia's signature, the Cunard ship would be in violation of quarantine law and subject to heavy fines upon arrival in the United States.

The prospect of a fine was sufficient to sway the Cunard officials, who allowed the inspection but refused to pay the doctor's fee. (La Guardia simply passed the bill along to the next Cunard ship seeking clearance for departure to New York.) The Cunard Line lodged a formal complaint in Washington, D.C., but La Guardia was never given official advice on the propriety or legality of his protocol. Consequently, he "kept right on with the practice." He reported, "Before long there was complete cooperation on the part of the steamship line, and we had a smaller percentage of rejections for health at Ellis Island than any other port in the world." In time, "the routine became well established. Inspection was speedy and efficient, and we saved many hundreds of innocent people from the expense of taking a trip all the way to New York only to be found inadmissible on health grounds and sent back."[25] In part, the efficiency of the examination abroad removed the burden of exclusion from the shoulders of PHS officers stationed at domestic immigration stations. Given that significant numbers of would-be immigrants chose not to make the voyage to America, the exam on the home front could be more responsive to the industrial imperative to discipline.

In other instances, the steamship companies themselves developed more elaborate mechanisms for inspection. The German steamship lines, for example, not only inspected immigrants at European seaports, but also entered into an agreement with the Prussian railway authorities to conduct immigrant medical inspections before transportation to the port cities. Paul Weindling notes that between 1880 and 1914, 5 million immigrants passed through Hamburg, Bremen, and Bremerhaven: "The increasingly stringent United States sanitary regulations exerted a powerful influence on German port health measures. The U.S. port health officials demanded rigorous medical controls on migration routes from the German-Russian borders to the U.S. medical and quarantine stations." The pressure intensified, resulting in "ever more elaborate routines of quarantine and disinfection." In response to a 1906 U.S. law requiring a six-day quarantine period before immigrants departed for the United States, Hamburg built new pavilions with dormitories to detain passengers, and more inspection stations with disinfecting facilities were built along the Russian border. The German government was certainly influenced by the U.S. port inspections, but it also acted to defend itself from infectious diseases that might be spread by transmigrants.[26] Thus, Henry Diedrich, the U.S. consul stationed at Bremen, reported that immigrants were inspected three times, "most carefully . . . for the third time here under the supervision of the United States consul."[27] In 1903 the German government rejected 5 percent of immigrants attempting to cross German borders en route to a seaport.[28]

Lack of precedent was not the only obstacle to establishing a formal system of medical inspection abroad. The actions of U.S. quarantine officials during the late nineteenth-century cholera epidemics in European, Asian, and South American ports left a legacy of deep resentment on the part of foreign governments.[29]

U.S. quarantine procedures during the epidemics of 1893 and 1900 resulted in interruptions of trade and other political and cultural conflicts. Thus, Belgium and the Netherlands flatly refused to allow PHS officials to inspect immigrants before departure.[30] Greece implied that it might be willing to consider the proposition if the United States would guarantee that immigrants inspected abroad would not be deported upon arrival in the United States—a compromise Americans were unwilling to entertain.[31] Austria and Hungary also refused, making the future possibility of such inspections contingent on receiving similar guarantees.[32] Great Britain, in contrast, offered limited, provisional support, stating that the inspections would be considered "purely optional so far as emigrants and ship owners are concerned."[33]

U.S. efforts to formalize immigrant medical inspections abroad ultimately brought an end to the official tolerance of informal systems. In China and Japan, the system collapsed when trans-Atlantic steamship companies raised objections, arguing that the PHS inspections abroad gave Pacific lines an unfair competitive advantage.[34] Although there was no consensus within the IS or PHS regarding the wisdom of discontinuing the exams, some officials felt that the PHS exam only prevented the government from exacting a fine from the steamship companies for transporting diseased or defective aliens.[35] The IS also believed that "substitution"—a fraudulent practice in which a healthy individual underwent the PHS and maybe even the steamship medical exam for an individual who either knew or suspected that he or she was diseased or physically unfit—foiled their best efforts anyway. After receiving clearance, the healthy individual would simply return the passport to the true immigrant, who would not have to undergo examination again until reaching the United States.[36] In January 1910 inspections in China and Japan came to an abrupt halt in spite of the objections of the surgeon general, the Pacific steamship companies, and the Japanese government.[37] After World War I Italy withdrew its consent to the work of PHS officers in Naples when the United States used official channels to attempt to extend the system to Genoa. In response, the IS warned inspectors at domestic ports and Atlantic steamship companies that

Italian immigrants should henceforth be inspected with "greater care and thoroughness."[38]

Nonetheless, ports abroad remained the most significant source of immigrant exclusions. Despite setbacks in Europe and Asia, the PHS and IS continued to press for a formal inspection system abroad, a goal they eventually achieved in 1924 following passage of the Immigration Restriction Act. (See the epilogue for a discussion of this system and the ways in which it served as a powerful means of excluding diseased and defective immigrants at the source.) With the goal of exclusion at the source as the objective, PHS officers stationed at domestic ports only ever represented a fail-safe.

If the federal government could not ensconce its own officers abroad, it could ensure that steamship companies would continue to conduct examinations sufficient to exclude immigrants with all but the most difficult-to-diagnose diseases (trachoma, venereal disease, tuberculosis) or the most minor conditions. Steamship lines paid at least five hundred thousand dollars in fines for rejected immigrants. When we add to that the costs of transportation, of meals and lodging for immigrants held at Ellis Island or another immigration station, and of legal expenses in contested cases, the financial burden on transportation companies was considerable. The goal, then, was to send back enough immigrants arriving on American shores to force steamship companies to maintain an adequate exam without hindering trade or interrupting the flow of cheap, healthy labor to the United States.

The medical exam abroad placed a greater emphasis on exclusion than did the exam at Ellis Island and other (but not all) American ports, and it was intertwined, particularly in the case of Germany, with the national interests of other countries.[39] At the same time, the exam abroad served disciplinary ends. Don Gussow described the cleansing and medical examination before departing for America as a totally foreign experience. The twelve-year-old had never had a shower and had only once before, in an orphanage, been totally naked. He explained, "For most of my life until then I had lived with lice on my body and in my clothing," like the rest of his family. While he welcomed the delousing, the rest of the experience had no precedent. Later he reflected, "I have often thought of this experience but particularly during the Hitler holocaust and the years that followed. As I read about the millions in the Nazi concentration camps who were marched in formation to stand in line for 'shower baths' and then led into the gas chambers for extermination, I could imagine—

almost feel—the choking sensation, the gasping for breath, the agony, the utter helplessness. . . . When I entered the shower room, however, I had no idea what to expect. Or what awaited me from the shower into the other room where the man dressed in white greeted us."[40]

With this in mind, it is misleading to look only at the number of immigrants actually excluded by the process of medical examination. Just as important, perhaps more so, was the number touched by this process: for twenty-five million immigrants, the moment of entry into the United States represented a profoundly consequential trial. Manny Steen, an Irish immigrant who arrived at Ellis Island in 1925, kept the moment of entry fresh in his memory for nearly seven decades. "I think, frankly, the worst memory I have of Ellis Island was the physical because the doctors were seated at a long table with a basin full of potassium chloride and you had to stand in front of them, follow me, and they'd ask you and you had to reveal yourself. . . . Right there in front of everyone! I mean, it wasn't private." To people acutely conscious of their individual and collective humiliation, the tense moments lasted an eternity: "You just had to stand. And the women had to open their blouse. And here, this was terrible. . . . I was as embarrassed as hell, you know, I had to open my trousers and fly and they would check you for venereal disease or hernia or whatever they were looking for. I don't know, when I had the physical. . . . I was in good shape, you know, but just the same I felt this was very demeaning. . . . There must have been some other way."[41]

Steen described not an examination but a public ordeal. Historians have disagreed over the significance of the immigrant medical exam for precisely this reason: it was not a thorough exam conducted in private, but an exam for all those present to witness (fig. 2-1). Even the more intensive examination of the estimated 10 percent to 20 percent of immigrants whom the PHS turned off the line was also a public event, though one segregated by gender, as illustrated in photographs taken at Ellis Island sometime after the turn of the century (figs. 2-2 and 2-3). Enid Griffiths Jones, examined at Ellis Island in 1923 at age ten, as an adult still felt the shock, the sting of the public medical exam: "And we went to this big, like an open room, and there were a couple of doctors there, and then they tell you, 'Strip.' And my mother had never, ever undressed in front of us. In those days nobody ever would. She was so embarrassed. And it was all these others, all nationalities, all people there."[42]

Eugene Lyman Fisk, medical director of the New York Life Extension Institute, drove home the point in the 1920s that adequacy of routine medical

FIGURE 2-1. Medical examination on the line at Ellis Island. (Photo:
National Park Service, U.S. Department of Interior)

inspections must be measured by the intent and not the extent of the exam.
Wrote Fisk, "An examination that attains its practical ends in its own particular field whether it be industry, life insurance, or military service, may be regarded as complete. It is idle to criticise such examinations or to say they are
not complete because they do not include a Wasserman test or an electrocardiogram or an X-ray of the chest."[43]

The public nature of the immigrant medical exam was the very source of its
strength, making it a vector of power: it served to communicate industrial values and norms in a public setting, and demonstrated the power to enforce
them by sending back a token number of immigrants. Here power worked
much as it does in a context where the aim is a punitive expression of strength
or mastery. A central feature of torture, Foucault writes, is that it "must be
spectacular, it must be seen by all almost as its triumph." In keeping with an

FIGURE 2-2. Women undergoing the secondary medical
examination at Ellis Island. (Photo: National Park Service,
U.S. Department of Interior)

older paradigm of power that rationalized public punishment or torture, the
spectacle of inspection on "the line" represented a "ritual recoding" to be "repeated as often as possible."[44] In accordance with the paradigm of power that
emerged when the purpose of discipline became correction or amelioration,
discipline represented immersion in a particular, routinized, ordered set of ex-
ercises or motions: waiting in line, moving in unison, stepping up to the medical inspector, moving forward, stepping up to the immigrant inspector, answering questions. The result was to establish habits through training, so it is

significant that individuals not only endured the inspection of themselves in public, but also witnessed the inspection of others. In this fashion, they were taught the rules, the repetitive, monotonous habits of industrial order.

This is not to say that the PHS was merely an agent of abstract or benign Americanization. It did not preach patriotism or teach English or civics or offer lessons in U.S. history to most of the immigrants inspected on the line. Although American industry, led by the Ford Motor Company, employed these methods as a means of reinforcing factory culture, the production line itself was a primary means of indoctrination, the first place in which workers were

FIGURE 2-3. Jewish immigrants undergoing the secondary medical examination at Ellis Island. (Photo: National Park Service, U.S. Department of Interior)

directly managed. Nor was the PHS the principal agent of any kind of industrial assimilation. It never explicitly articulated discipline as the purpose of the medical exam. Rather, PHS officers consistently framed their mission as one of defending the nation by excluding dangerous diseases and undesirable immigrants. Nonetheless, the PHS exam mirrored the kind of effort taking place on the floors of American factories. It became analogous to the factory operating under the principles of scientific management. It did not teach immigrants how to clean themselves, but instead bathed them—in public, if necessary. It sorted immigrants publicly, baring them if need be, in an effective if not entirely conscious demonstration of the principles of efficiency, order, and cohesion that were expected of them.[45]

The process of entry into the United States consisted of far more than the forty seconds in which the immigrant passed under the gaze of the PHS officer at the end of an inspection line in an American immigration station. It was part of the opening segment of the process of Americanization, not as an adjustment to American life, culture, and society but as an adjustment to the industrial working class. It was, moreover, one of many reinforcing moments in the new immigrant's life. As Mary Antin's account suggests, it was a process that began overseas, although there the exact form of the exam reflected not only the interests of the United States but also those of the countries from which immigrants departed. Thus Gussow's account can be read as implying that instruction in American industrial values began abroad "in institutionalized barracks life, organized and extremely efficient," in which he and his family lived for several weeks in Danzig while awaiting passage to New York. In some fashion, it foreshadowed the factory floor: "You got up, washed, dressed, ate, and moved around with people always around you. . . . And then came evening. Lights were put out. And you went to sleep."[46] Ellis Island was, in the words of La Sorte, part of "a seamless continuity" that began overseas "and ended somewhere in America."[47] The end was on the factory floor.

Herbert Gutman argues that "the changing composition of the American working class caused a recurrence of the 'preindustrial' patterns of collective behavior usually associated with the early phases of industrialization."[48] Imposing industrial culture on immigrants, therefore, carried special import after the 1870s, a period in which managers and not workers increasingly controlled and sought to control the means of production. As Stephen Meyer concludes, "the adaptation of a new industrial workforce involved a complex matrix of interrelationships between industrialization, social class, and culture." Above all, "the new industrial worker needed a new culture, i.e., a new set of attitudes, values, and habits for his survival and for his very existence in the factory and in industrial society."[49] The immigrant would have to bend to "Industry's iron will," "Industry's iron creeds."[50] Thus the immigrant medical inspection was part of a larger process of industrial assimilation.

INSTRUCTING BODIES: THE IMMIGRANT MEDICAL EXAM AT AMERICA'S GATE

The immigrant medical inspection is of a piece with the broader industrial ordeal of examination, management, and indoctrination that began before the immigrant departed for the United States and continued in everyday life and work. The official mandate of the Immigration Service was much broader than excluding immigrants with disease. IS officials ensured that each immigrant was eligible to enter under the law, and the law excluded, in addition to immigrants with disease, contract laborers, persons likely to become a public charge, youths under sixteen years of age and unaccompanied by a parent, people who had received financial assistance to come to the United States, polygamists, anarchists, prostitutes, pimps, criminals, Chinese and Japanese laborers, and those without proper passports.[51]

IS officers used these legal categories of exclusion as a guide to examining immigrants. Upon arrival, IS inspectors checked each passenger's identification tag and questioned him or her, checking information against that provided in the steamship's passenger manifest, which included the immigrant's destination, occupation, finances, literacy, age, and marital status. Upon arrival at the nation's largest ports, either customs or IS officials would pin an identification tag to the clothing of immigrants. Manny Steen, arriving at Ellis Island from Ireland in 1925, later described the inspection ordeal as a sort of flesh-and-blood assembly line: "The guards, as we called them there, the customs officers and immigration officials, they, first of all they slammed a tag on you with your name, address, country of origin, et cetera. . . . Everybody was tagged. They didn't ask you whether you spoke English or not, everybody was tagged. They took your papers and they tagged you. That was the first thing."[52] Thus, the ordering and sorting of immigrants began immediately upon arrival. It was an intimidating process, and one for which the steamship agents began to prepare immigrants long before arrival in America. La Sorte reports that "the agent made a point of rehearsing his charges in how to conduct them-

selves. . . . Each of the questions that would be asked by the American inspectors was carefully reviewed by the agents, and stock answers were re-
hearsed time and again until immigrants had them memorized. . . . The intensity with which the agents reviewed these instructions added to the
emigrants' feeling of apprehension."[53]

Next, it was the turn of the IS to begin to convey the rules of American society. Proceeding through each column of the manifest, IS examiners asked
immigrants a set of rote questions—"long, tiresome questions"[54]—about why
they had immigrated to the United States (had they been promised a job? were
they coming to join family?), whether they were morally fit (were they prostitutes, pimps, criminals, polygamists, or anarchists?), and whether they were
economically fit (what was their occupation? where did they plan to seek work?
did they have any prospects for employment? did they have in their possession
the requisite twenty-five dollars?).[55] According to Regina Tepper, who entered
through Ellis Island as a fourteen-year-old from Poland, the process was intimidating, terrifying: "And we finally get to Ellis Island and they take us in and
we go before this great big jury in this courtroom . . . And they ask us our names
and our ages and where we came from and the father's and the mother's name
and all the basic information and they tell us to sit down. And they cross-
examine us. You would think we were really spies."[56] Like the medical inspection, the interview with IS officials took place within sight and earshot of
other immigrants (fig. 2-4).

For Louis Adamic from Balto, Yugoslavia, it was the immigration inspection, rather than the medical inspection, that he most vividly remembered
from his 1913 stay at Ellis Island. All day he sat listening to hundreds of immigrants answering questions about their nationality, finances, parents, intentions, before he was called. "The examiner sat bureaucratically—very much in
the manner of officials in the Old Country—behind a great desk, which stood
upon a high platform. . . . The official spoke a bewildering mixture of many
Slavic languages. He had a stern voice and a sour visage. I had difficulty understanding some of his questions." After his examination, Adamic once again
sat on the benches in this strange court to await the friend who was to pick him
up. The friend, too, was interrogated, during which time Adamic's "heart
pounded." When it was all over, he reported, "I was weak in the knees and just
managed to walk out of the room, then downstairs and onto the ferryboat. I
had been shouted at, denounced as a liar by an official of the United States on
my second day in the country." For Adamic, as for Steen, adding to his chagrin

FIGURE 2-4. IS officers question immigrants using the ship's manifest.
(Photo: Culver Pictures)

was the constant awareness that both the mundane and humiliating portions of his ordeal took place "before a roomful of people."[57]

The typical immigrant was unable to distinguish doctors from guards from IS officials, all of whom, dressed in imposing uniforms, represented authority, judgment.[58] Immigration regulations required that all employees of the IS wear uniforms with buttons, caps, cap insignia, collar insignia, and service insignia (consisting of gold braids designating one or more year of service). Along the Mexican border, immigrants were also confronted with gun-toting soldiers and Border Patrol officials.[59] Waiting for her father at Ellis Island with her mother shortly after the turn of the century, Marie Jastrow grew panicky, thinking he would not come. Her mother slapped her, fearing that her crying would attract the attention of the immigration officers, who might mistake her for a lunatic. An immigrant would "[freeze] in his tracks at the very sight of a uniform—any uniform."[60] Marge Glasgow remembered undifferentiated authority figures: "I remember the Great Hall, and the desks there with men. I don't know if they were doctors, judges or what, questioning the people, you know. And that's when I was very scared, to be all alone in that big building be-

ing questioned."[61] Paulina Caramando, who entered Ellis Island from southern Italy in 1920 at age eight, noted that the facilities themselves had a military aura. She "remember[ed] wondering if Ellis Island was a place for soldiers."[62]

For many immigrants, such as Jack Weinstock, who emigrated from Austria with his family as a nine-year-old in 1923, the prison merged with the
stockade: "We felt like livestock."[63] Even though she was only three when she
arrived at Ellis Island, Margaret Wertle from Hungary vividly remembered the
prisonlike atmosphere in which people were akin to animals: "They put us in
this cage at Ellis Island. Wire fence all around. All these people shoved in
there. And I remember just holding on to that wire and looking around and
looking to see the people."[64] Recalled Rachel Shapiro Chenitz, who arrived at
Ellis Island in 1922 as a ten-year-old Palestinian immigrant, "the boat docked,
and they dropped us off on Ellis Island, but like a bunch of cattle or sheep, I'll
tell you, not like humans."[65] The immigrant was made to feel like Taylor's pig-
iron-handling gorilla—expected only to obey and not to question—even before setting foot on American soil. Indeed, the imagery of prisons, cages, and
stockades would reemerge as immigrants described the factory experience.
Songs of labor labeled the shop "Trade's concentration-camp," "Where, ox-
like, I must drudge until I drop."[66]

Although fourteen-year-old Bessie Kriesberg and her brother Dave had
traveled first class to New York on the *Lusitania*, they had lost the address of
the two brothers in Chicago who had preceded them. Bessie and Dave were
sent immediately to Ellis Island. Writing in the third person, Bessie wrote that
she "was put in a locked room. This frightened her. It made her feel as if she
were in jail. Besides this, they had taken their money away from them. She
cried because she had no way of knowing how long they might have to stay
locked up. She couldn't see Dave because he was in another section where the
men were. It was the first time in their lives they had been in locked quarters.
She was angry at the man who had been so strict with them. Whenever she
asked him about something he ignored her, and she felt bewildered." The experience—a short four days, compared to the weeks that many detained immigrants were held at Ellis Island, particularly those held for medical reasons—impressed on her the need "to obey the rules."[67]

For those immigrants actually "turned off the line" and medically certified
by the Public Health Service, the interrogation was far more daunting—
threatening, even.[68] Each received a hearing before an IS Board of Special In-

quiry. Here, a panel of three IS officers—never PHS officers, though they might testify—questioned the immigrant about his or her occupation, finances, and family residing in the United States. The questions asked of a suspect alien appearing before the BSI often had little or nothing to do with the nature of the disease, as is evident from the transcript documenting Samuel Nelkin's BSI hearing in 1926.

Nelkin, a sixty-year-old Russian "Hebrew," found himself before the BSI after the PHS certified him for "absence of teeth (partial). Partial ankylosis [stiffening or immobility] of right joint. Extensive scar of scalp." BSI examiners fired off a rapid stream of questions: Who paid for your passage? Have you other relatives in this country? Did you have any children? How have you supported your wife up to the present time?

Nelkin made a poor case for himself. He had been unable to pay for his trans-Atlantic trip himself. He had only a sister-in-law in Philadelphia and brother-in-law in St. Louis—the benefactor who had provided the steamship tickets. All three of his children had died. Although he had owned a tannery, after "the Bolcheviki came and took it away" from him, he had "not done anything." "[M]y relatives supported me," he reported. The examiners dug deeper. "Your wife do anything?" they asked. Nelkin, by now, saw the emerging picture of himself and struggled to recover. Ignoring questions regarding his wife Ester, he pleaded that he had found eight months of work in Southampton. Nelkin's brother-in-law, a citizen of thirty years' standing, owned a printing shop and had invited Samuel and Ester to America. He had even visited him at Ellis Island. Surely this suggested that he would employ Samuel, that the familial bond was strong.

The examiners turned, at long last, to the cause of Nelkin's interrogation—his health. "How long have you been in your present physical condition?" they asked. "I have had the hypertension the last 20 years," reported Nelkin, desperately trying to convey that his health had not yet interfered with his capacity to work. Nelkin had endured far worse things than hypertension, he reminded his interrogators: "Since 1905, I have been attacked by Russian Pogroms and my skull was broken."[69] Unsuccessfully, he tried once again to turn the tide, reiterating that Ester's brother Hersch was a bookbinder. He had a store.

The BSI unanimously voted to reject Samuel and Ester Nelkin as likely to become public charges.

While Nelkin's interrogation was typical, the outcome was unusual. In most instances the BSI overruled the medical certificate and did not reject the immigrant. In all instances, though, interrogations were intense, even hostile. The immigrant was on guard. BSI examiners did not seek to comfort or reassure. They sought to warn. For some the BSI inquiry ended with an oath, as in the case of Elizabeth Nimmo's mother, raising her children alone, who in 1920 "had to swear on that Bible" that "neither [she] [n]or anyone of we four kids would ever become a burden on this country."[70] Through the dual examinations—the medical exam followed by the IS interrogation—federal officials could impress on the immigrant and the hundreds or thousands of immigrant witnesses awaiting their turn on line behind him or her what was expected in this society, where each belonged, what each was worth, and how each must conform in order to stay. The medical exam conveyed the importance of health, of avoiding hospitalization; the subsequent IS exam warned the immigrant to find and hold a job, keep money in his or her pocket, obey the law, and earn a living. The immigrant medical inspection, then, was not so much an exclusionary exercise as a normative exercise. It was an expression of authority that was most fully realized when the immigrant was *not* turned away.

MANAGING BODIES: AFTER ADMISSION

The notion of power—to borrow from Andrew Mendelsohn's reflections on the idea of the "social"—"is a thing of too much use and too little dispute. Its murkiness makes it indispensable."[71] Social theorists and historians of public health, often addressing questions of social control, are fundamentally concerned with power and power relationships, but may give insufficient consideration to who wielded it and why. What was at stake? Who gained? Who benefited? Who controlled what? How extensive was that control? With regard to the practice and meaning of the immigrant medical exam and the ethos and ideology of the PHS and IS, what would constitute proof that these agencies successfully wielded power—particularly when they expressed power not by turning immigrants away, but by allowing them to pass? Available evidence strongly suggests that immigrants admitted to the United States absorbed the messages and warnings that were conveyed during the course of the immigration exams, pounded in by industry, and reinforced by myriad immigrant aid societies, health officials, and writers in the popular press.

Immigrant and native-born American alike maintained an intense interest in their personal health.[72] John Birge Sawyer, for example, worked for the Immigration Service and the Consular Service from the early years of the twentieth century through the 1930s. His diaries, though personal and bearing little direct relation to immigration work, underscore the great importance individuals—and, more important, IS officials—placed on health. Sawyer was very attuned to death and illness in his family. He describes the death of his wife's mother from Bright's disease in 1912 and the health of his father. In 1913 he reports that "our vacation was marred by Nancy [his daughter] having whooping cough which lasted for several months and by my having a severe attack of malaria with chills and fever regularly from June 21 to July 10. During the rest of the year all the family remained in good health." In 1912, he also reported an attack of gallstones. Sawyer's frequent remarks on health continue throughout all five volumes of his diary.

But health was more than an individual good, it was a social good. Health was part of a seamless continuum of personal, domestic, and national welfare that included earning a living wage, having enough to eat, benefiting from clean water and adequate sewage, and having decent working and living conditions.[73] Health, then, was not simply determined by wages or housing, but like them was considered a material social condition of the individual and of the community or nation. Immigrants, in particular, were exhorted to preserve their health. The North American Civic League for Immigrants advised the newcomer to "keep everything clean and sweet about your person, your home, and your street. This is your best protection against disease." For "in America as in other countries nothing is more necessary to the ambitious than good health."[74]

Margaret Wertle's mother boarded with her aunt while she established herself in America as a "cook for very wealthy people." Margaret underscored the immigrant's embrace of American health ideals when she contrasted her appearance with her mother's when her mother returned to Budapest to claim the three-year-old Margaret. "I remember the day she came. She stepped down from that carriage. . . . She saw me and started to cry. I was full of lice. I was dirty. . . . My mother was so immaculate, so clean. I never saw such a clean person. They took me to the doctor to see if I was all right. I was healthy, but I was a dirty little girl. They had to scrub me, because my aunt never washed my hair or anything."[75] She laughed when describing her condition decades later, as an elderly American citizen granting an oral history interview, yet the situation was clearly no laughing matter for her mother who anticipated the medical ordeal her daughter would undergo.

Mrs. Maurice, a thirty-eight-year-old Italian immigrant, gloated that in the years before her husband lost his job as a shoemaker, their first child "had everything that a child of moderate circumstances could have. I went to the hospital when she was born, had a private nurse and a doctor and not one time did I ever have to take her to a Free Dispensary, nor did any nurses with a blue coat and hat have to call at our home." In the lean years, however, after the Maurice family lost their home and Mr. Maurice barely made ends meet by working as a street cleaner, Mrs. Maurice was deeply humiliated when a police officer reported her family's squalid living conditions to a public health nurse.[76]

The federal process of socializing the immigrant did not end once he or she passed through the immigrant medical exam. Surveillance became an important tool not only for tracking immigrants' behavior but also for conveying society's expectations.[77] The Immigration Commission, for example, was keenly concerned with the living conditions of immigrant workers in different occupations. Far broader in scope than its survey of charity organizations and hospitals was its investigation into the distribution of immigrant groups in different industries around the nation and the manner in which they lived. The commission sent investigators into laborers' homes—predominantly immigrant homes—around the nation, armed with survey instruments to record the number of inhabitants in each dwelling and cleanliness and care of the home. Investigators rated the floors, ceilings, windows and doors, water supply and toilet conditions, lighting, and heat as either good, fair, bad, or very bad. They also rated the amount of dirt present in dwellings and the level of general tidiness.[78]

The federal government, using the immigrant medical exam and various forms of surveillance, was not the only social force acting to indoctrinate immigrants, to impress upon them the importance of adhering to American standards and norms. The efforts of the federal government mirrored the efforts of corporations, which in the early twentieth century brought a cadre of welfare workers, drawn from the ranks of YMCAs, settlement houses, and public health movements, into industry to train and supervise the immigrant work force. Corporate welfare workers sought to bring the immigrant in line with American practices of cooking, housekeeping, hygiene, and education. Companies felt they had to act with such "direct paternalism" because immigrants clearly were not capable of self-leadership. In this fashion, industry sought to apply the principles of scientific management outside of the factory walls, in the community itself.[79] In life, as in the workplace, immigrants required direction over their bodies. They were incapable of self-discipline or self-control.

At Ford Motor Company, for example, the Sociological Department evaluated the "habits, home lives, and attitudes of workers" to determine who was fit to share in company profits. In 1914 the company granted to 57 percent of its work force profit sharing amounting to an extra five dollars per day. Significantly, immigrants—mostly Russians, Romanians, and Austro-Hungarians—represented the vast majority of Ford workers. Ford's Sociological Department, therefore, developed profit-sharing standards that "pertained above all to the life-styles of the immigrants."[80] Ford required workers to meet high standards in "thrift, honesty, sobriety, better housing, and better living generally." The company duly admonished workers to "live in clean, well conducted homes, in rooms that are well lighted and ventilated. Avoid congested parts of the city. The company will not approve, as profit sharers, men who herd themselves into overcrowded boarding houses which are menaces to their health" and promote high rates of absenteeism and turnover. Perhaps Margaret Wertle's mother absorbed the company's advice to "use plenty of soap and water in the home, and upon [the] children, bathing frequently. Nothing makes for right living and health so much as cleanliness. Notice that the most advanced people are the cleanest."[81] The profit-sharing endeavor, like the immigrant medical inspection, was as much about disciplining the immigrant labor force as about increasing the efficiency of production and reducing worker turnover.[82]

For Frederick W. Taylor, as well as a host of scientific managers and a changing industrial leadership, the ideal worker was one who did "exactly as he's told from morning till night . . . and no back talk."[83] Industry believed that managing men—training them to fit the mold demanded of them by the nation, by the workplace—made them better men. Those subject to Taylor's experiments at Bethlehem Steel supposedly became "not only more thrifty but better men in every way; . . . they live rather better, begin to save money, become more sober, work more steadily."[84]

Just as scientific management eventually won the day and resulted in the profound transformation of the industrial workplace, evidence suggests that immigrants may have heeded the warning they heard at Ellis Island and the nation's other points of entry. The U.S. Immigration Commission in 1909 found that southern and eastern European immigrants, who dominated the

manufacturing and mining industries in the eastern and middle western
United States, "represent the economic failures—the derelicts" of American
society. Still, investigators were surprised by the degree to which these workers tried to emulate American standards of living: "While instances of extreme
uncleanliness were found, the care of the households as regards cleanliness
and an attempt to live under proper conditions was usually found unexpect-
edly good, about five-sixths of all families visited in the poorer quarters of these
large cities keeping their homes in reasonably good or fair condition."[85]

Even organized labor adopted the language of health, using it to justify a
strong labor movement (and viewing ill health as an indictment of working
conditions in the United States). The International Ladies Garment Workers'
Union and Joint Board of Sanitary Control (responsible for the sanitary inspection of New York State's garment industry), proclaimed: "In sanitation and
health as well as in economics, the salvation of the workers depends on the
working class itself."[86] The ILGWU merged the language of health and the
language of the factory in promoting health inspections to its members, suggesting that each "Inspect and Overhaul Your Machinery." The rationale was
simple: "Your bodily health, your muscular strength, your general bodily and
mental equipment are your only assets and capital."[87] In general, immigrant
communities well understood that health was a form of social currency. They
used this currency, when they could, for their own direct benefit.[88]

Immigrants became ardent advocates of health and independence. Indeed,
the Chicago Foreign Language Press Survey—a collection of indexed and
translated articles representing twenty-one ethnic groups, drawn from the
foreign-language press of Chicago from 1861 to 1936—contains no articles
complaining of immigrant medical inspection or immigrant restriction.
Rather, immigrants championed health and demanded preventive sanitary
services. The draft of a report of the Foreign Language Information Service,
for example, quoted what it called a typical immigrant health "catechism" frequently printed in foreign-language newspapers (see the epigraph to this chapter, above).[89] Such a view was shared across the spectrum of the ethnic press:[90]
"To have and to hold health is our divine heritage. There is no greater wealth
than health."[91] One Croatian writer proclaimed, "A nation, which wants to
continue as a nation and to progress as such, must be healthy."[92] Polish immigrants in Chicago internalized the shame of dependence as a community.
Wrote one Polish Jew, "We are good citizens and good Jews, and, therefore,
we do not burden the community with our helplessness; we maintain charity

institutions of our own."[93] Concludes historian Lizabeth Cohen, "America had clearly taught ethnic communities to support charitable institutions as patrons."[94]

Yet while immigrant aid societies, settlement houses, the popular press, the foreign-language press, and various other sources might work to create both a burning sense of shame regarding filth and disease, and an imperative to shine, the immigrant examination ordeal that took place on the nation's shores was the only normative force with dual powers. IS and PHS officers wielded the authority not only to exclude immigrants, but also to admit them. The IS and PHS saw themselves as granting admission to immigrants who earned it. Immigrants themselves understood the PHS and IS examinations as an expression of power. Wrote one Chinese immigrant on the walls of Angel Island,

> I thoroughly hate the barbarians because they do not respect justice.
> They continually promulgate harsh laws to show off their prowess.
> They oppress the overseas Chinese and also violate treaties.
> They examine for hookworms and practice hundreds of despotic acts.[95]

Individual immigrants might explicitly acknowledge and write in defiance of American power, but most acknowledged it subtly as they yielded to it. Nayan Shah, for example, describes how the Chinese community in San Francisco "reinterpreted their identity and their relationship to the nation through racially coded languages of hygiene and health. . . . The outcome of the Chinese American community's claim to citizenship and cultural belonging depended on the performance of normative hygiene."[96]

This is not to suggest that immigrants were docile conformists naively absorbing different forms of socialization. As James Barrett observes—echoing the work of Rudolph Vecoli, Herbert Gutman, and other historians who challenged Oscar Handlin's notion of the immigrants as "uprooted"[97]—many different settings, from the shop floor to trade unions, "provided immigrants with alternatives to the world view and the values advocated" from other sources of Americanization, such as scientific managers or immigrant medical inspectors.[98] Montgomery similarly argues that employees adopted not simply "the values and habits welfare plans sought to inculcate," but also "working-class mores."[99] That workers described themselves as "machines" or "just like a horse and wagon" in their songs and poems of labor was no indicator of conformity.[100] Such descriptions, after all, were intended to galvanize workers to organize and resist, to assert their own power.

Daniel Bender argues that Jewish clothing workers learned to view "the bodies of workers emerging from garment shops with the same critical medical eye that had examined them at Ellis Island or Castle Garden." In the case of garment workers, adopting the perspective that "garment labor meant a confrontation with disease . . . helped create a coherent set of issues that consolidated and empowered the labor movement of men and women, especially after the transformative strikes of 1909–1910."[101]

Workers in several industries mounted a successful campaign against the adoption of scientific management. American Locomotive, a company made up of ten works in the United States and Canada, attempted to introduce the stopwatch and incentive pay in 1907, but the effort was thwarted by resistance on the part of boilermakers and machinists. Opposition to Taylorism next spread to the metal trades: "The very appearance of stopwatches, time cards, or measurements of machine cutters, beds, or T-bolts that so much as hinted at standardization," Montgomery writes, "was enough to trigger anxious caucuses of craftsmen, strikes, or beatings of those who seemed to be collaborating with the systematizers."[102] Workers managed to have Taylorism banned in several government industries. But even if "Taylorism was liquidated, 'scientific management' was not by any means."[103]

Americanization was fundamentally a coercive process, widely conceived in terms of transformation in a "crucible" or "melting-pot," which spoke directly to the physical hardship that immigrants endured both in passage to the United States and in their early years of industrial employment, if not for the rest of their lives.[104] Mary Antin prefaced her autobiography with the dramatic declaration, "I was born, I have lived, and I have been made over. . . . I am just as much out of the way as if I were dead, for I am absolutely other than the person whose story I have to tell."[105] Edward Steiner echoed that sentiment, explaining, "I dimly felt what it meant, but I did not realize how new was the life which awaited me, or how completely I was being severed from my past and my former self."[106] Steiner and Antin were certainly partners in their own transformation, in the construction of that new identity.[107] Yet both convey that something was done to them, something involving—if not the sharp edge of violence—physical and emotional adversity. Adoption of a new identity was not entirely negative, to be sure. Indeed, Steiner and Antin were both proud of their new identities, and many immigrants endured the hardship of coming to and working in America precisely because they faced greater hardships abroad. But there remained inescapable elements of coercion that we cannot discount or play down. Antin draws an analogy to death; Steiner evokes images of laceration in explaining that he was "severed" from his past. Historians such as Gary Gerstle argue that there were many possibilities for immigrants to shape their own identities, but they also acknowledge that they faced many constraints.[108] Gerstle concludes that "the nation is itself a structure of power that, like class, gender, and race, necessarily limits the array of identities available" to the working class. Therefore, "any analysis of Americanization, past and present, must accord coercion a role in the making of Americans."[109] That workers protested their working conditions, argues Montgomery, the very terms they used to describe the dehumanizing effect of the factory floor, revealed that power did not lie in the hands of the worker.[110]

Industry and government created and promoted an industrial work ethic and established the political and social context that would shape workers' responses.[111] Power, according to Foucault, "is not exercised simply as an obligation or a prohibition on those who 'do not have it'; it invests them, is transmitted by them and through them; it exerts pressure upon them, just as they themselves, in their struggle against it, resist the grip it has on them." Power, as something "exercised rather than possessed," results in "innumerable points of confrontation, focuses of instability, each of which has its own risks of conflict, of struggles, and of an at least temporary inversion of the power relations."[112] What is perhaps most critical is that industry and government set the terms of confrontation.

Erving Goffman observed that "every institution has encompassing tendencies." That is, every institution provides "something of a world" for its members. Goffman, of course, was interested in what he called "total institutions"—those places that sought to control all the different spheres of an individual's life. "The handling of many human needs by the bureaucratic organization of whole blocks of people," he wrote, "is the fact of total institutions."[113] An immigration station was not a total institution, nor were the stations at the port of departure abroad, the ships on which immigrants traveled, the tenements in which many industrial workers lived, or the places in which they were employed. Viewed individually, they either failed to (or did not seek to) control all spheres of the immigrants' life or did so for only a very brief time. Each of these experiences, however, worked toward a common end in a nation that sought to serve as a kind of total institution for immigrants.

Immigrants, consequently, drew on the language of total institutions—prisons, cages, guards—to describe their immigration experience. On the one hand, this represented an Old World view, in which they saw authority in terms of tyranny and oppression, dictators and soldiers.[114] On the other, it was an acknowledgment of a new industrial order. In both instances, immigrants—who developed a language for conveying their experiences long after arrival in the United States—were not just describing the PHS or the IS, the steamship inspection, or the nature of factory or sweatshop work. They found the imagery of total institutions appropriate for explaining the entire ordeal of entering America—an ordeal that lasted not forty seconds, or two weeks, or even two months, but years and years.

CHAPTER THREE

THE MEDICAL GAZE

Science in Industrial-Era America

The medical profession has a great opportunity and a great duty. No one
is better qualified than the physician to say which immigrants are desirable. . . . [U]nder the law the most important feature of immigrant inspection devolves upon physicians.
>   —Dr. Alfred C. Reed, assistant surgeon, U.S. Public Health
>   and Marine Hospital Service, New York City

OFFICIALS CONDUCTED immigrant medical inspection swiftly so as not
to interfere with shipping and trade. The sheer numbers of immigrants arriving in New York and other ports prevented extensive examination in most
cases. Ellis Island officers were known to examine several thousand immigrants a day.[1] As one Ellis Island physician recalled, "The invaders were arriving in such numbers that individual physical examination with our meager staff
was out of the question. A snap diagnosis which stood a reasonable chance of
proving correct had to be made in a few seconds."[2] And Ellis Island had am-
ple staff and resources compared to most stations, where officers worked under poor conditions, in dilapidated facilities, with inadequate equipment. In
Boston, for example, there was "no single station or shed where steerage passengers are examined, as at New York, Philadelphia, or Baltimore." Instead,
"saloon [first class], second-cabin, and steerage passengers must be examined
either on shipboard or in the wharf sheds where the ships tie up. Large passenger ships are frequently passed by quarantine after sunset, and while under such circumstances the examination of the steerage passengers is usually
deferred until the next day, it is often necessary to attempt the examination of
the saloon and second cabin late at night."[3] As late as 1924, the surgeon in
charge still complained for want of proper facilities: "You can appreciate making heart and lung examinations in a room over an electric water pump, ad-
joining a dry dock and railway terminal. Noise of machinery and so much dirt

and dust that it is uncomfortable to have the windows open in the summer season. For dark room examinations of eye and ear cases we make use of a disused toilet."[4]

The microscope—the most basic laboratory tool, the emblem of bacteriology—was unavailable at many immigration stations across the nation. Stations began requesting microscopes and other laboratory equipment beginning in the 1890s.[5] The surgeon stationed at Niagara Falls in 1913, exasperated by requests denied, purchased his own microscope.[6] In 1893, after his first request for a microscope was denied, Dr. Preston Bailhache asked how he was to perform his duties without one: "By what means shall they [PHS officers examining immigrants] determine the presence of certain pathological conditions, which up to this time can be discovered only by the use of the microscope?"[7]

Here was a question that might have rocked the rationale of immigrant medical inspection. In the 1880s and 1890s medical schools successfully began to incorporate bacteriology and basic science—the hallmarks of a new scientific medicine—into their curricula as a means of transforming medical education and increasing the status of the medical profession.[8] Scientific medicine promised to shore up the professional authority of the physician and fortify him against irregular medical practitioners, such as homeopaths, herbalists, osteopaths, or eclectics. Adopting the accoutrements of scientific medicine paved a road to status and reputation in an intensely competitive environment.[9] The PHS officer was one of a new cadre of public health professionals who readily embraced bacteriology and championed the laboratory.[10] The Public Health Service opened the nation's first bacteriological laboratory—the Hygienic Laboratory on Staten Island—in 1887.[11] Its commissioned officers were skilled in laboratory technique.[12]

The PHS envisioned its commissioned officers as renaissance men of sorts.[13] Many were southerners educated at the University of Virginia, which promoted "a very strong esprit de corps."[14] Dr. John Heller, who graduated from Emory University School of Medicine and entered the PHS as acting assistant surgeon in 1931, became a commissioned officer by examination in 1934. Later in life he reminisced that, even into the 1930s, the PHS "could be regarded as something of a nineteenth century Corps rather than the modern concept of the twentieth century, and it was somewhat of a throwback to the old days when we lived in an elite rather than arrogant, snobbish type of society, very proud of it, and it was a very fine medical organization."[15] Two to four physician applicants were accepted into the coveted positions of commis-

sioned officers of the PHS each year, after undergoing days of "grueling" examinations. In his memoir of a career in the PHS, Dr. Victor Heiser described
the two-week examinations required of commissioned officers. The first several days were spent answering four questions each day, none of which the
applicant could get wrong. Each morning, unsuccessful applicants were dismissed before the next day of exams. This round was followed by oral examinations in history, philosophy, economics, and literature. Those candidates
able to demonstrate fluency in all of these fields were then taken to a hospital
to examine and diagnose six patients. Finally, "the same technique was followed in the laboratory, where we were required to analyze specimens and
identify bacteria and parasites under the microscope; many of these slides had
been prepared to confuse us."[16]

But despite the frustrations of some officers within the PHS—and many
stakeholders outside it—regarding access to technology, specifically, and the
nature of the exam, more generally, we should resist framing the immigrant
medical exam as a problem because examiners lacked either the equipment or
time to conduct a rigorous exam. To do so would be to ignore the conflicting
demands and interests creating tension within the United States when it came
to the immigrant question. The PHS medical exam was situated at the cross-
roads of many contending interests. The factors that we often think of as having prevented a more effective exam—vast numbers of arrivals, few resources
devoted to exclusion—were themselves the consequence of a decision to admit a tremendous and relatively unimpeded flow of immigrant labor.[17] This
decision necessitated a processing or disciplining rather than a screening regimen at the nation's borders. And in such a regimen the laboratory was not necessary.

In 1891 Surgeon General Walter Wyman flatly stated that he had no intention of establishing bacteriological laboratories at immigration stations.[18]
There was little further elaboration on the intentions of the PHS until the
agency published its first set of regulations guiding the immigrant medical
exam, more than a decade after the passage of the 1891 law mandating immigrant medical inspection. By this time the PHS had already perfected a system suited to the mass processing of industrial laborers in a society that had
begun thinking "in terms of a complex social technology, of a mechanized and
systematized factory."[19] The PHS regulations first issued in 1903 never explicitly defined the agency's mission as one of disciplining the laboring class.
Indeed, they defined it rather narrowly, in terms of preventing the entrance of

disease into the nation. PHS officers, however, interpreted their job far more liberally. In their eyes, the goal was to prevent the entrance of undesirable peoples—those "who would not make good citizens."[20] In the context of industrial-era America, this meant immigrants who would not make good industrial citizens, who would, instead, wear out prematurely and require care and maintenance that the nation could ill afford. Despite the rhetoric, the regulations outlined an inspection procedure that made possible a search for only a limited range of conditions visibly affecting the ability of the immigrant to work. The regulations and the manner in which doctors interpreted and acted on them indirectly but powerfully testify to the extent to which the American industrial context shaped and directed the uses of science and medicine.

THE "LINE"

Immigrant medical examination centered on the "line." Although there were important regional variations that figure in the story of small numbers, the line is central to the story of large numbers, the story of inclusion. The line became shorthand for techniques and procedures for quickly examining thousands of immigrants, and represented a direct and meaningful analogy to the industrial assembly line.

Ellis Island, where roughly 70 percent of immigrants entered the United States, set the standard for examination on the line. For arriving immigrants, the PHS inspection was part of a long, exhausting ordeal that began overseas and continued after admission to the United States. Telescoping the immigrant medical exam to the few seconds in which the immigrant actually passed under the eye of the PHS officer opens the door to misreading the exam and its function.

Rose Cohen, for example, after a two- to three-day train ride, was smuggled beneath "ill-smelling" hay at the bottom of a hot, dusty wagon from the Jewish Pale of Settlement into Hamburg. Here she and her aunt, with whom she traveled, waited for a week. Days they spent waiting for their names to be called confirming their passage on the steamship; nights were passed in quarters "where many dirty, narrow cots stood along the walls." The already exhausted travelers soon learned that here "sleep was out of the question. The air in the room was so foul and thick that it felt as if it could be touched. From every corner came sounds of groaning and snoring. But worst of all were the insects in the cot." When they finally embarked for New York, they were "deathly seasick the first three days." Upon arrival they "had [not] had enough to eat in a month."[21] Lena Karelitz Rosenman's food ran out after the first day of her ocean voyage. For the next eight days she had nothing to eat but bits of onion and a morsel of nonkosher food, which provoked the harsh criticism of other Jewish passengers.[22]

Although they had been bathed and deloused just before departure, passage in the steerage compartment again reduced immigrants into a miserable state. One immigrant wrote that "in the dark, filthy compartments in the steerage . . . [s]easickness broke out among us. Hundreds of people had vomiting fits, throwing up even their mother's milk. . . . I wanted to escape from that inferno but no sooner had I thrust my head forward from the lower bunk than someone above me vomited straight upon my head. I wiped the vomit away, dragged myself onto the deck, leaned against the railing and vomited my share into the sea, and lay down half-dead upon the deck."[23] While seasickness would eventually pass, conditions remained difficult. A midwestern clergyman confirmed the appalling conditions in 1906: "Crowds everywhere, ill smelling bunks, uninviting washrooms—this is steerage. The odors of scattered orange peelings, tobacco, garlic and disinfectants meeting but not blending." Even as late as 1910 conditions were little improved: "Everything was dirty, sticky and disagreeable to the touch," an investigator reported to a congressional committee charged with investigating conditions. The basin in the washroom "served as a dishpan for greasy tins, a laundry for soiled handkerchiefs and clothing, and a basin for shampoos without receiving any special cleaning. It was the only receptacle to be found for use in the case of seasickness." The toilets simply "baffle description as much as they did use." These filthy "open troughs . . . were apparently not cleaned at all."[24] After traveling on the White Star liner *Teutonic*, Edward Steiner summarized, "The steerage was crowded and the nine days at sea were nine days in prison."[25]

After an arriving ship passed the quarantine inspection in New York Harbor, IS and PHS immigrant examiners boarded and examined all first- and second-class passengers as the ship proceeded up the harbor.[26] When the ship docked, PHS officers transferred steerage or third-class passengers to Ellis Island by barge. Weary not only from the ocean voyage but also from the earlier trek to their European port of departure, immigrants were by this point in their journey often malnourished.[27] They stank, carrying the odor of sweat and powerful disinfectants, which one immigrant reported combined to create the smell of "a terrible sewer."[28] Those smells mingled with other sorts of stench:

stomach acid and partially digested food, "various vegetables, the smell of sheep-skin coats and of booted and unbooted feet."[29] Samuel Chotzinoff offered a similar description, stressing the smell of "bilge" and "less identifiable putrescences," which "settled on one's person, clothers, and luggage and stayed there forever, impervious to changes of habitat, clothing," or cleaning.[30] In this state immigrants began the next phase of the inspection procedure.

Proceeding one after the other and lugging heavy baggage, prospective immigrants entered the often-congested immigration station and proceeded slowly through a series of gated passageways resembling cattle pens (fig. 3-1). The winding passage toward the PHS officers who awaited at the end ensured that each person could witness the inspection of dozens of others ahead. As they reached the end of the line, immigrants slowly filed past one or more PHS officers who quickly surveyed them for a variety of serious and minor diseases and conditions, finally turning back their eyelids with their fingers or a buttonhook to check for trachoma (see fig. 2-1). "Were they ready to enter? Or would they to be sent back?" wondered each immigrant with faces "taut, eyes narrowed" throughout the ordeal.[31]

We see in the medical inspection line the analogy to the factory. The two most prominent symbols of Taylorism were the stopwatch and the camera.[32] The camera, especially, helped the scientific manager to reduce work to its discrete parts. While speed was essential on the medical inspection line, at the same time the individual medical officer took on the penetrating responsibility of the camera. He carefully yet swiftly analyzed the immigrant from various angles, deftly factoring in how he or she might figure into the new industrial economy. At the same time, the ordeal of the line told the immigrant what to expect in the nation's factories. The worker had to be open to the uses to which his or her body would be put on the industrial assembly line. The immigrant, therefore, had to patiently stand on line at Ellis Island, watch the repetitious medical examination of each one before him or her, and follow the commands of the medical managers of the immigration station even as they demanded public humiliation.

The regulations for inspecting immigrants promulgated by the surgeon general described the diagnostic protocol that PHS line officers were to follow. The protocol emphasized the physician's "gaze" or "glance." The similarity to the "medical gaze" described by Foucault is striking. And, as with Foucault, the gaze idealized by PHS officers was one unmediated (initially) by diagnostic technology.[33] The 1910 *Book of Instructions for the Medical Inspection of*

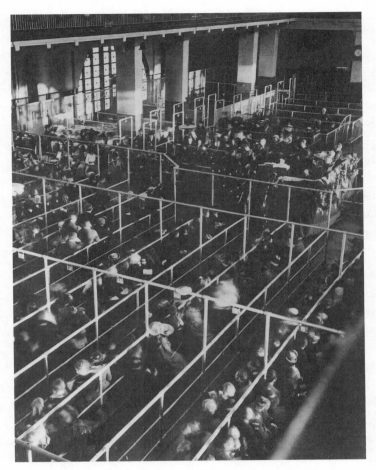

FIGURE 3-1. Immigrants proceed through a series of gates
as they are processed at Ellis Island. (Ellis Island Archives,
Box 19, Folder 6, image 19.16#3)

Aliens underscored that the forty seconds or so an officer spent assessing an immigrant was hardly a perfunctory, assembly-line procedure. Indeed, only a long passage from the regulations can begin to convey the sense of the knowledge, expertise, and judgment that the PHS presumed its officers could exercise in a flash:

As the alien approaches, he should be rapidly but thoroughly surveyed from his feet up, as the latter are the first to leave the examiner's visual field. . . . As the

gaze sweeps upward it will note any undue prominence in the region of the
crotch suggestive of hernia, hydrocele, tumors, and abscesses. The hands should
next be carefully scrutinized, as they furnish important indicia of the general
physical development, disease of the respiratory or vascular system, disordered
or impaired nutrition, enfeebled or defective mentality, nervous and cutaneous
disease, besides local defects and deformities. The abdomen is surveyed with a
view to detecting undue protrusion, as of ascites, splenic enlargement, pregnancy and abdominal tumors in general, the chest for marked asymmetry, undue prominence, and defective development, and the back for spinal disease and
deformities. The neck is inspected for goiter, abnormal pulsation of the cervical
blood vessels, enlarged glands, tumors, and other diseased conditions. Due note
is to be taken of the existence of abnormalities of the head, such as unusual
shape, deformity, disproportion, and marked asymmetry affecting the bones of
the face and the skull. The possible existence of disease of the ears, as well as cutaneous and local diseases, must not be forgotten.[34]

Descriptions of the line inspection drive home the importance not only of
the visual inspection of aliens, but also of touch in the early years of immigrant
inspection. PHS officers palpated, pushed, and pulled on the immigrants they
were inspecting: "Not only sight but the sense of touch and smell play a part.
The hand against the forehead gives an idea of the presence of fever at the
same time sight is taking in conditions about the mouth. . . . [T]he sense of
smell simultaneously may arouse suspicion of uraemia, ozaena, favus, foul discharge from ear, abscesses or ulcers, concealed by clothing."[35]

The conviction that the PHS officer could consider such a volume of information in a matter of seconds and accurately sift the diseased from the healthy
hinged on the assumption that disease was apparent, that it was written on the
body. Dr. Albert Nute, while stationed in Boston, argued that "almost no grave
organic disease can have a hold on an individual without stamping some evidence of its presence upon the appearance of the patient evident to the eye or
hand of the trained observer."[36] Exemplifying this notion, PHS regulations encouraged officers to place a chalk mark indicating the suspected disease or defect on the clothing of immigrants as they passed through the line: the letters
EX on the lapel of a coat indicated that an immigrant should merely be further examined; the letter C, that the PHS officer suspected an eye condition;
S, that senility was suspected; and X, insanity [37]

Laymen, like the PHS officers, believed that disease displayed itself on the body—that immigrants and laborers could be judged on the basis of their appearance. Declared Terence Vincent Powderly, the first commissioner general of immigration, "Vice may come in the cabin or the steerage, in rags or fine raiment, and escape detection, but . . . diseases . . . proclaim their presence and are their own detectors."[38] Nonmedical immigration officials, however, claimed no such skills for themselves. Rather, they respected greatly the ability of the PHS officer to detect that which the lay observer could not see: the PHS officer could detect the presence of "loathsome or dangerous contagious diseases . . . without causing the aliens to undress or without laboratory tests."[39]

PHS officers were unwilling to concede that any but the experienced officer could accurately diagnose disease at a glance. The only trained observer was the PHS officer. Assistant Surgeon General Cofer bragged in 1912 that he had "personally seen the most wonderfully qualified physicians and surgeons make the most ludicrous mistakes in diagnosis at Ellis Island." He particularly delighted in telling the story of one "specialist in nervous diseases" who pulled an immigrant off the line, declaring that his gait was a sure indicator of paraplegia. The experienced line officer almost instantly diagnosed, quite correctly, a simple abscess on the immigrant's thigh. "I say with a great deal of pride that our novitiates," concluded Cofer, "are able to baffle with their skill the average physician who has not made a study of the possibilities of the quick detection of cases on the immigrant line."[40]

The medical gaze, then, remained an art that could not be taught, but only learned through experience: "You saw what the doctors were doing—there was no training, except a few words of instruction."[41] Looking back on a long career in the PHS that began on the line at Ellis Island, Dr. Samuel Grubbs recalled his sense of admiration for the diagnostic wizardry of the experienced PHS officer: "I wanted to acquire this magical intuition but found there were few rules. Even the keenest of these medical detectives did not know just why they suspected at a glance a handicap which later might require a week to prove; but I lost money when I began giving odds on the field, so to speak, against their hunches."[42]

Most physicians valued the medical gaze as highly as intensive physical and laboratory examination. Dr. Victor Safford, who served at both Ellis Island and Boston, insisted that "defects, derangements and symptoms of disease which would not be disclosed by a so-called 'careful physical examination,' are often recognizable in watching a person twenty-five feet away." He concluded that

"a man's posture, a movement of his head or the appearance of his ears, requiring only a fraction of a second of the time of an observer to notice, may
disclose more than could be detected by puttering around a man's chest with
a stethoscope for a week." But, significantly, Safford added that after an immigrant was pulled off the inspection line for more intensive examination, "a
week might be required to demonstrate what was really wrong with the
man."[43] The microscope was complementary yet subordinate to the gaze
within the PHS officer's diagnostic armamentarium.

Even if the trained medical officer was the only one who could detect disease at a glance, the PHS believed that immigrants had knowledge of their own
health status and were aware of eye infections, defective vision, hernias, varicose veins. The PHS faced the immigrant as an adversary of sorts, for the PHS
officers encountered "the shrewdest evasion and concealment."[44] Immigrants,
aware of some of the conditions for which PHS officers searched, did attempt
to hide deformities of the arms and hands, and to mask disease either physically or pharmacologically. Favus seems to have been the disease immigrants
most regularly attempted to conceal. An immigrant with favus would often
"have had his scalp cleaned, scrubbed, perfumend [sic], etc., in the hope that
his condition will thus escape the scrutiny of the medical examiner." Beginning in the 1920s immigrants would receive x-ray treatment abroad, resulting
in hair loss and removal of the outer layer of skin.[45] "Under these conditions
it is impracticable to make a diagnosis, and these 'bald heads' have had to be
held . . . . The routine procedure at [Ellis Island] after the growth of hair has
occurred is to apply three sweat caps at intervals, after which microscopic
examinations were made. By these means only could failures in treatment be
detected."[46] In 1921 Ellis Island reported holding sixty such cases for observation in one week.[47] Dr. Mullan advised that in the search for favus, "pom-
padours are always a suspicious sign. Beneath such long growths of hair are
frequently seen areas of favus."[48] Immigrants also attempted to hide ringworm infection and trachoma. In the early years of the twentieth century, for
example, San Francisco papers described the topical use of adrenaline to
temporarily mask the symptoms of trachoma among Japanese and Chinese
"coolie" laborers.[49] Immigrants had fewer tactics to avoid laboratory detection
of some diseases, but laboratory examination was no fail-safe solution. In one
instance, San Francisco officials discovered that an alien with hookworm had
borrowed part of the stool of another alien. Alas, the loaned specimen was also
infected.[50]

The immigrant onlookers may not have been able to decipher the code that the PHS officers inscribed on their clothing, but their awareness of disease and its significance was such that the meaning of the writing was clear as each marked immigrant was turned off the line, separated from friends, family, and fellow passengers, and directed into cagelike areas (figs. 3-2 and 3-3). Indeed, the medical gaze could serve a socializing function—instructing the immigrant about the expectations for a laborer in a highly fluid industrial society—only if it demonstrated the power of the federal government to exclude in a fashion that other immigrant witnesses could understand. In practice the PHS focused on those diseases and conditions that were transparent not only to highly experienced medical examiners but also to ordinary immigrants. Everyone could see that the elderly were turned aside for further inspection. Everyone could see a stooped back or a pregnant woman traveling alone. Everyone could see the attention that was given to eyes, and could thus gauge the importance

FIGURE 3-2. Immigrants who have been turned off the line wait either for further medical exams or for a BSI hearing. (Photo: New York Public Library)

FIGURE 3-3. Turned off the line at Ellis Island. This image,
centering our attention on the emotions of one family, is taken
from a larger picture entitled "U.S. Inspectors examining eyes
of immigrants, Ellis Island, New York Harbor."
(Photo: Library of Congress)

of vision. Everyone could see that the young (except those too young to work),
the muscular, the robust were *not* turned aside. To make the immigrant medical examination work as a socializing tool, the PHS had to focus on those diseases that could convey a message about industrial citizenship.

Outspoken nativists, eugenicists, politicians, and labor leaders continued to
see the immigrant medical exam as a means of defending the nation, of excluding unwanted, undesirable immigrant groups. Accordingly, the immigrant
medical exam became the subject of frequent attack. Members of Congress,
the press, and physicians often claimed that it was cursory and that stations like

Ellis Island were understaffed and overburdened. But while the exam sometimes became the subject of investigation, as it did in New York following the
typhus epidemic of 1892, a congressional committee investigating the immigration protocol failed to find fault with "the inspection procedures, bills of
health, or other practical matters concerning the flow of immigration into the
United States."[51] Indeed, the phrase "concerning the flow of immigration into
the United States" was most telling, for it indicated that the aim was not to impede this flow but rather to direct it. The exam worked as it was needed to
work. It was an important and consistent part of the emerging spectrum of routine medical examinations of supposedly healthy individuals in a society dependent on immigrant labor.

The concept of examining supposedly healthy people emerged during the
mid to late nineteenth century at the same moment that bacteriology created
tension within the medical profession.[52] The new cadre of physicians who pi-
oneered routine medical examinations focused on particular populations and
specific diseases. Health department officials examined schoolchildren for
contagious diseases; firemen, soldiers, policemen, and railroad workers were
examined to determine whether they met health and fitness standards. The
National Association for the Prevention of Tuberculosis advocated preventive
chest x-rays in the early years of the twentieth century. A campaign for routine
examinations of the general population in order to promote health and prevent disease blossomed in the 1920s, after the concept of routine health exams
received the endorsement of the American Medical Association in 1922.[53]

Early advocates of routine medical examinations in the late nineteenth and
early twentieth centuries stressed, not technological diagnostic procedures,
but rather a holistic approach to understanding the history of individuals, their
family relations, habits, and environments. George Gould, who was among the
first to promote routine examination of the general public, acknowledged the
availability of "instruments of diagnosis" but emphasized the collection of
hereditary information, anthropometric measurements, intellectual capacity,
and personal and family history.[54] Within the life insurance field and industry
at large, however, the development of the concept of routine medical examinations created new medical diagnostic expectations. Life insurance companies were among the first to develop and promote the routine medical exam
and to insist on the universal application of technology. Like the routine exam
developed in industry after 1910, the life insurance exam was designed to detect disease in supposedly healthy populations.[55]

In both life insurance and industry, technology was ideologically important because the physician and the employee were now in an adversarial relationship, since employees had incentives to hide their physical condition. Industrial medicine worked to dispel the idea that physical appearance was useful. According to the Conference Board on Safety and Sanitation, "robust-looking fellows are usually hired for jobs demanding brawn, active-looking persons for work requiring agility and speed, keen-eyed craftsmen for operations that need close and careful attention, steady-looking men and women for tasks that call for special endurance. But this practice of judging persons merely by their physical appearance is often misleading and unsatisfactory."[56] As in life insurance exams, the instruments and tests most commonly used included the stethoscope and perhaps the sphygmomanometer, otoscope, and urinalysis. Some larger firms developed extensive laboratories and clinics approaching those of hospitals, using x-ray machines and fluoroscopes.

Even more than industry, the military, in its examination of potential recruits, provided the closest parallel to the conduct, development, and rationale of the immigrant medical examination. Both the immigrant and military exams were designed to process large numbers of individuals rapidly.[57] Both the PHS officer and the military surgeon were on the lookout for concealed defects.[58] In each setting, the examiner made important decisions about how to weigh the seriousness of disease and disability: "the difficulty arises in deciding whether or not a certain physical condition is serious enough to be a cause for rejection."[59] The charge of the military medical examiner resembled the PHS officer's mission to protect the nation from dependency: "you must remember," gently chided a military physician, "that a soldier is forced to undergo trials and hardships which men in civil life are seldom called to face. Therefore, a generally robust physique is essential if we are not to have an army of weaklings, increasing the sick report, filling the hospitals, entailing extra burdens on an already overworked medical department, and, by depleting an army at a critical period, inviting disaster."[60]

Physical requirements for military recruits were clearly articulated, culminating in the advice that the soldier "must be intelligent and have the stamina of manhood in its prime, to bear with triumphant fortitude the hardships of service, neither inclining to the tenderness of youth, nor yet to the submission of age."[61]

In some respects, the military examination—perfected between 1898 and 1917—was more intense than that conducted at Ellis Island. All military recruits were disrobed; the examining surgeon auscultated and examined the chest with a stethoscope, examined the genitals for hernia and other diseases and conditions, and carefully inspected the feet, searching particularly for flat feet. The recruit also briefly exercised for the surgeon, who then examined his heart and lungs; finally, the surgeon carefully examined the eyes, nose, and throat of each recruit in a dark room.[62] The military surgeon and civilian physician stationed at local exemption boards exercised great discretion in accepting or rejecting military candidates; thus, rejection rates ranging anywhere between 20 percent and 75 percent were standard.[63]

The physical examination of recruits conducted by the military was not all that different from the medical examination of arriving immigrants. For example, bacteriological tests, x-rays, and ophthalmoscopic examination of the eyes were not routine, but were used only when specific diseases or defects were suspected. Sputum tests and x-ray examinations for tuberculosis were not typical screening procedures, even among standing troops.[64]

The instructions for military officers examining recruits were remarkably similar to those formulated for PHS officers inspecting immigrants. Although the military placed a much greater emphasis on touching and closely examining the body, the medical gaze was also important. During the military exam, "the features, physique, stamina, bearing, and general desirability are . . . observed and a rapid comprehensive *glance* taken to detect any glaring or prominent disqualifications."[65] After this initial glance, each recruit was weighed and measured, and stood ready for completion of the exam. The fundamental similarity to instructions given the PHS officer is striking:

> The applicant now stands directly in front of the examiner who, while passing the hand over the head in search of parasites and other diseases of the scalp, scars, depressions or lumps, asymmetry of the skull, malformations or loss of the ears, enlarged glands or goitre or mumps, asks, 'Have you ever been struck or injured on the head, been knocked unconscious, or had fits or fainting spells; do you use alcohol, tobacco or other drugs?' The face is then examined, while he is replying for gross occular [sic] variations. . . . the nose for deformities, rhynophyma, acne rosacea, lupus and scars. The mouth is examined for hair lip and cleft palate, deforming scars, caries of the jaws or ankylosis of the temporomaxillary joint, and the ears for loss of pinna, atresia or evidences of operations or disease. The head and face should be symmetrical. Any unsightly scars, discolored marks, tumors or marked ugliness, though not the evidence of active or

latent disease, are cause for rejection, as such bring unpleasant notoriety to the man and are therefore subversive of discipline. The throat is examined for goitre, branchial cysts or fistulae, enlarged glands, aneurysms and scars. The chest is then inspected for asymmetry, deformities such as pigeon and funnel breast, mammary development and all evidences of operations, wounds, accidents or disease of skin, chest wall, or lungs. The thoracic organs are examined by the usual diagnostic methods, meanwhile asking the man if he has ever had a persistent cough, spat up blood, had lung fever, or rheumatism, or sore throat, and whether any of his family died or are sick with consumption. The X-ray may be called in, and the microscope, to decide questions of doubt.[66]

In both the military exam and the immigrant medical exam, a fundamental goal was to impose order and a sense of hierarchy, though for the military exam this was a stated organizational goal rather than a social consequence. Both exams were public displays of power that served at once to shore up state power in a display of medical authority and to convey a sense of rules and expectations to those examined.[67] The doctors of the Public Health Service, after all, were commissioned officers in a quasi-military institution headed not just by a chief surgeon, but by a surgeon general. Like the military, the PHS had the potential for serving as a "total institution," though it had no conscripts or inmates outside of the seamen and immigrants who came temporarily under its authority.[68]

OFF THE LINE AND IN THE LAB

For the immigrant medical exam to serve a disciplinary function, it had to incorporate an expression of power. Exclusion, therefore, was a real possibility. Thus, at Ellis Island, PHS officers immediately transferred immigrants bearing chalk marks—typically 15 percent to 20 percent of arrivals—to either the physical or mental examination rooms after passing through the line. In these rooms, "men are divested of their clothing and are thoroughly examined with the aid of stethoscope, thermometer, and Snelling test-types [eye charts]. The height and weight in certain cases are taken. When the examination is complete, an OK card or appropriate medical certificate, setting forth the physical condition, and signed by three medical officers, is issued to the alien." The process would then begin anew: "After twenty or thirty aliens have been examined and disposed of, the big doors of the empty room are again opened to receive another batch of immigrants. This process of filling the room with immigrants, examining them, and disposing of them is repeated over and over again throughout the day." But this more intensive exam still served normative functions; like examination on the line, it was chiefly a spectacle. As figures 2-2 and 2-3 showed, even the more intensive examination of immigrants who had been turned off the line made a public display. At Ellis Island and some other stations, officers conducted the business of definitively diagnosing and certifying disease among immigrants in semi-private examining rooms where groups of immigrants witnessed each other's close examination.

Laboratory diagnosis, however, was not ideally suited for discipline, for it took the examination process out of a public setting. Immigrants could not witness diagnosis in the lab, nor could they understand it. Still, detention at Ellis Island and at the nation's other immigration stations served to discipline and warn the confined immigrant. Ellis Island did so in part by virtue of its location: at the mouth of the East River that led to the prison, almshouse, workhouse, lunatic asylum, and isolation hospitals on Blackwell's Island (also known as Welfare Island) and the infectious disease hospitals on North Brother Island used to quarantine those with infectious disease who, like "Typhoid Mary," proved "recalcitrant" by refusing to acknowledge and abide by social norms. Those immigrants confined, often for many months and sometimes years, in the isolation units in the southernmost wing of Ellis Island gazed out onto Lady Liberty's backside, as she seemingly spurned them with her heel while turning her face away to the east. Angel Island, San Francisco's immigration station, served a similar symbolic purpose, standing in the icy waters of the bay, beyond even the military prison on Alcatraz. Detained immigrants— the ones at greatest risk of failing as industrial citizens—required extended socialization if they were to be deemed fit to enter the nation.

Although the lab itself did not serve to discipline immigrants, it did help to back up the warning conveyed on the line with actual exclusion, by enhancing the ability of the PHS to confirm disease. The PHS encouraged its officers to spend as much time as necessary to make an accurate diagnosis of those immigrants "turned off the line."[69] Dr. George Stoner noted, "I can well wish it were possible for us to furnish stronger aid in deporting all such undesirables. But we are not in position to certify contagion where it does not exist."[70]

As access to bacteriological testing increased, at least at Ellis Island, PHS officers readily incorporated the lab into the secondary immigrant medical inspection. Thus, while the PHS at Ellis Island performed only 1,223 laboratory exams in 1915, by 1916 the number of procedures performed increased by more than 300 percent, to over 5,000.[71] Significantly, the total number of immigrants examined decreased during this period (falling from 242,722 in 1915 to 176, 461 in 1916). The number of laboratory exams performed increased almost twenty-fold by 1921 (still surpassing the 82 percent increase in immigration from 1915 to 1921), the PHS reporting having performed over 20,000 procedures. The laboratory tests most commonly ordered in 1921 were "bacteriological examinations" (42%), blood tests (18.16%), and urine tests (17.61%).[72]

By the World War I years the PHS considered laboratory testing an integral part of the secondary medical examination at Ellis Island and other major ports. Yet Ellis Island remained a bastion of the medical gaze. Some PHS officers, like other mainstream medical practitioners, lamented the rise of laboratory testing, casting it as a challenge to the physician's art. Dr. J. G. Wilson in 1911 expressed the sentiments of more than one officer when he wrote, "In these days of laboratory experimentation and the use of refined methods of diagnosis, the value of simple inspection of the patient has gradually been lost sight of, and the art of snap shot diagnosis has been left almost entirely in the hands of the charlatan." Without denying the great value of the laboratory, Wilson could not help but regret that "there is a tendency developing to rely too much on the opinion of the laboratory diagnostician and to let his findings be considered as final, even when they contradict the verdict of common sense."[73] Dr. A. J. Nute echoed similar sentiments several years later, in 1914: "Since the days of the laboratory and other aids to diagnosis we have tended to lose sight of the value of observation and what it may tell us."[74]

Thus, within the context of the immigrant medical inspection, the importance of the laboratory should not be overstated. The structure of the laboratory was open to question in the late nineteenth century, on the grounds that it was not "naturally" suited for the diagnosis of disease. Although the PHS was the early champion of the laboratory in the 1880s, when the Hygienic Laboratory (renamed the National Institute of Health in 1930) was first established on Staten Island, it was not a diagnostic facility. Joseph Kinyon and Milton J. Rosenau, the laboratory's first two directors, created an institution that produced diphtheria and tetanus antitoxin; improved methods of vaccination; studied allergic reactions to antitoxins; sought to discover and isolate the causative organisms of disease, such as tularemia (a plaguelike disease of squirrels to which humans were susceptible) and Rocky Mountain spotted fever (for which the lab also developed a vaccine); and tested milk supplies for purity and rats for plague. In short, the laboratory was originally a research institution serving a broad public health function and not strictly a medical diagnostic or medical screening facility.[75]

The laboratory as a diagnostic tool was slow in arriving at even the largest stations, reflecting the delicate balance between the social and the scientific, between the exclusionary and normative social imperatives to which the exam responded. El Paso, for example, did not establish a laboratory service until it became the nation's second largest immigration station in 1924.[76] Yet here the primary medical inspection was far more thorough than that conducted at Ellis Island, for PHS officers along the Mexican border faced greater disciplinary challenges and exclusionary mandates. The laboratory also remained dependent on finances and PHS priorities. Some small ports managed to set up labs.[77] Most, however, continued to lack adequate laboratory facilities well into the 1920s.[78] As late as 1921, the surgeon in charge at Detroit complained of "the lack of sufficient office equipment . . . to make a complete and conscientious physical examination of clients." He requested, among other things, a microscope, a stethoscope, an examining table, a screen, a bottle of methalene blue (for staining the tuberculosis bacillus), and one set of gram stain.[79]

The mental exam and x-ray show the ways in which the PHS balanced the social and the scientific, and demonstrate that a careful weighing was required when medical judgment was confronted with other diagnostic "advances." The mental examination, in particular, provokes and intrigues, for it powerfully evokes thoughts of the prejudices and injustices of the eugenics movement. It also potentially opened the door for the PHS to employ a civic rather than industrial notion of citizenship.[80] But in practice mental examinations remained primarily within the purview of civilian "alienists" or psychiatrists who worked outside the boundaries of the PHS.[81] To be sure, PHS officers actively experimented with different diagnostic tests. They maintained that, like organic disease, mental illness proclaimed itself clearly on the immigrant body.[82]

Although the PHS did not dismiss the importance of mental illness, the agency remained skeptical of the value of the diagnostic tools then in vogue for use in an immigrant population. PHS officers screened immigrants with industrial labor in mind. Industry wanted workers who would perform repetitious, monotonous tasks without questioning.

> I am the shop's dictator hand,
> For all must do what I have planned,

Without disputes which waste but time.
In business this is more than crime.[83]

Frederick W. Taylor insisted that often the worker ideally suited to a particular job is "so stupid and so phlegmatic that he more nearly resembles in his mental make up the ox" or a gorilla than a man.[84] In place of intelligence tests, then, PHS officers preferred to rely on their professional judgment.

The secondary examination for mental conditions at Ellis Island was done in three stages. Those immigrants marked with an X (indicating a suspected mental defect) or an X with a circle around it (indicating definite signs of mental defectiveness) were immediately taken to the "mental room," where a PHS officer asked them simple questions, such as their name or age, and gave them simple tests, requiring manipulation of cubes or puzzles or interpretation of events depicted in photographs. Those immigrants still suspected of mental defects were held overnight for further examination twenty-four hours later. After they had rested and bathed, suspected immigrants were again questioned and tested. Those still thought to be mentally defective after day two were given a second question and testing period lasting twenty minutes to one hour. In an effort "to size up the immigrant from all angles," the examiner might inquire "into the home life, customs, schooling, occupation, voyage, and intentions of the subject. When necessary, questions are put in order to bring to light the whys and wherefores regarding the immigrant's attitude, emotional states, habits, interests, and health. In addition to the psychological tests and questions a neurological examination and test of vision are occasionally made."[85] Dr. Robert Leslie explained that officers would ask "social" questions about relationships with children and spouses: "I would evaluate their whole idea of life. We call it psychiatry today, but it wasn't that."[86] The third and determining examination, conducted the following day, was almost identical. The intensive examinations were conducted with the aid of an interpreter, and each time a different physician conducted the testing.

At the request of individuals like psychologist and eugenicist Henry Goddard, director of the laboratory for the study of mental deficiency at the Vineland, New Jersey, Training School for Feeble-Minded Boys and Girls, who believed that intelligence tests could help to exclude inferior southern and eastern European races, the PHS did allow experiments with the Simon-Binet intelligence tests. However, the agency did not insist on the use of such testing instruments and did not place great faith in them.[87] Howard Knox, for example, criticized Goddard and his staff for the "fallacy" of classifying "nearly all peasants from certain European countries . . . of the moron type."[88] (Knox and other Ellis Island physicians were also reacting to the sting of Goddard's assertion that PHS line officers made accurate assessments of mental defectiveness in less than 50 percent of cases, compared to the 80 percent accuracy of his test team.)

Dr. Eugene Mullan, one of the very few PHS officers specifically interested in refining the mental examination of immigrants, understood that an immigrant's ability to comprehend most tests, especially those requiring interpretation of drawings, was substantially influenced by his or her culture and life experiences. A particularly interesting test developed by Dr. Mullan involved asking an immigrant questions about a drawing called "Last Honors to Bunny" (fig. 3-4). This item is prominently displayed today at the Ellis Island Museum in New York, as an example of the testing protocol for mental conditions. Mullan himself, however, recognized the limitations of this particular test, in which the examiner showed immigrants a picture of a boy and girl with a dead rabbit; the boy dug a hole and the girl held flowers. Mullan observed that most of the immigrants did not understand that a burial was taking place until well into the questioning. He concluded, "In order to interpret a picture correctly, the immigrant must be familiar with pictures of various kinds. He must be also experienced in the customs or events portrayed in a picture. In addition, he must possess the power of constructive imagination." Consequently, "the poor showing which immigrants make in this test is due more often to a lack of experience with pictures and the scenes which they depict than to a lack of constructive imagination." Mullan understood only too well that "these immigrants had never seen pets treated well. They had never seen them treated with signal honors. Many are not accustomed to see rabbits used as pets. Peasants do not see well dressed children and sheep grazing at the same time. . . . In certain places it is not customary to cover the grave with flowers. Even if an immigrant has had some experience with pictures, pictures of this kind are hard to interpret."[89] Moreover, the PHS was convinced that racial types exhibited mental defects differently. This belief caused the value of standardized tests to pale in comparison to the experience and judgment of PHS officers, who considered themselves experts on the manifestations of mental and physical disease in different races. Thus, the PHS recognized the vastly different experiences and backgrounds of different immigrant groups, and this recognition made the agency comparatively less eager than alienists or eugenicists

FIGURE 3-4. "Last Honors to Bunny," part of a mental test designed
by Dr. E. H. Mullan and administered to immigrants at Ellis Island.
(E. H. Mullan, "Mentality of the Arriving Immigrant," *Public Health
Bulletin* No. 90 [Washington D.C.: GPO, 1917]:119)

to certify an immigrant for a mental disorder or to value the results of an "objective" test over its officers' own experience or intuition.[90]

The Public Health Service did rely on diagnostic technology when the "inscrutable" Asian body obscured the gaze and particular diseases made such
races unfit for citizenship. Yet the successful exclusion of Asian immigrants
sometimes demanded that the PHS abandon innovative diagnostic technology.
Although the PHS regularly used diagnostic instruments like the stethoscope,
thermometer, and microscope throughout the entire period covered in this
study, it appears that PHS officers only judiciously incorporated the full
panoply of diagnostic technology into the process of the secondary medical
exam.[91] Available data indicate that the x-ray was rarely used for diagnostic
purposes until the 1920s.[92] The immigration station at Ellis Island began to report on the use of x-rays in 1920. Here the x-ray machine was used primarily
to detect pulmonary conditions (especially tuberculosis), as well as to confirm

diagnoses of bones and joints.[93] Yet the most instructive use of the x-ray was that to which it was *not* put.

The Chinese Exclusion Act of 1882 prohibited the entrance of all Chinese people except those born in the United States, merchants, students, diplomats, and often their wives and minor children. When the Immigration Service began administering the Chinese Exclusion law in 1900, IS officials worried that Chinese immigrants would fraudulently claim exempt status.[94] After the San Francisco earthquake of 1906, during which the city's municipal records were destroyed by fire, it became easier for Chinese immigrants to claim a relation to an immigrant who was exempt from the law by forging a birth certificate. IS officials, fully aware of the "paper son" system, frequently asked PHS officers for a medical assessment of age when the age of the immigrant did not seem to correspond to the age stated in the birth certificate.[95]

In the 1910s, after the PHS had over twenty years of experience examining immigrants on the line and when the bacteriological laboratory was featuring more heavily in definitive diagnosis of immigrants, attorneys of aliens whose age came into question began to champion x-ray examinations of the bones and joints as definitive determinants of age. PHS officials, however, consistently repudiated x-ray evidence. In 1911, for example, IS officers in Boston granted a Chinese immigrant claiming to be the seventeen-year-old minor son of a Chinese merchant authorization to obtain outside opinion on his actual age. The *Boston Sunday Post* reported that "by means of an X-ray machine, Chang Hong . . . has been able to refute the contention of the immigration officials at this port that he is not a minor."[96] Dr. Safford, however, maintained that he was no minor, insisting that "an X-ray examination is not necessary." Safford formed his conclusion based on the "general appearance of maturity and the facial expression, the comparative flexibility of the breast bone and ribs through their attachments thereto, the degree of pigmentation of the skin in certain places and especially the development of the jaw bones and teeth and the amount of wear which some of the latter may show."[97] Safford doubted that the x-rays presented as evidence were really from the immigrant in question. He was also skeptical that one could compare the development of a "Chinaman's bones" to a "White man's bones." Safford ruled, "I still cannot escape the belief that he must be 22 years of age if not older."[98]

The PHS rejection of x-ray evidence represented not only skepticism about the usefulness of the results, but an affirmation of its own authority in the face of challenge. Safford indignantly protested, "I can conceive of no legitimate reason for still further extending the privileges already accorded a diseased immigrant by recognizing the right of an irresponsible private physician to take part in our medical examination no matter how high his professional attainments or personal integrity and even if such a course be consistent with law. A medical certificate issued in conformity to our Regulations means something more than an ordinary professional opinion."[99] Consequently, not until 1930 did the PHS agree to consider providing x-ray exams to supplement the judgment of the examining officer (and this only after the demise of the medical gaze, as discussed in the epilogue).[100]

The medical laboratory and other diagnostic technologies, like the mental examination or the use of x-ray, did not revolutionize medical diagnosis, taking it definitively out of the terrain of the social. Bacteriology had relatively few diagnostic tests to its credit. Between 1880 and 1898 the microbial causes of only twenty diseases had been discovered. Among these, only one was remotely relevant to the immigrant medical inspection—tuberculosis.[101] Even after the turn of the century, when laboratory detection of several venereal diseases became possible, the potential for using the laboratory to systematically screen immigrants for excludable diseases remained limited. Thus, in 1916, while Wassermann tests accounted for 20 percent of the exams performed, throat cultures for diphtheria represented 27 percent of the lab tests performed, followed by urine tests (15%), and sputum tests for tuberculosis (8%).[102]

Further, the diseases of chief concern to the PHS were not primarily infectious, for the overarching purpose of the exam was to control rather than to exclude. Indeed, the diseases on which the PHS primarily focused required no laboratory confirmation. By the turn of the century the immigrant body no longer represented a discrete unit of production, but rather a cog—ideally a sound cog but ultimately a replaceable one—suited for work in a variety of machines.[103] Just as new methods of production necessitated a new view of the laboring body, they also required a new view of disease. In the immigration arena, the PHS took on the critical function of defining the relationship between laboring bodies and disease, between health and dependency. The emphasis fell on diseases that disabled the laboring body, as shown in figures 3-5 through 3-10, displaying certification patterns using the PHS nomenclature for New York (fig. 3-5), Boston (fig. 3-6), San Francisco (fig. 3-7), New Orleans (fig. 3-8), El Paso (fig. 3-9), and stations along the Canadian border (fig.3-10).[104]

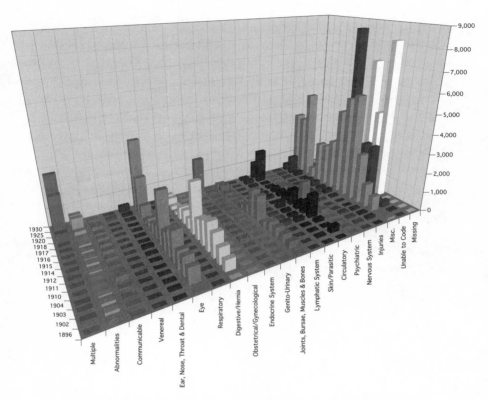

FIGURE 3-5. Immigrants certified by PHS nomenclature, New York,
selected nonconsecutive years.

Although there are interesting regional variations, an overall pattern tended
to repeat itself: most immigrant diagnoses fell into "diseases of the eye and annexa" and "miscellaneous diseases and conditions." Within the category of diseases of the eye and annexa, trachoma typically represented roughly 70–90
percent of eye diseases until World War I; thereafter, trachoma represented
20–50 percent, while defective vision and blindness represented the majority
of eye diseases and conditions. Most surprising is the prominence of senility
in the miscellaneous category. Meaning simply "old age," senility typically represented approximately 90 percent of miscellaneous certifications. Observed
an unemployed laborer in Cleveland, Ohio, in the late 1920s, "Factory no want
older men; men of forty-five not wanted—me [looking ashamed] fifty-two."[105]
Alter Abelson, in "The Designer," spoke on behalf of many workers when he

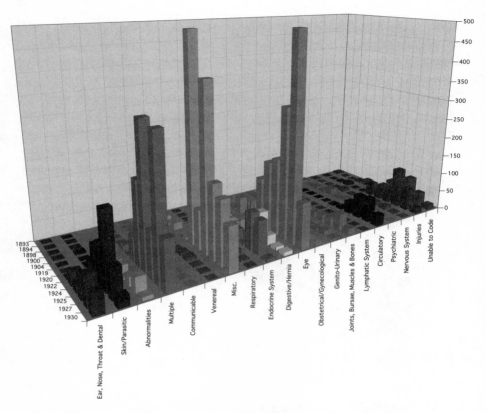

FIGURE 3-6. Immigrants certified by PHS nomenclature, Boston,
selected nonconsecutive years.

wrote, "We all are only furniture here, / To be discarded when our face / Reveals that age has slowed our pace."[106]

The one major exception to this general disease pattern was in San Francisco, where miscellaneous diseases and conditions were rarely certified and,
sometime after 1911, parasitic infections began to represent well over 50 percent of all certifications. But even parasitic infectious were very much tied to
questions of fitness for industrial citizenship, for the immigrant infected with
hookworm or river fluke threatened to degrade and undercut the American
worker. Pregnancy (the most prominent condition among the obstetrical and
gynecological diseases and conditions diagnosed with relative frequency in San
Francisco and along the Mexican border), hernia, joint stiffness, and curvature

of the spine (prominent in New York and along the Canadian border) represented the second tier of diseases on which the PHS focused. Pregnancy represented dependency and potentially signaled a lack of sexual control; hernias, bent backs, and immobile joints prevented an immigrant from engaging in heavy labor or repetitive assembly-line tasks requiring dexterity. In certifying immigrants, the PHS mapped out a set of diseases that defined the immigrant's ability to participate in an industrial society.

The PHS medical exam was of a piece with an emerging industrial ideology of scientifically classifying and managing the working body. In certifying blindness, hernia, old age, vericose veins, and pregnancy, the PHS was helping to define what made people susceptible to unemployment, to dependency. In certifying immigrants with disease, the PHS was in effect asking not, "Do you have a contagious disease representing a threat to the health of the nation?"

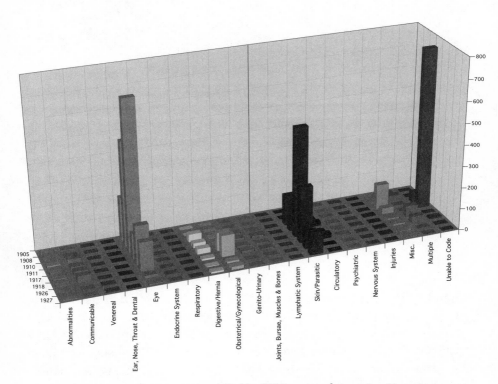

FIGURE 3-7. Immigrants certified by PHS nomenclature, San Francisco, selected nonconsecutive years.

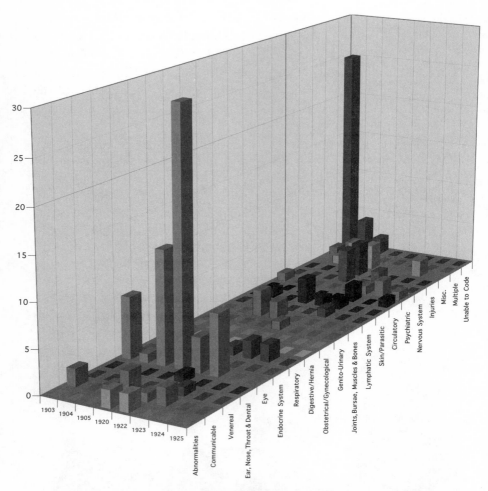

FIGURE 3-8. Immigrants certified by PHS nomenclature, New Orleans, selected nonconsecutive years.

but rather, "Are you capable of working?" It was affirming that the new working class would have to fit into a new industrial work regimen—one in which the body was dispensable and replaceable, a working part that would make its admission a good investment. Significantly, over time the PHS increasingly deported immigrants who had previously been granted entry to the United States but who had become dependent within one to five years of their initial landing (fig. 3-11). The conditions that the PHS began to identify among arriving immigrants represented a serious obstacle to performing heavy labor or gaining and retaining employment of any kind. The seriousness of each was mediated by the immigrant's gender, race, social class, and employment prospects in the United States.

The Office of the Surgeon General remained firmly supportive of the traditional practice of immigrant medical inspection—the medical gaze. Officials considered the value of possible alterations to the role of the gaze in the exam in light of both their effectiveness and the overarching intent of the exam. During World War I Ellis Island physicians experimented with conducting a more intensive physical examination; although the station experienced staff reductions as a result of the war, because of the considerable depression in immigration it actually enjoyed something of a surplus of medical officers for the first time in its history.[107] With more time on their hands, the staff began to

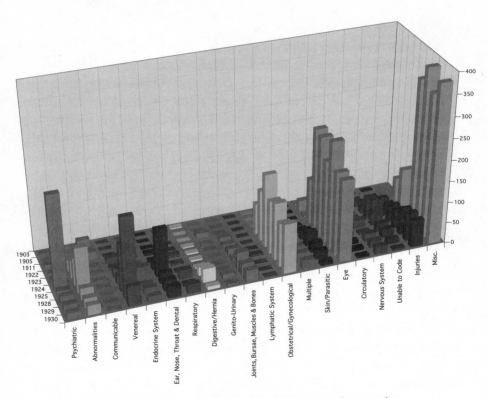

FIGURE 3-9. Immigrants certified by PHS nomenclature, El Paso, selected nonconsecutive years.

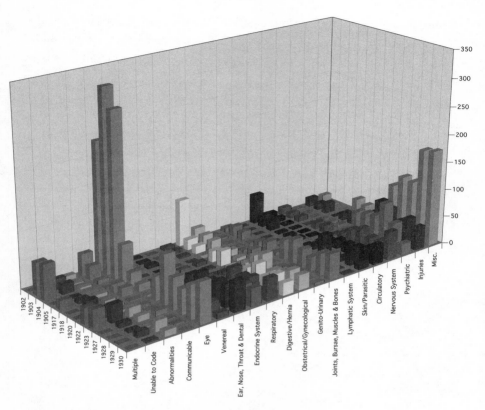

FIGURE 3-10. Immigrants certified by PHS nomenclature,
Canadian border, selected nonconsecutive years.

turn as many immigrants as possible aside into more private examination rooms; some days they were able to intensively examine all arriving immigrants.[108] The chief medical officer at Ellis Island reported that the certification rate rose from 2.29 percent in 1914 to 5.37 percent in 1915. When all the immigrants were turned aside for an intensive examination, the certification rate rose to 9.37 percent.[109]

The World War I experiment with intensive examinations was brought to an abrupt halt by the exigencies of war. The military took over Ellis Island for the treatment of injured soldiers on 8 March 1918, forcing PHS officers to conduct all their work on the decks of arriving ships.[110] After the PHS returned to its island stronghold in 1920, the surgeon general rejected proposals to make intensive examination routine for all arriving immigrants.[111] True, the certification rates had increased, but he underscored that this amounted to a mere handful of additional rejections: "it should be clearly evident from a study of the . . . figures and the methods employed during past years in the medical examination of immigrants that the procedure as carried out is reasonably satisfactory for the purpose for which it is employed."[112] He concluded that some insignificant number of immigrants would continue to slip by, no matter what measures were taken.

The surgeon general was not implying that these numbers did not justify the extra effort. Rather, he intended to emphasize that the PHS examination of immigrants was quite effective: in excluding immigrants with diseases that conveyed the proper message to other immigrants witnessing the medical exam on the line, the agency fulfilled a socializing responsibility. The very structure of the exam conveyed meaning. The goal was not to use every available means to catch every single diseased immigrant. Rather, the PHS exercised professional medical and social judgment to communicate social and industrial norms by underscoring in a public setting those diseases and conditions that compromised an immigrant's capacity for industrial citizenship.

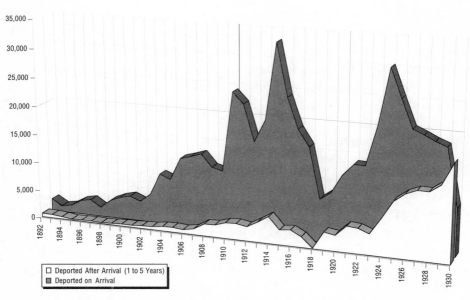

FIGURE 3-11. Comparison of immigrants excluded on arrival and immigrants deported within five years of arrival, 1892–1930.

SCIENCE AND INDUSTRY

Robert Wiebe argues that during the Progressive era, science ceased to be regarded as a set of "principles that men could comprehend and apply." Under the new bureaucratic regimes of the growing industrial nation, "science had become a procedure, or an orientation, rather than a body of results."[113] As such, it required individuals not merely trained in technique, but committed to a coherent set of underlying principles that made science, and therefore society, work. Taylor had recognized that grasping and applying the tools and methods of scientific management was "by no means a formidable undertaking"—indeed, it could "be accomplished by ordinary, every-day men without any elaborate scientific training." Still, it was imperative not to "mistak[e] the mechanism for the true essence. Scientific management fundamentally consists of certain broad general principles, a certain philosophy, which can be applied in many ways, and a description of what any one man or men believe to be the best mechanism for applying these general principles should in no way be confused with the principles themselves." Undertaken in "the wrong spirit," scientific management would inevitably "lead to failure and disaster."[114]

Within the industrial setting, the right spirit or philosophy entailed an ideological understanding of the proper relationship between management and worker, and a commitment to select workers toward the end of efficiency of production and national prosperity. So, too, in the immigration arena, the line represented far more than a technique or process. Inspection on the line drew on a philosophy or ideology, embodied in the notion of a medical gaze. Recognition of this helps us account for variations in the immigrant medical inspection in other parts of the country. PHS officers mastered both technique and an industrial orientation at Ellis Island, which provided a sort of proving ground. Many new officers served their first months or years at Ellis Island before being transferred to other domestic or international ports. Noted the assistant surgeon general in 1912, "When a medical officer has proven to his colleagues at Ellis Island that he is able to be trusted on the line, or in other words, that he has acquired the standard, he is pronounced 'a safe man.'"[115] The safe man understood the immigrant medical examination within the larger context of national industrial efficiency, economic prosperity, and careful classification of the laboring body.

Just as the needs of industry shaped the immigrant medical exam, they shaped medicine more broadly, for in the nation with the world's most dan-

gerous workplace, health became increasingly linked to achieving and maintaining order.[116] The immigrant medical examination foreshadowed the wave
of employee evaluations to determine fitness that would come in the 1910s and
1920s.[117] When first proposed at the turn of the century, physical examinations
were conceived as part of an emerging system of classifying the human body—
from military recruits to criminals—that would render "a genuine and all-
encompassing science of anthropology based upon all the data, morphologic,
physiologic, and pathogenic, of the entire individual life" intended to bring civilization to perfection.[118]

   By the 1910s, when the doctrine of scientific management was familiar to,
if not yet widespread practice in, American industry, health examinations became solidly cast in the language of industrial production and efficiency.[119] The
Norton Company in Worcester, Massachusetts, became one of the first to conduct routine examinations for tuberculosis, for example, with "the point of
view of increasing the efficiency of their force."[120] James Tobey wrote in 1924,
"A health examination is a thorough appraisal of an individual, made by a competent doctor of medicine in order to detect physical and mental impairments
and faulty habits of hygiene with a view to their correction. It has been called
an audit of one's assets and liabilities." Its intent was to determine whether the
"human mechanism has been functioning at its highest rate of efficiency." The
benefits of carefully maintaining the "human machine" could be measured in
terms of "return of the investment" and "profit" to both companies and workers in "the business of life."[121]

   The immigrant medical examination, when viewed as a product of industrial society, has implications for how we understand the past and the present.
Industrial labor concerns had a profound impact not only on the form and conduct of the immigrant medical exam, but also on the practice of medicine and
its meaning in society.[122] While the immigrant medical inspection led the way
into the United States, disciplining and managing the work force with medicine became a more far-reaching social effort.

PART II

NUMBERS SMALL

Immigrant Medical Inspection
as an Exclusionary Tool

CHAPTER FOUR

THE SHAPE OF THE LINE

Immigrant Medical Inspection from Coast to Coast

I had 17 transfers in my first 15 years. Hospital in Chicago, plague work in New Orleans, hospital in St. Louis, rural sanitation in Greenville, South Carolina, Ellis Island, Coast Guard, Ellis Island, X-Ray School at Cornell, Ellis Island, Leavenworth, Kansas, Washington, Ellis Island, Hygienic Laboratory, Fort Stanton Hospital, New Orleans, Pittsburgh, New Orleans. The last 16 years, only four stations. Actually we didn't mind this too much, as we knew what the Service was like before entering it. Also, we were nearly always with fellow officers, frequently already known, or with mutual friends. It made for a certain clannishness with us.

—Dr. T. Bruce H. Anderson, 1977

FROM 1892 TO 1930 Ellis Island was the nation's largest port of entry for immigration; roughly 70 percent of all immigrants to the United States attempted entrance there. The PHS and IS recorded some 25 million people arriving at immigration stations along the Atlantic coast—where New York was the dominant port—compared to nearly 2 million along the Pacific coast, 1 million along the Gulf coast, 3.5 million along the Canadian and Mexican borders, and almost 1 million in the U.S. island possessions (fig. 4-1). The Atlantic coast was responsible for the vast majority of certifications and led the nation in deportations for all causes. In 1920, for example, ports along the Atlantic coast certified a total of 35,749 immigrants, compared to a high of only 2,735 for the Pacific coast (1913), 2,076 for the Gulf coast (1919), 8,509 for the Canadian border (1919), 5,351 for the Mexican border (1929), and only 545 for the island possessions of the United States (1903) (figs. 4-2 and 4-3).[1]

For these reasons, the history of an immigrant's arrival in the United States has most often been told as a sentimental story of Ellis Island.[2] With few exceptions, histories of immigrant medical inspection have similarly focused almost exclusively on the "Island of Hope and Island of Tears."[3] Immigrants have emblazoned it in the national memory as the place where "people used to come in crying and crying. Your heart would break."[4]

As Ellis Island received the vast majority of immigrants from 1891 through 1930, it set the standard for medical inspection on the line, and also for discipline. Still, Ellis Island was only one port and can provide only one perspective on the conduct and meaning of the immigrant medical examination. The number of doorways to the United States began to increase substantially after the turn of the century; on average, immigrants arrived at sixty-two ports across the nation from 1891 to 1930. Speaking in 1912 before the American Academy of Medicine, Assistant Surgeon General L. E. Cofer explained that the practices of immigrant medical inspection "may be divided into those in vogue at the large stations, where immigrants are examined on the line, and those at small stations, where each immigrant is given, for practical purposes, a personal examination by the medical officer. It may be said, therefore, that at the smaller stations the examiners follow no fixed routine."[5] Variations in the medical examination of immigrants from station to station could, as Cofer indicated, be a function of size. But the broader regional variations in the practice of immigrant medical inspection—in the shape of the line—and in the subsequent patterns of certification and, more important, exclusion suggest that variations were less a function of size than of specific regional needs and concerns relating race and labor to the national industrial welfare. The function of the examination varied from region to region, and the balance between discipline and exclusion shifted accordingly.

Figures 4-1, 4-2, and 4-3 represent immigrant certification and deportation in absolute numbers and are important in illustrating the difference in the magnitude of immigration and certification in different regions. But we must turn to certification as a percentage of all immigrants inspected or all immigrants arriving to begin to uncover qualitative differences in the inspection procedure.[6] Figures 4-4 and 4-5 present a very different picture of certification and deportation in the United States. From about 1904 to 1913, the Pacific coast certified more than 5 percent of the immigrants who came through its ports, and in 1908, more than 20 percent. The certification rate did not exceed 5 percent along the Atlantic coast until after World War I and never exceeded 5 percent along the Gulf coast, along the Mexican border (with the exception of a peak of almost 20 percent in 1906), or in the U.S. island possessions. Along the Canadian border the percentage of immigrants certified was consistently high, often exceeding 10 percent.

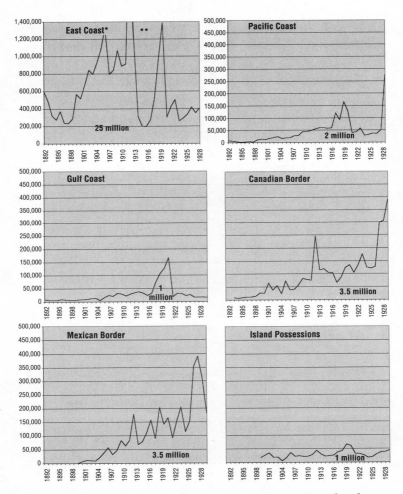

FIGURE 4-1. Immigrants arrived by region, 1892–1928. Note that data were recorded beginning in 1892. (°Scale for East Coast is 1,400,000; scale for all other coasts is 500,000. °°Arrivals for 1913 for the Atlantic coast were 2,423,096.)

THE ATLANTIC COAST: THE STAMP OF CLASS

In 1911 Immigration Service Commissioner William Williams wrote that, as "cabin passengers are as a rule less apt to be ineligible than steerage passengers, their inspection can proceed more rapidly and far fewer are held for special inquiry. As to this, however, much depends on the character of the pas-

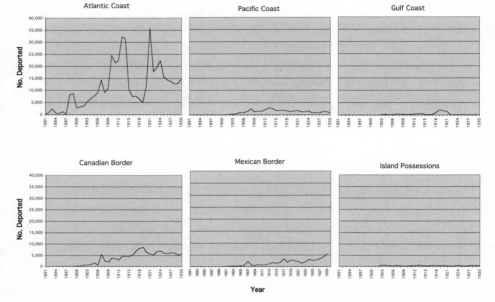

FIGURE 4-2. Number of immigrants medically certified, by region, 1891–1930.

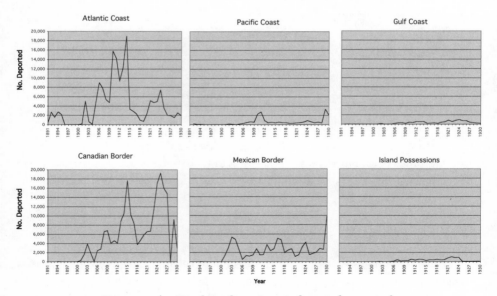

FIGURE 4-3. Number of immigrants deported on arrival
for all causes, by region, 1891–1930.

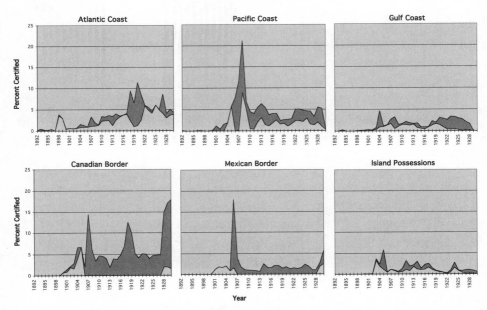

FIGURE 4-4. Estimates of proportion of immigrants medically certified, by region, 1892–1928. (The actual percentage certified, based on available data, probably lies somewhere within the shaded band, which represents the higher and lower estimate of percentage certified calculated using either Total Arrivals or Number Examined as the denominator. Because of missing data, though, even the upper limit of this band may underestimate the actual percentage certified for most years.)

sengers. . . . Experience shows that those from southern, south-eastern and eastern Europe are more likely to be ineligible than those from northern Europe."[7] Southern and eastern Europeans primarily filled the ranks of unskilled labor in industrial-era America. They represented the bulk of immigrants traveling steerage during this period. Thus, for the PHS to associate southern and eastern European origin with higher rates of disease was also to draw a class distinction. Although there was certainly an overlap between class and race in the minds of immigration officials, they discussed the problems of medical examination associated with European immigration to the East Coast almost exclusively in terms of class. Officers working in the Northeast faced a particular set of concerns that worked to structure the exam around the immigrant's social class.

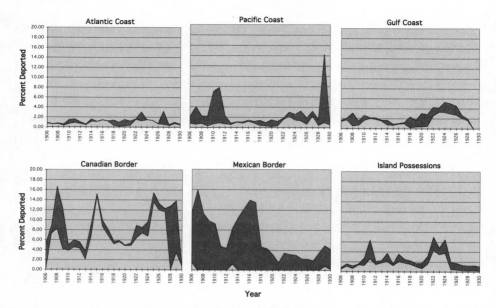

FIGURE 4-5. Estimates of Proportion of Immigrants Deported for All Causes
by Region, 1906 to 1930. (The actual percentage certified, based on available data,
probably lies somewhere within the shaded band, which represents the higher
and lower estimate of percentage certified calculated using either Total Arrivals
or Number Examined as the denominator. Because of missing data, though, even
the upper limit of this band may underestimate the actual percentage
certified for most years.)

The medical force at the port of New York was divided into the Boarding
Division, which examined first- and second-class cabin passengers on board
ship, and the medical examiners stationed at Ellis Island, who examined third-
class, or steerage, passengers transported by barge to the immigration station
on the island after the steamship docked at one of the many wharves in New
York City. Along the Atlantic coast, "class" corresponded to both the portion
of the ship in which an immigrant traveled and an individual's station in life,
which were considered equivalent. Class thus appeared as something of an inherent quality that could be easily inferred by looking at an immigrant and
judging his or her "natural sphere."[8] The surgeon general in 1906 described
"the scrutiny of the cabin passengers" as "close," explaining that it was limited
to "circulating among them, and thus observing them," for such passengers did

not represent the nation's industrial labor force and therefore did not need to learn the rules and values of the factory. Although an officer might, on occasion, send a first-class passenger to Ellis Island for further examination, he searched primarily not for physical but for social aberration.[9] "If a passenger is seen in the first cabin, but his appearance *stamps* him as belonging in the steerage or second cabin, his examination usually follows."[10] Class, then, served almost as a fixed biological characteristic akin to skin color, body type, or skull shape.

When first-class passengers were caught in the medical net, they complained bitterly, for the PHS regarded disease as a potential indicator that the immigrant might be of a lower social class than his or her ticket and accommodations would indicate. In 1906, for example, a British physician traveling to New York in the first cabin protested to his consul general of his treatment upon arrival. William H. Howitt claimed full awareness of "the existence of the law that forbids the landing of aliens afflicted with certain diseases at any port of the United States. As soon, therefore, as a physician of the Emigration Department appeared, I told him I was myself a medical man, and that I was not afflicted with any ophthalmic disorder, whatsoever,—expecting that the word of a gentleman and a physician would be sufficient." According to Howitt, "in spite of this, however, he proceeded, without asking my permission, and in an exceedingly rude and violent way, to pull down with his unwashed fingers, my lower eyelids, and not content with this, which, unless it disclosed signs of disease, ought to have sufficed,—he next gouged his thumbs in a most painful manner into the sockets of my eyeballs, in order to evert the upper lids." Exercising remarkable restraint, Howitt claimed that he "made not the slightest resistance at all this, out of consideration for the ladies that accompanied me. The severity of the treatment received may be judged . . . [when I tell you that for] . . . the best part of a week . . . my eyes have not ceased to ache and water from that moment to the present, so much so, indeed, that I intend, as the symptoms have not, as I hoped, worn away by this time, to consult a specialist."[11]

Dr. William Ward, the offending medical examiner, offered a different account of events, claiming that Howitt never identified himself as a physician and that his wife ultimately prevented the eversion of Howitt's eyes by shoving a muff into his face at the crucial moment. "I would have never approached Mr. Howitt in the first place, if on a cursory inspection I had not seen that his eyes were congested and that his lids seemed thickened. Frequently in my work I may inspect five or six, or even more, vessels with first cabin passengers and neither speak to nor touch any of this class, because none show any signs of disease." Clearly, Ward's mission was not to stop individuals such as Howitt. "This cabin inspection is only made to prevent persons suffering from diseases rejectable by the United States Immigration Laws from smuggling themselves into the country as they would do if they knew there were no examination made of this class."[12] According to Ward, then, the cabin inspection was aimed not at the cabin passengers, but rather at those who might travel in the cabin to try to avoid medical examination. For example, Ludwig Hofmeister, who emigrated from Germany in 1925 as a child of five or six, recalled that he had traveled from Hamburg to New York in the first class because the ship's agent told him that he had a heart murmur and would "never go through Ellis Island."[13]

Boston, the home of the immigration restriction movement, took this logic one step further. Henry Cabot Lodge, Massachusetts senator and class-conscious co-founder of the Immigration Restriction League (1894), was one of the first to notice the shift in the source of immigration in the 1890s, away from northern and western Europeans and in favor of southern and eastern Europeans. The cornerstone of Lodge's restriction campaign was the "literacy test," which he sought to require for all arriving immigrants beginning in 1891. While the literacy test would come to have racial overtones in the years before World War I and passage of the measure in 1917, it was originally conceived not for racial selection, but for class selection. Historians of several generations have thus found issues of social class and status at the heart of Progressive-era anxieties and the nativist movement of the 1890s and beyond.[14] Along the Atlantic coast local PHS and IS officers gave lip service to the importance of race. Yet it was social class—particularly at the rigidly class-structured port of Boston—that most profoundly shaped the immigrant medical exam.[15] In the heavily industrial Northeast, it was individuals who attempted to misrepresent their place in the industrial order who posed the greatest threat to society.

Boston based the medical inspection on the conviction that the second-class passenger posed a greater threat than did the steerage passenger. Medical inspectors reasoned that diseased steerage immigrants would attempt to travel in the second cabin in the hopes of avoiding a thorough medical exam. Boston's Dr. Victor Safford, who also served as a medical inspector at Ellis Island, wrote in his annual report for 1911 that "the ancient practice of regarding steerage passengers and immigrants as synonymous is without justification under present conditions of ocean travel. For the entire year 6 per cent of the steerage passengers arriving at Boston were United States citizens, and over three-fourths of the second-cabin passengers were aliens." More to the point, "it is also to be noted that about 25 per cent of the aliens arriving at Boston come as cabin passengers, and their proper medical examination is often made exceedingly difficult from the fact that a long-established custom dictates that the medical inspection of cabin passengers must be made somehow on shipboard whenever they may arrive, day or night, and that they can not be removed to a suitable place ashore for the purpose, as is done in the case of the steerage passengers."[16]

Although other ports along the Atlantic coast shared the conviction that diseased immigrants preferred to travel in the second cabin, only Boston carefully documented certification rates by class.[17] Year after year the certification rate for second-class immigrants was, without exception, higher than that for immigrants traveling in first class and steerage.[18] Although the differences in certification were greater for Class C conditions (5% or 6% of second-class immigrants were usually certified, compared to about 3% of steerage and less than 1% of cabin passengers), by 1915 an increased proportion of second-class passengers were certified for Class A and B conditions.

In 1915 and 1917, although up to 12 percent of all immigrants were certified as diseased, close to one-quarter of second-class passengers were certified for some disease or defect. This nearly matched the certification rates for San Francisco, where the PHS conducted its most aggressive examinations. There is some evidence to indicate that shipping companies did shift diseased or otherwise ineligible passengers from steerage to second cabin in an effort to carry more passengers while avoiding fines for providing passage to diseased passengers. However, it is likely that the more intensive inspection given these second-class immigrants accounts for the difference in certification rates between second class and steerage.

The structure of the immigrant medical exam along the Atlantic coast was not only a product of class-based thinking regarding the industrial labor force, though it consistently reflected the notion of the immutable nature of class. Boston's concern with class was closely related to the rates charged by shipping companies. Baltimore, for example, perceived no problem with diseased passengers trying to travel second class, because the first- and second-class rates were nearly the same; officials reasoned that "true" steerage passengers would be unable to afford the second-class fare. In Boston and New York, in

TABLE 4-1. Certification by Class, Boston

Year	Class	No. Immigrant Arrivals	Certified Class A or B		Certified Class C		Total Certified	
			No.	%	No.	%	No.	%
1910	First cabin	1,198	8	0.67	5	0.42	13	1.09
	Second cabin	11,244	149	1.33	586	5.21	735	6.54
	Steerage	48,922	412	0.84	1,584	3.24	1,996	4.08
	Total	61,364	576	0.94	2,183	3.56	2,759	4.50
1911	First cabin	1,324	10	0.76	11	0.83	21	1.59
	Second cabin	12,297	160	1.30	716	5.82	876	7.12
	Steerage	41,122	367	0.89	1,287	3.13	1,654	4.02
	Total	54,743	537	0.98	2,014	3.68	2,551	4.66
1912	First cabin	1,153	3	0.26	1	0.09	4	0.35
	Second cabin	12,742	170	1.33	767	6.02	937	7.35
	Steerage	35,320	276	0.78	1,288	3.65	1,564	4.43
	Stowaways	49	18	36.73	0	0.00	18	36.73
	Total	49,264	467	0.95	2,056	4.17	2,523	5.12
1913	First cabin	1,179	3	0.25	3	0.25	6	0.51
	Second cabin	15,192	227	1.49	787	5.18	1,014	6.67
	Steerage	50,422	449	0.89	1764	3.50	2,213	4.39
	Stowaways	19	1	5.26	0	0.00	1	5.26
	Total	66,812	680	1.02	2,554	3.82	3,234	4.84
1915	First cabin	425	3	0.71	7	1.65	10	2.35
	Second cabin	1,579	125	7.92	249	15.77	374	23.69
	Steerage	11,639	273	2.35	746	6.41	1,019	8.76
	Stowaways	71	11	15.49	6	8.45	17	23.94
	Total	13,174	412	3.00	1,008	7.35	1,420	10.35
1917	First cabin	278	1	0.36	4	1.44	5	1.80
	Second cabin	2,266	185	8.16	376	16.59	561	24.76
	Steerage	9,863	332	3.37	660	6.69	992	10.06
	Stowaways	39	5	12.82	1	2.56	6	15.38
	Total	12,446	523	4.20	1,041	8.36	1,564	12.57

SOURCE: CGAR, 1911, 1912, 1913, 1915, and 1917.
   NOTE: In 1906, Boston certified 2.6% of first cabin immigrants, 8% of second cabin immigrants, and 5% of immigrants traveling in steerage. These figures were similar for 1909, with 2.6% of first cabin, 8% of second cabin and 5% of steerage being certified. CGAR, 1906, 1909.

contrast, "there was very little difference in the prices . . . of second-class and steerage, and . . . as a result of that . . . there were large numbers of aliens who *really* were steerage passengers that were coming second-class purposely to avoid a more rigid inspection both medical and immigration than they got at Ellis Island."[19] Thus, the business interests and practices of steamship companies were also an important factor influencing the nature of the medical exam. The United States feared the influence that shipping lines might have in preventing countries like Mexico from legislating the medical examination of immigrants.[20] In some instances, however, shipping practices worked to intensify the medical exam. Along the Pacific coast, the IS and PHS moved to make the medical exam as rigid as possible at all ports to prevent shipping companies from preferentially taking their business to ports where the exam was most lenient.

In the East, however, shipping interests worked to prevent rigid medical examinations for certain classes of immigrants. In New York, shipping interests proved a critical factor in preventing more rigorous inspection of second-class passengers. Based on the success of second-cabin inspections at Boston, in January 1916 the Department of Labor launched an investigation into the possibility of examining second-class passengers arriving at New York who were destined for Boston, Philadelphia, or Baltimore, along with steerage passengers. In 1913 New York had unsuccessfully attempted to transfer all second-class passengers to Ellis Island for examination. Transportation companies and civic organizations quickly lodged complaints, in part arguing that removing second-class passengers via barge to Ellis Island was altogether humiliating, and the practice was almost immediately suspended pending an investigation of whether better facilities could be provided either on the ships themselves or on the piers.[21] The Department of Labor in 1916 concluded that any change in the examination of second-class passengers along the Atlantic coast was unworkable. The inspection of second-class aliens was again attempted in New York in 1924.[22] Along the Atlantic coast only Boston ever successfully established the examination of second-class aliens on the docks.[23] This, however, was not accomplished until 1917.[24]

Perhaps class was perceived by the PHS as an immutable concept, as an inherent quality of the immigrant, precisely because of the belief that race, biologically understood, was also immutable. As we will see in chapter 6, the PHS was fond of "typing" immigrants, not primarily racially but according to other seemingly fixed or immutable features that sometimes overlapped with notions of biological race or nationality. Thus, PHS officers spoke of "steerage types."[25] Some even referred to "trachoma types."[26] Such language may have carried explicit racial coding, but to the extent that class or disease did overlap with southern or eastern European origin among immigrants to this region of the United States, the language of the PHS also suggests that these groups were deemed "natural" industrial citizens, ideally suited to serve as America's unskilled labor force. Inspection efforts were not, after all, directed at achieving higher certification rates among particular groups. Rather, they were aimed at preventing those groups from disrupting the social or industrial hierarchy by attempting to "pass" as something other than what they were, by attempting to travel in the "wrong" class. That the PHS prioritized the lan-

guage of class over race is significant because if class was the language of inclusion in the industrial work force, race was the language of exclusion—not
only from industrial participation specifically but, at least in the minds of nativists and eugenicists, from civic participation more generally.

This argument is reinforced by the nature of the diseases that the PHS certified in the East. Reflecting the assumptions regarding the types of diseases
that made immigrants unsuitable as laborers, which underlay the immigrant
medical inspection in the East, when PHS officers did certify immigrants with
disease, they focused primarily on conditions affecting the economic potential
of the body: eye diseases, hernia, heart conditions, and diseases of the bones
or joints.

Physicians in New York began certifying across a wide range of disease categories quite early. Only in the category of venereal diseases did the PHS officers consistently make no certifications; they also rarely diagnosed immigrants with multiple conditions until 1920. By 1925, as a result of immigrant
medical inspections abroad, certifications in New York were once again confined to a relatively narrow range of conditions. (See fig. 3-5 in chapter 3.)

In New York, eye diseases typically represented less than 10 percent of certifications after 1910, increasing again only in 1925. Among eye diseases, trachoma typically represented more than 70 percent of diagnoses through 1912.
In 1914 that percentage fell to 46 percent, with blindness and defective vision
representing 50 percent of certifications. Trachoma continued to account for
up to 60 percent of certifications for eye diseases until 1920; thereafter, it represented only 10 percent of eye diseases—this before it was reclassified as a
parasitic disease in 1921 (meaning that the predominance of blindness and defective vision thereafter was not simply an artifact of classification). Psychiatric
diseases and conditions never represented more than about 5 percent of total
certifications. Miscellaneous or other conditions, however, were prominent in
all years except 1902, representing between 25 percent and 60 percent of certifications from 1915 to 1920. For New York, as for other ports, this category
largely represented senility, which was commonly understood to mean old
age.[27]

The concept of fit industrial citizens, whether understood in terms of race
or class, was not the only factor informing the outcome of the immigrant medical exam. At various times several categories of disease received special consideration at Ellis Island. Digestive diseases were prominent in 1902–4, representing approximately 15 percent to 25 percent of certifications. In the

period from 1910 to 1915, digestive diseases represented a smaller percentage of certifications (about 7% of all certifications). With the exception of 1896, when cirrhosis of the liver was the only digestive condition diagnosed, hernia represented 96 percent to 100 percent of the digestive conditions. From 1916 to 1918, hernia continued to represent about 10 percent of certifications, before falling off in 1920.

From 1896 to 1915 diseases of the circulatory system and diseases of the joints, bursae, muscles, and bones each represented about 5 percent of certifications. These diseases bore a more straightforward connection to an immigrant's ability to be a useful worker. Though representing a relatively small percentage of the total, such conditions were certified more often in New York than in any other port. After 1916 these two categories continued to represent roughly 5 percent of certifications. Circulatory diseases were also conceptually linked to the usefulness of unskilled laborers. Among circulatory diseases, valvular heart disease represented 78 percent to 100 percent of diagnoses through 1904. By 1910 arteriosclerosis, valvular heart disease, and various conditions of the veins (varicose veins, varix, or varicocele) constituted nearly 100 percent of certifications in the circulatory category. Though arteriosclerosis and varicose veins or other conditions of the veins continued to represent almost 100 percent of circulatory diagnoses, from 1914 to 1920 valvular heart disease once again accounted for 70 percent to 90 percent of cases. In 1925 and 1930 valvular heart disease represented approximately 50 percent of cases, joined once again by varicose veins and varicocele (a condition of the veins of the spermatic cord).

Among diseases of the joints and bursae, curvature of the spine and ankylosis (immobility or consolidation of a joint) typically constituted the bulk of diagnoses, with curvature of the spine predominating until 1917. By 1920 ankylosis represented between 75 percent and 95 percent of joint diseases or defects. Finally, in 1925 and 1930 diseases of the endocrines (mostly goiter or an endocrine imbalance) were added to the diminishing number of diseases.

Of course, diseases affecting industrial citizenship could also serve the ends of exclusion. However, when diseases such as trachoma were made to serve exclusionary ends, the context was one in which immigration was explicitly discussed in racial terms.

The Pacific Coast: Inscrutable Races
and Abandonment of the Line

After the California Gold Rush, San Francisco was a major city with a sizable immigrant population. The majority of its ethnic residents did not arrive
at the seaports of the West. Rather, they entered the United States at Atlantic
coast ports and traveled across the continent after having spent some time living in the East.[28] Asian immigrants, in contrast, entered San Francisco and
other Pacific coast cities by sea.[29] Although discrimination toward southern
and eastern Europeans was not unknown in San Francisco and other Pacific
coast cities, San Francisco was perhaps the nation's first cohesive multinational
city. Charles Wollenberg argues that nativism in San Francisco "was less ve-
hement and had less impact than in most other American metropolitan regions
with very large and diverse foreign-born populations. Overt job discrimination
against European immigrants was rare in the Bay Area. Moreover, multinational working-class districts, rather than exclusive ethnic neighborhoods,
were the general rule. While eastern employers often exploited national and
religious differences to break unions and strikes, the powerful Bay Area labor
movement was largely the creation of immigrants from many nations who
showed a remarkable ability to cooperate and maintain solidarity." Wollenberg
attributes mild anti-European sentiments to the relative newness of the area,
to which all groups had recently come, and to the experience of most European immigrants, who arrived and lived in the East before moving to California.[30]

A history of intense workplace competition between native-born whites and
Asian "coolie" labor worked to focus nativist heat on the Chinese and Japanese, in particular.[31] Anti-Asian sentiment in California was instrumental to
the passage of the Chinese Exclusion Act of 1882, which prohibited the entrance of Chinese laborers. In 1907 the U.S. government forged a Gentlemen's
Agreement with the Japanese government to prohibit the entry of Japanese
and Korean laborers. In 1900 the IS took on responsibility for enforcing Chinese exclusion.

The spate of legislation restricting the immigration of Asians helped to limit
the scope of immigrant medical inspection at San Francisco for nearly two
decades. Before 1903 the separation of Asians under the law was so complete
as to prevent the PHS and IS from even regarding them as immigrants. That
is, Chinese immigration was handled under the Chinese Exclusion Act rather

than under the Immigration Act. As a consequence, Dr. Hugh Cumming, convinced of the thoroughness of the quarantine exam, saw little need for a designated medical inspector in San Francisco. Cumming, accordingly, inspected
all arriving immigrants—both Asian and European—for quarantinable diseases only: cholera, yellow fever, smallpox, plague.[32]

In 1903 the IS ruled that, contrary to practice up to that time, Chinese immigrants were subject to the medical inspection and were to be "debarred as
other aliens." This prompted Commissioner General of Immigration Frank
Sargent to push to have a medical officer assigned to San Francisco specifically
for the medical examination of immigrants. Before 1903 Chinese immigrants
had been examined and debarred only for Class A conditions. The surgeon
general's annual report for 1903 stated that "the importance of this ruling is
shown by the figures. For the ten months preceding the ruling an average of
7.9 persons per month were certified and 3 deported, while for the two months
during which Chinese were examined an average of 38.5 per month were certified and 24 per month were deported."[33] To handle the increased workload,
the PHS transferred Dr. Lord from Galveston to San Francisco. For the next
several years Lord conducted the medical inspection of immigrants as part of
the quarantine examination, though his focus was on the "immigrant" diseases
specified in the law and PHS regulations.[34] Immigrants were "examined as
rapidly as possible" and those detained were visited again in the Pacific Mail
Dock "Shed" within two to five days when "a thorough physical examination"
was made and the alien certified. Those certified with Class A conditions were
sent before the Special Board of Medical Inquiry for verification of the diagnosis.[35] Although Lord noted that he took advantage of the glandular or groin
examination made by the quarantine officers to detect hernia, varicose veins,
and other diseases or conditions hidden by clothing, he was not satisfied with
the arrangements. The problem lay in the completion of the medical examination, for he was unable to detain for observation those immigrants he suspected of disease, particularly those with suspected eye and skin diseases.[36]

Not until the opening of the immigration station on Angel Island on 22 June
1910 did the medical examination of immigrants at San Francisco begin to
develop along the lines of Ellis Island. Two officers now examined first- and
second-class passengers on shipboard. They transferred all steerage passengers to the immigration station on Angel Island for a more thorough inspection.[37] (Beginning in 1912, all second-class passengers were also transferred
to Angel Island for the primary medical inspection.)[38]

Inspection at Angel Island, however, hardly resembled the line at Ellis Island. Medical officers stationed in San Francisco conducted a primary physical examination on all second-class and steerage passengers involving "inspection of the stripped body, stethoscopic examination of the chest, eversion of the eyelids, a saltatory test for the detection of beriberi, and microscopic examination of the centrifugalized feces."[39] Officers separated Asian men and women for the initial inspection, as did physicians at all ports when performing an intrusive immigration examination. Asian women, a relatively small percentage of arrivals, underwent a less rigorous medical exam; PHS officers did not ask women to disrobe unless they detected specific signs of disease.[40] In contrast, they "stripped [men] to the waist" before taking them "individually behind a ward screen" where they "completely stripped, in order that any existing abnormality below the waist might be observed."

San Francisco and other Pacific coast ports required that both men and women "furnish a specimen of feces for hookworm examination."[41] The hookworm examination required at least one night's stay at Angel Island, for it entailed both time and paperwork. The PHS provided each immigrant with a common tin basin with a loose wooden cover at night and created a memorandum for each immigrant that matched his name to his basin. The next morning, attendants collected basins and centrifuged the stools. Officers would then review the results for each awaiting immigrant. Individuals who came up negative were free to leave, provided they passed the IS interrogation. Those aliens who had failed to provide "a satisfactory stool" or who the PHS suspected had borrowed another's stool sample, were held to endure the procedure yet again.[42]

The function of the exam along the Pacific coast was not discipline. Rather, consistent with other federal legislation, the goal was exclusion of a group of immigrants deemed wholly unsuited for either civic or industrial citizenship. Here disease interacted with race to undercut class distinctions prevailing in most other regions.

In 1923, for example, two sons of a Chinese government official complained of their certification with trachoma, protesting that they should not have been transferred to Angel Island for the medical examination. The Immigration Service was unsympathetic. Wrote the assistant commissioner general to San Francisco's immigration commissioner, "if they are actually high-class Chinese [we would think] they would be glad to learn of their being afflicted with a dangerous or loathsome contagious disease, such as trachoma, and doubtless take treatment therefore at their own expense while in detention at Angel Island." Their protest suggested instead a sorry attempt to "[becloud] other issues. . . . Aliens found to be afflicted with communicable diseases and who . . . should be segregated, should be treated the same, irrespective of the status claimed by them."[43]

Because the immigration law worked to eliminate the immigration of laboring-class or "low-class" Asians, the social class of all arriving Asians was suspect. The IS constantly guarded against Asian laborers claiming to be merchants or businessmen. PHS and IS officials alike condescendingly referred to the "claimed" status of Asians, which was not easily "read" in the "inscrutable" Asian immigrant, particularly the Chinese. In 1904, for example, one author explained that the "Chinaman . . . always remains the same incomprehensible Asiatic he was when he first landed in America."[44] Consistent with this stance, the PHS professed no ability to detect disease at a glance among Asian immigrants. In 1911 Dr. W. C. Billings made clear that PHS officers could not detect hookworm at a glance, and could not even penetrate the Asian body with their medical gaze: "at least when dealing with oriental races, it is quite impossible to detect on primary examination anywhere near all the cases of hookworm, and a microscopical examination is therefore indicated."[45] In his annual report for 1914 the surgeon general noted, "The medical officers of the station so fully appreciate the impossibility of accurate decision without the microscope that on primary examination aboard ship their attitude is not to say to themselves, 'I am sure from his appearance that this alien has hookworm disease, therefore he must be taken to the hospital for further examination.'" In a critical reversal of the principle of the medical gaze, the officer stationed along the Pacific coast had to remind himself, "'I am not sure that this alien has not hookworm disease, therefore he must not be released until I can determine positively that he is not infected.'"[46]

San Francisco's inspection procedure for parasitic infection raised eyebrows both at home and abroad, prompting questions regarding different inspection procedures for different immigrant groups at different ports. In August 1911 Japan sent two physicians from its immigration service to observe U.S. inspection procedures on the Pacific coast. Dr. Trotter told the surgeon general, "I have been confidentially informed that the object of these gentlemen was to ascertain why there should be such a radical difference in the medical examination of immigrants at the ports of Seattle and San Francisco and why in particular, should Japanese be examined for hookworm, with consequent delay to the alien, at San Francisco and not Seattle."[47]

The PHS and IS scrambled to achieve uniformity along the Pacific coast. To prevent foreign shipping companies from selectively taking their business to ports other than San Francisco, the surgeon general directed Dr. B. J. Lloyd in Seattle to bring his examinations into conformity with those of San Francisco, instructing him to have an attendant collect stool samples from all Chinese immigrants.[48] Mexican, Central American, and Japanese immigrants were to be tested for intestinal parasites only "if appearances are suggestive."[49] Although this directive was ostensibly based on San Francisco protocol, it is noteworthy that San Francisco defended its procedures on the grounds that all second-class and steerage immigrants, regardless of race, were required to provide stool samples. Thus, the instructions failed to address the original complaint: different inspection procedures for different ethnic or racial groups at different ports. Indeed, the Department of Labor saw no contradiction, blithely insisting in 1912 that "this examination is not ordered for any particular nationality, and in its conduct there is no racial discrimination. The examination is conducted on the Pacific coast upon the Chinese, East Indians and Japanese, upon the Mexico-Texas border upon Mexicans and Syrians, and at other ports, such as Ellis Island, N.Y., upon such persons arriving from a region where the disease is prevalent or who present evidence of anemia not referable to any obvious cause."[50]

In 1912 the surgeon general warned all officers at stations along the nation's other land and sea borders to stay alert to hookworm infection but not to conduct routine examinations: "It is not the intention of the Bureau to issue an order for the general examination of arriving aliens for the detection of uncinariasis [hookworm], for the combined reason that this would impose an unnecessary hardship on a large number of people and would prove a task either difficult or impossible to accomplish at some of the large stations."[51] Nonetheless, San Francisco set an important precedent not only for the Pacific coast, but also for other ports receiving Asian immigrants, ensuring that the interaction of Asians and hookworm would always work to subvert class, even at those ports where the inspection typically turned on class. Although Boston's inspectors recognized the presence of hookworm infection in some immigrants by 1917 and examined immigrants for hookworm beginning in 1923, they did not consider such conditions dangerous.[52] Officers at Boston did not consider

routine feces examination necessary until Boston was designated a port of entry for the Chinese.[53] Indeed, for other immigrant groups, Boston officials
commented that "infection with such parasites, including the hookworms,
does not appear to be necessarily incompatible with apparent good health."[54]

The Canadian Pacific Railroad's plans to bring the first "shipment" of Chinese from St. John, New Brunswick, to Boston in the early part of 1923 found
Boston unprepared to conduct examinations according to San Francisco protocol. Dr. Albert Nute explained to the surgeon general that Boston was wholly
unprepared to examine Chinese immigrants. Nute insisted that "the examination here should be just as rigid as at Angel Island, otherwise the transportation companies will be able to avoid the annoyance of Angel Island by bring-
ing the aliens over the Canadian route. . . . To do this work properly these
aliens, men, women, and children must be under constant supervision until
the stools are collected."[55] The Boston station lacked both the facilities to detain Chinese immigrants and also the laboratory equipment to diagnose hookworm and other parasitic infections on the scale now required, for stool examinations were not standard at any other port along the Atlantic coast.[56]
Nevertheless, the inspection routine for Chinese immigrants was firmly established by 1930, when Boston reported that "a considerable number of this
race are medically examined. . . . In addition to the routine physical examination, stool examinations are required for ascertaining the presence or absence
of hookworm and other intestinal parasites."[57]

The Angel Island parasitic exam was gradually scaled back. By 1923 the
PHS took only steerage passengers to Angel Island.[58] By 1925 it transferred to
the immigration station only those immigrants suspected of being diseased or
defective when examined on board ship.[59] This change, however, came in the
wake of immigration restriction beginning in 1921. That is, it did not result
from the PHS or the IS rethinking the need to inspect Asian immigrants. Indeed, the PHS maintained strict standards of examination among Asians: because the "sanitary antecedents" of immigrants arriving from Asian countries
were "subnormal," immigrants were subjected to an intensive examination
whenever possible.[60] In the view of the PHS, Asian immigrants continued to
"present an intestinal fauna of perhaps unsurpassed richness and variety," allowing the stool examination at Angel Island to remain the staple of the medical examination and the "most productive feature of the laboratory work."[61]

In the East the language of class was mapped onto the language of race,
which not only gave class a fixed biological quality but also created a pre-

sumption of fitness for industrial citizenship. In the West it was primarily race, not class, that was "stamped" on the immigrant. This racial stamp, however, made it impossible for the PHS to read the class of the Asian immigrant—a necessary step in enforcing the Chinese Exclusion Act of 1882, which prohibited the immigration of laborers. Effectively, then, the PHS subjected not only all Chinese but also all Asians to a test for admissibility, by transferring all second- and third-class passengers to Angel Island for inspection.

Since the law made clear that all Chinese laborers were ineligible even for industrial citizenship, the PHS did not seek to prevent Asians from traveling under the wrong guise—that is, as first- or second-class immigrants when they were really third-class. In refusing to distinguish second-cabin from steerage immigrants in terms of treatment in the West, in subjecting all Asian men to a more revealing physical and bacteriological exam, and in claiming the inability to read class in the Asian immigrants, the PHS allowed the concept of race to preempt the concept of class. In so doing, the agency revealed an exclusionary intent broader than simply enforcing the particulars of the Chinese Exclusion Act. Medical certifications for all diseases in this instance were intended to serve the ends of exclusion. And disease in this situation came to have strong and explicit racial overtones. The language of class effectively fell out because viewing immigrants primarily in class terms implied that such immigrants were appropriate for the industrial working class.

In San Francisco (see fig. 3-7 in chapter 3) certifications fell primarily into three disease categories, which reflect on the PHS's supposed inability to read disease in the Asian body and to focus its efforts on exclusion rather than discipline. Through 1910 eye diseases accounted for the vast majority of certifications; from 1911 to 1926 eye diseases represented roughly 10 percent to 20 percent of cases. Trachoma accounted for between 65 percent and 99 percent of certifications for eye conditions until 1918. Significant here is the virtual absence of certifications for defective vision or other seemingly innocuous eye diseases or conditions, which suggests that the PHS failed to look for the diseases of industrial citizenship among Asians because it did not think of that group in those terms. Why search for blindness or defective vision among a group of immigrants who had no place in American political *or* industrial life?

In contrast to the situation at most other ports, obstetrical and gynecological conditions were important in San Francisco, representing roughly 10 percent to 15 percent of total certifications after 1911 (they surpassed eye diseases by 1927). All these were cases of pregnancy, which suggests a particular concern about race and citizenship. The Fourteenth Amendment to the Constitution made citizenship a birthright, setting up the paradox that although Asian immigrants were excluded from citizenship, their children were not. A pregnant Asian immigrant, whether traveling alone or with a husband, signified not only an unwanted industrial citizen, but an unwanted and inherently unsuitable participant in American public life.

But it was in the instance of parasitic infections that the PHS worked most actively as an agent of exclusion. Certifications due to parasitic and skin conditions became significant by 1917. With the exception of 1918, parasitic and skin conditions represented more than 50 percent of the San Francisco certifications, approaching nearly 75 percent in 1926 and 1927.[62] As we shall see, parasitic conditions became strongly and explicitly linked to racial unsuitability for participation in the industrial work force.

THE GULF COAST: ERASING THE LINE

In some instances, neither discipline nor exclusion was at issue. Public Health Service officers inspected immigrants along the Gulf coast under the poorest of conditions. The Immigration Service provided no immigration station at most Gulf coast ports, and both PHS and IS officers often made cursory inspections at night.[63] The primary medical inspection was usually made aboard ships or on wharves at ports along the coast. For first- and second-class passengers, it might take place in the dining or drawing room of a ship in space too small to allow immigrants to stand in line. The PHS inspectors had to examine third-class immigrants in the cramped steerage quarters or corridors below deck. Officers complained that "the light throughout the ships is bad, but it is much worse in the corridors and steerage where the aliens are inspected, and renders the detection of trachoma cases and other diseases . . . impossible." The IS inspector in Tampa confirmed the miserable conditions under which PHS inspectors worked, concluding that "the aliens come at times in such a crowd that in the narrow spaces on board the ships it is impossible to have them stand with uncovered heads and walk past the doctor at a distance as required by regulations. All the medical officer can do at the present time is to go among them and look them over in the dim light."[64]

Most inspectors at Gulf coast stations, because of the smaller volume of immigrant traffic, were responsible for making examinations at multiple ports. The inspector stationed at Port Arthur, Texas, for example, also inspected immigrants at Sabine, Sabine Pass, Beaumont, Port Neches, and Orange—cities thirty to forty-five miles away.[65] In New Orleans, one of the largest ports along the Gulf, the PHS officer was faced with a twenty-five-mile stretch along either side of the Mississippi where ships could dock: ships would enter the river at all times of the day and night, dock at a wharf, and then the crew and sometimes the passengers would disembark before the PHS and IS inspectors arrived. World War I exacerbated this situation, for increased shipping traffic meant that ships more frequently moved from one dock to another.[66]

Yet New Orleans was perhaps the best equipped and manned of the southern ports. Its history illuminates some of the factors that contributed to undercutting the immigrant medical examination, preventing it from serving either disciplinary or exclusionary ends. The IS opened one of the Gulf coast's premier immigration stations at Algiers in 1913. There the PHS was to combine the quarantine and medical examination of immigrants. Nonetheless, even at Algiers space to conduct the combined exam remained inadequate and public. Indeed, the station's opening did not significantly alter the medical inspection of immigrants, which PHS officers continued to conduct on the deck of the ship or in its saloon as late as 1918. Only those immigrants suspected of being diseased at quarantine were ever transferred to the immigration station.[67]

Years after Algiers opened, Dr. Scott, the PHS physician stationed in New Orleans in the 1910s, complained of the conditions under which he inspected immigrants and of a lack of support on the part of the IS. Although the IS insisted that he call in twice a day, Scott traveled long distances between vessels and found it difficult to drop whatever he was doing to locate a phone. Consequently, he was often misinformed, either by the IS or the shipping company agent, about the times or locations of ship arrivals; not infrequently, he arrived hours too early or too late. IS officials blamed Scott, who simply retorted, "If there is any delay in inspections of seamen or alien passengers it is not my fault."[68] Delay, however, was not the most frequent result of late arrival on Scott's part. Once examined by the IS officers, passengers had little incentive to await medical inspection, and IS officers—influenced by local leadership—were not inclined to hold them.

The IS chastised Scott not only for failing to appear to inspect immigrants on time, but also for conducting an overly thorough examination of first-class passengers. Wrote S. E. Redfern, commissioner of immigration in New Orleans, to Scott, "Comment has been made by a number of first cabin passengers arriving at this port relative to the difference in medical inspections under the Immigration Law here from such examinations at the Port of New York." Redfern, accordingly, encouraged Scott to exercise restraint: "The idea I wish to convey is that first cabin passengers should be treated with every possible consideration and courtesy so far as the law will permit. . . . Commerce and trade relations between the port of New Orleans and Central and South America need stimulating and encouraging."[69]

The PHS, concerned about the implications of this message, instructed Scott's supervisor, Dr. Fairbanks, that "it is against the desire and policy of the Bureau to have any Service work performed in a perfunctory manner."[70] Nevertheless, the immigrant medical inspection at New Orleans remained superficial. Local officials heralded the new immigration station not as a facility to ensure the proper control of immigration, but as one that promised to attract more immigrants to the state. In 1927 the *Times Picayune* proclaimed it "the Antithesis of Ellis Island."[71] Algiers was a place of comfort, of humanity, of welcome—not of discipline or exclusion. That the contrast was acknowledged supports the notion that the PHS exam worked to very different ends in different regions and indicates that the disciplinary and exclusionary functions, though they might not be explicitly stated, were widely perceived.

In the first decade of the twentieth century, when American business and industry eagerly sought the labor of immigrants, southern states generated the greatest demand. Cotton mills and railroads spread across the south. As African American labor moved northward, southern leaders reactivated immigration bureaus and sent agents to European ports in the hopes of enticing immigrant labor beyond the Mason-Dixon line.[72] Like many other southern leaders opposing immigration restriction, Justin Denechaud, a member of a prominent New Orleans family who served as secretary of the Immigration Division of the Louisiana State Board of Agriculture and Immigration from 1910 to 1924, promoted "white" immigration to the South. So great was his state's demand for labor that he actively campaigned against immigration restriction during the period of his tenure with the state of Louisiana.[73]

For Denechaud and other Louisiana officials, concerns regarding the race, class, and diseases of incoming immigrants took a back seat to promoting the state as a healthy one. The State Board of Health and the Immigration Division expended considerable effort in combating not only malaria, cholera, and yellow fever, but also the perception that Louisiana was plagued by such diseases of an "uncivilized" society: "Conservation of health as a natural resource or asset, is one of the most advanced of modern ideas. Health begets wealth."[74] Louisiana was not unconcerned with the quality of immigration to the state, explicitly stating that it sought to attract "a good class of immigrants and home-seekers to the State by showing them just what conditions are in the state of Louisiana."[75] The state was more concerned with the image of disease it pro-jected. In 1913, for example, a prospective settler from Kansas wrote to the Louisiana State Board of Health, having heard that "the people are suffering with Ague and Malaria, and are carrying quinine in their pockets, as we do to-bacco here."[76] Similarly, Mr. Olaf Huseby of the Empire Land Company in Moorhead, Minnesota, who was interested in buying land for colonization by northern farmers, wondered about conditions in the South, asking why south-erners still lagged behind northerners in production. Perhaps, he suggested, the southerner is really not naturally lazy and ignorant, but simply affected by the climate: "I am afraid that the low prices of your land, the laziness, the ig-norance etc must have something to do with the climate (the same as in South-ern and Southeastern Europe). . . . Is there any book or pamphlet provided telling the truth about mosquitoes, malaria and other diseases[?] . . . I do not expect you to sit down on your own state; do not want it either. What I am af-ter is the plain scientific truth."[77] Even in the early nineteenth century, one lo-cal physician observed that "an opinion has gone abroad, and attained, as is be-lieved, a very general currency, that in consequence of its sickliness, especially its liability to pestilential fever, New Orleans can never become a great empo-rium of business, the seat of an extensive commercial capital, and of direct im-portation from European ports." This misconception "rests on a single posi-tion, namely, the presumption, that, from its climate and topography, the city of New Orleans is necessarily and irremediably unfavourable to health."[78]

Denechaud spearheaded an effort to "correct" such false impressions of Louisiana. In 1913 he wrote to parish officials demanding that they comply with reporting vital statistics related to health: "You are also no doubt aware that the people living outside of our State have an idea that health conditions in Louisiana are fare [sic] from warranting them to come here either as home-seekers or investors, and if we can furnish them with authentic information as to health conditions, we can correct this erroneous impression and thus induce many to locate in our State."[79] A common assurance given to prospective homesteaders was that Louisiana and its lands are "very fertile and health con-ditions are excellent."[80]

The PHS medical inspection in Louisiana might have followed the model of the East and served a disciplinary function, for the state and region were trying hard to attract immigrant labor that would presumably require discipline. In this case, however, health was associated not with the industrial fitness of immigrant labor, but with the attractiveness of Louisiana. The city of New Orleans, like the state of Louisiana, was almost entirely uninterested in the process of immigrant medical inspection. It cared about the quarantine services provided by the PHS.[81] In 1909, after having stationed an extra officer in New Orleans to aid with the detection of yellow fever and cholera during the warm months, as a practical matter the PHS discontinued his services during the winter.[82] This move touched off a strong protest in New Orleans, which wanted to keep all possible quarantine services running at all times. The New Orleans Board of Trade and the United Fruit Company, the biggest shipping line serving Louisiana, both tried to persuade the PHS to leave an extra physician in the employment of the quarantine service, fearing that if "innocent parties" were blamed for epidemics, shipping and commerce in the city would suffer.[83]

The economic base of Louisiana differed substantially from that of the Northeast. Although the state began industrial development after 1900, Louisiana maintained a predominantly agricultural economy, with sugar, tobacco, and rice as its staple crops.[84] Sharecropping, rather than scientific management, provided a model for worker discipline.[85] Consequently, neither the PHS nor the IS felt a strong imperative to inspect immigrants to Louisiana rigidly. Indeed, the overriding concern was to remove as many obstacles to immigration as possible— that is, to facilitate the arrival of immigrants. In supporting the quarantine exam, moreover, the immigrant medical exam not only worked better to prevent the spread of highly communicable diseases but also helped buttress the argument that Louisiana was as civilized as the rest of the nation.

New Orleans, accordingly, provides a window for understanding the agencies' reluctance to funnel resources for medical inspection into southern states. Although PHS officers stationed along the Gulf coast perceived a great need for improved conditions, the PHS and IS withheld particular types of resources that would have facilitated immigrant medical inspection, while providing others, such as quarantine inspection, as a means of satisfying southern leaders, who sought to promote rather than restrict immigration to the region. Where resources were not withheld, as in New Orleans, both federal and local officials effectively constrained PHS officers, subverting the medical inspection of arriving immigrants.

Perhaps reflecting a regional reluctance to examine immigrants, data for the Gulf coast are scarce. Sufficient data were available for New Orleans, and these were limited (fig. 3-8). For example, no data were available for the period 1911–20. In New Orleans the PHS rarely made certifications in more than a handful of different categories of disease in any year. Eye diseases were predominant, generally representing more than 10 percent of certifications and typically more than 40 percent of certifications. These were almost all cases of trachoma. As in the West, the absence of certifications for blindness or defective vision reflects a lack of interest in assessing immigrants as industrial citizens. Beginning in 1924 blindness and defective vision replaced trachoma. Miscellaneous (primarily senility) and respiratory (primarily tuberculosis) diseases and conditions are the only other categories that were more than sporadically prominent.

The Canadian Border: Maintaining the Line at Vulnerable Points

The IS and the PHS took their most rhetorically vigilant stance along the Canadian and Mexican borders. They assumed that diseased or defective immigrants—particularly those turned away at Atlantic or Pacific coast ports—would attempt entry across the long land borders that separated the United States from Canada to the north and Mexico to the south, in anticipation of a more lenient examination.[86] Although discipline was not abandoned, especially along the Mexican border, exclusion was at stake along the land borders just as it was along the Pacific coast.

Special Immigrant Inspector Robert Watchorn, later to become the commissioner of immigration at the U.S. immigration station in Montreal, insisted that "there can be no doubt" that "a very large number of people from Europe who are either diseased or likely to become public charges, as a subterfuge," attempt to cross the Canadian border. One had only to look at the lot seeking entry across Canada to conclude that "they are the very essence of all that is undesirable in the way of acquisitions to our population from foreign countries. . . . I unhesitatingly assert no human beings who ever came under my observation presented a more forlorn and hopelessly unimprovable appearance than those who have attempted to enter the United States via the Canadian border." Making the situation near desperate, Watchorn noted, "the Canadian route to the United States is known to every unscrupulous agent in

Europe, and is by that means made known to the very dregs of society."[87] Less dramatically, Dr. Victor Heiser described the Canadian border as yet "another leak in the health dam." Like Watchorn, Heiser claimed that the ease of entrance by the Canadian route was well known, citing a steamship circular that read, "For sickly or defective passengers who want to avoid to land in the United States ports, I recommend my new Canadian line to St. John, where passengers are freely landed, without any examination. From St. John passengers can get within a few hours ride by rail to any place in the United States."[88]

The perception of disease and danger, however, did not translate into remarkable differences in the nature of the examination along the Canadian border, for two key reasons. First, the United States took a series of steps to plug the health dam in Canada. The first was to station IS and PHS officers at the Canadian ports of Quebec, Halifax, and St. John to examine immigrants destined for the United States. Immigrants, however, were not obliged to inform U.S. officials of their destination.[89] The Canadian government, unwilling to hinder immigration, resisted examining immigrants destined either for Canada or the United States for anything but quarantinable diseases. So the PHS and IS negotiated directly with the Canadian steamship companies and the Canadian Pacific Railroad to help prevent the passage of diseased immigrants. Railroad and steamship companies signing on to the Canadian Agreement, forged in 1901, granted U.S. PHS officers authority to examine all immigrants destined for the United States who arrived in and sought to leave Canada. Any companies not participating risked devastating business losses, for the United States flatly refused to admit immigrants arriving on steamship lines that failed to sign the Canadian Agreement and that supposedly "dumped" undesirable immigrants in Canada to avoid the American fine of one hundred dollars per diseased passenger. In the case of railroad lines, the United States threatened to halt all incoming passenger trains whose owners did not allow U.S. medical inspection in Canada and hold them at the border while officials sorted out and examined all immigrants.[90]

Second, and more important, the Canadian Agreement, which left Canada stuck with American "rejects," prompted the Canadian government to mandate the medical inspection of arriving immigrants for the benefit of Canada. The 1902 amendment to the Canadian Immigration Act prohibited the entrance of "any immigrant or other passenger who is suffering from any loathsome, dangerous or infectious disease or malady, whether such immigrant intends to settle in Canada, or intends to pass through Canada to settle in some other country." The Canadian medical inspection service was thus created, and physicians were stationed in Quebec, Halifax, St. John, Montreal, and Winnipeg.[91] Thereafter, Canadian officials first inspected immigrants destined for the United States to guarantee that they would be admissible to Canada, before sending them to U.S. PHS and IS officials.[92]

Despite U.S. complaints that Canadian officials conducted a lax medical inspection, Canadian policy provided an important buffer because the Canadian philosophy of immigration control was consistent with that of the United States, though it emphasized selection rather than discipline. Alan Sears notes that Canadian government officials regarded medical inspection as a tool for selecting out immigrants likely to impair "national efficiency." Canada was particularly concerned with societal "degeneration," but this concern was not specifically linked to immigrant genetic inferiority. Rather, it was linked to the behaviors, habits, and social and economic condition of arriving immigrants. Consequently, in Canada the medical inspection, ostensibly modeled on U.S. practice, was intended to assess the "usefulness" of immigrants; quarantine would serve to screen out diseased immigrants. As explained by the chief medical officer of Canada's provincial Board of Health, P. H. Bryce, the idea behind medical inspection was to ask "is that man going to be of use to Canada?"[93] Although the immigrant inspection in Canada resulted in a lower overall rate of rejection for all causes, medical rejections consistently represented a much larger percentage of overall rejections than in the United States. From 1902, the year Canada began medical inspection of its arriving immigrants, through 1911, Canadian health officials rejected a total of 0.46 percent of arriving immigrants for all causes.[94] Medical rejections accounted for over 40 percent of total rejections through 1916.[95]

Although the Canadian border was consistently cited as a vulnerable point in America's defenses, the IS invested few resources in immigration stations along the three-thousand-mile stretch of land, with the exception of those stations near the Pacific coast that received Asian immigrants, where practice was in keeping with that of San Francisco. The conduct of the examination likely varied widely along the Canadian border, which comprised both land stations and seaports. In Eastport, Maine, for example, Dr. John Brooks examined, presumably on either the ship decks or shipping docks, passengers arriving on large steamers and sailing ships, or on small craft that could land on any part of the island "and scatter every where." Brooks assumed that immigrants traveling on small boats sought to avoid examination.[96]

Virtually nothing is known of the inspection regime undertaken at stations receiving immigrants by rail. Inspectors write of following the Ellis Island system for disease classification, but there is no evidence to indicate how inspections were conducted, whether immigrants disembarked from trains and stood on line for an examination, or whether PHS officers boarded cars and moved among the passengers, examining them in the manner of the first- and second-class medical exam at many coastal stations. There is no evidence that the inspection along the Canadian border resembled that along the Mexican border, where immigrants arriving by train, foot, or other means were relatively intensely inspected en masse.

Because of the expanse that had to be guarded, immigration stations, especially before 1910, often maintained irregular hours of operation, were isolated, and were typically manned by a single physician and a mere handful of other IS officers or attendants.[97] To compensate for such conditions, many PHS physicians inspecting immigrants were local practitioners who maintained their private practices and served as civil surgeons or acting assistant surgeons rather than as commissioned officers. In many instances such arrangements appear to have worked well, but a great deal of IS and PHS correspondence was devoted to complaints about such physicians. Resource investment on the part of the IS and PHS, however, did not reflect on vigilance along the Canadian border as it did along the Gulf coast. Both the IS and PHS aimed sharp criticism—with no regard for scarce resources or impossible working conditions—at medical officers whom they perceived to represent weak links in the northern border defense.

The PHS did not maintain exacting standards for part-time, noncommissioned officers working along the Canadian border at insignificant immigration stations.[98] Of his full-time officers working along the border, however, whether commissioned or noncommissioned, the surgeon general demanded both competency and unwavering devotion to duty. Where a PHS physician did occupy a full-time position along the Canadian border, the work tended to be quite burdensome.[99] It became even more so as traffic through these ports increased.[100] In 1912 the surgeon general described the duties of the physician in charge at Buffalo, New York. The PHS officer reported to the immigration office at 10:00 A.M. or earlier, if called. He examined immigrants until 1:30 or 2:30 P.M., when he could leave for lunch. The 4:15 train brought the PHS officer back to the immigration station until around 6:00, when he was again free for dinner. His final shift of the day kept him at the immigration office from 7:30 to 10:30 P.M. or much later. "Often it is 12 to 12.30 before he leaves the office for the night. Outside of these hours he is obliged to hold himself in readiness to be called at any other time during the day and night. These conditions, coupled with the fact that he is called upon for the examination of aliens who become public charges in the local hospitals and for the care and examination of those aliens detained in the county jail by the immigration service, gives him little or no time to himself."[101]

So overwhelming was the work that Dr. Albert Nute, stationed in Port Huron, threatened resignation if he were not transferred to "a better place."[102] He argued, "I do not really believe that [Dr. Wyman] understood the conditions here when the appointment was made." The pay, in Nute's estimation, was too low (only sixty dollars per month) and the workload too great. Dr. Nute eventually received his transfer to Boston, but others stationed in Port Huron were not so lucky.[103] Dr. J. J. Siffer, who replaced Nute, resigned in 1906 after the PHS denied him a raise, despite an increase in the workload.[104] And in 1912 Dr. B. L. Schuster, battling allegations on the part of the IS that he was failing to make examinations when requested, protested that "I have never refused to meet . . . any . . . train or ferry boat, or failed to examine an alien when requested." He also noted that the IS expected him to make night inspections. If his superiors truly expected him "to be on duty both day and night, I will do the best I can; but I would be the only one at this port of entry who would be placed in the position of remaining on duty from 8:30 A.M. until any and all hours of the night."[105] The surgeon general's response, however, was fairly unsympathetic, indicating that he was more concerned with inconveniencing many passengers than one PHS officer. The surgeon general concluded by reprimanding Dr. Schuster: "if you find that you cannot perform this work as is proposed the Bureau will of necessity be compelled to take steps to fill your place."[106]

Patterns of certification along the northernmost border, then, were most similar to those of the East Coast. Immigrants crossing from Canada were in practice not regarded with an eye to exclusion, particularly after the Canadian Agreement and Canada's own medical inspection legislation. Along the Canadian border, few conditions exceeded 10 percent of certifications (fig. 3-10). The PHS typically made no certifications in two categories only: diseases of the lymphatic system and venereal diseases. As early as 1902 PHS officers were certifying across a wide range of diseases, though eye diseases represented approximately 50 percent to 70 percent of certifications. Between 85 percent and 99 percent of eye conditions were attributable to trachoma or nontrachomatous conjunctivitis. After 1917 eye diseases represented 5 percent to 10 percent of total certifications. Though trachoma still accounted for roughly 20 percent to 30 percent of certifications for eye diseases, blindness and forms of defective vision began to predominate at immigration stations along the Canadian border. Defective vision and blindness in one or both eyes represented about 70 percent of eye diseases certified from 1922 to 1927. While the spectrum of eye diseases diagnosed broadened from 1928 to 1930, blindness and defective vision continued to represent about 50 percent of cases.

By 1917 PHS officers were making certifications fairly evenly over a wide spectrum of diseases. Though in any single year one category—eye diseases, miscellaneous conditions, psychiatric conditions—might stand out, no single category was dominant. In 1917 and 1918, senility represented about 60 percent of miscellaneous conditions. Beginning in 1922, senility consistently represented more than 80 percent of miscellaneous conditions. As it should for other ports, then, the miscellaneous category should largely be read as senility.

Certifications also indicated that America regarded immigrants crossing the Canadian border in terms of their fitness to participate in both industrial *and* civic life. In 1917 feeblemindedness accounted for 50 percent of the certifications. In 1918 feeblemindedness accounted for 33 percent and insanity for 50 percent. In 1922 chronic psychopathic inferiority (also referred to as CPI or simply psychopathic inferiority) appeared on the scene, representing 48 percent of the psychiatric cases. Chronic psychopathic inferiority continued to represent between 30 percent and 40 percent of certifications, though certifications for various forms of dementia precox began to account for about 20 percent of certifications in 1928, and various forms of psychosis accounted for 25 percent of cases in 1930. Overall, it appears that PHS officers along the Canadian border showed shifting certification preferences, moving from feeblemindedness to chronic psychopathic inferiority, to dementia precox, and finally to psychosis, providing an index of sorts for tracing trends in psychiatric terminology. The only other region of the country to place any such emphasis on psychiatric conditions was the East Coast. This indicates that European immigrants could largely be regarded as having a proper place in both American industrial and public life, for only individuals from groups with the inherent capacity for self-government could lose this capacity to disease. Moreover, the PHS, as we shall see, drew no links between mental illness and race, despite outside pressure to do so.

THE MEXICAN BORDER: FORTIFYING THE LINE

In sharp contrast to Canada, the Mexican border remained a particularly vulnerable point in the nation's defenses. In 1906 the commissioner general of immigration lamented that "the very worst elements of the foreigners enter by that route. Aliens who are so diseased, or of such frail physique, or so apparently paupers, as to convince even the interested steamship companies of the risk attendant upon trying to enter them at Atlantic seaports are sent by the steamship agents to Mexico, and from there enter surreptitiously across the border."[107] All immigrants seeking entry along this "devious route" invited immediate suspicion that they were "diseased and pauperized."[108] Exclusion was foremost in the minds of immigration officials. The intent of exclusion caused both the PHS and IS to invoke the language of race and class, but the language used was far more complex than in any other region of the country. The Mexican border became a region in which different, supposedly biological races, nationalities, and types—labeled Mexicans, Syrians, peons, coolies, native Mestizos, and Indians—could be mapped onto one another and then altogether remapped, depending on the reading of class, which carried connotations not only of proper position in the work force but also in civilization. The imperative to exclude was mixed with the need to discipline, which took a far more rudimentary, almost colonial form that sought to civilize backward peoples rather than socialize industrial laborers.

Mexico presented problems of a different order than America's "civilized" neighbor to the north. The Mexican approach to public health differed sharply from that of the United States. For example, Dr. Eduardo Licéaga, president of the Mexican Board of Health, was a consistent advocate for liberalizing quarantine laws, even to the point of doing away with them. At the Second International Convention of the American Republics in 1906, he maintained that the international community should "sacrifice the word quarantine, let us strike it out of our actual vocabulary. . . . Let us adopt a new flag for the battle against the transmissible sicknesses; let us inscribe on same the motto, 'To safeguard the interests of the public health without impairing, or impairing the least possible, the interests of the commerce and of the free communication of men.'"[109] In 1907 Licéaga made a similar plea to allow "goods to have free transportation."[110] Although a plan similar to the Canadian Agreement would have facilitated commerce and passage between the United States and Mexico, the United States was never able to convince the Mexican railroads, which were government-owned, to allow PHS officers to inspect immigrants destined for the United States at their various points of origin in Mexico.[111]

But the Mexican border problem was more than one of international cooperation or trade relations. Mexican philosophy, like that of other Latin American countries, stressed the importance, not of the health of the immigrant in shaping the nation, but of the nation in shaping the health and productivity of the immigrant. The power of the state to transform and mold its inhabitants was the mark of a "civilized" society: "For many centuries the peoples of the earth have paid attention only to their defense against exotic diseases, without taking into consideration that it would be much more sensible and logical to prevent any disease from originating in their respective country." Thus, argued Licéaga at the Third International Sanitary Conference of the American Republics in 1907, those cities that provided adequate water, removed human and animal waste, kept streets paved and swept, regulated the light and ventilation in dwellings, and prevented overcrowding "will never be invaded by transmissible diseases." Licéaga concluded, "Let us try to obtain a healthier and more liberal view of matters, and not defend our legitimate interests by wounding those of others; let us be more practical and more humane; let us not look upon our unfortunate brethren as enemies, whatever may be their title or nationality, through the mere fact that they are sick."[112]

Such philosophy flew in the face of the PHS philosophy and function. Wyman, as a rebuttal to Licéaga, argued that the diseased immigrant was destined to live in "an environment of comparative poverty and squalor. He is unable to contribute his due portion to the industrial and mental development of the state, and his very presence is, therefore, a factor in reducing the standard of the civilization of the country to which he comes."[113] While the state might socialize the immigrant—teach him to uphold industrial values and norms—it could not "cure" the unsuitable industrial citizen. Discipline, in the United States, was so important precisely because the onus fell on the immigrant to maintain health, to serve as a reliable cog in the machinery of industrial production.

In 1908 Mexico passed immigration legislation mandating the medical inspection of immigrants.[114] Although the Mexican law was seemingly based on

U.S. legislation, a different philosophy of disease and immigration continued to separate the two countries. Licéaga, touting the new immigration law at the Fourth International Sanitary Conference of the American Republics in 1910, argued that "it will be seen that the law to which we refer not only protects the Mexican Republic against transmissible, acute or chronic diseases, but also in providing for its own defense, protects the nations which maintain relations with us by land or sea."[115] Trachoma and other diseases with which the United States was concerned were mentioned in the Mexican law, but the law gave priority, not to diseases like trachoma or hernia that promised to undermine the immigrant's ability to contribute to the industrial labor force, but to the acute contagious diseases that, in the United States, were covered under the quarantine law. Therefore, though the language was similar, the emphasis was very different. Further, convincing the United States that the Mexican immigration law was merely trumped-up quarantine legislation, the Mexican immigration law placed disease in immigrants and travelers on equal footing. Wyman, speaking at a previous sanitary conference, explicitly stated that quarantine law was not immigration law: "Immigrants are not like other passengers. I request very severe laws for immigrants, but not for passengers."[116] At stake was not the spread of disease but the suitability of immigrants for the U.S. labor force. The United States, consequently, never regarded the Mexican medical inspection as remotely comparable to that conducted by Canadian officials. Its own officers remained the primary defenders of the nation.

The United States maintained its major immigration stations for the Mexican border along the Texas line. Texas, unlike other Gulf coast states, remained indifferent to immigration during this period. Indeed, the state maintained an immigration committee only in 1887.[117] Immigrants to Texas and other states of the Southwest, particularly Mexican laborers, found employment primarily in agriculture, not industry. Some Mexican laborers found employment building railroads and working in coal and metal mines along with Chinese and Japanese laborers. Texas was keenly concerned with quarantinable diseases. The threat of epidemic diseases, such as smallpox, typhus, or yellow fever, arriving from Mexico was almost constant.[118] Diseases bearing on industrial citizenship, then, were of less concern in this region. For the PHS, civilization was at stake. Therefore, a different and more pejorative brand of discipline was applied in this agricultural region, which had neither a strong industrial base nor a strong sharecropping tradition, as in Louisiana, for it had never been one of slavery's strongholds.

The Mexican Revolution, which began in 1910, brought chaos and disorder to Mexico. For the PHS inspectors, conditions symptomatic of disorder and lack of control were sexual in nature. Despite government regulation of prostitution along the Mexican border, the PHS found a higher prevalence of sexually transmitted diseases, along with a relative preponderance of pregnancy and other obstetrical and gynecological conditions. That the PHS focused on these conditions in this region underscored its concern with reestablishing order. It was also interested in establishing standards of civilization, ensuring that immigrants conformed to a rudimentary notion of "civil" citizenship—"civilized" citizenship.

At the Mexican border, then, the interests of the region and the PHS allowed the unique combination of the quarantine and immigration functions of the PHS, producing a cleansing and scrutinizing process that was unparalleled in any other region. The PHS, after all, was not preparing these immigrants for the factory floor, but for simple cohabitation in the nation. The PHS and state officials were also concerned with quarantinable disease in Louisiana, but that state sought primarily to attract "white" immigrants and was primarily conscious of its own reputation. With the exception of measures taken during a typhus scare along the Atlantic coast in 1921, such quarantine procedures were never regularly performed at any other domestic port or station.[119] And there they never carried the overtones of immigrants coming from primitive countries, as they did when performed along the Mexican border.

Immigrants crossing the Mexican border were required to pass through the bathhouse and disinfecting plant, where they were entirely disrobed and examined by trained attendants or a medical officer. Inspectors examined the naked immigrants for "vermin infestation, for eruptions of any sort, enlarged glands or any abnormality." Inspectors also searched for lice, shaving the heads of men found to be infested and burning their hair. Women with head lice sat for one hour with a solution of kerosene and acetic acid applied to the scalp and hair. "After [the lice treatment] the person is passed on to the shower baths, where the bathing process is supervised by an attendant and then passed into a rear room in which the clothing is received back through an opening in the wall after having been disinfected by steam and dried in a vacuum."[120] Such procedures had been well established along the Mexican border since at least 1916.[121]

The immigrant medical exam on the Mexican border was reminiscent of that conducted in many German border control stations and ports of depar-

ture where, after immigrants undressed, they were "rubbed down with a dis-
infectant for a compulsory shower. Then came screening for skin diseases and
fevers while naked; clothes and possessions were returned after disinfection."[122]
Because disinfection of clothing, baggage, and other possessions took longer
than personal disinfection, immigrants had to remain disrobed in waiting
rooms, covered only with blankets. As European historian Paul Weindling
explains, "Sanitary authorities dragooned eastern populations into expecting
disinfection as part of the experience of migration. New techniques of disinfection with bactericidal antiseptics like carbolic acid, Lysol, and paraffin
emulsion were deployed—although particularly effective against lice, the disinfectants could cause skin reactions and were not fully effective against the
tenacious nits (the louse eggs)."[123]

All incoming travelers from Mexico were subject to the quarantine disinfection and inspection, yet PHS officers exercised considerable discretion in
determining who was an appropriate subject for disinfection, bathing, and inspection. This gave rise to conflicts between the PHS and IS over how to distinguish immigrants from travelers and decide which immigrants actually required the full medical exam. In 1916 the IS formulated new rules for the
medical inspection of different classes along the Mexican border. Immigrants
who "corresponded to the steerage passengers at seaports" were never to be
allowed to enter the United States without medical inspection.[124]

A problem encountered in Brownsville was that immigrants would seek to
enter the United States across the bridge connecting Mexico and Texas after
PHS and IS hours of operation. Those corresponding to steerage would have
to return to Mexico and apply the next day. Those corresponding to first-cabin
passengers at seaports, however, could enter Brownsville and come to the
immigration station the next day at their leisure if, in the judgment of the IS
inspector on duty at the bridge, they were *clearly admissible.*"[125] To aid in
determining which immigrants were to be considered first class and which
steerage, the IS specified, "Aliens not of the laboring classes are usually considered as 'first-class.'"[126] There were no second-class immigrants along the
Mexican border, only first-class immigrants and laborers. Officers along the
Gulf coast experienced similar problems with modifying the exam by class of
the immigrant. Shipping lines along the Gulf coast did not uphold the same
class distinctions that other regions had come to expect. In 1911 Dr. Scott, stationed in New Orleans, observed: "A larger number of aliens arrive from Central American and West Indian ports in transit, or on a short visit. This class of

immigrants object to a medical examination of any kind. The situation might be improved if the transportation companies would assist to the extent of classifying their passengers. This they can not or will not do, claiming that owing to the short passage it would not pay them to give a steerage rate; and again many of their boats are too small to carry any but first-class passengers. As a consequence, it is necessary to examine all passengers alike. Only the Southern Pacific Co. and Italian line make a distinction between classes on vessels. The steamers from England carry only cabin passengers."[127]

While there were written rules for modifying the examination by *class*, these rules were, again and again, confounded by unwritten rules for modifying the examination by *race*—rules that involved biology but also nationality, complexion, and occupation—along the Mexican border. In theory, a medical examination was not required for first-class "Mexicans."[128] Although Mexican laborers were required to carry bath certificates and undergo disinfection once a week, travelers from Mexico could skirt the disinfection and examination procedure if they appeared "obviously clean and not louse-infested."[129] In practice, however, the rules for race were much more complex, expressing a strange ambivalence toward Mexicans. Having lived side by side with Mexicans as citizens since the end of the Mexican-American war, the United States, on the one hand, was able to treat new immigrants from Mexico, in true colonial form, as conquered natives in need of the tutelage of a civilized nation.[130] On the other hand, the racial status of Mexicans continued to perplex America, so that immigrants from its neighbor to the south were regarded as foreign.

Underlying the procedures regarding Mexicans were quite diverse racial attitudes. Local attitudes toward the Mexican laborer heightened the sense that such individuals required disinfection: "The peon is a fatalist and believes implicitly that all the affairs of his life are ordered according to an inexorable destiny. . . . [T]his peculiar belief makes the peon difficult to handle when quarantines and other sanitary measures are being enforced. Whether he contracts a disease, or dies from it, according to his belief, is a matter entirely in the control of deity, and nothing that he could do would, in the least, alter the circumstances. He therefore, disbelieves in quarantines and vaccination; and he not only fails to cooperate in their enforcement, but frequently stubbornly resists them. For this reason, diseases are prevalent and the mortality rate high among the peon class."[131] That this discussion was framed in terms of the quarantine law suggests a wholly different, and narrower, notion of fitness for citizenship, in which the idea of a pre-industrial society connoted backwardness. Accordingly, views of the Mexican in American society turned prevailing notions regarding dependency, poor physique, and race suicide on their heads: "Generally they [Mexicans] enjoyed good health, and I never heard the doctor complain even once of a Mexican taking advantage of free medical service and making a nuisance of himself by asking for what he really did not need. I always thought that their smallness of stature and limited waistline was due to their extreme poverty and consequent lack of sufficient nourishing food. Their general good health I attributed to the fact that only the strongest survived. All the weaker ones died in infancy or early childhood."[132] Almost counterintuitively, the underdeveloped Mexican immigrant is neither destined for dependency nor necessarily unhealthy. Evoked is an older notion of survival of the fittest based on an animal rather than human model of evolution, in which the weakest do not survive or thrive. Yet Mexico also presented an environment in which the fittest were tested on the harshest of terrains. It is not clear that the Mexican, if grown on American soil, would be read as fit.

Likewise it is not clear if the PHS and IS officers recognized the class distinctions they specified among Mexicans. At the very least, evidence confounds the notion that any Mexicans could be universally regarded as "clearly admissible." For example, Dr. Woodall, stationed in Hidalgo, responded to a complaint that he had been abusive to a Mexican immigrant during the medical exam.[133] He explained that the case "must have been one of those few of the so called higher cast[e] of Mexicans who appear to think they should enjoy immunity from the examination and the treatment accorded to all arriving aliens in conformity to existing regulations."[134] The assistant secretary of labor supported Woodall's assessment, further suggesting that even "the lower classes of Mexicans are very resentful of any treatment which would tend to reflect discredit on their habits of personal hygiene, and they not infrequently employ very abusive language."[135] Dr. Fairbanks in Brownsville objected to using class to determine the rigor of the inspection where "there are no first and second-class passengers to indicate a difference. Some of the poorest people are the cleanest and vice versa some of the rich are filthy. If we try to separate the clean from the dirty, there are all degrees of cleanliness and all degrees of filthiness, and where can a line be drawn?" The PHS opted not to comment on Fairbanks's observation, but merely told him that no one was exempt from inspection or disinfection.[136] This advice stood in sharp contrast to the PHS's stated policy along the southern border.

The discrepancies were certainly tied to confusing issues of racial diversity, for Mexicans were a group consisting of Spaniards, "Indians," and mestizos of "mixed" blood. The U.S. secretary of labor wrote that "the Mexican people are of such a mixed stock and individuals have such a limited knowledge of their racial composition that it would be impossible for the most learned and experienced ethnologist or anthropologist to classify or determine their racial origin."[137] Law failed to provide a convenient solution, as it did for Asians. The 1848 Treaty of Guadalupe Hidalgo made all Mexicans living in territory ceded to the United States after the Mexican-American war U.S. citizens. A federal court in 1897 upheld this right to naturalized citizenship, despite persistent confusion over the proper classification of the man whose admissibility to citizenship came into question.[138]

Nor did policy help with the Mexican question. Characteristic of American confusion regarding the racial status of Mexicans, but also reflecting the importance of foreign relations with Mexico and the power of business interests in the American Southwest, Mexicans were not explicitly included in the quota laws of the 1920s. In her intriguing and powerful account, Mae Ngai argues that, as the political economy of the American Southwest was dramatically reorganized in the late 1920s and the nature of Mexican labor became migratory, the United States mapped a new racial category of "illegal" immigrants onto Mexicans. The United States would formalize this categorization of Mexicans as "a disposable labor force" in 1930.[139] At that time the U.S. Census would enumerate first- and second-generation Mexicans as an ambiguous yet separate race, "not definitely white" but also not " Negro, Indian, Chinese, or Japanese."[140]

Whether to classify Mexicans as "white" was not the only factor complicating immigration along the border. Distinctions among groups on the part of PHS and IS officials were made based both on complexion and on social class.[141] This played into the apprehensions of the PHS and IS regarding the diseased Chinese and Syrian immigrants whom they feared would attempt to cross over the Mexican border and the concerns they expressed about differentiating not only among Mexicans but between Mexicans and, say, Syrians. Other races of immigrants, particularly Syrians and Chinese, were compared to Mexican "peons" in assessing their desirability.

Reporting on the influx of Syrians into Mexico, for example, Inspector Seraphic of the Immigration Service noted that Syrians represented no problem for Mexico: "The Syrians endeavor to, and succeed very readily, in becoming Mexicanized. . . . Their mode of life in Syria is not dissimilar to that of the low Mexican and as both the Arabic and Spanish abound in insincere expressions of politeness, bordering on servility, they assimilate with Mexicans easier than all other races."[142] The implication was, however, that they did present a problem to the United States. In 1907 Special Investigator Marcus Braun, sent to report on the Mexican border situation, described Syrians as endowed with a physique that "would positively be found below the standard set by our laws."[143]

Fears of disease among Syrian and Chinese immigrants were instrumental in rousing the United States to exert pressure on Mexico to adopt its own immigrant medical inspection legislation in 1908. U.S. concern with the illegal passage of Syrian and Chinese immigrants across the Mexican border dated to at least 1903.[144] By 1907 it had become clear that the railroad companies could not inspect passengers they transported across the Mexican border without authorization from the Mexican government.[145] The United States began warning Mexico of "immigrant" diseases not covered under the quarantine law (or sanitary law in Mexico), particularly trachoma, among arriving Syrian, Greek, Chinese, and Japanese immigrants.[146]

Fear of diseased Syrian and Chinese immigrants also explain the shape of the medical exam in this region where Mexicans—who were considered only marginally desirable—crossed the border with relative freedom since they were not necessarily considered immigrants under the law, yet were interspersed with Syrian and Chinese immigrants who might "pass" as a local peon. Mexican officials regarded such "slag of humanity" from abroad as "no worse than our own peons."[147] Indeed, IS officials feared that a diseased Syrian or Asian immigrant would cross the border "dressed as a Mexican" knowing that "no one would question" him.[148] Thus, the immigrant medical exam, which necessitated complete disrobing of the immigrant, was also a response to fears of being unable to distinguish among the different races. Clothing worn among the "Mexicanized" races might deceive the medical inspector. He therefore had to insist on its removal. This complex mapping foreshadows the complexity of the term *race* and all that figured into it.

Given the different emphasis of the immigrant medical examination along the Mexican border, some of the typical immigrant diseases, such as trachoma, did not predominate, while others, such as venereal disease (which signaled a different, more decadent, perhaps even savage way of life) did. In El Paso, for example, eye diseases did not represent quite so large a proportion of total certifications as they did in most other regions, yet, with the exception of 1911, they consistently represented at least 10 percent of certifications. By 1922 eye diseases consistently represented approximately 20 percent of all certifications. Reflecting the relative importance of venereal diseases, gonorrheal ophthalmia was, after trachoma, the most important eye disease through 1911. Beginning in 1922, El Paso followed the pattern of other ports, with blindness and defective vision predominating after trachoma was reclassified in the PHS nomenclature (but not in immigrant nomenclature) as a parasitic disease. In addition, obstetrical and gynecological conditions (almost exclusively consisting of certifications for pregnancy), venereal diseases (gonorrhea and syphilis), and injuries stand out. Injuries never represented more than 10 percent of certifications and, as at the other ports, loss or amputation of limbs or other appendages or body parts typically accounted for 50 percent to 70 percent of this category. Miscellaneous conditions—a major category of certification—consisted almost entirely of senility.[149]

THE NATIONAL PICTURE

While each region had unique needs, those regional labor needs and priorities served as the lens through which the PHS viewed race and class, and applied those concepts to questions either of industrial citizenship, civic citizenship, or levels of civilization by working variations in the line. The picture emerging from this regional discussion can help us understand the ways in which the PHS considered and acted on different diseases and how, in tandem with the IS, it began to define who was to be included and excluded from the industrial labor force. The key point is that immigrants were marked as distinct in different ways, using different criteria.

Painting with the broadest brush, we find that class preempts the language of race in the North and East, that race trumps class in the West, and that along the Mexican border the two are mingled. Adopting a regional perspective necessarily brings up questions of specific diseases, as the exam in different regions was tailored to target certain conditions. Yet when we consider more carefully the interaction of disease and region—making disease the focal point—we begin to understand the complexity with which notions of race and class combined.

CHAPTER FIVE

AT THE BORDERS OF SCIENCE

Diagnostic Technology at the Intersection of Race, Class,
Disease, and Industrial Citizenship

After . . . many trials and endeavors, on my part, after interviews with
Dr. Fairbanks, your Local Representative, I succeeded in having my
husband, Manual Marron, re-examined, the diagnosis of the Dr., while
not made public, I understand to be favorable, with the exception that
he informed me my husband had a sore in his rectum, which would
have to be cured and this he stated would be a very simple matter. . . .
[Several Mexican and American doctors agree] that the trouble was not
of a serious or contagious nature and could be easily treated. . . . My
husband is interested in an Estate in Spain and I myself am interested
in property at Brownsville. It is necessary that we both are admitted into
the United States. My husband is not a pauper, is able to care for himself and is not infected with "syphilis" or any contagious disease. . . . Is
there not someway [sic] to secure justice at your hands? Is the action
of Dr. Fairbanks final or may my husband have an opportunity of being
examined by a Board of Surgeons appointed from your office? The situation is desperate, very humiliating and embarrassing to me and my
family . . .

> —Letter from Mrs. Victoria Fernandez Marron
> to Surgeon General Rupert Blue, 9 October 1918

WHEN THE PUBLIC HEALTH SERVICE certified and the Immigra-
tion Service excluded individual immigrants with disease, the process served
as an expression of power, as a means of shoring up authority by demonstrating that failure to adopt and conform to industrial norms and values carried
severe consequences. The PHS could also use its power more proactively, in
an attempt to certify not just individuals but groups of people, typically those
with one or two diseases that neatly characterized the nature of their unsuitability for industrial participation. Such targeted exclusion from citizenship at

the nation's borders, whether civic or industrial, involved questions of unchangeable biological characteristics. It further required making a solid connection to a specific disease.

When exclusion of a group or class of people from industrial citizenship was in question, the value of a new set of scientific technologies, innovations, or understandings was inevitably at issue. The PHS, after all, relied on science to help it define and control the industrial working class, even if it did not depend on diagnostic technology in the immigrant arena.

But science was not static. Rather, as an emerging, developing way of understanding and ordering the world, it constantly worked to force the PHS to reconsider and redefine the type of authority—disciplinary or exclusionary—it would or should wield. The social context, in turn, shaped the ways in which the PHS used science toward different ends.

For the PHS to take an exclusionary stance, class, race, and disease all had to be brought to bear on questions of industrial citizenship. Class and race could combine in remarkably complex ways. Consideration of a different set of diseases—perhaps a disease like trachoma whose diagnosis was not entangled with questions of technology, but one that was connected to both race and industrial efficiency—might have revealed yet a different pattern of overlap between race and class. Class alone—as far as it implied unchangeable physical or biological traits—may have aligned with a disease that I did not examine to effectively define what amounted to a group of biologically unfit immigrants. Here I focus on diseases and conditions in which new scientific methods of diagnosis were at stake and questions of whether the PHS would serve a disciplinary or exclusion function were at issue. In these cases, I found class and race to be inextricably intertwined.

TUBERCULOSIS

In the early twentieth century social reformers were particularly concerned with the devastating effects of tuberculosis (TB) on families and employment. As in the case of the Orloff family in Boston, the disease could quickly make an entire family dependent on charity or assistance. Mr. Orloff came to the United States from Russia in 1912 and found steady work in a rubber factory by 1915. Not until 1927 did Mr. Orloff develop "a cough." The diagnosis was tuberculosis, which quickly resulted in the loss of his job. The family had seven hundred dollars in savings, but this was quickly depleted after Mrs. Orloff "be-

gan to show signs of increased poor health. An examination proved tuberculosis in her case, too, and she was sent to a tuberculosis camp." Within a year,
both parents were in the state sanatorium and the first of the couple's seven
children began to show signs of the disease, leaving thirteen-year-old Maria, a
candy factory worker earning ten dollars per week, as the family's primary wage
earner.[1] Immigration law, building on a growing connection between tuberculosis and immigrant communities, specifically prohibited the entrance of immigrants with tuberculosis beginning in 1907.[2]

The PHS, however, drew few direct links between tuberculosis and arriving immigrants. Although tuberculosis was considered a common cause of dependency among immigrants, the PHS believed that most were not exposed
to the disease until after arrival in the United States.[3] Because the agency saw
the disease as a consequence of industrial living and made no explicit connection between TB and any particular immigrant group, it was predisposed to favor a disciplinary rather than exclusionary stance.[4] However, the PHS did not
work alone at the nation's borders. The IS, which made the final decisions regarding deportation, tended to view the connections between the disease and
race through the lens of eugenics. Pressure from the IS tested the extent to
which the PHS was grounded in a model of industrial citizenship that did not
draw on shared cultural assumptions about TB and race, and also the extent
to which industrial imperatives informed its decision making.

The PHS first encountered problems with diagnosing tuberculosis in immigrants in 1901, when the surgeon general labeled pulmonary tuberculosis a
dangerous and contagious disease.[5] Almost immediately, PHS physicians at Ellis Island issued a Class A medical certification to Thomas P. Boden. The IS
Board of Special Inquiry authorized his deportation. Boden and his family,
who were already living in the United States, were not without means or determination. In short order, Boden retained a lawyer who marshaled the aid
of two local physicians and appealed the decision directly to Theodore Roosevelt, with the hope that he would intervene in the case. Each physician submitted written testimony arguing that tuberculosis was not contagious. Boden's lawyer argued that "there is no scientific basis on which to classify
pulmonary tuberculosis among the dangerously contagious diseases." Specifically, Boden's lawyer argued that the bacilli could be destroyed with proper
disposal of sputum: "By excluding pauper immigrants, whether tuberculous or
not, the immigration authorities do their duty, and every loyal American citizen must approve of it." Surely the law was not intended to exclude individu-

als of a finer sort: "by excluding consumptive aliens of means, or at least such who can give evidence that they will not become a burden to the community, we must subject ourselves to retaliatory measures on the part of other governments and wealthy American pulmonary invalids may no longer be allowed to enjoy the hospitality of foreign health resorts."[6]

Although Boden's lawyer raised questions of how an immigrant's class and subsequent social behavior modified the threat he represented—questions that worked to modulate the ways in which the PHS balanced the social and the scientific in different circumstances and with different diseases—his challenge rested on the increasingly important distinction between "communicable" and "contagious" diseases.

In New York City, for example, classification of tuberculosis as communicable rather than contagious was a crucial and highly sensitive point underlying more explicit controversy over municipal encroachment on the authority of private practitioners. In 1897 the city Department of Health made notification of patients with tuberculosis mandatory. Private physicians resented the "offensively dictatorial" attitude of the Department of Health.[7] They also questioned the scientific rationale for this decision. Speaking at the New York State Medical Society meeting held in protest of mandatory notification, Dr. J. Blake White insisted that "there was high authority against the positive statement that tuberculosis was infectious and communicable." George Fowler, the city health commissioner, attempted to quell anxieties by insisting that the health department had not declared tuberculosis to be a contagious disease, but "had declared it to be among the infectious and communicable diseases, dangerous to the public health."[8] Hermann Biggs, director of the New York City Laboratory and soon to be city health commissioner, justified compulsory reporting of tuberculosis on the grounds that it was classified a communicable rather than a contagious disease: "I have always felt that much harm has been done by calling tuberculosis a contagious disease; it produces confusion in the minds of both the laity and the medical profession, because the conception of a contagious disease is always related to such diseases as scarlet fever, small-pox, etc., in which very limited contact or even simple proximity may result in their transmission." Rather, Biggs condescended, "Every intelligent person knows that tuberculosis is different in nature from these diseases, and I believe that this distinction should be made and kept clear and definite. Tuberculosis is communicable, but not contagious."[9]

The PHS, in sharp contrast, easily interchanged the terms *contagious* and *communicable.* Nonetheless, the PHS well understood that the question of whether tuberculosis was "contagious" or "communicable" was delicate and controversial. The agency quickly suspended temporarily all such certifications, except those made with the blessing of the PHS officer in command at an immigration station, until it resolved the Boden case.

The PHS convened a Board of Special Medical Inquiry to consider Boden's appeal on medical grounds. The PHS physicians who served on the medical board were neither divided on nor troubled by the question of contagion. To Boden's challenge that tuberculosis was not "contagious," Ellis Island physicians bluntly replied, "Tuberculosis of the lungs is undoubtedly a communicable disease."[10] The surgeon general, however, in a memorandum to the commissioner general considering the question in detail to aid him in his decision, switched back to the language of "contagion." Based on the consensus of scientific medical authorities in the United States and Europe, "there seems to be no doubt . . . that . . . tuberculosis of the lungs is a contagious disease. Now, is it a *dangerous* contagious disease?" On this point, there could be little debate: "So fatal is it and so prevalent that it has been termed the 'great white plague.' There is a world-wide movement now in progression for its suppression, and that it is a dangerous contagious disease is shown with penalties for their violation to prevent spitting in public conveyances and places of public assemblage, and in some instances upon the streets, the reason of these ordinances being that the dried sputum from consumptives may be inhaled and the disease contracted." Concluded the surgeon general, "a number of Western States have proposed, through their boards of health, to prevent the ingress of consumptives within their own borders. Consumptive patients in most city hospitals are segregated from other patients and kept in special wards."[11]

The PHS maneuvered around distinctions between contagious and communicable diseases by suggesting that, as the agent of the law, it would not assess the ease of transmission or the likely behavior of the tuberculous immigrant: "It is true that consumption may lose some of its dangerous character when the subject is made to observe certain restrictions which involve his personal habits. So, also, it may be said that most contagious diseases may be robbed of a large part of their danger by restrictive measures." Still, "the law has not taken these restrictive measures into account. It does not say that an immigrant with a contagious disease may be admitted provided he will carry a sputum flask with him and use it and disinfect it, or provided he will reside under good hygienic conditions and not mingle in closely crowded tenements."[12] Such discussion suggests that the PHS may have implicitly framed understandings of TB in class-based behavior, perhaps even inherent class-based behavior. Additionally, the argument by Boden's lawyer that the intent of the law was not to exclude immigrants "of means" reveals the deep-rooted assumptions regarding disease and class underlying the broader debate and discussion.

Based on the surgeon general's report, the commissioner general upheld the PHS assessment of pulmonary tuberculosis and ruled that Boden was excludable if infected with tuberculosis. Although the IS allowed Boden—to the chagrin and dismay of the surgeon general—to be examined by a group of private practitioners, PHS officers confirmed tuberculosis by bacteriological examination, and he was deported.[13] Nevertheless, the import of the Boden decision revolves around the PHS resistance to nailing down precise language of exclusion, ultimately leaving meaning vague and loose—never conceding whether or how class might figure into the ways in which TB should be treated at the borders—and maximizing its ability to maneuver medically.

While PHS officers were united on the question of how to classify pulmonary tuberculosis, other forms of tuberculosis raised complicated questions. In particular, after Congress named tuberculosis an excludable condition in the immigration law of 1907, PHS officers found themselves divided over the issue of which forms of tuberculosis should make an immigrant's exclusion mandatory. Between October 1907 and January 1908 they arrived at consensus, concluding that mandatorily excludable tuberculosis was limited to pulmonary disease.[14] All other forms—including "physical signs of [pulmonary] tuberculosis" that could be subjected to the medical gaze—remained Class B conditions.[15]

The PHS most commonly described nonpulmonary tuberculosis as "poor physique." The 1903 regulations for the medical inspection of immigrants defined the clinical symptoms of "poor physique" as "cases of so-called 'chicken breast,' especially those having some of the physical signs of pulmonary tuberculosis, but in which the tubercle bacillus can not be found in the sputum."[16] By 1906 certifications for "poor physique" had increased to such an extent "that the said medical term has become perhaps one of the most important employed."[17] In Boston, for example, PHS officers used the term quite liberally. In 1905—the first year in which the term made a significant showing—"poor physique" was the third most common cause for medical certifi-

cation and represented 20 percent of medical certificates (problems with eyesight and various deformities were the two most common causes of certification). Nevertheless, while suspected cases of nonpulmonary tuberculosis
were often certified as poor physique, the IS noted that the precise meaning
of the term was "not perhaps always clearly understood by those not connected
with the Immigration Service"—or, as the PHS might have argued, by those
within the IS.[18] The IS, consequently, requested a more complete definition
of "poor physique" from Dr. J. W. Schereschewsky, stationed in Baltimore.
Schereschewsky produced for the agency a definition that rested on a broad
notion of citizenship, incorporating not only fears about industrial fitness, and
thus class fitness, but also and more directly fears about racial unsuitability for
participation in a democracy.[19]

There was considerable overlap among bacteriology, sanitarianism, and
eugenics in the early twentieth century. The PHS was generally careful to
maintain a distinction between bacteriology and eugenics. According to
Schereschewsky's definition, however, "poor physique" meant that "the alien
concerned is afflicted with a body not only but illy adapted to the work necessary to earn his bread, but also poorly able to withstand the onslaught of disease." He "is undersized, poorly developed, with feeble heart action, arteries
below the standard size; that he is physically degenerate, and as such not only
unlikely to become a desirable citizen, but also very likely to transmit his
undesirable qualities to his offspring should he, unfortunately for the country
in which he is domiciled, have any." So serious was the diagnosis in
Schereschewsky's mind that "of all causes for rejection . . . 'poor physique'
should receive the most weight, for in admitting such aliens not only do we increase the number of public charges by their inability to gain their bread
through their physical inaptitude and their low resistance to disease, but we
admit likewise progenitors to this country whose offspring will reproduce, often in an exaggerated degree, the physical degeneracy of their parents."[20]

Although Schereschewsky himself never associated poor physique or TB
with Jewish immigrants, an image of the tubercular Jew was drawn in terms of
physique and would become more clearly defined as the century progressed.
Nativists like E. A. Ross defined the Jew as hollow-chested, wasted. "On the
physical side," he wrote, "the Hebrews . . . are undersized and weak muscled."
Madison Grant described their "dwarf stature."[21] During this period, even
reform-minded physicians, such as Maurice Fishberg, who sought to dispel
the notion of the biological inferiority of the Jews described the tubercular na-

ture of this group.[22] They spoke in particular of the "flat" chests and "inferior capacity" of the Jewish garment workers.[23]

Enormously pleased with Schereschewsky's definition of "poor physique," the IS immediately distributed it to officers at each of the nation's borders.[24] It was well received, too, by nativists such as Harvard's Robert DeC. Ward, who believed that "aliens of 'poor physique' are not the kind of fathers and mothers whom we need in this country; they are not the kind of people whom we need to do a hard day's work, for they cannot do it. They are not the kind of people who will maintain a high standard of health." Within the PHS, Schereschewsky's definition was seen as drawing too heavily on racial imagery and unduly emphasizing questions of degeneration rather than industrial citizenship. As a result, it did not set well with the most senior PHS officers, who attempted to place it on firmer clinical footing without undercutting its import for a conception of the useful laborer. Dr. George W. Stoner reported to the commissioner general that while there was "no specific demonstrable disease sufficient to warrant" a "poor physique" certification, a more precise clinical definition could be developed. In the opinion of Stoner and his fellow officers, a host of defects served as clinical indicators. The four most important included: "(a) Malformation of the thorax and lessening respiratory expansion. (b) Deficient muscular development. (c) Feeble circulation. (d) Lack of correlation between height and weight." All agreed with Schereschewsky that poor physique would "unfit the alien to earn a living at manual labor and lessen his power to resist disease."[25] But they put questions of procreation aside, for they were concerned with the labor force as disposable in some sense.

Even the revised definition of "poor physique" did not ease the concerns of high-ranking PHS officials. It departed too much from the original conception of "poor physique" as a form of nonpulmonary tuberculosis. By 1907 the surgeon general had become wary of the phrase altogether. He insisted that it was "not a diagnosis"—meaning that "the term does not imply a clinical or pathological entity"—and advocated abandoning the phrase entirely: "The term 'poor physique' is perhaps a misnomer, and we can do away with it, and what constitutes poor physique should be inserted instead. I think that term has been considered by the board of inquiry as a stronger term than was ever intended by the medical branch."[26]

In Boston the term had fallen out of use by 1920; in its place appeared the notion of "poor physical development," which appeared in only 1.5 percent of certifications in 1920. In New York, the term made its appearance in 1904, when the PHS certified only 1 percent of cases as representing "poor physique." The terms "poor physical development," "poor muscular development," and "lack of physical development" replaced "poor physique" by 1910. In that year "poor development" represented approximately 2 percent of cases. From 1911 to 1916 the diagnosis was fairly prevalent, representing from 8 percent to 27 percent of all certifications, but by 1918 typically fewer than 5 percent of cases were certified as involving some form of poor development.

With the elimination of "poor physique," several forms of tuberculosis were redefined as Class A conditions. In its 1910 regulations the PHS designated tuberculosis of the respiratory tract, intestinal tract, and genito-urinary tract as conditions for which exclusion was mandatory. Significantly, however, tuberculosis was designated neither a "loathsome contagious disease" nor a "dangerous contagious disease." Rather, tuberculosis stood on its own in Class A, somewhat obscuring the question of whether those with TB represented a class or group warranting not discipline but exclusion. The categorization suggested that exclusion was mandatory but failed to use the strong, suggestive language of loathsome or dangerous contagious diseases.

Although the PHS officially excluded "poor physique" from its 1910 regulations, both the term and Schereschewsky's definition of it continued to carry broad appeal within the IS. In 1910 and again in 1914 the commissioner general of immigration drew on the essence of Schereschewsky's definition to argue not only that unfit immigrants should be excluded from the United States, but also that the country should demand a high level of physical fitness from its immigrants. "Any measure that will tend . . . to raise the standard of physical excellence ought to meet with the approval of all citizens who are anxious to preserve and improve the American race." For the commissioner general, "this is not only a question of the present; it is more distinctly a matter of grave concern for the future. The strength of a nation is the combined strength of its individual members. Can we expect, if we continue to inject into the veins of our nation the blood of ill-formed, undersized persons, as are so many of the immigrants now coming, that the American of to-morrow will be the sturdy man he is to-day?"[27] Even as late as 1920 William Williams, commissioner of immigration at Ellis Island, produced a report (in his capacity as chair of the Sub-Committee on Immigration of the Committee on National Affairs of the National Republican Club of New York) identifying the "poor physique" of arriving immigrants as requiring Congress to raise and make more specific the physical standards for arriving immigrants.[28]

In the hands of the IS, the term "poor physique" expanded to mean something more than nonpulmonary tuberculosis. The language in which the IS discussed the term and the racial assumptions on which it drew suggested that for the IS, discipline may have been secondary to exclusion. To the PHS, however, the term now clearly smacked of eugenics, a social and scientific movement seeking to achieve "better breeding" through application of the laws of heredity, tied to notions of national identity and exclusion.[29] Where the leadership of the IS embraced the eugenical aspects of "poor physique" and its positive counterpart, "fitness," all available evidence indicates that the PHS shied away from eugenics and the implied business of improving or guarding America's genetic stock. Not until 1930 did the PHS specifically comment on proposals to expand the list of Class A conditions to include not only "contagious" or "communicable" diseases but also hereditary conditions "transmissible" to an immigrant's offspring. The PHS officially rejected the proposal, made by one of its own officers, on the grounds that while the change was "advisable from a eugenic standpoint," hereditary conditions were not contemplated by the law and, moreover, were not "as objectionable" as either mental conditions or loathsome and dangerous contagious diseases.[30] Unofficially, however, the PHS was troubled by the prospect of embracing eugenics and questioned "just how far we want to go in such matters."[31] Ultimately, the agency wanted no part of advocating for including hereditary defects in the law or of enforcing such a law.

Although the PHS was not interested in using TB or a looser notion of "poor physique" to weigh in on questions of exclusion related to civic or racial fitness, the agency would do so when it perceived that questions of class or labor and race were at stake. Use of the term "poor physique" (as well as "poor development" and other related terms occasionally used in later years), for example, was intimately linked to race and the labor market.[32] Along the Gulf coast, available data show that it was never a major cause for certification. New Orleans—a major entrance point for a labor-hungry region—certified only 3 percent of cases in 1903 and 4 percent of cases in 1905 as displaying "poor physique." The port never adopted other terms to describe the condition and made no other similar diagnoses in subsequent years. Likewise, ports along the Canadian border found little use for the diagnosis. Yet in areas receiving groups perceived as making poor industrial citizens and, accordingly, as requiring exclusion, the term retained value and meaning. In San Francisco and El Paso, where large numbers of Asian immigrants were received or where of-

ficials feared a large number of Asian immigrants would attempt to illegally
cross the border, the term "poor physique" persisted long after it was officially
abandoned and fell out of use at most other ports.[33] Diagnoses for "poor
physique" represented a fairly large portion of cases (12%) only in San Francisco in 1911. Marcus and Van Buren, cities along the Canadian border, used
the term in the period from 1917 to 1920, but certifications were very low,
reaching only 3 percent in Marcus from 1917 to 1920 and 8 percent in Van Buren in 1920. Marcus, notably, was in Washington and, like San Francisco, received Asian immigrants. By 1917 "poor physique" represented less than 1
percent of certifications at this port—typically only one case per year. Noteworthy is that the term did not persist at Ellis Island or other ports along the
East Coast, which received the vast majority of immigrants.

This is not to say that the category of "poor physique" achieved Asian exclusion in the West, but it demonstrates how race and class could combine in
a fashion to give a disease a different meaning. The coastal differences are
noteworthy because they reflect a willingness to retain the term for a largely
"nonwhite" Asian population—unsuited for both civic and industrial citizenship, and therefore requiring exclusion as opposed to proper industrial socialization. We find the same pattern in the case of hookworm, discussed below.

Yet the importance of TB and related diagnoses like "poor physique" even
at ports where the PHS revealed an interest in exclusion as opposed to discipline should not be overstated. The disease simply remained too difficult to
diagnose, both clinically and bacteriologically, and its classification remained
confusing to most officers, severely limiting its usefulness as a means to effect
the exclusion of groups or classes that would not make good industrial citizens.
For instance, along the Canadian border, in Sault Ste. Marie, Michigan, an acting assistant surgeon, Dr. Wesley Townsend, resigned in disgrace in 1910 after certifying a Canadian Indian boy, Dewey Wigwas, as having "'tubercular
glands of the neck,' a dangerous contagious disease," having made the diagnosis while allegedly in a state of obvious inebriation.[34] The boy, his father insisted, merely suffered from a mosquito or fly bite on the neck.[35] More serious than the lapse in professional conduct, Townsend had mistakenly given the
boy a Class A rather than a Class B certificate, throwing his entire diagnosis
into question.

The physician called in to reevaluate the case—though not impaired by al-
cohol—was no better equipped to make an accurate diagnosis. Dr. A. S. Mc-
Craig, a Canadian official sent to the reservation where Wigwas lived, offered

three different medical certificates on three different occasions. McCraig concluded in August that Dewey presented an "enlargement of the right lobe of the Thyroid gland, and slight infection of the Posteriour Cervical glands on both sides."[36] Presumably, the infection was not tuberculous. Five days later, McCraig described the boy's condition as "an enlargement of the right lobe of Thyroid gland and slight *tubercular* infection of the posterior cervical glands which may ultimately develop into a dangerous and contagious disease."[37] By September McCraig was convinced that the boy's glands were slightly enlarged specifically from tubercular infection but that the condition was neither dangerous nor contagious.[38] Certainly McCraig, a Canadian official, struggled to provide a diagnosis in accordance with U.S. immigration law. No doubt this task was complicated by the legal status of native peoples in the United States. Many tribes were formally barred from citizenship, along with their children born in the United States, until 1924.[39] But McCraig also struggled to diagnose Dewey's affliction precisely.

Precise diagnosis of tuberculosis by any means was difficult.[40] Bacteriological confirmation of the presence of the tuberculosis bacillus was not necessarily more reliable than clinical identification of the disease. As William Osler explained in the eighth edition of his *Principles and Practice of Medicine* (1916), sputum examinations were complicated: "the difficulty . . . is that it requires a long series of examinations to exclude positively the presence of tubercle bacilli. Time and again with suspicious cases, or in pleurisy with effusion, I have asked a clinical clerk day by day 'Any bacilli yet?', and in one instance there were none found until the twentieth examination!" Such problems made the sputum test impracticable in many settings. "Of course," concluded Osler, "in private practice this is impossible, but it is well to bear in mind that one or two negative examinations are not sufficient. Various methods of digesting the sputum and examining the centrifugalized sediment are important when few bacilli are present."[41]

PHS regulations stipulated that a diagnosis of pulmonary tuberculosis required laboratory identification of the bacillus from a sputum sample.[42] PHS officers inspecting immigrants expressed distinct unhappiness with this requirement. Not only did the test itself take time, but as one officer explained, "It may take a week to get a satisfactory specimen of sputum from an unwilling subject."[43] Dr. Alfred Reed of Ellis Island complained that the requirement of bacteriological confirmation was too limiting. He argued, "The tubercle bacillus rarely appears in the sputum until the disease is well advanced and there has been a certain degree of destruction of the lung tissue. To limit the diagnosis to such cases alone as show the bacillus allows numerous cases to pass free in which the clinical diagnosis is practically certain." Concluded Reed, "it would be more effective to hold for hospital observation all cases presenting clinical evidence of pulmonary lesions, and to allow diagnosis in such cases as after careful and repeated examination showed a definite lesion, perhaps using the tuberculin reaction as an aid in selected cases. In other words, if the diagnosis of pulmonary tuberculosis could be made by a competent and careful physician, even though there were no bacilli in the sputum, the case could be certified as tuberculosis."[44]

In 1917 the PHS resolved both classification and diagnostic difficulties: all forms of tuberculosis were considered Class A conditions. (Table 1-1 in chapter 1 traces the classification and reclassification of tuberculosis and other immigrant diseases in the PHS regulations published between 1903 and 1930.) At the same time, it became easier for PHS officers to certify an immigrant with tuberculosis. The PHS discontinued the practice of certifying "physical signs of tuberculosis" as a Class B condition when the symptoms were indicative of excludable disease but no bacillus could be isolated.[45] Rather, although "the finding of tubercle bacilli in the discharges and excretions from the body is conclusive evidence," officers were allowed to exercise their judgment in cases where no bacillus could be detected but where "pronounced clinical symptoms and physical signs of tuberculosis" indicated that "a certification of that disease is warranted."[46]

In essence, the PHS freed TB from the laboratory and took it back into the realm of the medical gaze. Because the medical gaze was a tool for discipline rather than exclusion, certification for tuberculosis never increased substantially despite the more lenient certification standards.

SYPHILIS

Congress, physicians, and the public believed that immigrants arrived on America's shores carrying a heavy burden of venereal infection.[47] In the years before the development of the Wassermann reaction for the detection of syphilis in 1906, early efforts to detect venereal infection among immigrants never turned up much disease. In 1903, at the request of the Immigration Service, the medical officer in command at Ellis Island briefly experimented with targeted exams for syphilis.[48] For more than a week Dr. George Stoner had his officers strip and examine all unmarried males, yet they found only two cases of venereal disease. Stoner accordingly "directed that the wholesale stripping method be discontinued and the usual form of examination be resumed as per the book of instructions, by which always a certain portion of the arriving aliens are examined in sufficient detail to determine the existence of any marked form of disease." Stoner concluded, "Syphilis is one of the rarest diseases amongst immigrants. For example, of the 3,427 arriving aliens admitted to the Immigrant Hospital at Ellis Island during the year ended June 30, 1903, only two were found to be suffering with Syphilis."[49] Surgeon General Walter Wyman supported Stoner's decision, agreeing that "the number of cases of venereal disease discovered in aliens . . . under this system of stripping them was too small to justify such a procedure: besides there were other reasons which rendered it objectionable."[50] More important, with regard to syphilis the PHS lacked either the critical connection to labor that would justify an attempt to use syphilis as a means of effecting broad-based exclusion or a link to race that would justify using it to achieve selective exclusion.

The syphilis question lay dormant for the next decade and a half, until laboratory advances were made in the diagnosis and treatment of syphilis and gonorrhea.[51] The times changed, too, with the anxieties of World War I.[52] There were increasingly vociferous accusations that the immigration exam for syphilis was a "farce."[53] As a result of these factors, the PHS stepped up efforts to detect venereal disease among immigrants.[54] The wartime shortage of labor and spate of labor strikes, in particular, opened the door to using a diagnosis of syphilis to exclude not a racial group, but a group of radical, uncontrolled immigrants who were clearly unsuited to industrial citizenship. As a sexually transmitted disease, syphilis could potentially embody the type of labor disorder and upheaval plaguing the nation's industries, for such diseases implied a lack of control.

Despite the PHS's wartime willingness to ferret out cases of syphilis and achieve a higher rate of exclusion, evidence that immigrants represented a significant venereal threat remained thin. Although at Ellis Island officers claimed a 900 percent increase in venereal cases, this increase represented only a very few cases. Wrote Dr. Kerr to the surgeon general, "There is no question but that a greater number of venereal diseases can be detected if they are looked for. This is plainly evident from our recent experience. During the period February 13th to 28th 5,577 passengers were examined, partly disrobed and 5,062 by ordinary line inspection and secondary inspection with the result that 28 cases [0.5%] of venereal disease were found among former group and 3 [0.1%] among the latter."[55]

PHS officers stationed on the nation's northern and southern borders—those places where both the PHS and IS imagined the threat of diseases like syphilis to be the greatest—produced similar results but mustered much less enthusiasm. Dr. Fairbanks, the PHS officer in charge at Brownsville, Texas, concluded in his annual report to the surgeon general in 1919, "A more strict observation for venereal disease has been carried out than heretofore but has not resulted in finding those diseases very prevalent. The lower class Mexicans were quite free from venereal disease." Fairbanks attributed the low level of disease among prostitutes to the rigid system of supervision, examination, and treatment carried out in Mexico.[56] The surgeon in charge at Montreal, Dr. F. Faget, similarly wrote in his annual report for 1924 that the routine inspection for venereal diseases in immigrants had not "resulted in any marked increase in the number of venereal diseases certified."[57] But as a consequence of the more intense examination conducted along the Mexican border and greater concern with diseases signifying disorder and connoting a less evolved civilization, PHS officials along the nation's southern land boundary—unlike those along the Canadian border or any other U.S. border—were already certifying the largest number of cases of venereal disease, which consistently represented one-quarter to half of all of Mexican border certifications. Thus, in this region there was little room for improvement.

In the immigration arena, the most important outcome of renewed interest in syphilis during World War I was linked to diagnostic testing. After the war, for example, Dr. Kerr at Ellis Island described the Wassermann reaction as the "sine qua non" in the diagnosis of syphilis.[58] In practice, however, increased reliance on the Wassermann reaction complicated the diagnosis and certification of syphilis.[59] In 1915, for example, PHS officers in Boston turned Luis de los Santos, a forty-seven-year-old man from Gibraltar apparently suffering from chronic malnutrition, off the line. His condition was indicative of syphilis. Although they could find no external signs of syphilis or anything else "to account for the alien's poor physical condition," a Wassermann reaction proved positive. Such cases of a positive Wassermann reaction in the absence of clinical disease, Safford claimed, were becoming more common, creating a problem of certification: were such cases syphilis or not?[60] The question proved thorny because it was more than a medical question of how to interpret the complex Wassermann reaction.[61] It was also a question of what the

PHS officer should communicate to Boards of Special Inquiry, which considered medical certificates and made final decisions regarding deportation on arrival.[62]

Dr. Lavinder, stationed at Ellis Island and serving on the surgeon general's committee to resolve the syphilis problem, maintained that a positive Wassermann was "a very important indication of the existence of an infection" and that any immigrant with a positive Wassermann should receive a Class A certificate.[63] Dr. Williams, in contrast, argued that a positive Wassermann with no clinical symptoms was to be certified as a case of "latent syphilis," affecting an immigrant's ability to earn a living—a Class B condition.[64] Victor Safford in Boston, who had raised the original question of the meaning of a positive Wassermann reaction in the absence of clinical disease, bristled at the term "latent syphilis": a Board of Special Inquiry (BSI), he felt, would have no idea what this meant.[65] IS officers, he insisted, specifically needed to know whether a disease was contagious in order to determine whether the condition mandated exclusion. Only a Class A rating would suffice, because it was otherwise impossible to assure a BSI that there was no chance that the individual would not transmit the disease at some time in the future. Medical precision, according to Safford, would prevent inappropriate lay interpretation of a medical certificate. "Whether rightly or wrongly," wrote Safford to the surgeon general, "the immigration officials deem the possible future contagiousness of any condition, whether syphilis, tuberculosis, or anything else, or even the existence or an unwarranted public belief in the contagiousness of such a condition to be a factor in determining an individual's future relations with society of sufficient importance to entitle it to careful consideration in deciding whether an alien in question may have a defect affecting his ability to earn a living or may be likely to become a public charge."[66]

As the director of the PHS Hygienic Laboratory saw it, the question was really one of how to tell a BSI that a positive Wassermann indicated the presence of the spirochete of syphilis and that although in the absence of open lesions the disease was not likely to be transmitted, it would probably lead to further health problems affecting ability to earn a living. Moreover, it was a question of how to communicate this information without giving a diagnosis of "syphilis," since "such boards do not like to go on record as having placed any case designated 'syphilitic' in any other category than in the deportable class."[67]

Ultimately, these officers agreed that in an instance where an immigrant tested positive for syphilis without clinical symptoms, officers would issue a certificate reading: "An abnormal systemic condition characterized by a positive Wassermann reaction in the blood serum, which may affect ability to earn a living. (This reaction is indicative of syphilitic infection.)" More important, the board clarified the role of laboratory results: "it is deemed of primary importance that medical officers, who may have occasion to draw deductions from reports as to Wassermann reactions, have impressed upon them the fact that the Wassermann reaction is not standardized and that reports are to be regarded as trustworthy only when it is known that they are done under proper conditions and by competent men." In order to certify an alien with an abnormal systemic condition, an officer needed two full, positive Wassermann reactions taken at least two days apart.[68] In consequence, the 1917 PHS regulations recommended only guarded laboratory testing for syphilis: "Whenever feasible a Wassermann test should be made. The result of this test is to be taken with caution and only as part of the evidence upon which the diagnosis rests."[69] Although the PHS informally allowed its officers to certify for TB without a positive sputum test, in 1917 syphilis became the only "loathsome contagious disease" for which the PHS officially did not require laboratory confirmation (table 1-1). Equally important, the PHS managed to blur the categorization of syphilis, defining it as both a Class A and a Class B condition, for it never specified how officers should certify a positive Wassermann test.

PHS officers in the field pushed the limits of their authority in cases of suspected syphilis. Although they did certify some immigrants with a positive Wassermann as Class B, officers proved profoundly unwilling to certify syphilis as anything less than a loathsome or dangerous contagious disease.[70] Their social and moral assessment of immigrants proved extremely powerful in influencing how they chose to interpret and weigh clinical and laboratory evidence of syphilis. For example, on 15 January 1917 the Office of the Surgeon General instructed Dr. G. D. Fairbanks, the medical inspector stationed in Brownsville, Texas, to "exercise especial care in the examination of arriving aliens, for the purpose of detecting lesions or conditions which indicates [sic] syphilitic infection."[71] Dr. Fairbanks took his charge seriously, and in 1918 he certified Mexican businessman Manuel Marron with primary syphilis, a Class A condition mandating exclusion under the U.S. immigration laws. Mrs. Marron took up her husband's case with Surgeon General Rupert Blue, insisting, "I feel certain that the Certificate of Dr. Fairbanks was made based on malice and hatred, rather than on the facts." In this letter she enclosed a certificate from the Texas Department of Health Laboratory in Houston, certifying that her husband did not, in fact, have syphilis.[72]

Dr. Fairbanks dismissed this certificate on the grounds that both Mr. and Mrs. Marron were "swindlers . . . and very much anti-American." Moreover, he wrote, "This man has been commonly known here for years as an 'old syphilitic' and when he was presented for examination showed old scars all over his body, nodes on the tibia, some enlarged glands, and subluxation of one (left) hip." Marron claimed to have fallen in Houston and dislocated his hip. Fairbanks was convinced, however, that "the history and physical signs [are] sufficient to justify a diagnosis of syphilis."[73] As for the diagnosis of the Health Department: "I believe there are other designing persons at the bottom of this complaint besides Mrs. Marron. Several times before, very unjust criticisms have been sent to the Bureau against this office which there is every reason to believe originate with the present State Health Officer located here." Although Surgeon General Blue sympathized, he informed Fairbanks that Marron's case sounded like one of tertiary, not primary, syphilis. Blue concluded that Marron "should have been certified as having a condition affecting ability to earn a living."[74]

Fairbanks remained undaunted. Although he reexamined the case, as Mrs. Marron had insisted he do, he was determined to prevent the entry of Manuel. Once again, he elaborated on Marron's clinical presentation: "an external pile of the anus with a red area a little smaller than a dime surrounding it," "a scab (about the size of a pea) of a sore on the right buttock which has just healed," a "falsetto" voice indicating "naso-pharyngeal affection." Yet it was not the clinical evidence that clinched the case for Fairbanks. "The man is known to everybody as being syphilitic and if I had not certified him as such, some of my critical enemies would have been delighted to herald my dereliction of duty. On the other hand I knew there would be a big kick from the same parties if he was excluded."[75]

The surgeon general, still inclined to believe that Fairbanks had misdiagnosed the case, straddled the line between supporting Fairbanks and following the letter of the law: "The only interest of the Bureau in the premises is that your certificate shall be made with a full appreciation of the regulations involved. . . . If you certify this as a case of syphilis in a communicable stage, your certificate will receive official support."[76]

Although Fairbanks reversed his certificate, he remained defiant. As Mrs. Marron explained to the surgeon general in a letter of 9 October 1918 (see the epigraph, above), Dr. Fairbanks cleared her husband, with the exception of pointing out a treatable rectal sore. Fairbanks, however, would not permit treatment in Brownsville. He insisted that Marron receive his treatment across the border in Matamoros, Mexico. Mrs. Marron wrote of the humiliation and embarrassment the situation was causing for her and her family, and ended, "I again urge you for prompt action in the premises."[77] Surgeon General Blue, however, informed Mrs. Marron that the matter now rested with the Immigration Service, where she would have to take her complaints.[78]

Although Fairbanks ultimately checked himself, the Marron case illustrates how the combination of Mexican race and high social class gave syphilis a particular charge along the nation's southern border. Fairbanks made a social diagnosis where the medical evidence and certification rules were unclear. Yet, depending on the circumstances, laboratory results could combine with other factors to restrain medical authority. In 1922, for example, Joao Souza Bispo was certified for syphilis and placed under observation at Ellis Island after the examining officers found a primary sore. On the day Bispo was admitted to the Ellis Island hospital for further examination, a "dark field for Treponema Pallida was positive." Two days later, however, his Wassermann was negative; one week later, he tested strongly positive. Three subsequent Wassermann reactions were negative.

The PHS officers were incredulous, for the clinical evidence contradicted the laboratory results. Attempting to reconcile the evidence of the medical gaze with that of the laboratory, they gave Bispo a second medical certificate reading, "Syphilis, but is in non-communicable stage. A loathsome contagious disease." As the officers saw it, despite the negative test results, so little time had elapsed that "it would hardly be justifiable to simply say 'effecting [sic] ability to earn a living' as there was no surety that communicable conditions of some sort would not re-appear. It may be that this was not the strictest possible interpretation" of the law, but the commanding medical officer asserted his personal opinion that "inasmuch as the primary object of medical examination is to protect the country against communicable diseases . . . it was the proper choice as to classification."[79] Although members of his staff supported this second medical certificate, Surgeon General Cumming ultimately disallowed it.[80] Cumming maintained that the regulations were quite explicit in demanding that medical certificates serve as statements of fact that could be used as evidence in a court investigation. Therefore, the second certificate was invalid. Cumming advised his Ellis Island officers always to observe the distinction between medical certificates and professional opinion.[81]

Only five years earlier, in the case of Manuel Marron, Surgeon General Rupert Blue had sought to strike a balance between professional opinion and the strictest interpretation of the law. The contrast may simply illustrate the different philosophies of two surgeons general who took a fundamentally different stance on borderline cases. Yet the two cases also illustrate that the interplay between the authority of laboratory results, the weight accorded clinical judgment, and the proper classification of disease shifted to accommodate the regional imperatives to discipline or exclude, complicated by the ways that race and class could combine and the instability of these combinations.

The extent of medical authority was somewhat regionally determined, as we have seen, and the rigor of the medical exam varied greatly by region. Along the Mexican border, where the threat of disease was paramount and the imperative to exclude undesirable immigrants of particular nationalities was in play, the authority of the PHS was maximized and certifications for venereal diseases were higher. Along the Atlantic coast, where the PHS was most likely to face community and business pressures, officers were accorded less discretion in how they certified a positive Wassermann. But, perhaps most important, in the case of Manuel Marron, Dr. Fairbanks was quite explicit in his assessment of the immigrant and his wife as "swindlers . . . and very much anti-American." His diagnosis challenged the notion that high social class—for Marron was clearly a man with both business and political connections in Texas—along the Mexican border was a sufficient criterion for admission. Fairbanks was more comfortable with the "peon-type" Mexican.

Bispo's examiners, in contrast, offered no commentary on the wisdom or desirability of admitting him to the country, merely stating the need to "protect the country against communicable diseases." Perhaps Bispo, an immigrant presumably from Mexico or South America, evoked unspoken concerns regarding disease, race, and perhaps even social class. But if he did, the officers in this case left them unexpressed and, significantly, framed their argument squarely in terms of preventing the entrance of disease, not of a person or a race or a class. The Ellis Island officers lacked an explicit imperative to protect the nation from immigrants unfit for either civic or industrial citizenship.

Hookworm and Clonorchiasis

While additions or subtractions were regularly made to the list of Class A and Class B conditions, only one disease other than TB was ever reclassified (table 1-1). In 1917 the PHS changed the classification of uncinariasis, or hookworm, from Class A to Class B, meaning that exclusion of immigrants with hookworm was no longer mandatory. Along with extrapulmonary tuberculosis and chronic malaria, it was the only communicable infection ever explicitly categorized as Class B.

The rationale for the change was that hookworm was easily treatable, but PHS officers warned that "this argument can not be pushed too far however, or else it will affect other diseases in Class A like Chancroid, for example, or oriental sore."[82] That the PHS identified "oriental sore" was perhaps coincidental, but hookworm, significantly, was a disease linked almost exclusively to Asian, particularly Chinese, immigrants. Dr. Victor Safford, in particular, was concerned that in the case of hookworm a reclassification to Class B would limit the ability of the PHS to require treatment for these groups.[83] While the IS increasingly allowed immigrants afflicted with other diseases, particularly trachoma, to apply for treatment until cure (provided they could afford it and provided that deportation was deemed an unnecessary hardship), increasing success in treating other diseases did not result in reclassification.[84] Thus, the fundamental question was not whether the PHS *could* treat a condition, but whether it *would*.

Effective, though potentially dangerous and certainly unpleasant, treatment of hookworm with thymol—which loosened the worms from the intestinal wall and allowed them to be expelled—had been standard since the 1880s. Willingness to offer universal treatment of hookworm hinged on a change in the perception of the threat of hookworm. John Ettling, the chronicler of hookworm in the American South, establishes that the campaigns of northern evangelical reformers to eradicate hookworm, the "germ of laziness," touched on sensitive questions of race, class, and economic, cultural, and political hegemony. The appearance of its undersized, anemic, and lethargic victims served as "a peculiar badge of an unfortunate economic and social class."[85]

In the case of Asian immigrants, the PHS turned the meaning of hookworm on its ear. First, "Asian" immigrants were believed not to manifest outward signs of infection. All, consequently, underwent laboratory examination for the disease. Second, though hookworm was linked to "coolie" labor and thus to social class, as it was in the South, the real threat presented by the "coolie" was not that he was lazy or indolent, but that he threatened to underbid, outwork, and thus pauperize and socially degrade the white American worker, who maintained higher, more civilized standards of living.[86] Like the term *peon* along the Mexican border, *coolie* carried connotations about both race and class.[87]

Thus, while the PHS and IS repeatedly cast hookworm infection as both a "public health" and "economic" threat, it represented an economic threat of a different order than did hookworm in the American South. Asians with hookworm did not represent a drain on the economy, as did immigrants with other Class B and even Class A conditions; rather, they represented an alteration of the economy. In effect, the Asian "coolie" and not the hookworm was the parasite in need of eradication. Reaction to East Indian "Hindus"—who fell into the broadly conceived notion of Asian "coolie" laborers and were also linked explicitly to hookworm—illustrates the connections made among labor, standard of living, the economy, and disease. Reflecting the sentiments expressed by many other labor councils affiliated with the American Federation of Labor to the Immigration Service around 1913, E. C. Berry of the Vallejo Trade and Labor Council in California strongly opposed all "Hindu" immigration on the grounds that those immigrants had a low standard of living, were unfit and inferior laborers, and lacked appropriate standards of family or morality. Unable and unwilling to reduce himself to this low state, the white worker would be pushed out of California. The letter concluded, "We don't want the manner of living of the Hindoo; we dont want his religion; we dont want his diseases; we dont want his filth."[88] This was one of dozens of similar letters written to the IS during this period by AFL-affiliated trade councils in California. In 1907 the Asiatic Exclusion League called "Hindu" workers "dirty, lustful, and diseased" and focused on "their lack of cleanliness, disregard of sanitary laws, petty pilfering, especially of chickens, and insolence to women."[89]

It is significant, then, that hookworm was only reclassified in the wake of the 1917 immigration law, which created the Asiatic Barred Zone, strictly prohibiting the entry of all Asian laborers.[90] Although the Chinese were described as "automatic engines of flesh and blood" with "muscles like iron"—seemingly the perfect industrial citizens—their racial status made them ineligible for industrial citizenship and civic participation alike.[91]

The story of clonorchiasis (liver fluke infection) further underscores the limited relevance of actual ability to treat disease and the importance of perception in reclassification decisions. After hookworm was downgraded to a Class B condition, the PHS found itself pressured to do the same for clonorchiasis.[92] In 1917 and again in 1922 the PHS refused to reconsider the classification of clonorchiasis on the grounds that the disease was resistant to all treatments. Concluded the surgeon general, "Any infection which produces a chronic, incurable disease, should be excluded from the country."[93] By 1924, however, immigration from China and Japan had been almost entirely choked off.[94] At that time the assistant surgeon general recommended—despite a lack of progress in treatment—that clonorchiasis be reclassified as Class B but that its description be left under the list of dangerous contagious diseases: "it is believed that any danger that may have existed previously will be very materially lessened because the type of aliens now admissible is a much higher sanitary type and not nearly so likely to have clonorchiasis as the laborer type that formerly came in." This solution, he continued, "does not put the Service in the embarrassing position of receding from its former classification of 'dangerous, contagious disease' but recognizes that the mandatory provisions of the law have removed the sanitary hazard."[95] Here, the "laborer type" was also a racial type—the "coolie." The "sanitary type," therefore, was "white," or at least European, since nonwhites were now formally excluded.

For both hookworm and clonorchiasis, then, legislative alterations in the flow of Asian immigration, rather than a change in the understanding, treatment, or diagnosis of disease, precipitated a change in the seriousness with which parasitic infection was regarded. But more important than treatment was the fact that both diseases were detectable only through laboratory diagnosis. The PHS felt that the Asian body was impenetrable by the medical gaze, so it went to great lengths to ensure that all Asian laborers would undergo laboratory scrutiny for the purpose of limiting the entrance of inherently unfit industrial citizens.

In the case of hookworm, the diagnostic reclassification from Class A to Class B provoked a furious clash between the PHS and IS at Angel Island. As a result, the IS had to yield to the scientific authority of the PHS and change its approach to hookworm certificates. Upon the reclassification of hookworm, the commissioner of immigration at San Francisco, Edward White, immediately revised exclusionary policy at Angel Island. He informed his officers that because hookworm "has been removed from the list of loathsome and dangerous contagious diseases . . . the practice recently followed at this Station of

uniformly excluding all aliens certified for hookworm, regardless of the attendant circumstances and collateral conditions, is incorrect. . . . If the Board is
of the opinion that the alien's ability to earn a living is not materially affected
by the cause of the medical certificate . . . he should be admitted." White
capped off this order with final instructions to "abolish the procedure heretofore followed of requiring treatment and cure before landing is permitted."[96]

White was a bureaucrat primarily concerned with the efficient operation of
his station and his own image in the popular press.[97] He viewed the reclassification of hookworm—a significant disease at his port—as an opportunity to
expedite and simplify his administrative chores. Such an outlook challenged
the PHS, for in this instance the agency clearly felt a mandate to hinder the
immigration of Asians with parasitic infection, for the combination of class,
race, and disease signaled unsuitability for industrial citizenship. Questions of
whether medical authority ultimately rested with the IS or with the PHS, then,
lay at the heart of White's order.

Dr. W. C. Billings, the chief medical officer stationed at San Francisco, flew
into a rage when he learned what White had ordered. White had blatantly dis-
regarded Billings's best professional advice regarding how to proceed in light
of the reclassification decision. Billings, therefore, took matters into his own
hands and appeared unsolicited before the BSI each time it considered a hookworm certificate. He subsequently convinced the BSI to exclude hookworm
as a Class B condition in at least one case, thus undermining White's order.[98]
In a thinly veiled threat, Billings "offered" to send a medical officer to all BSI
hookworm case hearings to testify "that in his opinion the disease most certainly did affect ability to earn a living."[99] When this tactic failed, Billings took
his complaint to the surgeon general, warning that if White's order were not
reversed, it would threaten the medical inspection of all Asian immigrants.[100]

Although the PHS had expressed some anxiety about the effects of the reclassification of hookworm before this incident, the San Francisco situation
prompted officials to clarify classification to underscore "the necessity for protecting the country against hookworm."[101] While not intending to undercut the
reclassification, the PHS issued a circular stressing that "the disease is communicable under certain insanitary conditions" and should be treated as a
communicable disease like typhoid or malaria. Thus, "any alien afflicted with
hookworm should either be excluded, or held until cured." While the PHS preferred exclusion, holding an Asian immigrant until cure had two beneficial
consequences. First, the prisonlike setting of Angel Island offered an oppor-

tunity for extended discipline over the course of the many weeks and sometimes months that immigrants had to undergo treatment. Second, it helped to
ensure that immigrants with hookworm had the financial resources to pay for
treatment—in effect putting up an additional barrier to the "low" type of
"coolie" labor that represented such an economic threat.

The IS concurred with the PHS circular and even worked an interesting
twist on the PHS qualifications: "aliens suffering with hookworm should not
be permitted to enter the United States. . . . [T]he disease affects ability to
earn a living, and renders the alien excludable on that ground alone." In other
words, the IS instructed its inspectors to consider hookworm a *mandatorily
excludable Class B condition* unless treated and cured.[102] Diseases affected
ability to earn a living not because of the nature of the disease, but because of
the economic status or prospects of the immigrant. The mandate to either exclude or treat "coolie" immigrants with hookworm, therefore, underscored the
extent to which race and class almost completely coincided in this instance.

White, for his part, refused to back down from his original order even after
acknowledging receipt of the PHS instructions. When called to the mat by the
commissioner general of immigration, White protested that he felt justified in
his persistence because he had received no explicit orders from the IS. Shifting attention from himself, he proceeded to request the transfer of Dr. Billings
"as he is now persona non grata."[103] Ultimately, White fished for the IS to af-
firm that PHS officers at Angel Island served only at his pleasure—that he exercised authority over them.[104] The commissioner general, however, gently
reprimanded White for defying the PHS and the IS (which had informed
White and all other commissioners of PHS instructions regarding hookworm)
and for baiting his chief medical officer.[105] Although the PHS transferred Dr.
Billings to Louisiana, ostensibly as a reward for his years of service, the hookworm saga ended with the triumph of the lab as an exclusionary device.[106] It
also represented a very different mapping of race and class than we found in
the instance of syphilis along the Mexican border.

THE CONVERGENCE OF RACE, CLASS, AND DISEASE

In the case of tuberculosis, the necessary ingredients for exclusion never
coalesced. When they did threaten to combine in such a fashion as to shift the
mandate of the PHS from discipline to exclusion, they did so by combining
questions of citizenship with race and disease. Issues of class, to the extent that

they existed in the larger culture, remained deep below the surface in discussions within the Public Health Service. For the Immigration Service, in the instance of TB, the issue was how the immigrant would affect the racial stock of the United States. As shown in table 6-12 in chapter 6, the PHS singled out no immigrant group for certification for TB, nor did the IS single out any group for exclusion.

Syphilis represents a transitional case of sorts. In most regions, race, class, and disease failed to coincide in a fashion that enabled PHS officers to cross the threshold from discipline to exclusion. Along the Mexican border, however, notions that venereal disease represented disorder and loss of control combined with ambivalent feelings about Mexicans and a concern about how the immigrant laborer would affect American civilization. But officials along the Mexican border were able to address these concerns with intensified efforts at primary examination, rendering superfluous further efforts to catch cases of syphilis. When the concerns were expressed, however, we find not "low" social class but "high" social class combining with race to make an immigrant unsuitable for admission.

In the case of questions involving parasitic diseases, race, class, and disease came into alignment in a different fashion. Although the overlap of these ideas in the case of syphilis along the Mexican border also speaks to these issues, the convergence of race, class, and disease in the case of Asians, in particular, helps underscore how disease was being used to map out the new contours of the nation. The work of scholars such as Mae Ngai, Ian Haney Lopez, Henry Yu, and Susan Craddock, among others, suggests the need to connect, though not necessarily equate, the exclusionary aspects of immigration policy to the segregation of African Americans—the more domestic side of difference, its construction, and its exclusion.[107] Rogers Smith notes that when the rise of Jim Crow "is seen in the context of all the changes in the nation's civic statuses during these years . . . second-class black citizenship looks in no way exceptional. It was instead the cornerstone of a general legal elaboration of ascriptive hierarchies."[108]

African Americans, once enslaved forced laborers confined mostly to the South, were given geographic mobility and a new and complex civic and industrial status following Emancipation and the passage of the Fourteenth and Fifteenth Amendments to the Constitution. The migration of African Americans to the North before World War I shared elements of the European migration from abroad. Despite the perception that the prewar migration represented the movement of the most skilled and educated, the majority of African American migrants were single, unskilled laborers who competed with southern and eastern European groups for the lowest paying, backbreaking industrial jobs. More than ninety thousand African Americans, most born in the South, lived in New York City by 1910. New York was second only to Washington, D.C., as the largest African American urban center. Philadelphia and Chicago were also important industrial centers of urban concentration.[109] World War I, in choking off European immigration while increasing production, also dramatically altered the position of African Americans as industrial citizens, as some half a million migrated from the South and left sharecropping for industrial, factory-based jobs. African Americans described the move in terms of the biblical Exodus and envisioned the move from South to North as akin to a journey to a new country.[110]

As a group, they were closely tied to the new immigrants and their position in the emerging industrial economy. Gwendolyn Mink argues that a "Chinese-black analogy was quite prominent in the debate over Chinese exclusion, both among abolitionist and radical Republicans and among Democrats," for both the "Chinese and blacks offered equivalent forms of labor."[111] Samuel Gompers, for example, along with the nation's political leaders, easily equated "Negroes, Chinamen, [and] Japs."[112] In the world of international affairs, Africans and Asians were also equated. The Paris Peace Conference in 1919 maintained the principle of self-determination for European nation-states, while what Colonel Edward House, President Wilson's adviser, termed the "backward countries" in Africa and Asia remained British and French colonies.[113] Peoples originating from both African and Asian nations, like their countries of origin, concludes Henry Yu, were objectified, treated not as persons but as objects or possessions.[114]

But although Chinese, Japanese, and other Asian immigrants were denounced, along with African Americans, as unfit industrial citizens, the role of racism made the status of African Americans as industrial citizens uniquely complex. On the one hand, racism could work to limit industrial opportunities—particularly employment in higher paying jobs—and to exclude African Americans from the labor movement.[115] Daniel Letwin notes that the coalfields of the South were racially mixed, with the proportion of black to white workers evening out by the 1890s and with black workers dominating after the turn of the century. Although blacks represented mainly unskilled mine labor, they were also established in skilled, supervisory positions.[116] In more urban industrial settings, African Americans, like the Chinese immigrants, may have had fewer industrial opportunities. Thomas Maloney and Warren Whatley observe that in Detroit, between 1916 and 1940, Ford employed almost every African American autoworker in the city; indeed, by 1922 Ford employed nearly 30 percent of the African American male population in Detroit. Maloney and Whatley suggest that the concentration of black male workers at Ford indicates the unavailability of other high-wage positions for African Americans, though blacks were also concentrated in the most hazardous foundry jobs beginning in the mid-1920s.[117] On the other hand, racism could serve the interests of industry.[118] African Americans, along with southern and eastern European immigrants, could benefit industry in the pivotal role of the strikebreaker, as they did in the mining industry, particularly before the turn of the century, in the slaughter and meat-packing strikes around the turn of the century, and in the packinghouse and steel strikes following World War I.[119] They became seen by industry as useful means of eroding ethnic solidarity on the shop floor.[120] In short, as Arnesen makes clear, there existed "no monolithic approach toward African Americans."[121]

The main difference, it seems, between African Americans and Asians lay in the ways the state contained the groups within the nation's borders. African Americans, unlike Asian immigrants, were granted citizenship under the Civil Rights Act of 1866 and then the Fourteenth Amendment to the Constitution, and were given the right to vote under the Fifteenth Amendment. These advances, however, were all undercut with the extension of Jim Crow. The development of Jim Crow segregation began in the South in the 1890s, following *Plessy v. Ferguson* (1896), in which the Supreme Court ruled that state-based racial segregation was constitutional. It later developed in other parts of the nation. Jim Crow found its counterpart in Asiatic exclusion, both of which worked to strip the group in question of many of the political and economic rights of citizenship. Following the Chinese Exclusion Act of 1882, which not only barred laborers but also declared Chinese immigrants ineligible for citizenship, but preceding the Immigration Restriction Act of 1924, came the Supreme Court decisions in the cases of *Ozawa v. United States* (1922) and *United States v. Thind* (1923). The decisions in these cases effectively made all Asian immigrants as defined by the 1917 Immigration Act ineligible for citizenship, on the grounds that Asians were not "white." These Supreme Court decisions were followed in 1923 by decisions upholding California and Washington laws prohibiting agricultural land ownership by Asians on the grounds that they were ineligible for citizenship, and then in 1924 by the Immigration Restriction Act, which finally banned the immigration of all peoples—not simply laborers—ineligible for citizenship. With an effective set of laws in place limiting the rights of all people of African descent—whether they were former slaves or new immigrants—there was little imperative to develop a strong rationale or mechanism for controlling immigrants from African nations at the borders. As mechanisms of controlling Chinese immigration were strengthened, the centrality of medical exclusion lessened. As we saw in this chapter, the more sweeping immigration restriction legislation of 1924, which effectively prohibited all Asian immigration, altered the PHS's willingness to remove certain conditions from the list of diseases requiring mandatory exclusion.

The U.S. situation stands in sharp contrast to that of Canada, which explicitly stated its intentions to bar African Americans from crossing its southern border.[122] Neither the PHS nor the IS discussed African immigrants in agency correspondence or annual reports, although African immigration was on par with that from both China and Japan after 1916.[123] Although the PHS tended to certify Asian immigrants at rates higher than average, it uniformly certified African immigrants at rates very close to or even below average. The IS, however, excluded both Asian—particularly the Chinese—and African immigrants at rates above average.

Just as the exam on the East Coast was telling the immigrants to adopt industrial values, the exam in the West was telling Asians, like African Americans, that they were a people apart, that they had no place in American civic or industrial life. Just as discipline was a process that continued on the nation's factory floors, segregation continued in the West. Asian immigrants, after all, were not merely segregated at Angel Island, but also within Chinatown after they were admitted. Susan Craddock, in her study of the Chinese population in San Francisco, shows when and why public health imperatives intersected with social and economic imperatives and worked not simply to stigmatize the Chinese immigrants, but to define and create a pathological containment space for them within the city of San Francisco, creating the boundaries of Chinatown. In her analysis of the public health response to TB, plague, smallpox, and syphilis in San Francisco, Craddock underscores the power of science not only to employ and define race, but to control, through the use of space, participation in American cultural and civic life. It is this use of the police powers of public health, then, that distinguishes the experience of the Chinese in

America from that of many other immigrant groups. As Craddock notes, "the ascription of disease to particular bodies and places can only work as a containment measure . . . if bodies remain within their specified geographic locations and if they can be easily coded as to their sexual orientation, social practices, ethnic origin, or nationality."[124]

Asian immigrants were not only denied entry into the United States in larger numbers, they were also denied admittance to citizenship. The physical, geographic, and legal/political color lines in America were being drawn not through Europe, but around African Americans; immigrants from African nations; Chinese, Japanese, and other Asian immigrants; and, beginning in the 1920s, Mexicans. Just as the policies of immigrant inspection in the East were part of a larger national endeavor to create a disciplined working class, the effort to exclude was part and parcel of a much broader effort to shape a national Anglo-American identity.[125]

CHAPTER SIX

DRAWING THE COLOR LINE

Racial Patterns of Medical Certification and Exclusion

Findings. The board finds that the applicant is an alien, a subject of
Japan, 26 years of age, and coming to the United States to join her husband to whom she has been married according to the laws of Japan
under what is known as a proxy arrangement, and to whom she has expressed her willingness to marry according to the laws of this country.
The Board further finds that the applicant is afflicted with uncinariasis
(hookworm), a dangerous contagious disease and accordingly rejects
her on that ground. The Board also finds that were she not afflicted she
would, in its opinion, be admissible to the United States. Deportation is
ordered at the expense of the importing vessel.

Applicant recalled:
Q You have been refused admission to the United States on a medical
   certificate certifying you for uncinariasis and from this decision you
   have no appeal.
A I understand.
         —RG 85, Box 225, File No. 53598/30, NARA, San Bruno, Calif.

BEFORE THE BOARD OF SPECIAL INQUIRY in San Francisco rendered its findings to Komume Kinoshita, a picture bride from Japan certified
with hookworm who was immigrating to the United States to join her husband,
whom she had last seen when she was twelve years old, the examiners, the
woman, and her husband went through an almost ritualistic dance of questions
and answers:

Q CHAIRMAN: What is your name, age, nationality, race, and birthplace?
A ALIEN: Kinoshita, Komume, age 26, Japanese, subject of Japan, and I was
   born in Mitogawa Mura, Wakayama Ken, Japan.

Q Ever been known by any other name or names?

A Kinoshita was also my maiden name.

Q Have you a passport reading to the United States?

A Yes.

Q How do you identify it?

A By the description thereon *(Description agrees with that of applicant.)*

Q Any person traveling with you?

A No, I came alone.

Q What was your last permanent residence and how long have you lived there?

A At my husband's home in my native village for about three years.

Q Have you a copy of that family record?

A Yes. *(Record shows marriage of Kinoshita, Komume to Kinoshita, Tomisaburo registered April 26, 1912)*

. . . .

Q Are your parents alive?

A Both dead.

Q How many brothers and sisters have you?

A Three sisters and no brothers.

Q Are your husband's parents living?

A Father dead and mother living.

Q How many brothers and sisters has he if you know?

A Four brothers and two sisters.

Q Can you read and write?

A Yes.

Q On what ship did you arrive and from what port?

A On the S.S. *Inaba Maru* sailing from the port of Kobe, Japan.

Q Who paid your passage?

A My husband.

Q What was your occupation in Japan and what will you do in this country?

A I was working on the farm in Japan and in the United States I will keep house.

Q Ever been in the United States or denied admission?

A No, this is the first time.

Q What is your husband's address and business?

A A sawmill laborer at Mukilteo, Washington, P.O. Box 27.

Q How much money have you?

A $50.00.

Some of the questions were racially loaded. Examiners were wary of incestuous relationships that might connote that the true purpose of immigration was prostitution or perhaps domestic labor, not lawful marriage.

Q Are you related to your husband by blood?
A No.
Q How does it happen that your family names are the same?
A I don't know, it just happened to be that way.

. . . .

Q Is it your intention to marry your proxy husband according to the laws of this state?
A Yes.
Q Have you any purpose in view of coming here except to live with him as his wife?
A No.

Addressing her husband, Tomisaburo Kinoshita, the BSI examiners further probed into the couple's intentions:

Q Have you a home ready for her?
A Yes, I have a house prepared for her.
Q Will she cook for other persons besides you?
A No, just myself.

The examiners suggested that, surely, this woman knew she had hookworm before immigrating, as did the Japanese physicians and officials who sent her off:

Q How many times were you examined for hookworm?
A Once.
Q Who made that examination?
A The harbor doctor at Kobe.
Q Did he give you a certificate or anything?
A No.
Q Did you take any medicine either before or after that examination?
A No.
Q Were you examined on the voyage on the way over?
A I was examined on board but not specifically for hookworm.
Q What was the result of that examination by the harbor doctor at Kobe?
A They found out that I was not afflicted with the disease, or any diseases.

Q Was this harbor doctor an officer of the steamship company or a Japanese government officer?

A A Japanese government officer.

The rite of examination ended, after Komume was flatly denied entry, with her request for treatment. Clearly she knew what to expect and how to respond. She was granted treatment, and we can only assume that after a period of days, weeks, or perhaps months, she was cured and ultimately joined her husband.[1] For many Japanese hopefuls and selected other immigrant groups, the process of certification and examination, stylized and loaded with racial assumptions though it might be, carried very real consequences.

Some concept of race was built into many different disciplines: eugenics, bacteriology, anthropology, psychology. Each drew differentially on ideas about biological heredity, culture, physique, and disease, as well as on political principles, in order to express and constitute differences in the social body. The construction of race was not unchanging, however, and scholars have traced a broad shift from nineteenth-century conceptions of race that emphasized culture and heredity to a twentieth-century scientific notion that emphasized biology.[2]

Even the "popular" conception of race that was distinguished from scientific definitions of race in court decisions from *In re Ah Yup* (1878) to *Ozawa v. United States* (1923), over the question of who was and was not "white," emphasized "natural," "unmistakable," "unchangeable," "physical" differences between peoples.[3] Ian Haney Lopez astutely observes that after 1909 we find "a schism appear[ing] among the courts over whether common knowledge or scientific evidence was the appropriate standard" for determining "whiteness." Still, biology remained central to race thinking.[4]

I expected to find the same phenomenon in my research for this book—to see an emphasis on biology in the construction of race and difference in the context of immigrant medical inspection and exclusion. Interesting, though not entirely unexpected, is the finding that this twentieth-century conception of race or difference does not consistently dominate.[5] Rather, we find the persistence of nineteenth-century notions of race—of difference—that conflated religion, language, nationality, place of origin, occupation, and biology. Race represented a remarkable combination of fixed and fluid elements, making it a much broader, more pliable, more unstable concept than was previously thought. While the courts may have capitalized on the biological underpin-

nings of race in the application of scientific and popular knowledge, they acknowledged that race, particularly "whiteness," was a "very indefinite" concept
that, over the course of the Progressive era, remained "not clearly defined to
be sure."[6]

THE CONSTRUCTION OF CATEGORIES

In an environment where the biological elements of race were viewed as increasingly central, particularly as concerned public debate over immigration,
Public Health Service officers, as agents of science, were uniquely positioned
to make and act on powerful statements about racial differences in the human
population that could profoundly shape the lives of immigrants seeking entry
to the United States. Day in and day out, at the bustling or sleepy ports of the
nation's borders, the PHS screened every arriving immigrant of every nationality. PHS officers developed a coherent ideology and set of regulations for inspection, exclusion, and admission that was adaptable to different settings, circumstances, and arriving immigrant groups. PHS officers espoused the notion
that they could "read" the race of the immigrant at a glance and quickly calculate how racial characteristics subtly altered the display of disease on the immigrant body. Indeed, the biological language of race influenced the broader
PHS vocabulary; the agency understood class and sometimes even disease as
constitutional characteristics of the immigrant.

Surprisingly, in a period and arena where science held such authority and
where the PHS might have been ideally situated to shape understandings of
race, it was the Immigration Service, a bureaucratic organization of civil servants and political appointees, that took the lead in defining race explicitly. Although the IS did not reject the racial formulations of leading scientists and
politicians, in formulating a "practical" construction of race, it made race speak
to questions regarding industrial capacity rather than civic capacity or genetic
quality. It did so in a way that, while not discarding biology, emphasized a
panoply of mutable characteristics.

The fledgling bureaucracy of the IS employed hundreds of clerks, immigrant inspectors, "Chinese inspectors" (who interviewed and handled the paperwork for Chinese immigrants only), and translators.[7] Approximately 60
percent of the IS employees were stationed at Ellis Island. Many inspectors
had been picked up from the state when immigration was placed under federal jurisdiction in 1891. Most of these had been New York political machine

appointees, so that despite the advent of the civil service, the island was staffed by political appointees well into the 1930s. Many, too, were immigrants. Indeed, Ellis Island boasted the highest percentage of foreign-born employees of any city agency.[8]

Jacob Auerbach, for instance, originally worked as a clerk for the city of New York before coming to Ellis Island as an immigrant inspector for the IS in 1930. At the city, he was repeatedly passed over for promotion and blamed political corruption for this. "So I said to myself, 'I have experience as an immigrant. I know what it feels like. I know that this country, the United States of America, is built of immigrants. What can I do about it?'" He took the civil service exam to become an immigrant inspector, "figuring I may as well make use of my English and my knowledge of the languages: Yiddish and Russian, some Polish and German . . . I could speak French, but not too well." In 1930 he "became an immigrant inspector with a uniform—a grey-green uniform with brass buttons with a gold and blue badge on my breast and a military cap with the insignia of the United States Immigration and Naturalization Service."[9] While Auerbach insisted that IS officers were not "rough with" or "cruel" to immigrant arrivals, he acknowledged the power that he and other officers held over them and the fear that they experienced while undergoing the inspection ordeal.

The available evidence suggests that the IS attracted men from a variety of different backgrounds with quite different ambitions. Some came to the IS in search of an administrative career. John Birge Sawyer, for example, who left diaries documenting his life and years in the U.S. Consular Service and the Immigration Service, saw the IS as a means to improve his lot. The son of a religion professor, Sawyer studiously tracked his salary, taxes, and living expenses each year. He bought property when he could. Sawyer aspired to the upper levels of the IS bureaucracy. He began his career in the federal bureaucracy as a "Chinese inspector" attached to the U.S. Consular Office in Hong Kong. In 1916, after failing to get the position of commissioner of immigration in Shanghai, Sawyer was offered the position of "Chinese and immigrant inspector" in San Francisco. It was a hard decision to make because it represented a step down, but Sawyer desperately wanted to leave Hong Kong, primarily because "the problem comes up of getting through it without illness. [My wife] Grace and I are starting in well this year but [our daughter] Nancy causes us some uneasiness because she has so little color and will not eat. The second consideration is our desire for a different social atmosphere with less of formality and caste and more of family ties and enduring friendships. Although we understand this life better and adjusted ourselves to it in some degree we realize always that we do not belong to it naturally and never shall."[10] Sawyer took the San Francisco post primarily because he saw the IS as holding promise for career advancement; he felt he could quickly move up the ranks into an administrative position. Career opportunities and social and career advancement were the overriding theme of his diary—indeed, they were his obsession. When he failed to advance within the IS, he made another career switch, back to the bureaucracy of the Consular Service. Sawyer spent the rest of his career in civil service, from 1918 to the 1930s, always vying for status and authority within the Consular Service and IS.

Others saw the IS as offering the opportunity for temporary employment—as a stepping-stone to careers in politics or business. For example, Fiorello La Guardia, who became a Republican member of Congress and New York City mayor, worked as an IS inspector at Ellis Island while attending law school in New York.[11] Similarly, Hart Hyatt North, a Hastings Law School graduate with political ambitions who served as San Francisco's commissioner of immigration from 1898 through 1910, focused his professional energies on the development of the Shasta Iron Company. For La Guardia, IS work became a means of feeling the pulse of the people he would represent politically; for North, it served as a means of beginning to manage the labor force that would drive his mining ventures.

The workhorses of the IS were the career civil service officers and political appointees who received and processed arriving immigrants day in and day out, and they are mostly silent and anonymous in the historical record.[12] La Guardia characterized these men as a "hardworking lot, conscientious and loyal."[13] Many came to the IS because it represented the first available job. Clifford Alan Perkins, an "immigrant and Chinese inspector" turned Border Patrol officer in Texas, never expressed the bureaucratic ambitions of someone like Sawyer, the industrial ambitions of North, or the idealism of Auerbach or La Guardia. He simply came to Texas as a youth with tuberculosis looking for health and a job. The IS gave him a uniform and the opportunity to tote a gun. He took opportunities for advancement whenever they presented themselves, and ended his career in law enforcement, patrolling the U.S.–Mexican border when the Border Patrol was founded in 1924. He preferred a good brawl or prairie shootout to deskwork and described his companions within the Border Patrol as an equally restless group of cowboys and drifters.[14]

IS administrative officials characterized the core of their employees as a rather rough and unprofessional lot. The career IS officer or senior administrator occasionally received praise for either denying entrance to diseased immigrants or admitting them.[15] He might equally receive a slap on the wrist, for the same reasons.[16] Rank-and-file IS employees were most frequently criticized for rude behavior, graft, or for poor conditions or inspection methods at immigration stations.[17] Not until 1929 did the IS attend to the professional conduct of its officers. Weeding out corruption and misconduct on the part of IS officers was a matter of concern, yet in the face of complaints of rudeness on the part of IS officers, of equal concern was the need to establish and instill a code of professionalism among its officers. Their ambitions offer insight into the organizational ethos to which the IS aspired. A commissioner of immigration at Ellis Island, Benjamin Day, argued that the IS should be characterized by gentlemanly conduct, the most important features of which were civility and "a businesslike conduct of [immigrants'] examinations." He continued, "A quiet, dignified, expeditious, though complete exam is ideal."[18] The IS aspired to be an efficient bureaucracy.

As a group, the IS officers presented a strong contrast to the elite PHS officer. The commissioned officers of the PHS were largely native-born southerners of similar social class whose circles crossed as early as college and medical school. In some sense, however, the IS was more representative of the laboring class in the United States. Its officers were drawn from different groups and different sectors of society. Some had toiled as unskilled laborers. Robert Watchhorn, who became commissioner of immigration at Ellis Island, worked as an immigrant coal miner in the Pittsburgh area when he first came to America.[19] They were not now common laborers, for they had to be educated to pass the civil service exam, but they represented the nation's blue- and white-collar labor force. They were immigrants or the children of immigrants. They consequently had the most at stake in establishing and reinforcing a racial hierarchy that was grounded in questions of where the immigrant working class belonged and, significantly, who belonged.[20] They saw it as their prerogative to define the proper place in the social hierarchy of the new arrivals.

Beginning in 1899, after having reported immigration statistics by "nationality" since 1891, the IS adopted the language of "race," assigned by country of birth. In 1902 the IS altered its terminology slightly, presenting immigration statistics according to "race or peoples." Despite the transition from "nationality" to "race or peoples," each of the categorization schemes reflected almost identical information: "Russian" replaced "Russia," "Scotch" replaced "Scotland," "Korean" replaced "Korea," and so forth.

The taxonomic shift, however, was not merely semantic. It represented a deliberate attempt to conflate race and nationality. Carl Degler notes that up to the early nineteenth century, the notions of race, nation, and people were roughly equivalent, though even at that point the concept of race could convey a sense of physical as well as cultural difference. With increasing scientific attention to classification in the nineteenth century, race began to take on specific biological connotations.[21] The IS chose not to uncouple race or biology and nationality, but rather to retain a sense of categorization that aggregated both cultural and physical differences, though it did not weigh them consistently or equally, as the term "race or peoples" implies.

Captured in its categorization scheme, the IS believed, were all the elements that contributed to each group's social potential. The term *race* was used by the IS "in its popular rather than in its strict ethnological sense; so that, from an experience of the distinguishing occupations of each race, its moral, mental, and physical characteristics, and their development under American institutions, a basis may be formed for estimating its effect on the population and industry of the United States."[22] Of note is the idea that American institutions could potentially alter a whole range of characteristics, including physical ones.

The committee that devised the classification scheme adopted by the IS informed the commissioner general that "undue importance is attached to occupation as stated by the immigrants on arrival. It seems to be assumed that this indicates that the immigrants will continue the same occupation here. At present this is probably erroneous in four cases out of five, except for expert mechanics and some professions."[23] For example, in the case of Henry Ryan Kenny Burke, a twenty-year-old Irish immigrant certified with "loss of ⅔ functioning grinding teeth [and] Loss of left foot and leg six inches above ankle," the IS was almost dismissive of past employment history. At his BSI hearing, Burke testified that he had been a clerical worker in Ireland. Although Burke heartily disagreed, the IS assumed that Burke—a former civil servant—would fail to win a job in American commercial houses. He had, after all, lost his position in Ireland due to his disability, according to his own testimony. The PHS officer who testified at his hearing emphasized the potential for future dependency: "The loss of leg involves interference with locomotion in some degree. In case of certain emergency and threatened accident the patient would be less likely to escape injury and the resulting disability from the injury." The

BSI sealed his fate with a simple observation: "You couldn't do any work as a laborer or any other work than a clerk." Burke confirmed the verdict: "I couldn't do work as a laborer." The BSI swiftly excluded the young man on dual grounds: "as a person being certified by the medical examiners at this station to be afflicted with a physical defect which may affect your ability to earn a living, and as a person likely to become a public charge."[24]

For the IS committee on the classification of immigrants, the point was not just that occupation was an inadequate determinant of the immigrant's prospects in individual instances. Rather, the point was that much broader generalizations could be drawn: "We believe that the race or language of the peoples, together with the destinations by States, will furnish the nearest approach to tangible information in this regard."[25] The term *race,* in other words, provided "a clew to what will be [the immigrant's] immediate future after he has landed. It is merely a grouping together . . . of people who maintain recognized communities in the various parts of this country where they settle, who have the same *aptitudes or industrial capacities* or who are found here identified with certain occupations."[26] Here, however, the IS suggested that America or its institutions did not have the ability to alter biological or physical features, but could draw out those that were innate.

Rather than espousing a purely biological racial ideology, or even a consistent theory, the IS viewed race pragmatically. Race reflected elements of social or industrial class: where an immigrant was likely to live (urban ghettos, mining communities, rural areas), and which industries or trades were likely to absorb the immigrant successfully (a function of former occupation, relatives already living in the United States, and, critically, health or fitness). Race also determined how the immigrant would assimilate (a function of both language and destination). In describing the "racial characteristics" of Italians to the readers of *National Geographic,* former assistant commissioner of immigration Z. F. McSweeny consistently used the language of class and occupation: southern Italy "was already represented in the immigration twenty years ago, but was composed chiefly of artisans, barbers, restaurant keepers, fruit venders, etc. Now the majority of Italian immigrants enter the field of unskilled labor." The northern Italian, in contrast, "is a type which belongs to the older period of immigration, and has little to differentiate him in economic possibilities from the Swiss, French, or Germans."[27]

This style of racial description also characterized the popular eugenics literature on immigration, though eugenicists concentrated on the physical features, particularly the facial and cranial features, as well as the intelligence of immigrants.[28] For Edward Alworth Ross, as for other eugenicists writing during this period, it is interesting to note that the "social effects" of immigration—illiteracy, crime, disease, pauperism—were more easily explained than the "racial effects." Thus, with the exception of "race suicide," which was clearly explained as the overbreeding of undesirable populations combined with underbreeding on the part of the native classes, the precise threat that immigrants posed to the "blood" of the nation was typically vaguely articulated. Thus, the Eugenics Society's Committee on Immigration demonstrated a remarkably flexible notion of "public health" that embraced promoting better breeding, preventing the spread of infectious disease, and protecting the national well-being by reducing the populations of charitable and penal institutions.[29]

The notion of suitability for industrial citizenship took on meaning within this socioeconomic framework for classifying peoples. In the new era of industrialization and scientific management, workers were mere replaceable parts, controlled by a manager who acted as the "brains" of the new immigrant labor force, which was expected only to perform—not to think—for itself. This helped place a premium on immigrant health. While the political and scientific leaders of the nation were primarily concerned with the civic capacity of the immigrant, the IS grappled with seemingly more mundane concerns. Since immigration was primarily concerned with admitting fit laborers, there was room and precedent for conceiving of immigrant arrivals not as potential citizens, but as those to be governed. Such conception necessitated racial classifications that emphasized not intellect or civic capacity, but where and how the group would fit into the working class.

As in industry, the immigrant's unfitness for self-government was practically a given at the nation's ports of entry. Frederick W. Taylor boldly asserted that "the man who is fit to work at any particular trade is unable to understand the science of that trade without the help and cooperation of men of a totally different type of education, men whose education is not necessarily higher but of a different type from his own."[30] Indeed, selecting the ideal worker for any particular task might require finding the man "so stupid and so phlegmatic that he more nearly resembles in his mental make up the ox than any other type. The man who is mentally alert and intelligent is for this very reason entirely unsuited to what would, for him, be the grinding monotony of work" of a physical character.[31]

Such thinking had a broad reach. Boston's settlement workers, for example, sometimes viewed intelligence as a vice among the working classes. Concerning the city's Italian classes, the 1903 settlement study conducted by the residents and associates of South End House concluded, "They sometimes have more education than honesty, and manage to live as parasites on their inexperienced countrymen." The authors of the study valued the Italians who served as the city's "skilled workmen following a variety of useful and productive callings. It is these specialized workmen who form the more permanent and desirable part of the North End colony." More generally, local settlement and charity workers tended to emphasize, first, the sanitary habits of the individual immigrant and those of the group within which he was classified, and, second, the industries in which different groups primarily engaged—the "industrial civilization" of different groups. Finally, settlement workers were interested less in civic capacity than in civic preferences: what motivated and then won the loyalty of the immigrant vote?[32]

While scientific racists were concerned with ensuring that the nation's citizens should remain "well born," those concerned with the labor half of the equation insisted that this was not enough: the worker "must be trained right as well as born right." They sought to admit "no rare specimen of humanity, difficult to find and therefore highly prized." Scientific selection of men for the labor force merely required "picking out from among very ordinary men the few who are especially suited" to labor.[33] The scientific manager or new corporate executive, like the PHS or IS officer, wanted the immigrant only to be able to perform a clearly defined task and take direction from others. As David Montgomery explains, "The essence of scientific management was systematic separation of the mental component of commodity production from the manual. The functions of thinking and deciding were what management sought to wrest from the worker, so that the manual efforts of wage earners might be directed in detail by a 'superior intelligence.'"[34]

Like the IS, American businesses were concerned with finding the right fit between different immigrant groups and labor. A Philadelphia executive noted that his company examined applications with an eye not only toward the applicants' "fitness for the particular tasks, but with respect to their constitutional ability to harmonize with the ideals and underlying principles of the company they are to serve; they must be capable of loyalty as well as efficiency."[35] Finding this combination required a careful classification and evaluation of workers in keeping with the thinking of those like H. A. Worman at International

Harvester, who asserted, "Each race has aptitude for certain kinds of work."[36] For example, an *Iron Age* editorial exhorted steel mill managers to hire "certain races only."[37] A Pittsburgh company assessed the "racial adaptability of thirty-six different ethnic groups to twenty-four different kinds of work under twelve sets of conditions" as a means of guiding hiring.[38]

The IS was an agency fundamentally concerned with labor and issues of this nature. Although the PHS and IS inspected some 25 million bodies from 1891 to 1930, many hundreds of thousands of those arriving had no intention of remaining in the United States. Between 1906 and 1930, when the IS drew distinctions between true "immigrants" and mere "arrivals," they classified 75 percent of arrivals as immigrants. During this period, the percentage of arrivals ranged from 1 percent in 1914 to 48 percent in 1918. In periods of economic recession and depression, even those who had come with the intent of immigrating to the United States often returned home. In 1908, for example, the number of Italians and Austro-Hungarians leaving the country surpassed the number arriving.[39] Frank Thistlewaite in 1960 estimated that as many as 1 million out of the 33 million who arrived in America in the century preceding 1924 returned home.[40] In its work at the nation's borders the IS was truly managing laboring bodies, not future citizens; thus, it clearly separated the realm of naturalization from immigration.

Originally established in 1891 as part of the Treasury Department and known as the Office of the Superintendent of Immigration, the Immigration Service began to administer the contract labor laws in 1895 and the Chinese exclusion laws in 1900, and as it did so, the agency's mission shifted. It was renamed the Bureau of Immigration when it became part of the Department of Commerce and Labor in 1903. By 1906 the problems of immigration and naturalization had become distinct enough to warrant the creation of two separate divisions within the new Bureau of Immigration and Naturalization: the Division of Immigration and the Division of Naturalization. Congress formally segmented immigration from naturalization work with the creation of two new bureaus—the Bureau of Immigration and the Bureau of Naturalization—in 1913. Both were transferred to the newly created Department of Labor, where they remained until moving to the Justice Department in 1940. The breadth of the labor-management mandate of the IS is evidenced by the explicit charge, from 1910 to 1917, to ensure the "beneficial distribution of aliens admitted into the United States."[41] The goal was to prevent "the congestion in our larger Atlantic seaport cities that has attended the immigration of recent

years" and "to supply information to all of our workers, whether native, foreign born, or alien, so that they may be constantly advised, in respect to every
part of the country as to what kind of labor may be in demand, the conditions
surrounding it, the rate of wages, and the cost of living."[42] (With the pressures
created by the war, a new agency, the U.S. Employment Service, began to
oversee the distribution of labor in 1917.)[43]

The immigrant racial classification schema used by the IS, then, represented a broad bio-socio-cultural-economic composite. The concept of race
reflected where an immigrant would fit in the industrial work force, to what
extent he or she could maintain the health standards necessary to remain a fit
worker, and whether he or she could absorb the industrial values of the nation.
Indeed, it almost seems a throwback to older eighteenth- and nineteenth-
century notions of race—indistinguishable from nations or peoples—that
combined elements of language, culture, and physical appearance.[44] This is
not to say that the IS was unaware that the notion of race had scientific and
political meanings or that such meanings were not reflected in its use of the
term. But the IS intended to capture far more than biological, intellectual, or
civic differences in its definition of race. Thus, it did not find the scientific
racial categories in vogue particularly useful in the context of immigration: "In
a scientific or ethnological sense it may be that there are only five or six
races. . . . [These broad divisions] were, however, too large to be of service and
it was decided to select what may be termed historical races, or races popularly known as such."[45]

Such understandings of race were not unique. Rather, they emphasized one
of several common components of race thinking in America, putting industry
and the IS at one end of a spectrum. Madison Grant, for example, understood
that the idea of race had particular implications for an industrial nation, that
immigrants were distinctly part of an industrial hierarchy. Yet Grant questioned industrial values that placed new immigrant groups at the bottom of any
labor hierarchy and, accordingly, stressed the biological component of race.
He argued, "A race that refused to do manual work and seeks 'white collar'
jobs is doomed through its falling birth rate to replacement by the lower races
or classes. In other words, the introduction of immigrants as lowly laborers
means a replacement of race."[46] Grant, in effect, was challenging a broad conception of race suited for thinking of immigrants as industrial citizens.

The Immigration Commission used the IS racial taxonomy as the backbone
of its 1911 *Dictionary of Races and Peoples,* yet emphasized a different com-

bination of racial elements. The *Dictionary of Races and Peoples* also avoided exclusive reliance on biology, arguing that a sole focus on "physical character-istics . . . is manifestly impracticable . . . in immigration work or in a census. The immigrant inspector or enumerator in the field . . . has neither the time nor training" to determine into which formal classification of race any immi-grant fits and may not even understand the different categories.[47] The dictio-nary, as part of a continuum of race thinking and classification, occupied a place on the spectrum next to the thinking of the IS. The dictionary used an array of fixed and mutable elements that, though less broad than the IS think-ing, attempted to reach toward a popular understanding of race. The Immi-gration Commission created the dictionary not "for the ethnologist" but rather "for the student of immigration." While based on the IS classification of races and peoples, the dictionary was more expansive, treating "more than six hun-dred subjects."[48] It explicitly stated that while "physical race" and language were its primary categorization tools, it also considered political history and geography as well as statistics describing immigration to and settlement pat-terns within the United States as critical to its endeavor.[49] While the dictionary confined the majority of its discussion within these parameters, it also consid-ered characteristics that exceeded these bounds. Albanians, for example, were described as "warriors rather than workers" and Bulgarians as drawn to "states where unskilled labor is mostly in demand." The dictionary characterized "He-brews" as an urban people "interested in inducing their relatives to follow." In the case of "Hindus," religion was a salient feature, while the illiteracy and poverty of Italians gave special import to their "immense capacity . . . to pop-ulate other parts of the earth."[50]

At the opposite end of the spectrum, those who dominated debate before the House Committee on Immigration and Naturalization were concerned with the effect of the "admixture" of races and the proliferation of those who were genetically and physically inferior on the germ plasma or germ stock of the American population.[51] Eugenical constructions of biological race were clearly articulated during the 1921 and 1924 hearings over immigration re-striction. Dr. Harry Laughlin, of the Eugenics Records Office at Cold Spring Harbor, was most influential. Laughlin conducted four extensive surveys for the House Committee on Immigration and Naturalization from 1921 to 1928 that stressed the physical and genetic inferiority of southern and eastern Eu-ropean immigrants.[52]

The Public Health Service found itself positioned somewhere between these two poles—between the IS classification scheme emphasizing the social and the economic, and a more eugenical construction of biological race stressing genetics. The PHS combined a scientific understanding of group membership that prioritized biology, though not necessarily genetics, with one that weighed race in terms of culture and economics and industry. Although the PHS did not bring a clearly articulated theory of race or precise racial taxonomy to the immigration arena, it considered knowledge of racial types key to the inspection process. The PHS officer believed race modified the face of disease. And race, like disease and social class, was inscribed or "stamped" on the body for the PHS officer to read. In 1914 Dr. A. J. Nute of Boston wrote that the first task of the line officer was "to determine the race or type of the individual and to have a good knowledge of his racial characteristics."[53] Continued Nute, "It takes considerable experience to know what constitutes the healthy color in a given race. A healthy Gypsy might readily be suspected of having Addison's disease, a healthy Greek of suffering from malarial cachexia or malignant disease. A normal West Indian Negro sometimes has the peculiar pallor suggestive of tuberculosis and the temperate Alpine mountaineer often [has] dilated capillaries resembling those seen in chronic alcoholics."[54] In 1917 the PHS incorporated the notion that race was crucial to diagnosis into its regulations for the medical inspection of immigrants, stating that "knowledge of racial characteristics in physique, costume, and behavior are important in this inspection procedure."[55] Sometimes occupation came to the forefront, sometimes biological race, sometimes nationality, sometimes culture.

The PHS did not limit itself to thinking in terms of biological racial types, but scientific notions of race helped to shape its framework, allowing it to see many different characteristics as fixed. The PHS referred to its own officers as "high" types.[56] Most broadly it discussed the "immigrant type."[57] Sometimes it specified "steerage types";[58] at other times, "primitive types," "Mongolian and negroid types," and "inferior types."[59] It referred to different "physical and mental types" among immigrants.[60] Other terms used included "sanitary types," "laborer types,"[61] "trachoma types,"[62] and "female types."[63] Even if the PHS did not necessarily define a clear causative relationship between biological race and disease, there was something immutable, inherent, and fixed in thinking about immigrants in terms of type that was akin to the thinking of the eugenicists.

Unlike the eugenicists, but like the IS, the PHS consistently viewed immigrants, regardless of type, within a framework of labor and disease. It was not concerned with capacity for self-government or racial degeneration. In 1903, for example, Dr. A. J. McLaughlin sought to assess "the relative value of the different races." To accomplish this he used a series of tables showing the proportion of unskilled labor that each of seven races brought to the United States, the proportion afflicted with "loathsome or dangerous contagious diseases," the proportion receiving hospital treatment on arrival, the proportion afflicted with diseases liable to render them public charges, the proportion afflicted with minor defects, and the proportion of illiterates.[64] Other PHS officers were less empirical in drawing links between race and disease. The PHS officer stationed in Portal, North Dakota, in 1911 described the arriving immigrants as "Canadian citizens of the robust type of laborers engaged in agricultural pursuits seeking localities in the Northwest States, and these classes of aliens are, as a consequence, particularly free from mental or physical defects."[65] At El Paso, in addition to describing the diseases most prevalent among Mexicans, the PHS officer in charge wrote in his annual report to the surgeon general for 1928, "by and large the Mexican seeking admission to the United States is comprised in the farm labor class, who lead an arduous, outdoor life for the greater share of the year, and being rather vigorous and robust, are not subject to any more diseases in particular than mankind in general falls victim to."[66] Thus, the PHS grounded its understanding of immigrant groups or types and disease within the laboring body.

Disease could serve as a vehicle not only for cutting across strictly biological racial categories but also for reevaluating those categories. We can see this in the immigration debates of the 1920s, when PHS data on disease among immigrants provided compelling evidence of inferiority, fueling eugenicists' demands to restrict immigration.[67] Indeed, Laughlin built the case for immigration restriction most convincingly not on a theoretical platform of genetic or intellectual inferiority, but on a slab showing concretely the burden that immigrants placed on penal and charity institutions in the United States. Laughlin's first report, which surveyed the national origins of the inmates of state and federal penal and charitable institutions, used the following nine categories of "defects": deafness, blindness, deformity, insanity, criminality, feeblemindedness, dependency, tuberculosis, and epilepsy.[68] Notably, though Laughlin was determined to read into his data the inferiority of southern and eastern European immigrants, others were quick to use those same data to mute the differences between European immigrants from different regions. H. S. Jennings, professor of zoology at Johns Hopkins University in Baltimore, brought

out the finer points of Laughlin's data, but not with an aim to show that southern and eastern Europeans were unfit and that northern and western Europeans were fit. Rather, he argued that, depending on the year used to establish the quota (either 1890 or 1900), the United States would receive a
different mix of defective dependents occupying American institutions. Using
the census of 1890, he argued, would yield a greater proportion of insane and
epileptics and a "very much larger proportion of dependents." However, it
would also produce fewer criminals and tubercular individuals. Jennings concluded that "the 1890 basis would not change the number of defectives in our
institutions but would change the combinations."[69] The choice presented to
the committee, then, involved considering what type of dependency rather
than which races of immigrants they most wanted to exclude. Similarly, Laughlin himself concluded that when it came to defectives living at the expense of
the state, "the division of European immigrants into those from the north and
west on the one hand, the south and the east on the other, or into the 'older'
and 'newer' immigration, is not a sharply defined one." Indeed, "the heaviest
source of defectives is found in one of the 'older' set; while the group with the
lowest proportion of defectives belong to the 'newer' set."[70] Although disease
could work to destabilize and reorder racial hierarchies, an important point of
agreement among House members during the hearings on the Immigration
Restriction Act of 1924 was that disease was a critical marker of "social inade-
quacy."[71]

CATEGORIES EMPLOYED

Only a fraction of arriving immigrants felt the brute force of exclusion. Yet
that fraction offers eloquent testimony to the physical expression of power and
its relation to how officials thought of and acted on categories of racial or group
difference. Federal immigration statistics—perhaps impersonal but not indifferent to identity—illuminate broad patterns of discrimination and exclusion
in unique fashion. The statistics on certification and exclusion by the IS category of "race or peoples" tell how the PHS and IS worked together (and sometimes in opposition) to define who must be rejected from industrial citizenship, and thus how they began to map the contours of Anglo-American
identity.

Despite the small numbers of immigrants medically certified or excluded,
the relationship between immigrant group and disease was complex. There is

important group variation in rates of certification and deportation. (See tables 6-1 through 6-23, at the end of this chapter.) To aid in comparing certification or deportation rates for different groups, I have expressed all rates in terms of standard deviations (SDs) from the mean rate for each year in which I am comparing different races.[72] I call this measure the NSDMR (literally, the number of standard deviations from the mean rate). My method is explained in detail in the appendix. Here, I may say briefly that the NSDMR given for each group in the tables reflects the magnitude and direction of difference from the mean rate of certification or deportation. An NSDMR between zero and 1 indicates that the rate was not more than one SD above the mean; an NSDMR between zero and −1 indicates that the rate was not more than one SD below the mean; and an NSDMR of zero indicates that the rate was exactly equal to the mean rate. I interpret an NSDMR of greater than 1 or less than −1 as representing substantial deviation. That is, an NSDMR of greater than 1 or less than −1 indicates that a group was either certified or deported at a rate much higher or much lower than average, given the normal variation in the entire data set. I refer to NSDMRs greater than 1 as "strongly positive" or "inflated" and to those less than −1 as "strongly negative."

Interpretation of the NSDMR varies depending on whether we are considering medical certifications or medical exclusions. We have no basis for comparison to determine whether racial differences in the NSDMRs for medical certification were "real" or "manufactured."[73] We do not know the true rate of disease among different groups. PHS officers believed that many factors that they associated with race—biology, culture, disposition—altered the face of disease, and that officers had to correctly identify an immigrant's race in order to diagnose disease accurately. Variations in the conduct of the immigrant medical exam, moreover, were, in some regions, a function of the immigrant groups predominating, as was the case along the Pacific coast and Mexican border. Thus, while there is reason to believe that the PHS might have "targeted" some groups, these data cannot answer the question of whether the differential racial distribution of certifications for different diseases was a product of the exam, regional practices, attitudes toward particular immigrant groups, or actual epidemiology. The NSDMR tells us only which groups the PHS singled out for either social or medical reasons. The pattern of diseases certified discussed in previous chapters, however, suggests that when the PHS felt a need to exclude a particular racial group, it could find a disease that suited its purposes.

The NSDMR for medical exclusions, in contrast, tells us how the IS evaluated the PHS medical certificate in the broad context of the immigrant's place in society. After the PHS certified an immigrant, he or she was sent before an IS Board of Special Inquiry (BSI), which made a final decision regarding admission or deportation. The BSI did not make decisions about whether an immigrant was actually diseased or not; rather, it decided, given the immigrant's social and economic condition, whether exclusion was warranted. In some sense, the PHS leveled the playing field, for medical certificates gave the BSI little or no information about the severity of the disease in any given class (A, B, or C).[74] The PHS gave the IS the bodies with which to work. The IS could not deport more than the number of any group that the PHS sent to it, at least not for medical causes. It could, however, etch its own patterns of exclusion based on an ordering that prized industrial capacity somewhat more highly than did the ordering employed by the PHS.

If the BSI made such decisions based on the nature of the disease alone, we would expect a relatively uniform exclusion rate (the proportion of those certified then debarred from entry) for all groups. We would thus expect all NSDMRs for all immigrant groups to equal the mean rate or to deviate only slightly from it. If the IS deported any people at a rate higher or lower than expected, this would indicate that the judgment of the IS differed from that of the PHS; that for any particular disease, the race, as defined by the IS, of the immigrant group somehow mattered.

TOTAL MEDICAL CERTIFICATIONS AND EXCLUSIONS

Table 6-1 shows that between 1916 and 1930 the PHS certified Asian and some southern and eastern European immigrants at the highest rates. Within these regions, the NSDMRs for Chinese, Croatians and Slovenians, Slovaks, Roumanians, and Hebrews (the only group in which the distinction between biological race and nation was not blurred) were consistently strongly positive. The NSDMRs for northern Italians, notably, tended to be strongly negative, indicating that certification rates were much lower than average. With the exception of Germans, no northern or western European immigrant group was characterized by anything but a weakly positive NSDMR; but even the NSDMR for Germans was strongly positive only in 1919. Although it is not a stark pattern that draws a crude line between southern and eastern European immigrants and their northern and western counterparts, this is consistent with what we might expect based on prevailing scientific notions of racial inferiority that defined not only Asians but also southern and eastern Europeans as peoples apart. When Asians are removed from the analysis, the pattern for European groups remains unchanged: the PHS overcertified only a very few southern and eastern European immigrants. (This result holds true for all subsequent analyses, none of which is shown here.)

In table 6-2, if the IS were consistently deporting approximately the same proportion of all immigrants, we would expect to see no variation among groups. That is, rates of deportation among those certified would all be equal to each other and thus equal to the mean. But table 6-2 shows that the IS tended to deport several groups at rates much higher than expected based on the average: Asians, Mexicans, and "African Blacks," and, among Europeans, Czechs and Bulgarians, Serbians, and Montenegrins.[75] Among Asians, the IS deported Chinese immigrants and, to some extent, Japanese and Syrian immigrants at rates much higher than average. While the PHS certified both Mexican and African immigrants at near average rates, the IS deported these groups at rates much higher than expected. Disease among these groups mattered more than disease among other groups. That is, the IS assessment of disease among different groups was not equivalent to the PHS assessment. This suggests that biology, which the PHS prized more highly than did the IS, was not sufficiently powerful in drawing the color line in America, in determining whether a group as a whole was fit for industrial citizenship.

This difference in judgment is particularly evident in the case of southern and eastern Europeans. Although we see several pockets of strongly positive NSDMRs here, the IS consistently deported only certified Czechs and Bulgarians, Serbians, and Montenegrins at rates much higher than average (that is, in more than 40% of years). In contrast, Mexican, Chinese, and African immigrants were deported at rates much higher than average in more than 50 percent of all years. In other words, the IS, in employing a broader, lay understanding of race, erased the line that the PHS drew through Europe and redrew it around Europe to more clearly distinguish it from the world of Asia and Latin America.

The major limitation of these data is time span: they allow us to draw conclusions only for the period 1916–30. Fortunately, overall medical deportation rates—based on the total population arrived rather than on the population medically certified—are available from 1899 to 1930. Overall variation in deportation was a product of both certification and deportation practices. As table 6-3 shows for the period 1899–1930, the IS made the biggest difference where the PHS had certified groups at rates much higher than expected, either exaggerating racial differences (as in the case of Asians) or muting differences (as in the case of Hebrews, Croatians and Slovenians, and Slovaks). It also singled out groups that the PHS had certified at rates near average, particularly Syrians and East Indians.

Note, however, that these figures consider certifications and deportations for all medical reasons. In the case of loathsome and dangerous contagious diseases, PHS assessments of disease most strongly shaped the final picture of deportation; in the case of diseases affecting ability to earn a living, IS deportation decisions were most powerful. In other words, the PHS and IS were most likely to agree in the case of loathsome and dangerous contagious diseases, but to assess the importance of group membership and disease differently in the case of diseases affecting ability to earn a living.

LOATHSOME AND DANGEROUS CONTAGIOUS DISEASES

Certifications for loathsome and dangerous contagious disease present a striking racial pattern: Asians, particularly the Chinese, were certified at rates much higher than average (table 6-4). Indeed, the certification rate for the Chinese was so high that almost all other groups were certified at a rate slightly below average. Among a few groups—Turks, Croatians and Slovenians, Mexicans, and Pacific Islanders—we find at least one aberrantly high NSDMR. Nonetheless, with the exception of Pacific Islanders, all rates were nearly average. Likewise, the PHS certified all other groups at approximately the same rate.

The IS consistently excluded all immigrants with Class A certificates at a uniform rate. That is, it consistently acted to deport the vast majority of immigrants with Class A certificates regardless of group membership. Nevertheless, this pattern is not necessarily obvious when all Class A conditions are considered together. The law allowed very few exceptions to the deportation of immigrants with Class A certificates. Thus, the law resulted in average deportation rates close to 100 percent, the maximum possible deportation rate. Because the average was so close to the maximum possible deportation rate, the standard deviations from the mean are unusually small. Thus, aberrations of two or three NSDMRs may be less consequential than they appear. That is, the cap of 100 percent coupled with the high mean exaggerates small deviations from the mean. Thus, while table 6-5 shows many strongly positive NSDMRs, this does not mean that the IS "overdeported" the bulk of Europeans. Consequently, the final picture of deportation for loathsome and dangerous contagious diseases among the total arrived from 1900 to 1930, displayed in table 6-6, parallels that of PHS certifications for 1916–30 (table 6-4): the Chinese, Syrians, and Armenians were most consistently excluded for Class A conditions at rates substantially higher than average, reflecting a no-tolerance policy. With the possible exception of Turks and Mexicans, no other group was consistently deported at a rate substantially greater than the mean.

When selected Class A conditions—trachoma, parasitic infections, tuber-culosis, mental conditions, poor physical development, venereal disease, and pregnancy—are considered separately, we again find distinct patterns of agreement between the PHS and the IS. In the case of trachoma, the most im-portant Class A condition, the PHS certified Armenian, Chinese, and Syrian immigrants at a rate much higher than average (table 6-7). For selected other eastern European groups—Turks fairly regularly, and Croatians and Sloveni-ans more sporadically—the NSDMRs were very strongly positive. Once pre-sented with these immigrants, the IS almost without exception deported all groups at a fairly uniform rate (table 6-8). Thus, the overall rate of deportation for trachoma shows the marked effect of certification on Armenians, Syrians, and Asians, particularly the Chinese (table 6-9). It is telling that trachoma worked primarily as a tool to exclude Asian immigrants, especially in the pe-riod before rigorous laboratory testing of Asians for parasitic infection, as noted in chapter 5. But trachoma was never as conceptually well suited for Asian exclusion as were parasitic infections, for it was explicitly linked to de-pendency and only implicitly associated with industrial capacity. Parasitic in-fection carried profound cultural connotations. For these and, presumably, epidemiological reasons, PHS certifications for parasitic infection were effec-tively limited to Asian populations (table 6-10; data are available for certifica-tions only).

In the case of tuberculosis—a condition for which few certifications were ever made—there was no clear association between any immigrant group and certification rates, though excessive certifications were certainly concentrated among European groups, especially southern and eastern Europeans. Table 6-11 shows very few years in which the NSDMR was strongly positive, re-flecting relatively uniform certification rates for all groups. Similarly, the IS ex-cluded immigrants certified with tuberculosis rather uniformly (table 6-12);

the overall deportation rate reflected no particular association between deportation for TB and group, with the exception that some southern and eastern European groups were deported at rates higher than expected after 1926 (table 6-13). We see only random bursts of inflated deportation rates, and these do not obviously cluster within any particular group or region or the world. Again, as we saw in chapter 5, the PHS made no link between TB and any particular immigrant group, although there was potential to link the disease either to industrial capacity or to biological or genetic degeneration.

For certifications for mental conditions—of which there were relatively few—we find strongly positive NSDMRs among European groups (table 6-14). Within northern and western Europe, the rates for French, Irish, and, to some extent, English and Scotch immigrants were characterized by strongly positive NSDMRs; within southern and eastern Europe, this pattern was evident only among Ruthenians. Syrians, too, showed a pattern of strongly positive NSDMRs. As with tuberculosis, the IS made deportations for mental conditions at a rather uniform rate (table 6-15). The final result, shown in table 6-16, is that the French, Irish, and Germans were most consistently deported for mental conditions at rates much higher than expected. In general, though we see sporadic, strongly positive NSDMRs among northern and western Europeans before the 1920s, thereafter such sporadic rates tend to shift to southern and eastern European groups.

In the case of mental conditions the PHS came the closest to using disease to address questions of civic capacity. But note that the immigrants whom the PHS certified with mental illness were primarily northern or western European groups—those groups already deemed suited for civic participation—rather than the eastern or southern European groups that the PHS viewed primarily as fodder for industry.[76] Given their civic and intellectual compatibility with the nation, the PHS may have felt a greater imperative to evaluate the capacity of these groups for self-rule. In turn, those groups that the PHS saw as fueling the engines of industry were most likely to be turned aside for conditions most seriously hampering ability to work. As a consequence, as shown in tables 6-17 and 6-18, the PHS certified immigrants falling primarily within the middle of the IS racial scale with senility (or old age) and poor physical development (data are available only for PHS certifications for both sets of conditions). The IS racial hierarchy, after all, was a means of ordering society. Though they might also labor, those at the top were fit for participation in democratic society; those in the middle were the nation's key industrial citizens;

those at the bottom were unsuited even for industrial citizenship. Just as certifications for parasitic diseases were concentrated at the bottom of the racial hierarchy, so too were certifications for venereal disease (table 6-19; data available for certifications only).

DISEASES AFFECTING ABILITY TO EARN A LIVING

The PHS certified the vast majority of immigrants not for Class A conditions, but for Class B conditions. Thus, we might expect to find the greatest variation in certification rates by peoples and the greatest divergence of opinion between the PHS and the IS among Class B conditions. After all, the IS and PHS never resolved tension over how to determine whether a disease affected "ability to earn a living" or where the authority for this decision lay.

When certifying immigrants for both Class B and Class C conditions (separate data are not available), the PHS produced a distinct racial portrait (table 6-20). Several southern and eastern European immigrant groups exhibited strongly positive NSDMRs for the period 1916–30: Croatians and Slovenians, Hebrews, southern Italians, Lithuanians, Magyars, Poles, and Slovaks. Asian immigration was already restricted by class, for the Chinese Exclusion Act of 1882, the Gentleman's Agreement of 1907 with Japan and Korea, and the Immigration Act of 1917, which created the Asiatic Barred Zone, all prohibited the immigration of Asian laborers. We see only occasional aberrant NSDMRs among these groups. Disease among Asians, moreover, was indicative not of dependency but of a racial propensity to subvert the American standard of living by working longer and harder for lower wages and then living in substandard conditions.

Ultimately, as a consequence of the many factors that figured in the deportation of immigrants certified with Class B conditions (table 6-21), the many factors that figured in the construction and ordering of groups of people, and the differences in PHS and IS judgment, the overall picture of deportation illustrated in table 6-22 is characterized more by bursts of excessive deportation among all groups (though most heavily concentrated among particular southern and eastern European groups) than by consistent patterns of excessive deportation for select groups (other than East Indians). While we cannot argue that the IS erased the lines drawn by the PHS, in the case of Class B and C conditions the IS muted or at least blurred the line that the PHS drew through

Europe. IS deportation rates for Czechs and to some extent Bulgarians, Serbians, Montenegrins, and Turks were strongly positive, or higher than average.

There is much we cannot know about how the PHS and IS worked together in the case of Class B conditions. While the PHS used only a handful of Class A conditions, many of which we have data for, the agency classified many diseases and conditions as Class B. Unfortunately, we lack the data to look at PHS and IS patterns of certification and exclusion for different Class B conditions by group or people. The only data opening a window onto this question are PHS certifications for pregnancy (table 6-23). Pregnancy could signal both reproduction of the working class and, like sexually transmitted diseases, the lack of control necessary for living in a civilized society. Higher than expected rates of certification for pregnancy are mostly concentrated in the middle tier of the IS hierarchy of races and peoples, with rates for Syrian and Japanese immigrants standing out.

Categorization Reconsidered

The Immigration Restriction Act of 1924 represented the culmination of American nativism. The legislation limited immigration to 150,000 people per year and restricted immigration to 2 percent of the number of each "race" recorded in the U.S. Census of 1890. It represented a deliberate attempt to "correct" the quota system put forth in 1921, which was based on the 1910 census and allotted 45 percent of the quota to southern and eastern European immigrants—a proportion far too generous, in the minds of those favoring immigration restriction. The 1924 adjustment substantially reduced the quota for immigration from southern and eastern European countries, granting those countries 16 percent of the total, while allotting northern and western Europe 84 percent of the total.[77] John Higham accordingly argues, "Immigration restriction marked both the climax and conclusion of an era of nationalistic legislation."[78] Some historians now argue that it marked not the height but rather the fundamental transformation of American nationalism.

Eugenics figured prominently in the passage of the nation's immigration restriction laws in the 1920s. It served as the vehicle through which the cultural differences of southern and eastern Europeans were translated into a genetic threat to the blood of the nation. At the apex of eugenic influence in the United States, those favoring immigration restriction successfully argued that the new, culturally unfamiliar immigrants were inferior to the "native" American population; they threatened to pollute America's superior genetic stock with inferior "germ plasma."

Rudolph Vecoli has observed that the literature of American immigration characteristically turns on the restrictive legislation of the 1920s.[79] This is true of the literature of both nativism and eugenics, in which the Immigration Restriction Act of 1924 looms large. Questions of racial inferiority were central to the passage of this legislation, which selectively restricted immigration. But while eugenics is an appealing way to explain this legislative dénouement of immigration restriction, we must remember that 1924 is unique. Questions of immigration and genetic inferiority crystallized in a manner not necessarily representative of the period leading up to this moment in history; one particular and relatively fixed racial ideology briefly coalesced in an uncharacteristically coherent fashion. Thus, by taking 1924 as the end-point, historians tend to tell the history of immigration restriction, implicitly or explicitly, from 1924 backward, linking all policies and sciences of exclusion narrowly to scientific racism.

As the story of immigration restriction played out in the late 1920s, the impetus behind maintaining genetic purity dissipated and the coalition in favor of restriction began to fragment. The Immigration Restriction Act of 1924 was only ever a temporary measure. The legislation specified that in 1927 the 1890 census would no longer serve as the basis for determining the maximum number of immigrants from any given nation. Instead, quotas were to be based on the national origins of the population in 1920.[80] Yet establishing quotas based on national origins was, as it turned out, an extremely complicated ethnological and political task.

The 1920 census did not merely replace the 1890 census as the basis for the quotas. The 1890 census was the first that listed the country where an individual's parents were born. Nevertheless, the 1924 legislation required a more rigorous determination of "nativity." Robert Divine, in his analysis of the much ignored national origins debates of the late 1920s, explains in detail the different methods for determining the national origins of the U.S. population that the Quota Board, charged with finding a means to determine the racial mix of the U.S. population from the 1920 census, used in this task. In essence, the board determined national origin based on surnames, using either the census of 1790 or the 1909 census publication, *A Century of Population Growth,* to estimate the proportion of the population descended from "native stock" (defined by descent from immigrants arriving in the United States before 1790) and "immigrant stock" (defined as descent from immigrants arriving after 1790) and then to determine the national origins of those descended from immigrant stock. On top of these genealogical and ethnographic complications, World War I changed the political topography of Europe, making assessments of national origin more complex.[81]

After 1924, while the question of racial nationalism or racial homogeneity remained paramount, the issue was drastically reconfigured in a way not so easily explained in terms of eugenics or biological superiority and inferiority. When Congress took up the question of how to establish the new 1927 quotas, several schemes were considered that left the quotas for southern and eastern Europeans virtually unaffected. Instead, in many of these schemes, northwestern European immigrants suffered numerical "setbacks." Germany, the Irish Free State, Norway, and Sweden found their numbers substantially reduced from the quotas based on the 1890 census in the original 1924 legislation. Different Quota Boards worked different and complex alterations on the method of determining national origin, sparking debates over the usefulness of these figures.[82]

In the bitter debate that ensued, Senator Albert Johnson—who explicitly crafted the 1924 restriction act as a system of eugenically construed racial quotas—withdrew his support for the national origins system when faced with the prospect that the coalition supporting immigration restriction would splinter along national lines. Johnson instead endorsed "the gradual elimination of all immigration."[83] Most important, in the end, was limiting the number of immigrants rather than strictly limiting immigration from northern and western Europe; though it was still accorded the lion's share of immigration, northern and western Europe's share of the quota was ultimately reduced from 84 percent to 67 percent. Southern and eastern Europe's share rose from 16 percent to 33 percent.

Divine's study is really an analysis of how interest-group politics came to bear on different quota schemes. Mae Ngai takes his analysis a step farther, arguing that while the final quotas were manipulated in telling ways from 1924 to 1929 (the year they received presidential approval), from the outset the 1924 Immigration Restriction Act intended to distinguish Europeans from all other immigrants—to distinguish "whites" from "nonwhites," drawing a line around, not through, Europe.[84] She notes that the law not only mandated a quota system, but also explicitly excluded "non-white" peoples deemed "ineligible for citizenship or their descendants" as well as "descendants of slave immigrants," and the "descendants of American aborigines."[85] As the Quota Board—chaired by Dr. Joseph A. Hill, the chief statistician of the U.S. Census Bureau—attempted to establish the national origins of the population, it made key decisions that further reinforced the intent of immigration restriction to "[construct] a white American race, in which persons of European descent shared a common whiteness distinct from those deemed to be not white."[86]

Specifically, writes Ngai, the quota board "discounted from the population all blacks and mulattos . . . all Chinese, Japanese, and South Asian persons 'ineligible to citizenship,' including those with American citizenship by native-birth." The provision also excluded the Territories of Hawaii, Puerto Rico, and Alaska, which came under U.S. immigration law and whose natives were U.S. citizens. In other words, Ngai argues, "to the extent that the inhabitants of the continental United States in 1920 constituted a legal representation of the American nation, the law excised all non-white, non-European peoples from that vision, erasing them from the American nationality." The effect of "eliminating non-whites from the formula resulted in larger quotas for European countries and smaller ones for other countries." Ngai thus concludes that "while the national origins quota system intended principally to restrict immigration from southern and eastern Europe and used the notion of national origins to justify discrimination against immigrants from those nations, it did more than divide Europe. It also divided Europe from the non-European world. It defined the world formally in terms of country and nationality but also in terms of race." While Europeans as a whole gradually became accepted as "white" in the years following 1924, the "nonwhite" Asian and Latin American groups were cast as "foreign," "alien," "unassimilable." They thus became the objects of concern as the United States shifted the focus of its immigration policy from exclusion on arrival to deportation of "illegal" immigrants or immigrants who had proven themselves undesirable or who were biologically undesirable.[87]

The way in which the PHS and IS worked together complements and complicates this story of the remapping of the nation in the 1920s. As Matthew Frye Jacobson points out, science and the state represent the primary alchemists creating and enforcing social conceptions of difference among peoples.[88] In the context of immigration, the PHS and IS both uniquely grounded their construction of difference in concerns over the immigrant as laborer, yet emphasized different components or elements—the PHS emphasizing fixed elements, the IS prioritizing the more mutable, socioeconomic aspects of difference. They thus illustrate the ways in which science and the state could work in harmony and in opposition. But while both conceptions of group difference could result in drawing a line, in one instance, through Europe and, in another, around Europe, numerically speaking, the line drawn around Europe prevailed.

Working together, with a shared notion of the importance of the immigrant as laborer, the PHS and IS tended to draw a line around Europe as a whole, setting it apart from Asian and Mexican peoples (Chinese, Japanese, "Hindoo," and Syrians, "coolies" associated in the minds of the IS with either Mexicans or other groups of Asian immigrants). These were primarily the groups deemed unassimilable by reason of disease burden and work ethic, as in the case of Chinese and Japanese immigrants, who were associated with parasitic diseases and parasitic laborers who would sap the strength of the nation. These were either the groups or peoples that the gaze of the PHS could not penetrate (again, the Chinese and Japanese) or those that were not entirely distinguishable from other immigrant groups by appearance (Mexicans and Syrians). Most important, these were the peoples unsuited for industrial citizenship.

Depending on the question at hand, immigrants could be mapped in different ways. When questions of fitness for self-rule in a democratic society were at stake, the proper contours of the nation could be mapped very differently than when questions of industrial citizenship were at stake. If we look forward from 1899—the year for which we have the first statistics reflecting how the United States drew the color line at the nation's borders and the point at which the PHS, IS, and the state began to concern themselves with immigration and industrial citizenship—we find a similar mapping of the nation that Mae Ngai has identified in the 1920s. A color line distinguishing Europeans from Asians and Mexicans was evident in the statistics of the PHS and IS; it was evident in the type of exam the PHS conducted and the diseases it prioritized at the nation's different land and sea borders.

Does this mean that the Irish and Jews were always white, or does it mean that they became white earlier than we had imagined?[89] It may be that while the IS and PHS clearly saw Europe in terms of many distinct and unequal races, they never viewed any particular European group as "nonwhite." But the notion of whiteness, though rarely explicitly employed, did have some salience for the agencies and was useful in marking Europe off from other parts of the world.[90] For example, although the PHS used the term *white* just as frugally, it used the term not as a means of underscoring European differences, but of differentiating Europeans from Asians.[91] Similarly, writing to Secretary of Commerce and Labor Oscar Straus, who sought data on the number of immigrants who actually remained in the United States, Ellis Island commissioner Robert Watchhorn stated, referring to all European immigration, "If it is desirable that the whites should predominate in numbers as they excel in enterprise and executive capacity, this white immigration, which seems to fill some people with fearful forebodings, is very essential to the maintenance of numerical supremacy of the whites."[92]

Evidence relating to the PHS and IS will in no fashion resolve the emerging debate over whiteness, its meaning, and its usefulness as an analytic category, but it helps to underscore that there were different ways to map the nation, to map who legitimately belonged and who did not. There were different criteria by which to group and then weigh the value of immigrants. The immigrant medical examination represented an important means of mapping and defining the contours of legitimate industrial citizenship, of determining which groups were fit to contribute to America's industrial economy. It was a significant national endeavor that set European immigrants apart from Asians and from Mexicans and other Latin American peoples. These patterns were formalized in the more sweeping restriction legislation of the 1920s. But once the exclusionary intentions of the exam were cemented in the Immigration Restriction Act of 1924, this formative role of the exam would be eclipsed.

Tables for this chapter follow on pages 221–52.

TABLE 6-1. Deviations in Rate of Medical Certification by Race, 1916–1930, Expressed as Rate of Certification and Number of Standard Deviations from Mean Rate of Medical Certification (NSDMR)

Race or People	1916		1917		1918		1919		1920		1921		1922		1923		1924		1925		1926		1927		1928		1929		1930	
	%	NSDMR	%	NSDMR	%	NSDMR	%	NSDMR	%	NSDMR	%	NSDMR	%	NSDMR	%	NSDMR	%	NSDMR	%	NSDMR	%	NSDMR	%	NSDMR	%	NSDMR	%	NSDMR	%	NSDMR
English	3.7	(0.0)	4.8	(-0.1)	5.9	(-0.1)	2.5	(-0.1)	1.7	(-0.3)	2.1	(-0.7)	2.5	(-0.8)	2.1	(-0.6)	1.7	(-0.6)	4.2	(-0.2)	3.2	(-0.2)	5.0	(-0.1)	5.4	(0.0)	2.5	(-0.3)	2.7	(-0.2)
Irish	4.8	(0.0)	5.0	(0.0)	7.5	(0.6)	4.2	(0.5)	2.0	(0.0)	2.3	(-0.5)	3.7	(-0.4)	2.9	(-0.3)	2.1	(-0.4)	5.5	(-0.1)	7.9	(0.5)	5.2	(0.0)	3.8	(-0.2)	2.5	(-0.3)	2.6	(-0.2)
Scotch	4.0	(0.0)	4.6	(-0.1)	5.8	(0.1)	3.4	(0.2)	2.0	(0.0)	2.3	(-0.6)	2.8	(-0.7)	2.1	(-0.6)	1.9	(-0.5)	3.7	(-0.3)	2.8	(-0.2)	4.3	(-0.2)	4.2	(-0.1)	2.4	(-0.4)	1.7	(-0.4)
Welsh	3.3	(-0.1)	6.3	(0.3)	8.9	(0.9)	2.4	(-0.2)	2.9	(0.6)	2.2	(-0.6)	2.1	(-1.0)	2.5	(-0.4)	2.2	(-0.4)	4.8	(-0.2)	5.3	(0.1)	6.9	(0.2)	5.2	(0.0)	2.3	(-0.4)	1.1	(-0.5)
Spanish	2.0	(-0.2)	2.1	(-0.7)	2.1	(-0.8)	0.7	(-0.9)	1.0	(-0.9)	1.4	(-1.1)	1.5	(-1.2)	1.3	(-1.0)	2.0	(-0.5)	11.4	(0.6)	1.3	(-0.4)	9.0	(0.4)	10.3	(0.4)	1.4	(-0.7)	2.1	(-0.3)
Portuguese	4.4	(0.0)	6.6	(0.4)	3.2	(-0.5)	1.3	(-0.6)	1.3	(-0.6)	1.6	(-1.0)	3.1	(-0.6)	2.2	(-0.3)	1.8	(-0.6)	10.1	(0.5)	2.0	(-0.3)	10.1	(0.5)	6.4	(0.1)	1.8	(-0.6)	2.7	(-0.4)
Italian (North)	3.3	(-0.1)	3.8	(-0.3)	5.6	(0.1)	2.2	(-0.3)	1.5	(-0.4)	1.1	(-1.2)	1.2	(-1.3)	0.6	(-1.3)	0.7	(-1.0)	4.3	(-0.2)	1.4	(-0.4)	2.2	(-0.4)	3.6	(-0.2)	3.5	(-0.6)	**9.0**	**(1.1)**
Italian (South)	7.2	(0.2)	7.6	(0.6)	7.5	(0.6)	2.2	(-0.3)	3.1	(0.8)	4.2	(0.4)	7.3	(0.9)	5.3	(0.8)	4.4	(0.5)	4.9	(-0.2)	6.5	(0.3)	13.5	(0.9)	10.6	(0.4)	**6.6**	**(1.0)**	8.2	(0.9)
Dutch and Flemish	3.3	(-0.1)	3.5	(-0.5)	4.4	(-0.2)	1.7	(-0.5)	1.9	(-0.2)	1.8	(0.4)	2.4	(-0.9)	2.1	(-0.6)	1.6	(-0.6)	4.9	(-0.2)	3.0	(-0.2)	7.9	(0.3)	9.4	(0.3)	3.5	(0.0)	3.2	(-0.1)
Finnish	2.5	(-0.1)	2.7	(-0.5)	3.5	(-0.4)	2.1	(-0.3)	2.5	(0.3)	3.4	(0.0)	4.7	(0.0)	3.1	(-0.2)	3.1	(0.0)	6.3	(0.0)	2.4	(-0.3)	12.9	(0.5)	11.9	(0.5)	2.8	(-0.2)	2.3	(-0.3)
Scandinavian*	3.1	(-0.1)	3.7	(-0.3)	5.2	(0.0)	2.2	(-0.2)	1.8	(-0.2)	2.6	(-0.4)	3.7	(-0.4)	2.2	(-0.6)	2.7	(-0.2)	5.0	(-0.1)	3.2	(-0.2)	5.7	(0.0)	7.0	(0.1)	3.7	(0.1)	3.6	(0.0)
Lithuanian	3.6	(-0.1)	6.5	(0.5)	6.7	(0.4)	4.3	(0.6)	2.6	(0.4)	5.7	(1.3)	7.0	(0.8)	4.7	(0.6)	6.7	(1.6)	23.0	(2.1)	10.7	(0.8)	18.0	(1.4)	25.2	(1.6)	9.9	(1.9)	9.6	(1.2)
Russian	3.1	(-0.1)	3.1	(-0.5)	4.0	(-0.3)	2.6	(-0.1)	2.1	(0.0)	1.8	(-0.9)	2.8	(-0.7)	2.7	(-0.3)	3.3	(-0.1)	5.9	(0.0)	2.7	(-0.2)	7.8	(0.3)	6.4	(0.1)	3.0	(-0.2)	2.7	(-0.2)
Polish	4.2	(0.0)	4.9	(0.0)	5.3	(0.0)	3.8	(0.4)	0.6	(-1.2)	2.1	(-0.7)	6.0	(0.5)	5.6	(1.0)	6.0	(1.3)	13.2	(0.8)	9.4	(0.7)	12.6	(0.8)	12.5	(0.5)	11.2	(2.5)	6.1	(0.5)
German	4.0	(0.0)	4.6	(-0.1)	7.5	(0.6)	6.1	(1.3)	2.3	(0.2)	3.9	(0.3)	6.1	(0.5)	5.0	(0.7)	4.9	(0.4)	7.4	(0.1)	5.6	(0.2)	4.6	(0.1)	5.2	(0.1)	2.6	(-0.3)	3.1	(-0.1)
Magyar	5.3	(0.1)	4.3	(-0.2)	7.9	(0.7)	6.9	(1.4)	6.2	(3.2)	4.9	(0.8)	6.8	(0.8)	4.7	(0.6)	7.1	(1.7)	12.7	(0.7)	8.6	(0.6)	17.8	(1.4)	12.7	(0.6)	6.2	(0.9)	7.1	(0.7)
Slovak	3.3	(-0.1)	3.6	(-0.3)	5.3	(0.3)	4.1	(0.5)	2.0	(-0.1)	5.3	(1.0)	10.5	(2.1)	9.0	(2.5)	10.1	(3.0)	42.5	(4.5)	46.7	(5.6)	46.7	(4.8)	16.5	(0.9)	6.9	(1.1)	2.0	(-0.3)
Croatian and Slovenian	3.8	(0.0)	4.3	(-0.3)	13.2	(2.0)	10.9	(3.3)	2.0	(-0.1)	1.1	(-1.2)	4.5	(-0.1)	6.1	(1.2)	6.9	(1.6)	22.4	(2.0)	12.9	(1.1)	35.8	(2.5)	25.2	(1.6)	22.1	(2.9)	0.5	(-0.6)
Czech†	6.1	(0.1)	5.3	(0.1)	7.7	(0.6)	3.6	(0.3)	1.7	(0.0)	0.8	(-1.4)	1.1	(-1.3)	1.7	(-0.8)	4.1	(0.4)	1.0	(-0.7)	0.1	(-0.6)	0.4	(-0.6)	1.6	(-0.3)	0.5	(-1.0)	20.7	(3.5)
Dalmatian, Bosnian, Herzegovinian			3.0	(-0.5)		(0.0)	9.1	(2.5)			0.4	(-1.6)	5.7	(0.3)	1.4	(-0.9)	5.2	(0.3)										(0.0)	10.9	(1.5)
Ruthenian (Russniak)	4.5	(0.0)	6.4	(0.3)	10.1	(1.2)	2.4	(-0.2)	3.3	(0.9)	2.4	(-0.5)	2.6	(-0.3)	1.6	(-0.9)	1.8	(-0.6)	3.0	(-0.4)	1.9	(-0.3)	9.4	(0.4)	9.2	(0.3)	2.8	(0.0)	1.5	(-0.4)
Roumanian	5.1	(0.0)	6.8	(0.4)	4.1	(-0.3)	3.9	(0.4)	1.2	(-0.7)	2.5	(-0.5)	5.2	(0.2)	5.8	(1.1)	5.9	(1.2)	16.6	(1.3)	7.9	(0.5)	18.5	(1.5)	18.3	(1.0)	5.5	(0.7)	9.5	(1.2)
Bulgarian, Serbian, Montenegrin	4.6	(0.0)	4.3	(-0.2)	7.2	(0.5)	2.7	(0.0)	1.5	(-0.5)	1.2	(-1.2)	3.7	(-0.4)	3.4	(0.0)	1.8	(-0.5)	5.9	(0.0)	1.9	(-0.3)	6.8	(0.1)	6.8	(0.1)	1.5	(-0.6)	20.6	(3.4)
Greek	4.0	(0.0)	3.7	(-0.2)	6.1	(0.2)	1.9	(-0.4)	2.3	(0.4)	3.1	(-0.1)	4.6	(-0.1)	3.8	(0.4)	4.2	(0.5)	12.2	(0.7)	3.2	(-0.2)	8.5	(0.3)	6.1	(0.0)	3.2	(-0.1)	3.7	(0.0)
Armenian	6.9	(0.2)	7.0	(0.5)	5.0	(0.1)	1.8	(-0.4)	2.6	(0.4)	4.4	(0.6)	5.9	(0.4)	6.5	(1.4)	6.3	(1.4)	9.2	(0.4)	5.8	(0.2)	7.3	(0.2)	6.0	(0.0)	4.3	(0.3)	4.6	(0.3)
Hebrew	5.1	(0.0)	5.2	(0.1)	5.1	(0.0)	4.5	(0.7)	3.7	(1.3)	7.0	(1.9)	10.9	(2.3)	9.2	(2.6)	7.5	(1.3)	17.8	(1.4)	14.2	(1.3)	12.8	(0.8)	11.9	(0.5)	8.4	(1.6)	10.0	(1.3)
Syrian	10.6	(0.4)	7.5	(0.6)	14.2	(2.2)	2.8	(0.0)	3.4	(1.0)	4.9	(0.8)	4.3	(-0.2)	4.5	(0.5)	4.4	(0.6)	10.9	(0.6)	6.1	(0.2)	8.2	(0.3)	6.9	(0.1)	5.1	(0.5)	4.5	(0.2)
French	3.4	(-0.1)	3.4	(-0.4)	6.6	(0.3)	3.3	(0.2)	2.0	(0.0)	2.4	(-0.5)	2.8	(-0.7)	2.5	(-0.4)	2.0	(-0.5)	4.1	(-0.1)	2.9	(-0.2)	4.3	(-0.1)	4.1	(-0.1)	2.4	(-0.4)	1.8	(-0.4)
Turkish	7.9	(0.4)	3.2	(-0.4)	3.8	(-0.3)	1.9	(-0.4)	3.0	(0.7)	7.7	(3.2)	10.6	(2.2)	4.3	(0.4)	6.1	(1.3)	5.3	(-0.1)	2.9	(-0.2)	9.8	(0.5)	7.0	(0.1)	1.3	(-0.7)	4.0	(0.1)
Mexican	4.2	(0.0)	6.3	(0.3)	2.3	(-0.7)	5.0	(0.9)	1.6	(-0.4)	1.4	(-1.1)	1.2	(-1.3)	1.1	(-1.1)	1.4	(-0.7)	2.9	(-0.4)	0.9	(-0.5)	2.6	(-0.3)	3.2	(-0.2)	4.4	(0.3)	4.6	(0.2)
Spanish American	1.7	(-0.2)	0.8	(-1.0)	1.3	(-1.0)	0.6	(-0.9)	0.7	(-1.1)	0.6	(-1.5)	0.9	(-1.4)	1.0	(-1.1)	1.2	(-0.8)	2.9	(-0.4)	0.9	(-0.5)	2.5	(-0.4)	2.3	(-0.3)	0.8	(-0.9)	0.6	(-0.7)
Cuban	0.7	(-0.2)	0.8	(-1.0)	1.1	(-1.0)	0.2	(-1.1)	0.2	(-1.4)	0.2	(-1.7)	0.5	(-1.6)	0.7	(-1.3)	0.7	(-1.0)	4.3	(-0.2)	0.5	(-0.5)	2.3	(-0.4)	1.7	(-0.3)	0.6	(-1.0)	0.3	(-0.7)
West Indian	3.2	(-0.1)	2.0	(-0.7)	2.0	(-0.8)	0.7	(-0.9)	0.4	(-1.4)	0.2	(-1.7)	0.8	(-1.5)	0.5	(-1.4)	0.4	(-1.2)	1.1	(-0.6)	0.6	(-0.5)	1.6	(-0.5)	2.3	(-0.3)	0.5	(-1.0)	0.6	(-0.6)
Chinese	19.5	(1.0)	26.1	(5.0)	18.5	(3.3)	4.1	(0.5)	2.5	(0.3)	2.8	(-0.0)	5.8	(0.4)	6.4	(1.4)	6.0	(0.3)	19.2	(1.6)	3.4	(-0.1)	26.4	(2.4)	60.2	(4.4)	8.6	(1.7)	4.4	(0.2)
East Indian	8.1	(0.2)	11.6	(1.6)	11.3	(1.5)	0.2	(-1.1)	0.3	(-1.4)	1.4	(-1.0)	2.5	(-0.8)	4.1	(0.3)	3.3	(0.1)	7.7	(0.2)	2.4	(0.0)	5.9	(0.0)	5.3	(0.0)	2.9	(-0.2)	2.9	(0.0)
Japanese	7.3	(0.2)	10.3	(1.3)	9.4	(1.0)	4.8	(0.8)	5.7	(2.5)	5.1	(0.9)	2.9	(-0.7)	2.7	(-0.3)	3.2	(0.0)	14.7	(1.0)	0.8	(-0.5)	6.8	(0.1)	29.7	(1.9)	1.6	(-0.6)	1.4	(-0.5)
Korean	6.7	(0.1)	12.2	(1.7)	7.9	(0.7)	2.9	(0.0)	2.4	(0.3)	2.4	(-0.5)	2.8	(-0.7)	5.0	(0.7)	6.3	(1.4)			7.0	(0.3)	12.8	(0.8)	27.3	(1.7)	5.1	(1.5)	4.1	(0.1)
Pacific Islander	10.0	(6.2)	10.0	(1.2)	16.7	(2.8)	10.0	(2.9)	3.6	(1.2)	0.0	(-1.8)	8.0	(1.2)	2.3	(-0.5)	1.9	(-0.5)	10.0	(0.5)	1.9	(-0.3)		(0.0)	50.0	(3.5)		(0.0)	14.3	(2.2)
African (black)	3.8	(0.0)	2.7	(-0.5)	2.4	(-0.7)	1.0	(-0.8)	0.6	(-1.2)	1.1	(-1.2)	1.7	(-1.1)	1.3	(-1.0)	1.4	(-0.7)	8.3	(0.2)	1.8	(-0.3)	6.9	(0.2)	5.2	(0.0)	1.1	(-0.8)	1.1	(-0.5)
Other	4.4	(0.0)	4.5	(-0.1)	2.3	(-0.7)	3.1	(0.1)	3.2	(0.8)	2.0	(-0.7)	2.6	(-0.3)	5.3	(0.8)	4.2	(0.5)	10.6	(0.5)	4.5	(0.0)	11.4	(0.7)	11.2	(0.4)	6.9	(1.1)	8.2	(0.9)

Bold = 1 or more NSDMR below mean   = 1 to 1.99 NSDMRs above mean   = 2 or more NSDMRs above mean

* Norwegians, Danes, and Swedes.   † Bohemians and Moravians.

TABLE 6-2. Deviations in Rate of Medical Deportations of Those Certified, by Race, 1916–1930, Expressed as Rate of Deportation and NSDMR

Race or People	1916 %	NSDMR	1917 %	NSDMR	1918 %	NSDMR	1919 %	NSDMR	1920 %	NSDMR	1921 %	NSDMR	1922 %	NSDMR	1923 %	NSDMR	1924 %	NSDMR	1925 %	NSDMR	1926 %	NSDMR	1927 %	NSDMR	1928 %	NSDMR	1929 %	NSDMR	1930 %	NSDMR
English	19.5	-0.2	21.0	-0.1	12.2	-0.2	7.7	-0.1	7.5	-0.1	6.1	(0.1)	6.9	(0.0)	6.6	-0.1	7.1	-0.2	3.8	-0.2	3.9	(0.0)	2.6	(0.0)	0.2	(0.0)	0.4	-0.5	0.4	-0.1
Irish	21.7	-0.1	23.9	(0.0)	15.5	(0.0)	12.3	-0.1	6.3	-0.2	6.2	(0.1)	10.8	(0.4)	8.5	(0.1)	7.6	-0.1	4.1	(0.1)	1.3	-0.2	1.9	-0.1	0.3	-0.2	0.3	-0.6	0.7	-0.1
Scotch	17.3	-0.3	24.5	(0.1)	14.8	(0.1)	11.4	-0.1	7.2	-0.1	5.7	(0.1)	7.4	(0.0)	3.4	-0.5	5.4	-0.3	4.4	-0.1	5.8	-0.1	2.4	-0.1	0.7	-0.2	0.6	-0.6	0.7	-0.1
Welsh	23.5	(0.0)	13.2	-0.6	18.5	(0.1)	12.5	-0.1	3.1	-0.4	3.8	-0.2	15.2	(0.8)	3.4	-0.5	10.8	(0.2)	1.5	-0.4	2.5	-0.1	2.2	-0.1	0.0	-0.4	0.0	-0.7	3.1	(0.3)
Spanish	42.5	(1.2)	35.1	(0.7)	36.4	(1.0)	22.6	(0.4)	19.0	(0.9)	21.8	(2.3)	11.8	(0.5)	12.3	(0.5)	12.7	(0.4)	6.7	(0.1)	5.4	(0.1)	5.2	(0.2)	0.0	(0.4)	0.0	-0.7	2.6	(0.3)
Portuguese	14.4	-0.5	8.1	-0.9	16.0	(0.0)	21.4	(0.3)	2.8	-0.5	7.9	(0.4)	3.4	-0.4	7.1	-0.1	11.2	(0.2)	14.8	(0.9)	6.8	(0.1)	8.2	(0.4)	8.5	(3.5)	0.0	-0.7	3.0	(0.2)
Italian (North)	21.0	-0.1	14.9	-0.5	14.3	-0.1	18.5	(0.1)	3.7	-0.4	8.9	(0.5)	21.6	(1.5)	19.6	(2.4)	23.5	(1.4)	16.1	(1.1)	11.8	(0.4)	10.2	(0.6)	5.2	(0.7)	1.5	-0.1	2.0	(0.3)
Italian (South)	17.2	-0.3	17.4	-0.3	5.6	-0.5	15.8	(0.1)	2.2	-0.5	4.4	-0.1	6.4	-0.1	5.2	-0.3	6.5	-0.2	4.3	-0.1	1.2	-0.2	1.1	-0.2	1.1	(0.2)	0.3	-0.6	0.4	-0.2
Dutch and Flemish	15.1	-0.5	10.4	-0.7	7.0	-0.4	5.3	-0.4	4.3	-0.3	3.3	-0.3	5.8	-0.1	6.3	-0.2	8.6	(0.0)	5.1	(0.0)	0.9	-0.2	1.6	-0.1	0.7	-0.2	0.3	-0.4	0.0	-0.2
Finnish	28.3	(0.3)	18.6	-0.3	22.4	(0.3)	6.4	-0.4	5.2	-0.3	2.5	-0.4	5.7	-0.1	7.6	(0.0)	3.9	-0.1	4.3	-0.1	8.7	(0.2)	11.1	(0.7)	3.1	(0.2)	0.4	-0.4	0.0	-0.2
Scandinavian*	12.8	(0.3)	12.6	-0.3	8.5	-0.4	3.7	-0.5	2.1	-0.5	3.3	-0.5	6.0	-0.1	4.0	-0.4	3.9	-0.5	3.2	-0.2	3.4	(0.0)	2.0	-0.1	0.4	-0.3	0.3	-0.6	0.3	(0.0)
Lithuanian	12.5	-0.6	12.1	-0.6	9.1	-0.3	10.0	-0.2	0.0	-0.7	1.9	-0.5	8.9	(0.2)	5.6	-0.2	4.4	-0.4	3.3	-0.3	1.2	-0.2	0.0	-0.2	0.0	-0.4	0.2	-0.2	1.2	(0.0)
Russian	24.1	(0.1)	30.1	(0.5)	17.9	(0.1)	12.3	-0.1	14.5	(0.4)	8.1	(0.4)	10.1	(0.3)	6.0	-0.2	4.4	-0.4	11.2	(0.6)	7.1	(0.2)	5.2	(0.2)	0.0	-0.4	3.1	(0.6)	0.6	(0.0)
Polish	19.0	-0.2	24.4	(0.1)	11.1	-0.1	11.1	-0.1	7.5	(0.1)	3.3	-0.3	4.1	-0.3	4.8	-0.3	3.0	-0.6	1.6	-0.4	1.3	-0.2	0.7	-0.2	0.2	-0.4	1.8	(0.0)	0.4	-0.1
German	14.3	-0.5	15.3	-0.5	21.4	(0.3)	11.4	-0.1	11.2	(0.2)	3.4	-0.3	3.9	-0.3	3.2	-0.5	3.2	-0.5	2.0	-0.2	2.0	-0.1	1.0	-0.2	0.5	-0.3	0.3	-0.3	0.4	-0.1
Magyar	13.0	-0.6	15.0	-0.5	0.0	-0.8	0.0	-0.6	18.2	(0.6)	3.5	-0.2	4.6	-0.3	4.1	-0.4	5.8	-0.3	2.9	-0.2	1.8	-0.1	1.1	-0.2	2.1	(0.0)	1.0	-0.3	0.4	-0.1
Slovak	0.0	-1.4	11.1	-0.7	0.0	-0.8	16.7	(0.0)	2.0	-0.5	2.3	-0.4	5.4	-0.2	2.3	-0.6	4.9	-0.3	0.5	-0.5	0.5	-0.2	0.6	-0.2	0.3	-0.2	0.0	-0.7	0.0	-0.2
Croatian and Slovenian	36.4	(0.8)	28.6	(0.3)	0.0	-0.8	60.0	(2.0)	8.0	(0.0)	12.3	(1.0)	5.6	-0.2	3.2	-0.5	4.9	-0.4	2.1	-0.3	1.3	-0.2	1.4	-0.1	3.0	(0.2)	2.6	(0.4)	22.2	(3.9)
Czech †	5.0	-1.1	5.6	-1.0	42.9	(1.3)	30.0	(0.7)	10.0	(0.1)	0.0	-0.7	23.5	(1.6)	3.7	-0.5	2.2	-0.6	27.5	(2.2)	100.0	(5.4)	40.0	(3.0)	0.0	-0.4	8.3	(2.8)	0.0	-0.2
Dalmatian, Bosnian, Herzegovinian	25.0	(0.1)	66.7	(2.4)	0.0	(0.0)	0.0	(0.0)	0.0	(0.0)			20.0	(1.3)					0.0	(0.0)			9.5	(0.5)			0.0	(0.0)		
Ruthenian (Russniak)	16.2	-0.4	24.2	(0.0)	0.0	-0.8	100.0	(3.9)	26.7	(1.5)	10.3	(0.7)	4.2	-0.3	34.5	(3.0)	18.8	(0.9)	15.6	(1.0)	18.2	(0.8)	9.5	(0.5)	0.0	(0.0)	0.0	-0.7	0.0	(0.0)
Roumanian	27.8	(0.3)	35.0	(0.3)	28.6	(0.6)	33.3	(0.8)	13.6	(0.4)	5.9	(0.1)	7.2	(0.0)	6.1	-0.2	6.9	-0.2	7.0	(0.2)	1.4	-0.2	3.8	(0.1)	1.2	-0.1	1.4	-0.1	1.0	(0.0)
Bulgarian, Serbian, Montenegrin	50.3	(1.6)	43.4	(1.1)	50.0	(1.7)	15.4	(0.0)	20.9	(1.0)	11.7	(0.9)	10.8	(0.4)	5.1	-0.3	22.2	(1.3)	14.7	(0.9)	7.1	(0.2)	0.0	-0.2	16.7	(3.1)	4.3	(1.1)	0.0	-0.2
Greek	38.1	(0.9)	25.8	(0.1)	15.2	(0.9)	33.3	(0.0)	6.9	-0.1	9.3	(0.6)	4.8	(0.1)	6.3	-0.2	6.3	-0.2	6.5	(0.1)	2.8	-0.1	0.9	-0.1	2.3	-0.1	0.5	(0.5)	3.5	(0.4)
Armenian	40.0	(1.0)	36.7	(0.7)	33.3	(0.9)	0.0	-0.6	1.3	-0.6	5.4	(0.1)	8.4	(0.1)	4.4	-0.2	4.1	-0.5	2.1	-0.3	6.3	(0.1)	6.9	(0.3)	1.6	-0.1	1.8	(0.5)	8.3	(1.1)
Hebrew	12.9	-0.6	9.0	-0.8	15.1	(0.1)	9.7	-0.2	4.0	-0.4	1.5	-0.5	3.5	-0.4	4.4	-0.4	4.7	-0.5	15.6	(1.2)	2.0	-0.1	1.6	-0.1	0.4	-0.1	0.4	-0.4	0.7	-0.1
Syrian	40.0	(1.0)	56.3	(1.8)	60.5	(2.2)	28.6	(0.6)	9.4	(0.1)	9.9	(0.7)	11.1	(0.4)	14.4	(0.7)	11.7	-0.2	12.9	(0.7)	9.4	(0.3)	1.8	-0.1	2.4	-0.1	0.1	-0.6	6.6	(0.8)
French	24.8	(0.1)	23.0	(0.0)	12.4	-0.2	15.1	(0.0)	12.4	(0.3)	14.4	(1.3)	17.9	(1.1)	18.8	(1.2)	13.5	(0.4)	12.9	(0.7)	13.5	(0.5)	6.1	(0.3)	0.7	(0.1)	0.3	(0.0)	3.0	(0.3)
Turkish	61.1	(2.9)	73.3	(2.5)	100.0	(4.1)	0.0	-0.6	50.0	(3.4)	3.2	-0.3	0.0	-0.3	7.1	-0.1	0.0	-0.8	40.0	(3.4)	29.6	(1.4)	0.0	(0.1)	0.0	-0.4	0.0	-0.7	7.7	(1.0)
Mexican	51.5	(1.7)	66.2	(2.4)	43.5	(1.4)	35.1	(0.9)	28.8	(1.7)	21.5	(2.3)	36.0	(1.9)	25.0	(1.9)	30.7	(2.1)	13.2	(0.8)	10.9	(0.4)	4.5	(0.1)	8.8	(1.4)	9.1	(3.1)	6.5	(1.1)
Spanish American	15.2	-0.5	18.2	-0.3	20.7	(0.2)	6.3	-0.4	3.0	-0.4	10.4	(0.7)	10.5	(0.4)	16.4	(0.9)	17.8	(0.8)	7.0	(0.2)	17.2	(0.7)	11.1	(0.7)	6.2	(0.9)	6.3	(1.9)	13.2	(1.8)
Cuban	32.0	(0.5)	19.2	-0.2	15.4	(0.0)	0.0	-0.6	4.5	-0.3	15.5	(1.5)	4.0	-0.3	22.0	(1.6)	8.1	(0.0)	4.9	(0.0)	2.3	-0.1	2.3	-0.1	5.6	(0.8)	4.1	(1.0)	18.5	(2.7)
West Indian	32.3	(0.6)	14.8	-0.5	20.0	(0.2)	0.0	-0.6	0.0	-0.6	35.0	(2.9)	4.8	-0.3	6.7	-0.1	38.9	(2.9)	25.0	(1.9)	0.0	(0.0)	50.0	(3.9)	0.0	(0.0)	0.4	-0.7	15.4	(2.2)
Chinese	6.9	-1.0	6.7	-0.9	11.2	-0.2	19.9	(0.2)	25.0	(1.4)	23.5	(2.6)	38.4	(2.8)	36.3	(3.1)	43.4	(3.2)	37.4	(3.2)	55.7	(3.1)	41.1	(3.1)	8.4	(1.4)	0.4	-0.5	0.8	-0.1
East Indian	60.0	(2.2)	54.5	(1.7)	0.0	-0.8	0.0	-0.6	0.0	-0.7	14.3	(1.3)	12.5	(0.6)	18.2	(1.1)	23.7	(1.4)	25.0	(1.8)	16.7	(0.7)	0.0	-0.2	0.0	-0.4	0.0	-0.7	14.3	(2.0)
Japanese	10.5	-0.7	16.4	-0.4	8.2	-0.4	11.0	-0.2	5.2	-0.3	1.0	-0.6	3.2	-0.4	13.6	(0.6)	23.7	(1.4)	23.7	(1.8)	15.6	(0.7)	4.4	(0.1)	2.6	(0.1)	2.3	(0.2)	0.8	(0.1)
Korean	9.1	-0.8	33.3	(0.1)	0.0	-0.8	0.0	-0.6	0.0	-0.7	0.0	-0.7	25.0	(1.8)	14.3	(0.7)	9.1	(0.0)	23.7	(1.8)	16.7	(0.7)	33.3	(2.5)	16.7	(3.1)	0.0	-0.7	55.0	(3.6)
Pacific Islander	0.0	-1.4	0.0	-1.3	30.0	(1.7)	12.4	(0.1)	33.3	(3.3)	0.0	(0.0)	50.0	(4.1)	21.8	(1.5)	39.7	(2.1)	24.3	(1.9)	0.0	(0.0)	9.1	(0.5)	6.0	(0.9)	0.0	(0.0)	3.8	(0.4)
African (black)	53.7	(1.8)	37.4	(0.8)	32.7	(1.2)	12.4	(0.1)	23.3	(1.1)	21.0	(2.0)	21.9	(1.5)	21.8	(1.5)	30.7	(2.1)	14.9	(1.9)	15.9	(0.7)	9.1	(0.5)	6.0	(0.9)	4.0	(1.0)	3.8	(0.4)
Other	60.3	-0.2	37.4	(0.7)	50.0	(1.7)	25.0	(0.5)	8.5	(0.0)	16.2	(1.6)	15.6	(0.9)	14.1	(0.7)	13.6	(0.4)	14.9	(0.9)	12.3	(0.4)	8.9	(0.5)	3.7	(0.4)	1.6	(0.2)	2.3	(0.2)

Bold = 1 or more NSDMR below mean
* Norwegians, Danes, and Swedes.
† Bohemians and Moravians.

= 1 to 1.99 NSDMRs above mean
= 2 or more NSDMRs above mean

TABLE 6-3. Deviations in Rate of Deportation, by Race, 1899–1930, Expressed as Rate of Deportation and NSDMR

Race or People	1899	1900	1901	1902	1903	1904	1905	1906	1907	1908	1909	1910	1911
English	0.0 (1.0)	0.1 (-0.2)	0.1 (0.0)	0.0 (-0.3)	0.0 (-0.2)	0.0 (-0.2)	0.1 (-0.1)	0.1 (-0.1)	0.1 (-0.1)	0.2 (-0.1)	0.1 (0.0)	0.1 (-0.1)	0.2 (-0.1)
Irish	0.0 (0.5)	0.1 (-0.1)	0.0 (0.0)	0.0 (-0.2)	0.0 (-0.2)	0.0 (-0.2)	0.1 (-0.1)	0.1 (-0.1)	0.2 (-0.1)	0.3 (-0.1)	0.3 (0.0)	0.2 (-0.1)	0.4 (-0.1)
Scotch	0.0 (-0.5)	0.1 (-0.1)	0.0 (0.0)	0.1 (-0.1)	0.0 (-0.3)	0.0 (-0.2)	0.1 (-0.1)	0.1 (-0.1)	0.2 (-0.1)	0.2 (-0.1)	0.2 (-0.1)	0.2 (-0.1)	0.3 (-0.1)
Welsh	0.0 (-0.5)	0.0 (-0.2)	0.0 (0.0)	0.1 (-0.1)	0.0 (-0.3)	0.1 (-0.2)	0.1 (-0.1)	0.0 (-0.1)	0.1 (-0.1)	0.1 (-0.2)	0.1 (-0.1)	0.2 (-0.1)	0.6 (0.0)
Australian	0.0 (-0.5)	(0.0)	(0.0)	(0.0)	(0.0)	(0.0)	(0.0)	(0.0)	(0.0)	(0.0)	(0.0)	(0.0)	(0.0)
Spanish	0.0 (-0.5)	0.0 (-0.2)	0.0 (0.0)	0.1 (0.0)	0.2 (0.0)	0.0 (0.0)	0.1 (-0.1)	0.3 (0.0)	0.4 (0.0)	1.2 (0.2)	0.2 (0.0)	0.2 (-0.1)	0.1 (0.0)
Portuguese	0.0 (-0.5)	0.0 (-0.2)	0.0 (0.0)	0.1 (-0.1)	0.0 (-0.3)	0.0 (-0.2)	0.1 (-0.1)	0.1 (-0.1)	0.1 (-0.1)	0.4 (-0.1)	0.2 (0.0)	0.3 (0.0)	0.2 (-0.1)
Italian (North)	0.0 (-0.5)	0.1 (-0.1)	0.0 (0.0)	0.1 (-0.1)	0.0 (-0.3)	0.0 (-0.1)	0.1 (-0.1)	0.1 (-0.1)	0.1 (0.0)	0.3 (0.0)	0.2 (-0.1)	0.2 (-0.1)	0.4 (-0.1)
Italian (South)	0.0 (-0.1)	0.0 (-0.1)	0.0 (0.0)	0.1 (-0.2)	0.0 (-0.2)	0.0 (-0.1)	0.1 (-0.1)	0.1 (-0.1)	0.2 (0.0)	0.5 (0.0)	0.2 (0.0)	0.2 (-0.1)	0.7 (0.0)
Dutch and Flemish	0.0 (-0.5)	0.0 (-0.2)	0.0 (0.0)	0.0 (-0.3)	0.0 (-0.2)	0.0 (-0.2)	0.1 (-0.1)	0.0 (-0.1)	0.2 (-0.1)	0.5 (0.0)	0.1 (-0.1)	0.2 (0.0)	0.3 (0.0)
Finnish	0.0 (-0.5)	0.2 (0.2)	0.2 (0.0)	0.2 (-0.3)	0.4 (-0.3)	0.5 (0.4)	0.3 (0.1)	0.0 (0.0)	0.4 (0.0)	0.4 (-0.1)	0.2 (0.0)	0.3 (0.0)	0.3 (-0.1)
Scandinavian*	0.0 (-0.2)	0.2 (0.2)	0.2 (0.0)	0.3 (0.5)	0.3 (0.5)	0.7 (0.7)	0.5 (0.3)	0.7 (0.3)	0.4 (0.0)	0.7 (0.0)	0.5 (0.1)	0.5 (0.1)	0.6 (0.0)
Lithuanian	0.0 (-0.5)	0.1 (-0.1)	0.3 (0.1)	0.6 (0.5)	0.5 (0.4)	0.3 (0.1)	0.2 (0.0)	0.2 (0.0)	0.2 (0.0)	0.4 (0.0)	0.3 (0.0)	0.4 (0.0)	0.2 (0.0)
Russian	0.0 (-0.2)	0.1 (-0.1)	0.1 (0.1)	0.6 (1.9)	0.9 (1.0)	0.3 (0.1)	0.8 (0.6)	0.4 (0.1)	0.4 (0.1)	0.4 (0.0)	0.4 (0.1)	0.4 (0.0)	0.4 (0.0)
Polish	0.0 (-0.4)	0.1 (-0.1)	0.1 (0.1)	0.2 (0.3)	0.2 (0.1)	0.3 (0.1)	0.2 (0.6)	0.2 (0.0)	0.2 (-0.1)	0.4 (0.0)	0.3 (0.0)	0.3 (0.0)	0.5 (0.0)
German	0.0 (-0.5)	0.1 (-0.1)	0.0 (0.0)	0.1 (-0.1)	0.1 (-0.2)	0.1 (-0.1)	0.2 (-0.1)	0.2 (-0.1)	0.2 (-0.1)	0.3 (0.0)	0.3 (0.0)	0.3 (0.0)	0.3 (0.0)
Magyar	0.0 (-0.5)	0.0 (-0.1)	0.0 (0.0)	0.1 (-0.1)	0.1 (-0.1)	0.1 (-0.1)	0.1 (-0.1)	0.1 (-0.1)	0.2 (-0.1)	0.3 (-0.1)	0.2 (0.0)	0.2 (-0.1)	0.6 (0.0)
Slovak	0.0 (-0.5)	0.0 (-0.1)	0.0 (0.0)	0.1 (-0.2)	0.1 (-0.1)	0.1 (-0.1)	0.1 (-0.1)	0.1 (-0.1)	0.2 (-0.1)	0.4 (0.0)	0.2 (0.0)	0.2 (-0.1)	0.6 (0.0)
Croatian & Slovenian	0.0 (-0.5)	0.0 (0.0)	0.0 (0.0)	0.1 (-0.1)	0.1 (-0.1)	0.1 (-0.1)	0.3 (0.0)	0.2 (0.0)	0.2 (0.0)	0.4 (0.0)	0.3 (0.0)	0.2 (0.0)	0.6 (0.0)
Transylvanian (Siebenburger)	0.0 (-0.5)	(0.0)	(0.0)	(0.0)	(0.0)	(0.0)	(0.0)	(0.0)	(0.0)	(0.0)	(0.0)	(0.0)	(0.0)
Czech †	0.0 (-0.5)	0.1 (0.1)	0.0 (0.0)	0.0 (-0.4)	0.0 (-0.3)	0.0 (-0.2)	0.1 (-0.1)	0.0 (0.0)	0.1 (-0.1)	0.2 (-0.1)	0.2 (-0.1)	0.3 (0.0)	0.3 (-0.1)
Dalmatian, Bosnian, Herzegovinian		0.0 (-0.2)	0.0 (0.0)	0.0 (-0.4)	0.0 (-0.3)	0.0 (-0.1)	0.0 (-0.1)	0.0 (-0.1)	0.0 (-0.1)	0.4 (-0.1)	0.3 (0.0)	0.3 (-0.1)	0.4 (-0.1)
Ruthenian (Russniak)		0.0 (-0.2)	0.0 (0.0)	0.1 (0.1)	0.1 (-0.2)	0.3 (0.1)	0.1 (-0.1)	0.1 (-0.1)	0.2 (-0.1)	0.3 (-0.1)	0.2 (-0.1)	0.2 (-0.1)	0.5 (0.0)
Austro-Hungarian		(0.0)	(0.0)	(0.0)	(0.0)	(0.0)	(0.0)	(0.0)	(0.0)	(0.0)	(0.0)	(0.0)	(0.0)
Roumanian		0.0 (-0.2)	0.0 (0.0)	0.0 (0.0)	0.0 (-0.1)	0.1 (0.0)	0.2 (0.0)	0.2 (0.0)	0.2 (0.0)	0.3 (0.0)	0.3 (0.0)	0.3 (0.0)	1.0 (0.1)
Bulgarian, Serbian, Montenegrin	0.0 (-0.5)	0.0 (0.0)	0.0 (0.0)	0.3 (0.7)	0.1 (0.1)	0.1 (-0.2)	0.3 (0.1)	0.1 (-0.1)	0.2 (0.2)	0.5 (0.2)	0.8 (0.2)	0.4 (0.0)	0.9 (0.1)
Greek	0.0 (-0.5)	0.2 (0.3)	0.0 (0.1)	0.6 (0.1)	0.4 (0.2)	0.4 (0.2)	0.2 (0.2)	0.1 (0.0)	0.2 (0.0)	0.5 (0.0)	0.7 (0.1)	0.6 (0.2)	1.7 (0.4)
Armenian	0.0 (-0.5)	1.1 (2.4)	0.4 (0.2)	1.8 (1.8)	1.9 (1.9)	2.2 (2.2)	2.7 (2.4)	1.7 (1.0)	1.8 (1.0)	3.0 (0.9)	2.3 (0.6)	3.1 (2.0)	3.1 (0.8)
Hebrew	0.0 (-0.1)	0.2 (0.2)	0.1 (0.1)	0.6 (0.3)	0.3 (0.2)	0.2 (0.0)	0.3 (0.1)	0.2 (0.0)	0.3 (0.0)	0.3 (-0.1)	0.5 (0.1)	0.4 (0.1)	0.8 (0.1)
Arabian	0.0 (0.0)	(0.0)	(0.0)	(0.0)	(0.0)	0.0 (0.0)	0.1 (-0.1)	0.2 (0.0)	0.3 (0.0)	(0.0)	(0.0)	(0.0)	(0.0)
Syrian	0.0 (-0.5)	1.6 (3.6)	0.7 (0.3)	1.5 (5.3)	1.0 (1.2)	3.1 (4.2)	3.2 (2.9)	5.6 (3.7)	8.2 (3.7)	6.4 (2.1)	3.3 (1.0)	2.9 (1.9)	4.4 (1.2)
Turkish	0.0 (-0.5)	0.0 (-0.2)	0.0 (0.0)	0.0 (-0.4)	0.9 (1.0)	0.6 (0.6)	0.4 (0.2)	1.1 (0.6)	0.5 (0.1)	1.0 (0.2)	0.7 (0.1)	1.2 (0.6)	2.1 (0.5)
French	0.0 (3.0)	0.0 (-0.1)	0.0 (0.0)	0.0 (-0.3)	0.0 (-0.3)	0.2 (-0.2)	0.1 (-0.1)	0.2 (0.0)	0.2 (0.0)	0.2 (-0.1)	0.2 (0.0)	0.3 (0.0)	0.4 (-0.1)
Swiss	0.0 (0.0)	(0.0)	(0.0)	0.1 (0.1)	(0.0)	(0.0)	(0.0)	(0.0)	(0.0)	(0.0)	(0.0)	(0.0)	(0.0)
Mexican	0.0 (-0.5)	(0.0)	(0.0)	0.1 (0.1)	(0.0)	1.6 (2.0)	3.5 (3.2)	1.3 (0.7)	1.1 (0.4)	1.5 (0.3)	0.7 (0.1)	0.6 (0.2)	0.5 (0.0)
Central American (NS)	0.0 (-0.5)	(0.0)	(0.0)	(0.0)	(0.0)	(0.0)	(0.0)	(0.0)	(0.0)	(0.0)	(0.0)	(0.0)	(0.0)
South American	0.0 (-0.5)	(0.0)	(0.0)	(0.0)	(0.0)	(0.0)	(0.0)	(0.0)	(0.0)	(0.0)	(0.0)	(0.0)	(0.0)
Spanish American	(0.0)	(0.0)	(0.0)	(0.0)	(0.0)	(0.0)	(0.0)	(0.0)	(0.0)	(0.0)	(0.0)	(0.0)	(0.0)
Cuban	0.1 (3.2)	0.0 (0.0)	0.0 (0.0)	0.0 (-0.4)	0.0 (-0.3)	0.0 (-0.3)	0.2 (-0.3)	0.3 (0.1)	1.6 (0.6)	0.7 (0.6)	0.2 (0.2)	0.2 (-0.1)	0.2 (-0.1)
West Indian	0.0 (-0.5)	0.0 (-0.2)	0.1 (0.1)	0.0 (-0.4)	0.0 (-0.3)	0.2 (-0.2)	0.1 (-0.2)	0.1 (0.1)	0.2 (0.2)	0.2 (-0.1)	0.1 (-0.1)	0.0 (-0.2)	0.1 (-0.1)
Chinese	0.0 (-0.5)	1.3 (2.7)	0.0 (0.0)	0.7 (2.2)	0.2 (0.0)	0.9 (1.1)	3.8 (3.5)	5.0 (3.2)	9.2 (4.2)	10.7 (3.6)	1.7 (0.5)	1.1 (0.6)	1.3 (0.2)
East Indian	0.0 (-0.5)	0.0 (-0.2)	0.0 (0.0)	0.0 (-0.4)	0.0 (-0.3)	1.2 (1.4)	1.7 (1.7)	2.7 (1.7)	6.9 (3.1)	14.0 (4.8)	19.2 (6.2)	5.6 (5.6)	8.2 (6.1)
Japanese	0.0 (-0.5)	0.3 (0.5)	0.6 (0.3)	0.2 (0.3)	2.7 (3.6)	1.4 (1.7)	2.6 (2.3)	1.8 (1.1)	2.2 (0.9)	2.0 (0.5)	2.9 (0.8)	1.0 (0.5)	0.5 (0.5)
Korean	0.0 (-0.5)	0.0 (-0.2)	2.1 (1.1)	2.7 (3.6)	2.7 (2.5)	0.1 (-0.1)	0.4 (-0.1)	5.8 (3.8)	2.0 (2.0)	0.0 (0.0)	0.0 (0.0)	0.0 (0.0)	0.0 (0.0)
Filipino	0.0 (0.0)	0.0 (-0.2)	12.3 (6.6)	0.0 (-0.4)	0.0 (-0.3)	0.4 (0.0)	0.4 (0.0)	(0.0)	(0.0)	(0.0)	(0.0)	0.4 (0.0)	(0.0)
Hawaiian	0.0 (-0.5)	(0.0)	(0.0)	(0.0)	(0.0)	(0.0)	(0.0)	(0.0)	(0.0)	(0.0)	(0.0)	(0.0)	(0.0)
Pacific Islander	0.0 (-0.5)	1.8 (3.9)	0.0 (0.0)	0.4 (0.4)	1.9 (2.5)	0.0 (-0.3)	0.0 (-0.2)	(0.0)	0.0 (0.0)	0.0 (-0.2)	3.4 (1.0)	1.8 (1.8)	0.0 (0.0)
African (black)	0.0 (-0.5)	0.1 (0.1)	0.0 (0.0)	0.0 (0.0)	0.2 (0.2)	0.0 (-0.2)	0.2 (-0.2)	0.1 (0.0)	0.0 (0.0)	0.2 (-0.1)	0.3 (0.0)	0.3 (0.0)	0.3 (0.0)
(NS)	(0.0)	(0.0)	(0.0)	(0.0)	(0.0)	(0.0)	(0.0)	(0.0)	(0.0)	(0.0)	(0.0)	(0.0)	(0.0)
Esquimaux	(0.0)	(0.0)	(0.0)	(0.0)	(0.0)	(0.0)	(0.0)	(0.0)	(0.0)	(0.0)	(0.0)	(0.0)	0.0 (0.0)
Other	(0.0)	(0.0)	(0.0)	0.0 (-0.4)	0.0 (-0.3)	2.4 (3.2)	1.4 (1.2)	1.1 (0.6)	1.3 (0.4)	1.7 (0.4)	0.7 (0.1)	0.7 (0.3)	2.4 (0.6)

Bold = 1 or more NSDMR below mean

= 1 to 1.99 NSDMRs above mean

= 2 or more NSDMRs above mean

* Norwegians, Danes, and Swedes.

† Bohemians and Moravians.

NS = not specified

TABLE 6-3. Continued

Race or People	1912 %	1912 NSDMR	1913 %	1913 NSDMR	1914 %	1914 NSDMR	1915 %	1915 NSDMR	1916 %	1916 NSDMR	1917 %	1917 NSDMR	1918 %	1918 NSDMR	1919 %	1919 NSDMR	1920 %	1920 NSDMR	1921 %	1921 NSDMR	1922 %	1922 NSDMR	1923 %	1923 NSDMR	1924 %	1924 NSDMR
English	0.1	(-0.2)	0.2	(-0.2)	0.2	(-0.5)	0.3	(-0.2)	0.7	(-0.2)	1.0	(-0.1)	0.7	(0.0)	0.2	(-0.1)	0.1	(-0.1)	0.1	(-0.3)	0.4	(-0.2)	0.1	(-0.3)	0.1	(-0.3)
Irish	0.2	(-0.1)	0.3	(-0.2)	0.2	(-0.4)	0.5	(-0.1)	1.0	(0.0)	1.2	(0.0)	1.2	(0.0)	0.5	(0.1)	0.1	(0.0)	0.1	(-0.3)	0.4	(0.1)	0.2	(-0.1)	0.2	(-0.2)
Scotch	0.2	(-0.1)	0.3	(-0.2)	0.2	(-0.5)	0.4	(-0.1)	0.7	(-0.3)	1.1	(0.0)	0.9	(0.0)	0.4	(0.0)	0.1	(0.0)	0.1	(-0.1)	0.2	(-0.2)	0.1	(-0.3)	0.1	(-0.4)
Welsh	0.1	(-0.1)	0.4	(-0.1)	0.1	(-0.1)	0.5	(-0.1)	0.8	(-0.2)	0.8	(-0.2)	1.7	(0.5)	0.3	(0.0)	0.1	(-0.2)	0.1	(-0.7)	0.3	(0.0)	0.1	(-0.5)	0.2	(-0.1)
Australian		(0.0)		(0.0)		(0.0)		(0.0)		(0.0)		(0.0)		(0.0)		(0.0)		(0.0)		(0.0)		(0.0)		(0.0)		(0.0)
Spanish	0.2	(-0.1)	0.6	(0.0)	0.8	(0.0)	0.9	(0.0)	0.9	(-0.1)	0.7	(-0.3)	0.8	(0.0)	0.3	(-0.1)	0.2	(-0.1)	0.3	(0.9)	0.2	(-0.2)	0.2	(-0.3)	0.2	(-0.1)
Portuguese	0.1	(-0.1)	0.2	(0.0)	0.5	(0.0)	0.6	(-0.1)	0.6	(-0.3)	0.5	(-0.5)	0.5	(0.0)	0.3	(0.0)	0.2	(-0.3)	0.2	(0.0)	0.1	(-0.3)	0.2	(-0.3)	0.2	(-0.2)
Italian (North)	0.3	(-0.1)	0.5	(-0.1)	0.5	(-0.2)	0.7	(0.0)	0.7	(-0.3)	0.6	(-0.4)	0.4	(0.0)	0.4	(0.0)	0.1	(-0.3)	0.1	(-0.5)	0.5	(-0.1)	0.2	(-0.2)	0.2	(-0.3)
Italian (South)	0.5	(-0.1)	0.8	(-0.1)	1.1	(0.1)	1.3	(0.1)	1.2	(0.2)	1.3	(0.1)	0.4	(-0.2)	0.6	(0.1)	0.1	(-0.3)	0.2	(0.2)	0.5	(0.2)	0.3	(0.0)	0.2	(-0.3)
Dutch and Flemish	0.1	(-0.2)	0.2	(-0.2)	0.6	(-0.5)	0.6	(-0.2)	0.5	(-0.4)	0.4	(-0.6)	0.4	(-0.2)	0.1	(-0.2)	0.1	(-0.2)	0.1	(-0.8)	0.3	(-0.3)	0.1	(-0.3)	0.1	(-0.3)
Finnish	0.3	(-0.2)	0.2	(-0.2)	0.3	(-0.3)	0.2	(-0.1)	0.3	(-0.3)	0.5	(-0.5)	0.8	(-0.2)	0.1	(-0.2)	0.1	(-0.3)	0.1	(-0.7)	0.3	(-0.1)	0.1	(-0.1)	0.2	(-0.1)
Scandinavian*	0.1	(-0.2)	0.2	(-0.2)	0.2	(-0.5)	0.4	(-0.1)	0.5	(-0.5)	0.5	(-0.5)	0.4	(-0.2)	0.1	(-0.1)	0.1	(-0.2)	0.2	(-0.7)	0.2	(-0.2)	0.1	(-0.5)	0.2	(-0.4)
Lithuanian	0.5	(0.0)	0.5	(-0.1)	0.7	(-0.1)	1.2	(0.1)	0.4	(-0.5)	0.8	(-0.3)	0.6	(-0.1)	0.4	(0.0)	0.3	(-0.4)	0.1	(-0.5)	0.6	(0.5)	0.3	(0.0)	0.3	(0.1)
Russian	0.3	(-0.1)	0.3	(-0.1)	0.6	(-0.1)	1.7	(0.1)	0.7	(-0.2)	0.8	(-0.3)	0.7	(0.0)	0.3	(0.0)	0.3	(0.3)	0.2	(-0.2)	0.3	(-0.1)	0.3	(-0.3)	0.3	(0.1)
German	0.2	(-0.1)	0.3	(-0.2)	0.4	(-0.3)	0.3	(-0.2)	0.6	(-0.4)	0.7	(-0.3)	0.6	(0.4)	0.4	(0.2)	0.3	(0.0)	0.2	(-0.8)	0.2	(-0.1)	0.2	(0.0)	0.2	(-0.2)
Magyar	0.3	(0.0)	0.4	(-0.1)	0.7	(-0.1)	0.4	(-0.1)	0.8	(-0.4)	1.2	(0.0)	1.6	(0.4)	0.4	(0.2)	0.3	(0.0)	0.1	(0.0)	0.3	(-0.1)	0.3	(0.0)	0.3	(-0.2)
Slovak	0.4	(0.0)	0.4	(-0.1)	0.7	(0.0)	0.7	(0.0)	0.7	(-0.3)	0.6	(-0.4)	0.7	(-0.4)	0.0	(0.2)	0.3	(0.0)	0.2	(0.0)	0.3	(0.0)	0.3	(-0.2)	0.3	(0.0)
Croatian & Slovenian	0.5	(0.0)	6.0	(2.7)	0.8	(-0.2)	4.8	(1.1)	0.3	(0.3)	0.4	(-0.6)	0.4	(-0.4)	6.5	(3.6)	0.0	(0.0)	0.1	(-0.4)	0.6	(0.4)	0.2	(-0.2)	0.6	(0.7)
Transylvanian (Siebenburger)		(0.0)		(0.0)		(0.0)		(0.0)		(0.0)		(0.0)		(0.0)		(0.0)		(0.0)		(0.0)		(0.0)		(0.0)		(0.0)
Czech †	2.3	(1.1)	2.2	(0.8)	2.7	(1.5)	3.6	(0.8)	4.2	(2.7)	4.2	(2.3)	3.3	(1.3)	1.1	(0.4)	0.2	(0.0)	0.0	(-1.3)	0.3	(-0.1)	0.1	(-0.5)	0.1	(-0.4)
Dalmatian, Bosnian, Herzegovinian	0.4	(0.0)	0.4	(-0.1)	0.7	(0.0)	3.1	(0.6)	0.8	(-0.2)	2.0	(0.7)	0.0	(-0.4)	9.1	(5.1)	0.0	(-0.4)	0.0	(-1.3)	0.0	(-0.5)	0.0	(0.0)	0.0	(-0.6)
Ruthenian (Russniak)	0.5	(0.0)	0.6	(0.6)	0.9	(0.1)	1.0	(0.0)	0.7	(-0.2)	1.6	(0.3)	0.0	(-0.4)	2.4	(1.2)	0.9	(1.8)	0.3	(0.6)	0.1	(-0.3)	0.5	(0.7)	0.3	(0.1)
Austro-Hungarian		(0.0)		(0.0)		(0.0)		(0.0)		(0.0)		(0.0)		(0.0)		(0.0)		(0.0)		(0.0)		(0.0)		(0.0)		(0.0)
Roumanian	0.5	(0.0)	1.0	(0.3)	1.3	(0.4)	1.9	(0.3)	1.4	(0.4)	2.4	(0.9)	1.2	(0.2)	1.3	(0.5)	0.3	(0.2)	0.1	(0.9)	0.3	(0.0)	0.4	(0.2)	0.4	(0.3)
Bulgarian, Serbian, Montenegrin	0.7	(0.2)	1.3	(0.3)	1.1	(0.3)	2.4	(0.4)	2.3	(1.1)	1.9	(0.5)	3.6	(1.5)	0.4	(0.0)	0.2	(0.3)	0.1	(-0.3)	0.4	(0.1)	0.2	(-0.2)	0.2	(0.3)
Greek	1.2	(0.4)	2.2	(0.8)	2.0	(1.0)	1.8	(0.3)	1.5	(0.4)	0.9	(-0.2)	0.9	(0.1)	0.6	(0.1)	0.2	(0.0)	0.3	(0.9)	0.2	(0.1)	0.2	(-0.1)	0.4	(0.0)
Armenian	1.8	(0.8)	1.1	(0.2)	2.1	(1.1)	2.5	(0.5)	2.8	(1.5)	2.6	(1.1)	1.7	(0.5)	0.4	(-0.2)	0.2	(0.0)	0.2	(0.5)	0.5	(0.3)	0.3	(0.1)	1.4	(2.5)
Hebrew	0.6	(0.1)	0.6	(0.6)	0.8	(0.1)	0.7	(0.1)	0.7	(-0.3)	0.5	(0.8)	0.8	(0.5)	0.4	(0.0)	0.1	(-0.1)	0.1	(-0.5)	0.1	(0.4)	0.3	(0.3)	0.3	(0.1)
Arabian		(0.0)		(0.0)		(0.0)		(0.0)		(0.0)		(0.0)		(0.0)		(0.0)		(0.0)		(0.0)		(0.0)		(0.0)		(0.0)
Syrian	2.3	(1.1)	2.2	(0.8)	2.7	(1.5)	3.6	(0.8)	4.2	(2.7)	4.2	(2.3)	8.6	(4.1)	0.8	(0.2)	0.3	(0.4)	0.5	(2.3)	0.5	(0.2)	0.6	(1.0)	0.5	(0.6)
Turkish	1.4	(0.6)	1.2	(0.3)	1.6	(0.7)	2.1	(0.4)	4.8	(3.2)	2.3	(0.9)	3.8	(1.6)	0.0	(0.2)	1.5	(2.3)	0.2	(0.6)	0.0	(-0.5)	0.3	(0.1)	0.0	(-0.6)
French	0.2	(-0.1)	0.3	(0.6)	0.2	(-0.4)	0.3	(-0.2)	0.8	(-0.1)	0.8	(-0.3)	0.8	(0.0)	0.5	(-0.1)	0.2	(0.2)	0.3	(1.9)	0.5	(0.3)	0.5	(0.5)	0.3	(0.0)
Swiss		(0.0)		(0.0)		(0.0)		(0.0)		(0.0)		(0.0)		(0.0)		(0.0)		(0.0)		(0.0)		(0.0)		(0.0)		(0.0)
Mexican	0.4	(0.0)	1.9	(0.6)	1.5	(0.6)	1.7	(0.2)	2.1	(1.0)	4.2	(2.3)	1.0	(0.1)	1.8	(0.8)	0.5	(0.8)	0.3	(0.9)	0.3	(0.0)	0.3	(0.0)	0.4	(0.3)
Central American (NS)		(0.0)		(0.0)		(0.0)		(0.0)		(0.0)		(0.0)		(0.0)		(0.0)		(0.0)		(0.0)		(0.0)		(0.0)		(0.0)
South American		(0.0)		(0.0)		(0.0)		(0.0)		(0.0)		(0.0)		(0.0)		(0.0)		(0.0)		(0.0)		(0.0)		(0.0)		(0.0)
Spanish American	0.0	(-0.2)	0.4	(0.0)	0.2	(-0.5)	0.4	(-0.1)	0.3	(-0.6)	0.2	(-0.8)	0.3	(-0.3)	0.0	(-0.2)	0.0	(-0.4)	0.1	(-0.8)	0.1	(-0.4)	0.2	(-0.3)	0.2	(-0.1)
Cuban	0.0	(0.0)	0.2	(0.2)	0.1	(-0.6)	0.1	(-0.2)	0.2	(-0.7)	0.1	(-0.8)	0.2	(-0.3)	0.0	(-0.2)	0.0	(-0.4)	0.0	(-1.0)	0.0	(-0.5)	0.2	(-0.3)	0.1	(-0.5)
West Indian	0.2	(-0.1)	0.5	(-0.1)	0.1	(-0.5)	0.0	(-0.1)	0.3	(-0.6)	0.3	(-0.6)	0.4	(-0.4)	0.2	(-0.2)	0.1	(-0.4)	0.1	(-1.3)	0.0	(-0.5)	0.2	(-0.6)	0.2	(-0.3)
Chinese	0.9	(0.3)	2.4	(0.9)	1.6	(0.7)	0.8	(0.0)	1.3	(0.3)	1.8	(0.5)	2.1	(0.7)	0.8	(0.3)	0.6	(1.2)	0.7	(3.7)	1.6	(2.1)	2.3	(5.5)	2.6	(5.3)
East Indian	10.5	(6.0)	11.6	(5.5)	7.6	(5.5)	22.8	(6.2)	4.8	(3.3)	6.3	(3.0)		(0.0)		(0.0)		(0.0)		(0.0)		(0.0)	1.3	(1.3)	0.0	(-0.6)
Japanese	1.1	(0.4)	2.2	(0.8)	1.1	(0.8)	1.7	(0.4)	0.8	(-0.2)	0.4	(0.4)	0.8	(0.0)	0.5	(0.0)	0.3	(-0.4)	0.0	(-0.9)	0.3	(-0.4)	0.7	(1.3)	0.8	(1.1)
Korean	2.1	(1.0)	1.5	(0.7)	1.8	(0.7)	1.3	(-0.1)	0.6	(-0.3)	4.1	(2.2)	0.0	(-0.9)		(0.0)		(0.0)		(0.0)	0.7	(0.6)	1.2	(1.2)	0.6	(0.7)
Filipino		(0.0)		(0.0)	0.5	(0.0)	0.7	(-0.1)	0.1	(-0.8)	0.0	(0.0)	0.0	(-0.4)	0.0	(-0.2)	0.0	(-0.4)	0.0	(-1.3)	0.7	(-0.5)	0.0	(-0.7)	0.0	(0.0)
Hawaiian		(0.0)		(0.0)		(0.0)		(0.0)		(0.0)		(0.0)		(0.0)		(0.0)		(0.0)		(0.0)		(0.0)		(0.0)		(0.0)
Pacific Islander	0.0	(-0.3)	0.0	(-0.3)	0.0	(-0.6)	0.0	(-0.3)	0.0	(-0.9)	0.0	(-0.9)	5.4	(3.0)	0.0	(-0.2)	2.6	(4.1)	0.0	(-1.3)	4.0	(5.5)	0.0	(-0.7)	0.0	(-0.6)
African (black)	0.2	(-0.1)	0.6	(0.0)	0.3	(-0.4)	1.4	(0.2)	2.1	(0.9)	1.3	(0.1)	0.8	(0.0)	0.1	(-0.2)	0.1	(0.1)	0.2	(1.1)	0.4	(0.1)	0.4	(0.0)	0.4	(0.3)
(NS)		(0.0)		(0.0)		(0.0)		(0.0)		(0.0)		(0.0)		(0.0)		(0.0)		(0.0)		(0.0)		(0.0)		(0.0)		(0.0)
Esquimaux		(0.0)		(0.0)		(0.0)		(0.0)		(0.0)		(0.0)		(0.0)		(0.0)		(0.0)		(0.0)		(0.0)		(0.0)		(0.0)
Other	1.2	(0.5)	2.0	(0.7)	2.0	(1.0)	2.4	(0.4)	2.6	(1.4)	1.7	(0.4)	1.1	(0.2)	0.8	(0.2)	0.3	(0.2)	0.3	(1.1)	0.4	(0.1)	0.7	(1.3)	0.6	(0.7)

TABLE 6-3. Continued

Race or People	1925 %	1925 NSDMR	1926 %	1926 NSDMR	1927 %	1927 NSDMR	1928 %	1928 NSDMR	1929 %	1929 NSDMR	1930 %	1930 NSDMR
English	0.2	-0.1	0.1	-0.2	0.1	(0.0)	0.0	-0.1	0.0	-0.6	0.0	-0.2
Irish	0.2	-0.1	0.1	-0.2	0.1	(0.0)	0.0	-0.1	0.0	-0.6	0.0	-0.1
Scotch	0.2	-0.1	0.2	-0.1	0.1	(0.0)	0.0	-0.1	0.0	-0.6	0.0	-0.2
Welsh	0.1	-0.2	0.1	-0.2	0.2	(0.0)	0.0	-0.1	0.0	-0.7	0.0	(0.0)
Australian	(0.0)		(0.0)		(0.0)		(0.0)		(0.0)		(0.0)	
Spanish	0.8	(0.3)	0.1	-0.3	0.5	(0.2)	0.0	-0.1	0.0	-0.7	0.1	(0.1)
Portuguese	1.5	(0.9)	0.1	-0.1	0.8	(0.4)	1.2	(1.0)	0.0	-0.3	0.1	(0.1)
Italian (North)	0.7	(0.3)	0.2	-0.1	0.2	(0.0)	0.2	(0.1)	0.0	-0.7	0.0	-0.1
Italian (South)	1.0	(0.5)	0.1	-0.3	0.1	(0.0)	0.1	(0.0)	0.0	-0.5	0.0	-0.2
Dutch and Flemish	0.3	-0.1	0.0	-0.5	0.1	(0.0)	0.1	(0.0)	0.0	-0.4	0.0	-0.2
Finnish	0.8	(0.3)	0.2	(0.1)	1.4	(0.7)	0.4	(0.2)	0.1	(0.9)	0.0	-0.2
Scandinavian*	0.2	-0.1	0.1	-0.2	0.1	(0.0)	0.1	-0.1	0.0	-0.5	0.0	-0.2
Lithuanian	0.5	(0.1)	0.1	-0.1	0.0	-0.1	0.0	-0.1	0.1	(0.6)	0.1	(0.4)
Russian	0.7	(0.3)	0.2	(0.0)	0.4	(0.1)	0.0	-0.1	0.1	(0.4)	0.0	-0.2
Polish	0.2	-0.1	0.1	-0.2	0.1	(0.0)	0.0	-0.1	0.2	(1.5)	0.1	-0.1
German	0.4	(0.0)	0.2	-0.1	0.2	(0.0)	0.3	(0.2)	0.1	(0.0)	0.0	-0.1
Magyar	0.2	-0.1	0.2	(0.1)	0.3	(0.1)	0.0	-0.1	0.0	-0.7	0.0	-0.2
Slovak	0.2	-0.1	0.2	(0.1)	0.4	(0.1)	0.0	-0.1	0.0	(0.4)	0.0	(0.4)
Croatian & Slovenian	0.5	(0.1)	0.2	(0.0)	0.4	(0.1)	0.0	-0.1	0.1	(0.4)	0.0	-0.2
Transylvanian (Siebenburger)	(0.0)		(0.0)		(0.0)		(0.0)		(0.0)		(0.1)	(0.3)
Czech †	0.3	-0.1	0.2	(0.0)	0.2	(0.0)	0.0	-0.1	0.0	-0.2	0.0	-0.2
Dalmatian, Bosnian, Herzegovinian	1.3	(0.7)	0.4	(0.5)	0.0	-0.1	0.0	-0.1	0.0	-0.7	0.0	-0.2
Ruthenian (Russniak)	0.5	(0.1)	0.3	(0.5)	0.9	(0.4)	0.0	-0.1	0.0	-0.7	0.0	-0.2
Austro-Hungarian	(0.0)		(0.0)		(0.0)		(0.0)		(0.0)		(0.0)	
Roumanian	1.2	(0.7)	0.1	-0.2	0.7	(0.3)	0.2	(0.1)	0.1	(0.2)	0.1	(0.3)
Bulgarian, Serbian, Montenegrin	0.9	(0.4)	0.1	-0.1	0.0	-0.1	1.1	(1.0)	0.1	(0.1)	0.0	-0.2
Greek	0.8	(0.4)	0.1	-0.3	0.1	(0.2)	0.1	(0.0)	0.1	(0.4)	0.1	(0.5)
Armenian	1.4	(0.9)	0.4	(0.5)	0.5	(0.2)	0.1	(0.0)	0.3	(2.2)	0.4	(1.8)
Hebrew	0.4	(0.0)	0.3	(0.3)	0.2	(0.0)	0.0	-0.1	0.1	(0.0)	0.0	(0.0)
Arabian	(0.0)		(0.0)		(0.0)		(0.0)		(0.0)		(0.0)	
Syrian	1.4	(0.9)	0.6	(1.1)	0.1	(0.1)	0.2	(0.1)	0.1	(0.2)	0.3	(1.3)
Turkish	2.1	(1.4)	0.8	(1.9)	0.0	-0.1	0.0	-0.1	0.1	-0.7	0.3	(1.4)
French	0.2	-0.1	0.4	(0.6)	0.3	(0.1)	0.0	-0.1	0.1	-0.6	0.1	(0.1)
Swiss	(0.0)		(0.0)		(0.0)		(0.0)		(0.0)		(0.0)	
Mexican	0.4	(0.0)	0.2	(0.1)	0.1	(0.0)	0.3	(0.2)	0.4	(3.9)	0.4	(1.8)
Central American (NS)	(0.0)				(0.0)		(0.0)		(0.0)		(0.0)	
Spanish American	0.2	-0.1	0.2	-0.1	0.3	(0.1)	0.1	(0.0)	0.0	-0.1	0.1	(0.2)
Cuban	0.2	-0.1	0.0	-0.5	0.1	(0.1)	0.1	(0.0)	0.0	-0.4	0.1	(0.1)
West Indian	0.3	(0.0)	0.0	-0.5	0.8	(0.4)	0.0	-0.1	0.0	-0.7	0.1	(0.0)
Chinese	7.2	(5.5)	1.9	(4.9)	10.7	(5.9)	5.0	(4.5)	0.0	-0.3	0.0	(0.0)
East Indian	0.0	-0.3	0.4	(0.6)	0.0	-0.1	0.0	(0.6)	0.0	-0.7	0.4	(2.0)
Japanese	3.5	(2.5)	0.1	-0.2	0.3	(0.1)	0.8	(0.6)	0.0	-0.3	0.0	-0.1
Korean	0.0	-0.3	1.2	(2.8)	4.3	(2.3)	4.5	(4.2)	1.0	-0.7	1.0	(5.2)
Filipino	(0.0)		(0.0)		(0.0)		(0.0)		(0.0)		(0.0)	
Hawaiian	(0.0)		(0.0)		(0.0)		(0.0)		(0.0)		(0.0)	
Pacific Islander	0.0	-0.3	0.0	-0.5	0.0	-0.1	0.0	-0.1	0.0	-0.7	0.0	-0.2
African (black)	2.0	(1.3)	0.3	(0.3)	0.6	(0.3)	0.3	(0.2)	0.0	-0.2	0.0	(0.0)
(NS)	(0.0)		(0.0)		(0.0)		(0.0)		(0.0)		(0.0)	
Esquimaux	(0.0)		(0.0)		(0.0)		(0.0)		(0.0)		(0.0)	
Other	1.6	(1.0)	0.6	(1.1)	1.0	(0.5)	0.4	(0.3)	0.1	(0.5)	0.2	(0.8)

TABLE 6-4. Deviations in the Rate of Certifications for Class A Conditions, by Race, 1916–1930, Expressed as Rate of Certification and NSDMR

Race or People	1916 %	1916 NSDMR	1917 %	1917 NSDMR	1918 %	1918 NSDMR	1919 %	1919 NSDMR	1920 %	1920 NSDMR	1921 %	1921 NSDMR	1922 %	1922 NSDMR	1923 %	1923 NSDMR	1924 %	1924 NSDMR	1925 %	1925 NSDMR	1926 %	1926 NSDMR	1927 %	1927 NSDMR	1928 %	1928 NSDMR	1929 %	1929 NSDMR	1930 %	1930 NSDMR
English	0.1	-0.1	0.2	-0.2	0.2	-0.4	0.0	-0.3	0.0	-0.3	0.0	-0.3	0.1	-0.1	0.0	-0.2	0.0	-0.2	0.1	-0.1	0.0	-0.1	0.1	0.0	0.1	0.0	0.0	-0.2	0.0	-0.1
Irish	0.1	-0.1	0.2	-0.2	0.4	-0.3	0.1	-0.2	0.0	-0.3	0.0	-0.3	0.1	-0.1	0.1	-0.2	0.0	-0.2	0.1	-0.1	0.0	-0.1	0.1	0.0	0.1	0.0	0.0	-0.2	0.0	-0.2
Scotch	0.1	-0.1	0.2	-0.2	0.3	-0.3	0.1	-0.2	0.0	-0.3	0.0	-0.3	0.1	-0.1	0.0	-0.2	0.0	-0.2	0.1	-0.1	0.1	-0.1	0.1	0.0	0.1	0.0	0.0	-0.2	0.0	-0.2
Welsh	0.0	-0.1	0.1	-0.2	1.0	-0.1	0.1	-0.2	0.0	-0.3	0.0	-0.3	0.1	-0.1	0.0	-0.2	0.3	-0.1	0.0	-0.1	0.1	-0.1	0.0	0.0	0.1	0.0	0.1	-0.1	0.0	-0.2
Spanish	0.3	0.0	0.4	-0.1	0.5	-0.2	0.1	-0.1	0.2	-0.1	0.2	0.0	0.1	-0.1	0.2	0.0	0.1	0.0	0.0	0.0	0.2	0.0	0.5	0.1	0.5	0.0	0.1	-0.2	0.2	0.1
Portuguese	0.2	0.0	0.2	-0.2	0.7	-0.2	0.4	-0.2	0.3	-0.3	0.1	-0.1	0.0	-0.1	0.3	-0.1	0.3	-0.1	1.7	0.5	0.1	-0.1	1.3	0.3	0.8	0.0	0.2	0.0	0.2	0.0
Italian (North)	0.4	0.0	0.2	-0.1	0.5	-0.3	0.2	-0.2	0.1	-0.3	0.2	-0.1	0.0	-0.1	0.0	-0.1	0.1	0.0	0.1	-0.1	0.1	-0.1	0.2	0.0	0.2	0.0	0.0	-0.2	0.0	-0.2
Italian (South)	0.4	0.0	0.5	-0.1	0.6	-0.2	0.2	-0.2	0.1	-0.2	0.2	-0.1	0.9	-0.1	0.0	-0.1	0.0	-0.1	1.2	0.3	0.3	0.1	0.2	0.0	0.2	0.0	0.0	-0.2	0.0	-0.2
Dutch and Flemish	0.2	-0.1	0.1	-0.2	0.4	-0.3	0.2	-0.2	0.1	-0.3	0.1	-0.1	0.0	-0.1	0.2	-0.1	0.2	-0.2	0.2	0.0	0.1	-0.1	0.1	0.0	0.2	0.0	0.0	-0.2	0.0	-0.2
Finnish	0.4	0.0	0.3	-0.1	0.8	-0.1	0.2	-0.2	0.1	-0.2	0.3	0.0	0.3	0.0	0.2	0.0	0.1	0.0	0.4	0.0	0.1	0.1	1.4	0.3	0.7	0.0	0.3	0.0	0.1	-0.1
Scandinavian*	0.1	-0.1	0.3	-0.1	0.6	-0.2	0.1	-0.2	0.1	-0.3	0.1	-0.2	0.0	-0.1	0.0	-0.2	0.1	-0.1	0.1	-0.1	0.1	-0.1	0.0	0.0	0.0	0.0	0.0	-0.2	0.1	-0.2
Lithuanian	0.7	0.0	0.6	-0.1	1.2	0.0	0.9	-0.2	0.5	-0.1	0.5	0.4	0.7	0.1	0.4	0.1	0.5	0.0	0.5	0.1	0.4	0.0	0.4	0.0	0.3	0.0	0.1	-0.1	0.1	-0.1
Russian	0.5	0.0	0.4	-0.1	0.6	0.0	0.2	-0.2	0.2	-0.1	0.2	-0.2	0.3	0.0	0.5	0.0	0.2	0.0	0.7	0.2	0.2	0.1	0.3	0.0	0.1	0.0	0.1	-0.2	0.1	0.0
Polish	0.5	0.0	0.4	-0.1	0.4	-0.3	0.3	-0.2	0.2	-0.3	0.2	-0.2	0.4	0.0	0.2	-0.1	0.1	-0.1	0.2	0.0	0.1	0.0	0.3	0.0	0.1	0.0	0.2	0.0	0.0	-0.2
German	0.2	0.0	0.1	-0.2	0.5	-0.3	0.2	-0.2	0.1	-0.2	0.1	-0.2	0.1	-0.1	0.1	-0.1	0.1	-0.1	0.1	-0.1	0.1	-0.1	0.0	0.0	0.0	0.0	0.0	-0.2	0.0	-0.2
Magyar	0.9	0.0	0.0	-0.2	0.0	-0.4	0.0	-0.3	0.1	1.1	0.1	-0.2	0.4	0.0	0.2	0.0	0.2	-0.1	0.4	0.0	0.2	0.0	0.4	0.0	0.3	0.0	0.1	-0.1	0.0	-0.1
Slovak	0.3	0.0	0.4	-0.1	7.9	(2.9)	0.7	0.1	1.1	1.0	0.1	-0.2	0.4	0.0	0.1	0.0	0.2	-0.1	0.6	0.1	0.1	-0.1	0.6	0.1	0.1	0.0	0.1	-0.1	0.1	-0.2
Croatian and Slovenian	1.3	-0.1	0.6	-0.1	2.6	(0.4)	4.3	(2.1)	0.2	0.0	0.1	-0.2	0.1	-0.1	0.1	-0.1	0.2	-0.1	0.5	0.1	0.2	0.0	0.5	0.1	0.5	0.0	0.3	0.0	0.1	-0.1
Czech †	0.0	-0.1	0.0	-0.2	1.1	-0.1	0.4	-0.1	0.2	-0.1	0.1	-0.3	0.0	0.0	0.1	-0.2	0.1	-0.1	0.0	0.0	0.1	0.0	0.0	0.0	0.0	0.0	0.0	-0.1	0.1	-0.1
Dalmatian, Bosnian, Herzegovinian	0.0	-0.1	2.0	(0.3)		(0.0)	0.0	-0.3	0.0	0.0	0.1	-0.2	0.8	0.1	0.0	-0.2	0.0	-0.2		0.0	0.0	0.0		0.0		0.0		0.0	0.0	-0.2
Ruthenian (Russniak)	0.4	0.0	0.3	-0.2	0.0	-0.4	1.2	0.4	0.4	0.0	0.0	-0.1	0.0	-0.1	0.0	-0.1	0.0	-0.2	0.1	0.0	0.3	0.1			0.5	0.0	0.0	-0.1	0.3	0.3
Roumanian	0.4	0.0	0.2	-0.2	0.6	-0.2	0.6	0.1	0.2	-0.1	0.0	0.0	0.5	0.0	0.5	0.2	0.3	0.1	1.0	0.2	0.2	0.0	0.2	0.0	0.7	0.0	0.0	-0.2	0.6	0.6
Bulgarian, Serbian, Montenegrin	0.6	0.0	0.7	0.0	2.1	0.2	0.4	0.1	0.2	0.2	0.0	0.0	0.3	0.3	0.3	0.0	0.2	0.2	0.3	0.0	0.0	0.0	0.2	0.0	0.8	0.0	0.1	0.1	0.4	0.4
Greek	0.4	0.0	0.5	0.5	1.2	1.2	0.6	0.6	0.2	0.2	0.2	0.2	1.4	1.3	0.3	0.3	0.2	0.2	0.6	0.6	0.1	0.1	0.1	0.1	0.2	0.2	0.3	0.3	0.4	0.1
Armenian	3.3	0.3	3.1	0.5	1.3	0.0	0.3	0.2	0.1	0.1	0.5	0.3	1.3	0.2	0.6	0.3	1.2	0.8	2.3	0.7	0.6	0.4	0.5	0.1	0.5	0.0	0.3	0.0	1.0	(1.2)
Hebrew	0.5	0.0	0.4	-0.1	0.6	-0.2	0.3	-0.2	0.1	-0.1	0.3	0.3	0.5	0.0	0.5	0.0	0.6	-0.1	0.2	0.3	0.2	0.2	0.2	0.1	0.1	0.1	0.1	0.1	0.1	-0.1
Syrian	5.5	0.5	3.6	0.7	7.9	(2.2)	0.6	0.0	0.3	0.1	0.6	0.4	0.7	-0.1	0.5	-0.2	0.6	0.3	1.2	0.3	0.9	0.7	0.4	0.0	0.7	0.0	0.2	-0.2	0.4	-0.4
French	0.2	-0.1	0.1	-0.2	0.3	-0.3	0.2	-0.2	0.2	-0.2	0.2	-0.3	0.1	-0.1	0.2	-0.2	0.1	0.0	0.1	0.1	0.2	0.2	0.2	0.2	0.1	0.0	0.1	0.1	0.1	-0.1
Turkish	3.9	0.4	0.9	0.0	3.8	0.8	0.0	-0.3	0.3	-0.3	2.6	(2.3)	4.4	0.9	0.3	-0.1	0.0	1.0	1.1	0.3	1.2	(1.0)	0.9	0.2	0.7	0.0	0.5	0.5	1.5	(2.0)
Mexican	1.6	0.1	3.1	0.5	0.6	0.6	0.0	0.7	0.3	0.1	0.2	0.2	0.1	-0.1	0.3	-0.1	0.3	0.3	0.3	0.3	0.3	(1.0)	0.3	0.2	0.5	0.0	0.9	0.5	0.9	(1.1)
Spanish American	0.6	0.0	0.1	-0.2	0.2	-0.4	0.1	-0.3	0.2	-0.2	0.2	-0.1	0.1	-0.1	0.3	0.0	0.2	0.0	0.5	0.1	0.1	0.0	0.2	0.0	0.1	0.0	0.1	-0.2	0.1	0.0
Cuban	0.1	-0.1	0.1	-0.2	0.1	-0.4	0.0	-0.3	0.0	-0.3	0.3	-0.3	0.1	-0.1	0.1	0.0	0.0	0.0	0.1	-0.1	0.1	-0.1	0.1	0.0	0.3	0.0	0.0	-0.2	0.1	-0.2
West Indian	0.6	0.0	0.2	-0.2	0.3	-0.3	0.0	-0.3	0.0	-0.3	0.1	-0.2	0.2	-0.1	0.2	-0.2	0.1	-0.1	0.0	-0.1	0.1	-0.1	0.0	-0.1	0.3	0.0	0.0	-0.2	0.0	-0.2
Chinese	14.2	(1.5)	22.3	(5.2)	15.4	(4.6)	3.4	(1.6)	2.2	(2.3)	2.5	(3.0)	2.2	(0.4)	5.3	(4.7)	5.0	(4.0)	15.5	(5.5)	2.9	(2.6)	2.0	(5.4)	54.2	(4.5)	7.8	(5.5)	3.7	(5.2)
East Indian	2.4	(0.2)	7.4	(1.6)	6.6	1.3	0.0	0.0	4.5	(3.1)	1.2	(1.3)	0.9	0.1	0.8	(1.5)	0.8	1.9	3.1	(1.0)	0.4	0.6	2.0	(0.4)	51.2	(1.0)	1.3	(0.7)	1.7	(2.1)
Japanese	4.5	(0.4)	7.4	(1.6)	6.6	1.7	3.5	1.8	4.5	(3.1)	4.7	1.9	1.3	1.3	1.3	1.0	1.9	1.4	8.5	(2.9)	0.4	0.2	3.9	(0.9)	19.2	(1.6)	0.4	0.4	0.4	0.4
Korean	6.1	(0.6)	11.8	(2.6)	4.3	(1.0)	2.9	1.0	0.0	-0.3	3.8	(4.7)	1.4	0.2	1.3	3.9	5.7	4.6	5.8	(5.3)	5.4	(5.4)	12.5	(3.1)	19.2	(1.5)	24.1	3.6	1.0	(1.3)
Pacific Islander	55.6	(6.9)	0.0	-0.2	8.3	(2.3)	10.0	(5.2)	1.8	1.9	0.0	0.0	12.0	(2.7)	0.0	-0.2	0.0	0.0	0.0	-0.1	1.9	1.6	0.0	0.0	50.0	(4.1)	0.0	0.0	0.0	0.0
African (black)	1.5	0.1	1.0	0.0	0.8	-0.2	0.1	0.1	0.2	-0.1	0.2	0.0	0.5	0.0	0.4	0.1	0.3	0.3	1.2	0.3	0.6	0.4	2.0	0.4	2.1	0.1	0.2	0.0	0.3	0.2
Other	0.6	0.0	0.9	0.0	0.9	0.0	0.4	0.4	0.2	-0.1	0.2	-0.1	0.0	0.0	0.7	0.4	0.4	0.7	1.3	0.3	1.3	(1.1)	0.8	0.1	0.4	0.0	0.4	0.1	0.7	0.7

Bold = 1 or more NSDMR below mean

(shaded) = 1 to 1.99 NSDMRs above mean

(dark shaded) = 2 or more NSDMRs above mean

* Norwegians, Danes, and Swedes.

† Bohemians and Moravians.

TABLE 6-5. Deviations in the Rate of Deportations for Those Certified with Class A Conditions, by Race, 1916–1930, Expressed as Rate of Deportation and NSDMR

Race or People	1916 %	NSDMR	1917 %	NSDMR	1918 %	NSDMR	1919 %	NSDMR	1920 %	NSDMR	1921 %	NSDMR	1922 %	NSDMR	1923 %	NSDMR	1924 %	NSDMR	1925 %	NSDMR	1926 %	NSDMR	1927 %	NSDMR	1928 %	NSDMR	1929 %	NSDMR	1930 %	NSDMR
English	85.7	(1.0)	81.8	(1.1)	85.7	(1.6)	80.0	(0.8)	61.3	(0.9)	90.0	(1.7)	77.4	(1.0)	68.0	(0.6)	61.5	(-0.1)	61.5	(0.2)	70.6	(0.3)	41.7	(0.0)	3.7	(-0.5)	17.6	(0.2)	21.4	(0.2)
Irish	66.7	(0.3)	90.9	(1.5)	85.0	(1.5)	76.9	(0.8)	70.0	(1.2)	59.1	(0.7)	93.3	(1.4)	86.1	(1.3)	86.5	(0.4)	79.4	(0.8)	44.0	(-0.5)	33.3	(-0.1)	0.0	(-0.5)	0.0	(-0.5)	36.4	(0.7)
Scotch	85.7	(1.0)	92.0	(1.5)	86.7	(1.6)	100.0	(2.1)	100.0	(2.1)	85.7	(1.6)	85.7	(1.2)	86.4	(1.3)	100.0	(1.1)	63.0	(0.2)	65.0	(0.2)	30.0	(-0.2)	6.7	(-0.3)	33.3	(0.9)	33.3	(0.6)
Welsh	0.0	(0.0)	100.0	(1.8)	66.7	(1.0)	100.0	(1.2)	100.0	(2.1)	0.0	(-1.0)	100.0	(1.5)	0.0	(0.0)	100.0	(1.1)	0.0	(-1.9)	0.0	(-1.9)	40.0	(0.0)	0.0	(-0.6)	0.0	(-0.5)	0.0	(-0.5)
Spanish	81.3	(0.8)	94.6	(1.6)	51.1	(0.6)	71.4	(0.7)	53.7	(0.6)	63.5	(0.9)	41.7	(0.1)	40.0	(-0.5)	42.4	(-0.8)	75.0	(0.6)	18.2	(-1.3)	40.0	(0.0)	0.0	(-0.6)	0.0	(-0.5)	6.7	(-0.2)
Portuguese	62.5	(0.9)	52.6	(0.9)	58.8	(0.8)	25.0	(0.7)	50.0	(0.5)	56.3	(0.7)	0.0	(1.5)	75.0	(0.8)	50.0	(0.0)	53.3	(-0.1)	25.0	(-1.1)	27.3	(-0.3)	71.4	(1.9)	11.1	(0.0)	33.3	(0.6)
Italian (North)	84.2	(0.9)	85.7	(1.3)	50.0	(0.5)	100.0	(1.2)	50.0	(0.5)	63.3	(1.3)	100.0	(1.5)	100.0	(1.3)	100.0	(1.1)	100.0	(1.5)	25.0	(-1.1)	60.0	(1.4)	80.0	(1.4)	0.0	(-0.5)	33.3	(0.6)
Italian (South)	71.5	(0.5)	59.1	(0.1)	38.7	(0.2)	26.7	(-0.1)	32.1	(0.0)	57.8	(0.7)	22.2	(-0.3)	51.1	(-0.1)	67.6	(0.1)	45.5	(-0.4)	50.0	(-0.3)	48.4	(0.2)	45.0	(1.1)	100.0	(3.6)	50.0	(1.2)
Dutch and Flemish	90.0	(1.0)	40.0	(-0.5)	40.0	(0.2)	33.3	(0.0)	0.0	(-1.0)	0.0	(-1.0)	100.0	(1.5)	66.7	(0.5)	100.0	(1.1)	42.9	(-0.5)	50.0	(-0.3)	42.9	(0.2)	0.0	(-0.6)	50.0	(1.6)	0.0	(0.0)
Finnish	92.3	(1.2)	76.5	(0.9)	53.3	(0.6)	64.7	(0.6)	92.3	(1.9)	28.6	(0.4)	62.5	(0.6)	55.6	(0.5)	100.0	(1.1)	42.9	(-0.5)	50.0	(-0.3)	88.9	(1.1)	0.0	(-0.6)	42.9	(1.3)	0.0	(-0.5)
Scandinavian*	80.8	(0.8)	72.5	(0.7)	64.7	(0.3)	64.7	(0.6)	92.3	(1.9)	46.7	(0.4)	63.2	(0.6)	66.7	(0.5)	78.8	(0.4)	64.4	(0.5)	95.0	(1.1)	100.0	(1.4)	0.0	(-0.6)	22.2	(0.6)	0.0	(-0.5)
Lithuanian	40.0	(-0.5)	66.7	(0.5)	0.0	(-0.9)	50.0	(0.0)	0.0	(-1.0)	0.0	(-1.0)	69.2	(0.8)	44.4	(-0.4)	36.4	(-1.0)	50.0	(-0.2)	40.0	(-0.6)	50.0	(0.3)	0.0	(-0.6)	0.0	(-0.5)	100.0	(2.9)
Russian	70.4	(0.5)	69.2	(0.6)	90.0	(1.7)	80.0	(0.8)	50.0	(0.5)	33.3	(0.0)	75.0	(0.9)	45.5	(-0.3)	17.2	(-1.6)	45.5	(-0.4)	71.4	(-0.6)	25.0	(0.3)	25.0	(0.3)	28.6	(2.5)	33.3	(0.6)
Polish	79.2	(0.8)	66.7	(0.6)	90.9	(1.7)	80.0	(0.8)	0.0	(-1.0)	22.9	(-0.4)	23.3	(-0.3)	67.6	(0.5)	45.2	(-0.7)	45.5	(-0.8)	71.4	(-0.6)	25.0	(-0.3)	25.0	(0.3)	7.3	(2.5)	33.3	(0.6)
German	100.0	(1.4)	60.0	(0.3)	80.0	(0.8)	80.0	(0.8)	85.7	(1.6)	66.7	(1.0)	66.7	(0.7)	81.4	(1.1)	78.6	(0.4)	80.6	(0.8)	82.1	(0.7)	47.8	(0.2)	64.3	(1.6)	45.0	(1.3)	50.0	(1.2)
Magyar	66.7	(0.3)	66.7	(0.8)	66.7	(1.0)	66.7	(0.0)	50.0	(0.0)	100.0	(1.3)	28.0	(0.0)	64.7	(0.4)	69.2	(0.1)	50.0	(-0.2)	100.0	(1.3)	50.0	(0.0)	66.7	(1.1)	33.3	(0.0)	25.0	(0.4)
Slovak	0.0	(-1.8)	0.0	(0.0)	0.0	(0.0)	0.0	(0.0)	50.0	(0.5)	56.8	(0.7)	22.2	(-0.4)	50.0	(-0.1)	100.0	(0.7)	33.3	(-0.8)	0.0	(-1.9)	0.0	(0.0)	50.0	(1.1)	0.0	(0.0)	0.0	(0.0)
Croatian and Slovenian	100.0	(1.4)	100.0	(1.5)	0.0	(-0.9)	100.0	(1.2)	0.0	(-1.0)	0.0	(-1.0)	20.0	(-0.4)	80.0	(1.0)	85.7	(0.7)	0.0	(-0.8)	33.3	(0.0)	50.0	(0.0)	90.0	(1.1)	33.3	(0.9)	100.0	(2.9)
Czech †	0.0	(0.0)	50.0	(-0.1)	0.0	(-0.9)	100.0	(1.2)	33.3	(0.0)	0.0	(-1.0)	0.0	(3.0)	0.0	(-2.1)	0.0	(0.0)	0.0	(0.0)	0.0	(0.0)	0.0	(0.0)	0.0	(0.0)	0.0	(0.0)	0.0	(-0.5)
Dalmatian, Bosnian, Herzegovinian	0.0	(0.0)	0.0	(0.0)	0.0	(0.0)	0.0	(0.0)	0.0	(0.0)	0.0	(-1.0)	0.0	(-0.8)	0.0	(0.0)	0.0	(0.0)	0.0	(0.0)	0.0	(0.0)	0.0	(0.0)	0.0	(0.0)	0.0	(0.0)	0.0	(0.0)
Ruthenian (Russniak)	85.7	(1.0)	100.0	(1.8)	0.0	(0.0)	100.0	(1.2)	100.0	(2.1)	100.0	(2.6)	100.0	(1.3)	0.0	(1.9)	0.0	(0.0)	100.0	(1.5)	66.7	(0.0)	100.0	(1.4)	0.0	(0.0)	0.0	(0.0)	0.0	(0.0)
Roumanian	75.0	(0.6)	100.0	(1.8)	100.0	(2.0)	100.0	(1.2)	100.0	(2.1)	53.3	(1.3)	55.6	(0.5)	50.0	(-0.1)	71.4	(0.2)	100.0	(1.5)	50.0	(-0.3)	100.0	(1.4)	33.3	(0.6)	0.0	(-0.5)	16.7	(0.1)
Bulgarian, Serbian, Montenegrin	95.0	(1.3)	77.8	(1.0)	75.0	(1.2)	50.0	(0.3)	53.3	(1.6)	100.0	(2.0)	0.9	(-0.8)	33.3	(-0.8)	100.0	(1.1)	100.0	(1.5)	0.0	(0.0)	0.0	(0.0)	75.0	(2.0)	0.0	(-0.5)	0.0	(-0.5)
Greek	80.8	(0.8)	70.3	(0.7)	30.3	(0.0)	66.7	(0.6)	61.5	(0.9)	66.7	(0.6)	26.3	(-0.1)	42.9	(-0.4)	82.4	(0.6)	75.0	(0.6)	66.7	(0.2)	33.3	(0.0)	60.0	(1.5)	44.4	(1.3)	33.3	(0.6)
Armenian	63.6	(0.2)	75.0	(0.8)	66.7	(1.0)	66.7	(0.0)	33.3	(0.0)	36.5	(0.1)	31.4	(-0.1)	41.2	(-0.5)	100.0	(1.3)	62.5	(0.2)	66.7	(0.2)	100.0	(1.4)	20.0	(0.1)	50.0	(1.6)	40.0	(0.9)
Hebrew	38.4	(-0.6)	34.2	(-0.6)	52.0	(0.6)	100.0	(1.2)	44.0	(0.3)	17.6	(-0.5)	25.4	(-0.1)	58.0	(0.0)	94.0	(0.9)	80.0	(0.5)	74.1	(0.4)	38.5	(0.0)	17.6	(-0.6)	36.8	(1.4)	15.4	(0.0)
Syrian	59.1	(0.1)	92.3	(1.5)	87.0	(2.0)	100.0	(1.2)	78.6	(1.4)	67.6	(1.1)	56.3	(0.5)	72.7	(0.7)	75.0	(0.3)	71.4	(0.5)	56.1	(-0.1)	58.3	(0.0)	12.0	(-0.6)	6.3	(-0.2)	56.3	(1.8)
French	66.7	(0.3)	79.4	(1.0)	57.0	(0.9)	85.7	(0.9)	70.8	(0.9)	70.0	(1.1)	91.7	(1.3)	58.8	(0.2)	100.0	(1.1)	57.1	(0.0)	60.7	(0.2)	38.2	(0.0)	12.0	(-0.2)	6.3	(-0.5)	56.3	(1.4)
Turkish	11.1	(-1.5)	100.0	(1.5)	100.0	(2.0)	96.2	(2.0)	97.3	(2.0)	97.3	(1.9)	100.0	(1.5)	94.4	(1.6)	98.5	(1.1)	100.0	(0.7)	60.7	(1.3)	25.1	(-0.3)	27.1	(0.3)	24.0	(-0.5)	20.0	(0.2)
Mexican	100.0	(1.4)	98.6	(1.5)	96.3	(1.9)	96.2	(1.1)	97.7	(2.0)	33.3	(0.0)	100.0	(1.5)	29.4	(-0.9)	86.7	(0.7)	30.8	(-0.9)	100.0	(-1.3)	25.1	(-0.9)	100.0	(2.6)	20.0	(-0.3)	54.5	(1.4)
Spanish American	41.7	(-0.5)	100.0	(1.5)	100.0	(2.0)	33.3	(0.0)	0.0	(-1.0)	100.0	(2.1)	100.0	(1.5)	51.5	(1.1)	75.0	(0.3)	75.0	(0.7)	33.0	(-0.9)	100.0	(1.4)	100.0	(2.5)	20.0	(-0.3)	90.0	(2.2)
Cuban	60.0	(0.1)	100.0	(1.5)	100.0	(1.9)	33.3	(0.0)	100.0	(2.1)	100.0	(2.1)	100.0	(0.8)	51.5	(1.1)	100.0	(1.1)	30.8	(-1.9)	100.0	(-1.9)	100.0	(2.1)	100.0	(2.5)	0.0	(-0.5)	100.0	(2.9)
West Indian	60.0	(0.4)	33.3	(-0.8)	0.0	(-0.9)	33.3	(0.0)	26.0	(0.0)	25.9	(-0.3)	73.2	(0.9)	43.7	(-0.4)	52.0	(-0.4)	45.7	(-0.4)	63.6	(0.0)	47.7	(0.2)	9.3	(-0.3)	0.2	(-0.5)	1.0	(-0.4)
Chinese	9.5	(-1.5)	7.0	(-1.8)	13.4	(-0.5)	23.3	(-0.2)	6.1	(-0.8)	16.7	(-0.5)	0.0	(0.0)	40.0	(-0.5)	52.0	(-2.2)	45.7	(-0.4)	50.0	(-1.0)	0.0	(-0.3)	4.0	(-0.4)	2.8	(-0.4)	25.0	(-0.4)
East Indian	100.0	(1.4)	71.4	(0.7)	11.0	(-0.9)	13.7	(-0.4)	6.1	(-0.8)	1.3	(-0.5)	100.0	(1.5)	27.5	(-1.0)	39.5	(-0.8)	33.8	(-0.8)	20.0	(-1.0)	7.7	(-0.7)	25.0	(0.3)	2.8	(-0.4)	2.8	(-0.4)
Japanese	14.9	(-1.3)	22.2	(-1.2)	40.0	(-0.9)	0.0	(-0.6)	6.1	(-0.8)	1.3	(0.0)	33.3	(0.3)	27.5	(-1.3)	10.0	(0.0)	33.8	(0.0)	27.3	(-1.3)	33.3	(-0.1)	25.0	(0.3)	2.8	(-0.4)	2.8	(-0.4)
Korean	10.0	(-1.5)	34.6	(-0.8)	0.0	(0.0)	0.0	(-0.9)	0.0	(-0.6)	0.0	(0.0)	16.7	(-1.4)	16.7	(0.0)	10.0	(0.0)	0.0	(0.0)	20.0	(-1.3)	33.3	(0.0)	25.0	(0.0)	0.0	(-0.5)	0.0	(0.0)
Pacific Islander	0.0	(-1.8)	0.0	(0.0)	50.0	(0.5)	100.0	(1.2)	0.0	(2.1)	0.0	(-1.0)	0.0	(0.0)	0.0	(-1.2)	0.0	(-1.9)	0.0	(0.0)	0.0	(-1.9)	15.8	(-0.5)	15.0	(0.0)	10.0	(-0.1)	6.7	(-0.2)
African (black)	77.8	(0.7)	51.7	(-0.1)	50.0	(0.5)	64.7	(0.6)	41.7	(0.6)	52.8	(0.6)	34.0	(0.0)	24.1	(-1.2)	52.7	(-0.4)	33.3	(-0.8)	27.3	(-0.8)	15.8	(-0.5)	15.0	(-0.6)	10.0	(-0.1)	6.7	(-0.2)
Other	90.9	(1.1)	85.7	(1.3)	33.3	(0.1)	100.0	(1.2)	66.7	(1.2)	71.4	(1.1)	34.0	(1.5)	62.5	(0.3)	50.0	(-0.5)	75.0	(0.6)	17.6	(-1.4)	100.0	(1.4)	50.0	(1.1)	25.0	(0.5)	28.6	(0.5)

Bold = 1 or more NSDMR below mean   [shaded] = 1 to 1.99 NSDMRs above mean   [dark shaded] = 2 or more NSDMRs above mean

* Norwegians, Danes, and Swedes.
† Bohemians and Moravians.

TABLE 6-6. Deviations in Deportation Rate for Class A Conditions, by Race, 1900–1930, Expressed as Percent Deported and NSDMR

Race or People	1900 %	1900 NSDMR	1901 %	1901 NSDMR	1902 %	1902 NSDMR	1903 %	1903 NSDMR	1904 %	1904 NSDMR	1905 %	1905 NSDMR	1906 %	1906 NSDMR	1907 %	1907 NSDMR	1908 %	1908 NSDMR	1909 %	1909 NSDMR	1910 %	1910 NSDMR
English	0.0	(-0.2)	0.0	(0.0)	0.0	(0.0)	0.0	(-0.3)	0.0	(0.0)	0.0	(-0.2)	0.0	(-0.1)	0.0	(-0.1)	0.1	(-0.1)	0.1	(-0.1)	0.1	(-0.1)
Irish	0.0	(-0.1)	0.0	(0.0)	0.0	(-0.3)	0.0	(-0.2)	0.0	(0.0)	0.1	(-0.2)	0.0	(-0.1)	0.0	(-0.1)	0.0	(-0.1)	0.1	(-0.1)	0.1	(-0.1)
Scotch	0.0	(-0.1)	0.0	(-0.3)	0.0	(-0.1)	0.0	(-0.3)	0.0	(0.0)	0.1	(-0.1)	0.1	(-0.1)	0.0	(-0.1)	0.1	(-0.1)	0.1	(-0.1)	0.1	(-0.1)
Welsh	0.0	(-0.2)	0.1	(0.0)	0.1	(-0.1)	0.0	(-0.3)	0.1	(-0.2)	0.0	(-0.2)	0.1	(-0.1)	0.1	(-0.1)	0.1	(0.0)	0.0	(-0.1)	0.0	(-0.2)
Australian	0.0	(0.0)	0.0	(0.0)	0.0	(0.0)	0.0	(0.0)	0.1	(0.0)	0.0	(0.0)	0.1	(-0.1)	0.1	(0.0)	0.1	(0.0)	0.0	(0.0)	0.0	(0.0)
Spanish	0.0	(-0.2)	0.0	(0.0)	0.0	(-0.4)	0.0	(0.0)	0.2	(-0.1)	0.1	(0.1)	0.3	(0.1)	0.4	(0.1)	1.1	(0.3)	0.1	(-0.1)	0.0	(-0.1)
Portuguese	0.0	(-0.1)	0.0	(0.0)	0.0	(0.0)	0.0	(-0.3)	0.0	(-0.1)	0.1	(-0.1)	0.1	(-0.1)	0.1	(0.0)	0.3	(0.0)	0.1	(-0.1)	0.3	(0.0)
Italian (North)	0.0	(-0.1)	0.0	(0.0)	0.1	(-0.2)	0.0	(-0.3)	0.0	(-0.1)	0.1	(-0.1)	0.1	(-0.1)	0.1	(0.0)	0.2	(-0.1)	0.1	(-0.1)	0.2	(-0.1)
Italian (South)	0.0	(-0.1)	0.0	(0.0)	0.0	(-0.2)	0.1	(-0.2)	0.1	(-0.1)	0.1	(-0.1)	0.1	(0.0)	0.2	(0.0)	0.3	(0.0)	0.1	(-0.1)	0.2	(0.0)
Dutch and Flemish	0.0	(-0.2)	0.0	(0.0)	0.0	(-0.3)	0.0	(-0.2)	0.1	(-0.2)	0.1	(-0.1)	0.1	(0.0)	0.2	(0.0)	0.3	(0.0)	0.1	(-0.1)	0.1	(-0.1)
Finnish	0.2	(0.2)	0.0	(0.0)	0.2	(0.3)	0.4	(0.3)	0.5	(0.4)	0.3	(-0.1)	0.2	(0.0)	0.3	(0.0)	0.3	(0.0)	0.2	(0.0)	0.2	(0.0)
Scandinavian*	0.0	(-0.2)	0.0	(0.0)	0.0	(-0.3)	0.0	(-0.3)	0.0	(-0.2)	0.0	(-0.1)	0.0	(0.0)	0.0	(-0.1)	0.3	(0.0)	0.0	(0.0)	0.1	(0.0)
Lithuanian	0.2	(0.2)	0.0	(0.0)	0.3	(0.6)	0.5	(0.4)	0.7	(0.7)	0.3	(0.3)	0.7	(0.4)	0.4	(0.0)	0.6	(0.1)	0.5	(0.1)	0.5	(0.2)
Russian	0.3	(0.1)	0.3	(0.1)	0.6	(0.6)	0.8	(0.9)	0.3	(0.1)	0.7	(0.5)	0.4	(0.2)	0.2	(0.0)	0.3	(0.0)	0.6	(0.2)	0.3	(0.0)
Polish	0.1	(0.0)	0.1	(0.0)	0.2	(0.3)	0.2	(0.2)	0.3	(0.1)	0.2	(0.0)	0.2	(0.0)	0.2	(0.0)	0.3	(0.0)	0.2	(0.0)	0.3	(0.0)
German	0.1	(-0.1)	0.0	(0.0)	0.1	(0.1)	0.1	(0.1)	0.1	(-0.1)	0.1	(0.0)	0.1	(-0.1)	0.1	(-0.1)	0.2	(0.0)	0.2	(0.0)	0.3	(0.1)
Magyar	0.1	(0.0)	0.0	(0.0)	0.1	(0.2)	0.1	(0.0)	0.1	(-0.1)	0.0	(0.0)	0.0	(0.0)	0.1	(0.0)	0.2	(0.0)	0.2	(0.0)	0.2	(0.0)
Slovak	0.0	(-0.1)	0.0	(0.0)	0.1	(0.1)	0.1	(0.0)	0.1	(0.0)	0.0	(-0.1)	0.0	(0.0)	0.0	(0.0)	0.1	(0.0)	0.1	(-0.1)	0.1	(-0.1)
Croatian & Slovenian	0.0	(-0.2)	0.0	(0.0)	0.0	(-0.1)	0.1	(-0.1)	0.1	(-0.1)	0.0	(-0.1)	0.0	(0.0)	0.0	(0.0)	0.3	(0.0)	0.2	(0.0)	0.2	(0.0)
Transylvanian (Siebenburger)	0.0	(0.0)	0.0	(0.0)	0.0	(0.0)	0.0	(0.0)	0.0	(0.0)	0.0	(0.0)	0.0	(0.0)	0.0	(0.0)	0.0	(0.0)	0.0	(0.0)	0.0	(0.0)
Czech†	0.1	(-0.1)	0.0	(0.0)	0.0	(-0.4)	0.0	(-0.3)	0.0	(-0.2)	0.0	(-0.1)	0.0	(-0.1)	0.1	(-0.1)	0.1	(0.0)	0.1	(0.0)	0.2	(0.0)
Dalmatian, Bosnian, Herzegovinian	0.0	(-0.2)	0.0	(0.0)	0.0	(-0.4)	0.0	(-0.3)	0.1	(-0.1)	0.0	(-0.1)	0.0	(-0.1)	0.1	(-0.1)	0.3	(0.0)	0.3	(0.1)	0.3	(0.1)
Ruthenian (Russniak)	0.0	(0.0)	0.0	(0.0)	0.1	(-0.2)	0.0	(-0.2)	0.3	(0.1)	0.0	(-0.1)	0.0	(-0.1)	0.2	(0.0)	0.1	(0.0)	0.2	(0.0)	0.2	(-0.1)
Austro-Hungarian	0.0	(0.0)	0.0	(0.0)	0.0	(0.0)	0.0	(0.0)	0.3	(0.1)	0.0	(0.0)	0.0	(0.0)	0.0	(0.0)	0.1	(0.0)	0.0	(0.0)	0.0	(0.0)
Roumanian	0.0	(-0.2)	0.0	(0.0)	0.0	(-0.2)	0.0	(-0.3)	0.1	(-0.1)	0.0	(0.0)	0.2	(0.0)	0.2	(0.0)	0.3	(0.0)	0.1	(0.0)	0.3	(0.1)
Bulgarian, Serbian, Montenegrin	0.0	(-0.2)	0.0	(0.0)	0.3	(0.7)	0.1	(0.0)	0.1	(-0.2)	0.1	(0.1)	0.1	(-0.1)	0.2	(0.0)	0.3	(0.0)	0.6	(0.2)	0.3	(0.1)
Greek	0.2	(0.3)	0.1	(0.1)	0.0	(0.1)	0.2	(0.1)	0.4	(0.0)	0.2	(0.2)	0.1	(-0.1)	0.2	(-0.1)	0.4	(0.0)	0.4	(0.0)	0.3	(0.0)
Armenian	1.1	(2.4)	0.4	(0.2)	0.2	(0.5)	1.5	(1.9)	1.7	(1.7)	2.7	(2.4)	1.7	(1.1)	1.8	(0.7)	2.9	(1.1)	2.1	(0.9)	3.0	(2.1)
Hebrew	0.2	(0.2)	0.1	(0.0)	0.2	(0.2)	0.2	(0.2)	0.2	(0.0)	0.3	(0.1)	0.2	(0.0)	0.2	(0.0)	0.3	(-0.1)	0.3	(0.1)	0.3	(0.1)
Arabian	0.2	(0.2)	0.1	(0.0)	0.2	(0.0)	0.2	(0.2)	0.0	(0.0)	0.2	(0.0)	0.0	(0.0)	0.2	(0.0)	0.4	(0.0)	0.4	(0.1)	0.3	(0.3)
Syrian	1.0	(3.0)	0.7	(0.3)	1.0	(5.1)	1.0	(1.2)	3.1	(4.3)	3.2	(2.9)	5.6	(3.7)	8.2	(3.7)	5.8	(2.4)	3.2	(1.4)	2.9	(2.0)
Turkish	0.0	(-0.2)	0.0	(0.0)	0.0	(-0.4)	0.9	(1.0)	0.6	(0.6)	0.4	(0.2)	1.1	(0.6)	0.5	(0.1)	0.8	(0.2)	0.7	(0.2)	1.2	(0.7)
French	0.0	(-0.1)	0.0	(0.0)	0.0	(-0.2)	0.0	(-0.3)	0.0	(-0.3)	0.1	(-0.1)	0.1	(-0.1)	0.1	(0.0)	0.2	(-0.1)	0.1	(0.1)	0.1	(0.1)
Swiss	0.0	(0.0)	0.0	(0.0)	0.0	(0.0)	0.0	(0.0)	0.0	(0.0)	0.0	(0.0)	0.0	(0.0)	0.0	(0.0)	0.1	(0.0)	0.0	(0.0)	0.0	(0.0)
Mexican	0.0	(-0.2)	0.0	(0.0)	0.0	(0.0)	0.0	(0.0)	0.2	(0.2)	0.0	(0.0)	0.0	(0.0)	0.0	(0.0)	1.2	(0.3)	0.6	(0.2)	0.5	(0.2)
Central American (NS)	0.0	(0.0)	0.0	(0.0)	0.0	(0.1)	0.0	(0.0)	0.0	(0.0)	0.0	(-0.2)	0.0	(-0.1)	0.0	(0.0)	0.0	(0.3)	0.0	(0.2)	0.0	(0.0)
South American	0.0	(0.0)	0.0	(0.0)	0.0	(0.0)	0.0	(0.0)	0.0	(0.0)	0.0	(0.0)	0.0	(0.0)	0.0	(0.0)	0.7	(0.1)	0.0	(0.0)	0.0	(0.0)
Spanish American	0.0	(-0.2)	0.0	(0.0)	0.0	(-0.4)	0.0	(-0.2)	0.0	(-0.3)	0.2	(0.0)	0.3	(0.1)	1.6	(0.6)	0.7	(0.1)	0.2	(0.0)	0.2	(0.1)
Cuban	0.0	(-0.2)	0.0	(0.0)	0.0	(-0.4)	0.0	(-0.3)	0.0	(-0.1)	0.1	(-0.1)	0.3	(0.1)	0.4	(0.1)	0.2	(0.2)	0.0	(-0.1)	0.0	(-0.2)
West Indian	1.3	(2.7)	0.0	(0.0)	0.7	(2.2)	0.1	(0.1)	0.2	(0.2)	0.1	(-0.2)	0.4	(0.1)	0.4	(0.1)	0.2	(0.2)	0.0	(-0.1)	0.0	(-0.2)
Chinese	0.0	(-0.2)	0.0	(0.0)	0.0	(-0.4)	0.0	(-0.3)	0.9	(1.1)	3.5	(3.5)	5.0	(3.3)	9.2	(4.2)	10.7	(4.5)	1.7	(0.7)	1.1	(0.7)
East Indian	0.3	(0.5)	0.6	(0.3)	0.0	(-0.4)	1.2	(1.4)	1.4	(1.4)	2.0	(2.3)	1.8	(1.1)	12.3	(3.1)	9.0	(3.7)	5.9	(5.3)	7.4	(5.4)
Japanese	0.3	(0.5)	0.6	(0.3)	0.2	(0.2)	0.0	(0.0)	0.0	(0.0)	0.0	(0.0)	0.0	(0.0)	1.9	(0.9)	0.0	(0.0)	2.7	(1.2)	1.0	(0.6)
Korean	0.0	(-0.2)	2.1	(1.1)	0.0	(0.0)	2.7	(3.7)	0.0	(0.0)	2.6	(2.3)	1.8	(0.1)	2.2	(0.9)	1.9	(-0.2)	2.7	(1.2)	1.0	(-0.1)
Filipino	0.0	(-0.2)	12.3	(6.6)	0.0	(0.0)	0.0	(-0.3)	0.0	(-0.3)	0.4	(0.1)	5.8	(3.8)	2.0	(0.8)	0.0	(0.0)	0.0	(0.0)	0.4	(0.1)
Hawaiian	0.0	(0.0)	0.0	(0.0)	0.0	(0.0)	0.0	(0.0)	0.0	(0.0)	0.0	(0.0)	0.0	(0.0)	0.0	(0.0)	0.0	(0.0)	0.0	(0.0)	0.0	(0.0)
Pacific Islander	1.5	(3.9)	0.0	(0.0)	0.0	(-0.4)	1.9	(2.5)	0.0	(-0.3)	0.0	(-0.2)	0.0	(-0.1)	0.0	(-0.1)	0.0	(-0.2)	3.4	(1.6)	2.9	(2.0)
African (black)	0.0	(-0.2)	0.0	(0.0)	0.0	(-0.4)	0.2	(0.2)	0.0	(-0.2)	0.0	(-0.2)	0.0	(-0.1)	0.3	(0.0)	0.2	(-0.1)	0.3	(0.0)	0.2	(-0.1)
(NS)	0.0	(0.0)	0.0	(0.0)	0.0	(0.0)	0.0	(0.0)	0.0	(0.0)	0.0	(0.0)	0.0	(0.0)	0.0	(0.0)	0.0	(0.0)	0.0	(0.0)	0.0	(0.0)
Esquimaux	0.0	(0.0)	0.0	(0.0)	0.0	(0.0)	0.0	(0.0)	0.0	(0.0)	0.0	(0.0)	0.0	(0.0)	0.0	(0.0)	0.0	(0.0)	0.0	(0.0)	0.0	(0.0)
Other	0.0	(0.0)	0.0	(0.0)	0.0	(0.0)	0.0	(-0.3)	2.4	(3.2)	1.4	(1.2)	1.0	(0.6)	1.3	(0.5)	1.0	(0.3)	0.5	(0.1)	0.6	(0.3)

Bold = 1 or more NSDMRs below mean = 1 to 1.99 NSDMRs above mean = 2 or more NSDMRs above mean NS = not specified

* Norwegians, Danes, and Swedes. † Bohemians and Moravians.

TABLE 6-6. Continued

Race or People	1911 %	1911 NSDMR	1912 %	1912 NSDMR	1913 %	1913 NSDMR	1914 %	1914 NSDMR	1915 %	1915 NSDMR	1916 %	1916 NSDMR	1917 %	1917 NSDMR	1918 %	1918 NSDMR	1919 %	1919 NSDMR	1920 %	1920 NSDMR	1921 %	1921 NSDMR
English	0.1	(0.1)	0.1	(0.1)	0.1	(0.1)	0.1	0.1	0.1	(0.1)	0.1	-0.4	0.1	-0.3	0.2	(0.0)	0.0	(0.1)	0.0	-0.2	0.0	-0.5
Irish	0.1	(0.0)	0.1	(0.1)	0.1	(0.1)	0.1	-0.2	0.1	-0.1	0.1	-0.4	0.2	-0.3	0.3	(0.0)	0.1	-0.1	0.0	-0.2	0.0	-0.5
Scotch	0.1	-0.1	0.1	-0.1	0.1	-0.1	0.1	-0.2	0.1	-0.1	0.1	-0.3	0.2	-0.3	0.2	-0.1	0.1	-0.1	0.0	-0.1	0.0	-0.6
Welsh	0.1	(0.1)	0.1	-0.1	0.1	-0.1	0.1	-0.2	0.1	-0.1	0.1	-0.1	0.1	-0.3	0.2	-0.1	0.0	-0.1	0.0	-0.3	0.0	-0.6
Australian	0.0	(0.0)	0.0	-0.1	0.1	(0.0)	0.2	-0.2	0.2	-0.1	0.1	(0.0)	0.1	(0.0)	0.0	-0.1	0.0	-0.2	0.0	-0.3	0.0	-0.7
Spanish	0.1	-0.1	0.0	(0.0)	0.0	(0.1)	0.5	0.3	0.3	-0.1	0.3	-0.1	0.3	-0.1	0.4	(0.0)	0.1	-0.1	0.1	(0.0)	0.2	(0.6)
Portuguese	0.1	-0.1	0.1	-0.1	0.1	-0.1	0.1	-0.1	0.2	-0.1	0.1	-0.3	0.1	-0.3	0.3	-0.1	0.2	-0.1	0.1	-0.2	0.1	(0.6)
Italian (North)	0.2	(0.0)	0.1	(0.0)	0.1	-0.1	0.1	-0.1	0.4	(0.0)	0.1	-0.1	0.3	-0.3	0.3	-0.1	0.2	(0.0)	0.0	-0.2	0.0	-0.4
Italian (South)	0.3	(0.0)	0.2	(0.0)	0.2	(0.0)	0.2	(0.0)	0.5	(0.0)	0.3	-0.1	0.3	-0.2	0.2	-0.1	0.0	-0.1	0.0	-0.2	0.1	(0.1)
Dutch and Flemish	0.1	-0.1	0.1	(0.0)	0.1	-0.1	0.1	-0.2	0.1	-0.1	0.1	-0.3	0.2	-0.4	0.2	-0.1	0.0	-0.1	0.0	-0.2	0.0	-0.6
Finnish	0.2	(0.0)	0.2	(0.0)	0.2	-0.1	0.1	-0.2	0.5	(0.0)	0.4	-0.1	0.4	-0.2	0.4	(0.0)	0.1	-0.1	0.1	-0.1	0.0	-0.4
Scandinavian*	0.0	-0.1	0.0	-0.1	0.1	-0.1	0.1	-0.2	0.3	(0.0)	0.1	-0.4	0.2	-0.2	0.3	-0.1	0.1	-0.1	0.0	-0.1	0.0	-0.4
Lithuanian	0.5	(0.1)	0.3	(0.1)	0.3	(0.1)	0.4	0.2	0.9	(0.1)	0.4	-0.1	0.4	-0.1	0.5	-0.3	0.4	(0.4)	0.0	-0.3	0.0	-0.7
Russian	0.3	(0.0)	0.2	(0.0)	0.2	-0.1	0.3	(0.0)	1.4	(0.3)	0.3	(0.0)	0.4	-0.1	0.5	-0.1	0.2	(0.0)	0.1	(0.0)	0.0	-0.5
Polish	0.2	(0.0)	0.2	(0.0)	0.2	(0.0)	0.6	(0.0)	0.6	(0.0)	0.4	(0.0)	0.4	-0.2	0.4	(0.2)	0.2	(0.0)	0.1	(0.0)	0.0	-0.4
German	0.2	(0.0)	0.1	(0.0)	0.2	(0.0)	0.1	-0.1	0.1	-0.1	0.2	-0.2	0.1	-0.3	0.4	(0.1)	0.2	(0.0)	0.1	-0.1	0.0	-0.4
Magyar	0.2	(0.0)	0.1	(0.0)	0.1	(0.0)	0.1	-0.1	0.2	-0.2	0.6	(0.3)	0.1	-0.4	0.4	-0.3	0.2	-0.2	0.8	(2.4)	0.1	(0.1)
Slovak	0.2	(0.0)	0.2	(0.0)	0.2	-0.1	0.1	-0.1	0.2	-0.1	0.4	-0.5	0.4	-0.1	0.0	-0.3	0.2	-0.2	0.0	-0.2	0.1	-0.1
Croatian & Slovenian	0.2	(0.0)	2.0	(1.1)	0.1	(0.2)	1.5	(1.2)	4.5	(1.4)	1.3	(1.2)	0.3	(1.5)	0.8	-0.3	0.0	-0.2	0.0	-0.2	0.1	-0.1
Transylvanian (Siebenburger)	0.0	(0.0)	0.0	(0.0)	0.0	(0.0)	0.0	(0.0)	0.0	(0.0)	0.0	(0.0)	0.0	(0.0)	0.0	-0.3	0.0	-0.2	0.0	(0.0)	0.0	(0.0)
Czech †	0.1	(0.0)	0.0	-0.1	0.0	-0.1	0.1	-0.1	2.5	(0.7)	0.0	-0.5	0.0	-0.4	1.1	(0.5)	0.0	(1.1)	0.0	-0.3	0.0	-0.7
Dalmatian, Bosnian, Herzegovinian	0.1	-0.1	0.1	(0.0)	0.0	-0.1	0.0	-0.1	0.5	(0.0)	0.8	(0.7)	2.0	(1.3)	0.0	-0.3	1.3	(1.3)	0.0	-0.3	0.0	-0.7
Ruthenian (Russniak)	0.2	(0.0)	0.1	(0.0)	0.0	(0.0)	0.0	(0.0)	0.5	(0.0)	0.4	(0.0)	0.1	-0.3	0.0	-0.3	1.8	(2.2)	0.4	(1.1)	0.2	(0.7)
Austro-Hungarian	0.0	(0.0)	0.0	(0.0)	0.0	(0.0)	0.0	(0.0)	0.0	(0.0)	0.3	(0.0)	0.2	(0.0)	0.6	(0.1)	0.6	(0.6)	0.2	(0.2)	0.1	(0.3)
Roumanian	0.5	(0.1)	0.1	(0.1)	0.0	(0.0)	0.0	(0.0)	1.0	(0.2)	0.3	-0.1	0.2	-0.3	0.6	(0.1)	0.6	(0.1)	0.2	(0.3)	0.1	(0.3)
Bulgarian, Serbian, Montenegrin	0.3	(0.0)	0.2	(0.0)	0.5	(0.2)	0.3	(0.1)	1.8	(0.4)	0.6	(0.3)	0.6	(0.1)	1.5	(0.8)	0.2	(0.1)	0.3	(0.3)	0.1	-0.2
Greek	0.4	(0.0)	0.2	(0.2)	0.6	(0.2)	0.2	(0.0)	0.6	(0.0)	0.3	-0.1	0.3	-0.1	0.4	(0.0)	0.4	(0.3)	0.1	-0.1	0.1	(0.0)
Armenian	2.6	(0.8)	1.3	(0.8)	0.6	(0.2)	1.5	(1.2)	2.0	(0.5)	2.1	(2.4)	2.3	(1.5)	0.8	(0.3)	0.2	-0.2	0.2	-0.2	0.2	(0.8)
Hebrew	0.3	(0.0)	0.3	(0.1)	0.2	(0.0)	0.8	(1.5)	0.9	(0.1)	0.2	-0.2	0.1	-0.3	0.3	(0.0)	0.1	(0.0)	0.1	-0.1	0.0	-0.3
Arabian	0.0	(0.0)	0.0	(0.0)	0.0	(0.0)	0.7	(0.0)	0.0	(0.0)	0.0	(0.0)	0.0	(0.0)	0.0	(0.0)	0.2	(0.0)	0.1	(0.0)	0.0	(0.0)
Syrian	3.8	(1.2)	1.8	(1.1)	1.6	(0.9)	2.0	(1.7)	3.3	(1.0)	3.2	(4.1)	3.4	(2.4)	7.9	(5.2)	0.6	(0.6)	0.3	(0.6)	9.4	(2.5)
Turkish	1.3	(0.4)	0.8	(0.4)	0.8	(0.3)	1.1	(0.9)	1.8	(0.5)	0.4	(0.1)	0.9	(0.3)	3.8	(2.4)	0.1	-0.2	0.5	(1.3)	0.0	-0.7
French	0.2	(0.0)	0.2	(0.0)	0.1	-0.1	0.1	-0.1	0.1	-0.1	0.1	-0.3	0.3	-0.3	0.3	-0.1	0.1	-0.1	0.0	-0.1	0.0	-0.5
Swiss	0.0	(0.0)	0.0	(0.0)	0.0	(0.0)	0.0	(0.0)	0.0	(0.0)	0.0	(0.0)	0.0	(0.0)	0.0	(0.0)	0.0	(0.0)	0.0	(0.0)	0.0	(0.0)
Mexican	0.4	(0.0)	0.3	(0.1)	1.7	(0.0)	0.0	(1.1)	1.4	(0.3)	1.8	(2.0)	3.1	(2.1)	0.5	(0.1)	0.7	(0.7)	0.3	(0.7)	0.2	(1.2)
Central American (NS)	0.0	(0.0)	0.0	(0.0)	0.0	(0.0)	0.0	(0.0)	0.0	(0.0)	0.0	(0.0)	0.0	(0.0)	0.0	(0.0)	0.0	(0.0)	0.0	(0.0)	0.0	(0.0)
South American	0.0	(0.0)	0.0	(0.0)	0.0	(0.0)	0.0	-0.1	0.0	(0.0)	0.0	-0.1	0.1	-0.3	0.2	-0.1	0.0	(0.0)	0.0	-0.3	0.1	-0.2
Spanish American	0.2	(0.0)	0.0	-0.1	0.1	-0.1	0.1	-0.1	0.4	(0.0)	0.3	-0.1	0.1	-0.3	0.0	-0.2	0.0	-0.2	0.0	-0.3	0.1	-0.5
Cuban	0.0	(0.0)	0.0	-0.1	0.0	(0.0)	0.0	-0.2	0.5	-0.2	0.1	-0.4	0.1	-0.3	0.0	-0.3	0.0	-0.2	0.0	-0.2	0.1	-0.5
West Indian	0.1	-0.1	0.0	(0.0)	0.3	(0.0)	0.0	-0.2	0.5	(0.0)	0.1	(0.5)	0.1	-0.3	0.0	-0.3	0.0	-0.2	0.0	-0.3	0.1	-0.2
Chinese	1.3	(0.4)	0.9	(0.5)	2.4	(1.3)	1.6	(1.3)	0.8	(0.1)	1.3	(1.4)	1.6	(0.9)	2.1	(1.2)	0.8	(0.9)	0.6	(1.5)	0.6	(4.7)
East Indian	17.8	(6.2)	8.9	(6.1)	9.7	(6.0)	6.0	(5.7)	18.1	(6.2)	2.4	(2.9)	3.0	(1.0)	0.7	(0.2)	0.5	(0.5)	0.3	(0.6)	0.2	(1.0)
Japanese	0.5	(0.1)	1.1	(0.6)	2.2	(1.2)	1.0	(0.8)	0.7	(0.1)	0.7	(0.4)	1.6	(1.0)	0.7	(0.2)	0.5	(0.5)	0.3	(0.6)	0.0	-0.3
Korean	0.0	-0.1	0.0	-0.1	1.5	(0.8)	1.8	(1.5)	0.3	(0.3)	0.6	(0.4)	0.0	(0.0)	0.0	(0.2)	0.0	-0.2	0.0	-0.3	0.0	-0.7
Filipino	0.0	(0.0)	0.0	(0.0)	1.9	(1.1)	0.5	(0.2)	0.7	(0.1)	0.1	-0.3	0.0	-0.4	0.0	-0.3	0.0	-0.2	0.0	-0.3	0.0	-0.7
Hawaiian	0.0	(0.0)	0.0	(0.0)	0.0	(0.0)	0.0	(0.0)	0.0	(0.0)	0.0	(0.0)	0.0	(0.0)	0.0	(0.0)	0.0	(0.0)	0.0	(0.0)	0.0	(0.0)
Pacific Islander	0.0	-0.1	0.0	(0.0)	0.0	-0.1	0.0	-0.2	0.0	-0.2	0.0	-0.5	0.0	-0.4	4.2	(2.6)	0.0	-0.2	1.8	(5.4)	0.0	-0.7
African (black)	0.2	(0.0)	0.1	(0.0)	0.4	(0.1)	0.1	(0.1)	1.0	(0.2)	1.1	(1.1)	0.5	(0.0)	0.4	(0.0)	0.1	(0.0)	0.1	(0.0)	0.1	(0.3)
(NS)	0.0	(0.0)	0.0	(0.0)	0.0	(0.0)	0.0	(0.0)	0.0	(0.0)	0.0	(0.0)	0.0	(0.0)	0.0	(0.0)	0.0	(0.0)	0.0	(0.0)	0.1	(0.0)
Esquimaux	0.0	(0.0)	0.0	(0.0)	0.0	(0.0)	0.0	(0.0)	0.0	(0.0)	0.0	(0.0)	0.0	(0.0)	0.0	(0.0)	0.0	(0.0)	0.0	(0.0)	0.0	(0.0)
Other	1.1	(0.3)	0.4	(0.1)	0.7	(0.3)	0.8	(0.6)	0.7	(0.1)	0.4	(0.1)	0.8	(0.3)	0.3	-0.1	0.4	(0.3)	0.1	(0.1)	0.1	(0.4)

TABLE 6-6. Continued

Race or People	1922 %	1922 NSDMR	1923 %	1923 NSDMR	1924 %	1924 NSDMR	1925 %	1925 NSDMR	1926 %	1926 NSDMR	1927 %	1927 NSDMR	1928 %	1928 NSDMR	1929 %	1929 NSDMR	1930 %	1930 NSDMR
English	0.0	(-0.2)	0.0	(-0.3)	0.0	(-0.3)	0.0	(-0.1)	0.0	(-0.2)	0.0	(0.0)	0.0	(-0.1)	0.0	(-0.5)	0.0	(-0.2)
Irish	0.1	(-0.1)	0.1	(-0.1)	0.0	(-0.3)	0.1	(-0.1)	0.0	(-0.2)	0.0	(0.0)	0.0	(-0.1)	0.0	(-0.6)	0.0	(-0.1)
Scotch	0.0	(-0.1)	0.0	(-0.2)	0.0	(-0.3)	0.1	(-0.1)	0.0	(-0.2)	0.0	(0.0)	0.0	(-0.1)	0.0	(-0.4)	0.0	(-0.2)
Welsh	0.1	(0.0)	0.0	(-0.4)	0.0	(-0.2)	0.0	(-0.1)	0.0	(-0.3)	0.0	(-0.1)	0.0	(-0.1)	0.0	(-0.6)	0.0	(-0.2)
Australian	0.1	(0.0)	0.0	(0.0)	0.0	(0.0)	0.0	(0.0)	0.0	(0.0)	0.0	(0.0)	0.0	(0.0)	0.0	(0.0)	0.0	(0.0)
Spanish	0.0	(-0.1)	0.1	(-0.1)	0.1	(-0.1)	0.1	(0.0)	0.0	(0.0)	0.0	(0.1)	0.0	(-0.1)	0.0	(0.0)	0.0	(-0.1)
Portuguese	0.0	(-0.2)	0.1	(-0.2)	0.1	(-0.1)	0.7	(0.4)	0.2	(0.0)	0.2	(0.1)	0.6	(0.5)	0.0	(0.0)	0.0	(0.3)
Italian (North)	0.1	(-0.1)	0.1	(-0.2)	0.1	(-0.1)	0.4	(0.2)	0.1	(-0.2)	0.2	(0.1)	0.1	(0.1)	0.1	(-0.6)	0.1	(-0.1)
Italian (South)	0.2	(0.1)	0.1	(0.0)	0.2	(0.0)	0.9	(0.6)	0.4	(0.2)	0.1	(0.0)	0.6	(0.5)	0.0	(-0.5)	0.0	(-0.1)
Dutch and Flemish	0.1	(-0.1)	0.1	(0.0)	0.2	(0.0)	0.5	(0.3)	0.0	(0.0)	0.1	(0.0)	0.1	(0.0)	0.0	(-0.4)	0.0	(-0.1)
Finnish	0.2	(0.0)	0.1	(-0.3)	0.1	(-0.1)	0.4	(0.2)	0.1	(-0.1)	0.0	(-0.3)	0.0	(-0.1)	0.0	(-0.4)	0.0	(-0.2)
Scandinavian*	0.2	(0.0)	0.0	(0.0)	0.1	(-0.2)	0.1	(-0.1)	0.1	(-0.1)	1.3	(0.7)	0.0	(-0.1)	0.1	(1.8)	0.0	(-0.2)
Lithuanian	0.5	(0.5)	0.2	(0.1)	0.2	(0.1)	0.3	(0.1)	0.1	(-0.3)	0.0	(-0.1)	0.0	(0.0)	0.0	(-0.6)	0.1	(0.9)
Russian	0.1	(0.1)	0.1	(0.1)	0.2	(0.0)	0.3	(0.1)	0.1	(-0.1)	0.2	(0.0)	0.0	(-0.1)	0.1	(0.5)	0.0	(-0.1)
Polish	0.1	(-0.1)	0.1	(0.0)	0.1	(-0.2)	0.1	(-0.1)	0.1	(0.0)	0.1	(0.0)	0.0	(0.0)	0.2	(2.3)	0.0	(-0.1)
German	0.1	(-0.1)	0.1	(0.0)	0.1	(-0.2)	0.1	(-0.1)	0.1	(-0.1)	0.2	(0.1)	0.0	(0.1)	0.0	(-0.4)	0.0	(-0.1)
Magyar	0.1	(-0.1)	0.1	(0.0)	0.1	(0.0)	0.2	(0.0)	0.2	(0.2)	0.2	(0.1)	0.2	(0.1)	0.0	(-0.1)	0.0	(-0.1)
Slovak	0.1	(-0.1)	0.1	(0.1)	0.2	(0.2)	0.5	(0.3)	0.0	(-0.1)	0.0	(-0.1)	0.0	(0.0)	0.0	(-0.6)	0.0	(-0.2)
Croatian & Slovenian	0.0	(-0.2)	0.0	(0.0)	0.1	(0.2)	0.2	(0.0)	0.1	(-0.1)	0.2	(0.1)	0.4	(0.3)	0.0	(0.0)	0.1	(0.7)
Transylvanian (Siebenburger)	0.0	(0.0)	0.0	(0.0)	0.0	(0.0)	0.2	(0.0)	0.0	(0.0)	0.0	(0.0)	0.0	(0.0)	0.0	(0.0)	0.0	(0.0)
Czech †	0.0	(-0.1)	0.0	(-0.4)	0.1	(-0.4)	0.2	(0.0)	0.1	(-0.2)	0.1	(0.0)	0.0	(-0.1)	0.0	(-0.6)	0.0	(-0.2)
Dalmatian, Bosnian, Herzegovinian	0.0	(-0.2)	0.0	(-0.4)	0.0	(-0.4)	0.0	(-0.1)	0.4	(0.7)	0.4	(0.2)	0.0	(-0.1)	0.0	(-0.6)	0.0	(-0.2)
Ruthenian (Russniak)	0.1	(-0.1)	0.1	(-0.2)	0.1	(-0.2)	0.3	(0.1)	0.2	(0.2)	0.2	(0.2)	0.0	(-0.1)	0.0	(-0.6)	0.0	(-0.1)
Austro-Hungarian	0.0	(0.0)	0.0	(0.0)	0.0	(0.0)	0.0	(0.0)	0.0	(0.0)	0.0	(0.0)	0.0	(0.0)	0.0	(0.0)	0.0	(0.0)
Roumanian	0.3	(0.2)	0.2	(0.1)	0.2	(0.3)	1.2	(0.8)	0.1	(0.0)	0.7	(0.3)	0.2	(0.2)	0.0	(-0.6)	0.1	(0.7)
Bulgarian, Serbian, Montenegrin	0.2	(0.1)	0.1	(-0.1)	0.3	(0.3)	0.5	(0.3)	0.0	(-0.3)	0.0	(-0.1)	0.6	(0.5)	0.0	(-0.6)	0.0	(-0.2)
Greek	0.1	(-0.1)	0.1	(0.0)	0.2	(0.1)	0.5	(0.2)	0.1	(0.0)	0.0	(0.0)	0.0	(0.0)	0.1	(0.7)	0.1	(0.4)
Armenian	0.4	(0.4)	0.3	(0.3)	1.3	(2.5)	1.4	(1.1)	0.3	(0.5)	0.5	(0.2)	0.1	(0.0)	0.2	(2.3)	0.4	(3.4)
Hebrew	0.1	(0.0)	0.2	(0.1)	0.1	(0.0)	0.2	(0.0)	0.1	(0.1)	0.1	(0.0)	0.1	(0.0)	0.0	(0.1)	0.0	(0.0)
Arabian	0.0	(0.0)	0.0	(0.0)	0.0	(0.0)	0.0	(0.0)	0.0	(0.0)	0.0	(0.0)	0.0	(0.0)	0.0	(0.0)	0.0	(0.0)
Syrian	0.4	(0.4)	0.4	(0.6)	0.5	(0.7)	0.6	(0.5)	0.5	(1.2)	0.0	(-0.1)	0.0	(-0.1)	0.0	(-0.6)	0.3	(2.5)
Turkish	0.0	(-0.2)	0.0	(-0.4)	0.0	(-0.4)	1.1	(0.7)	0.8	(-0.1)	0.1	(-0.1)	0.0	(-0.1)	0.3	(-0.6)	0.3	(2.6)
French	0.0	(-0.1)	0.0	(-0.2)	0.1	(-0.2)	0.0	(-0.1)	0.1	(0.0)	0.1	(0.0)	0.0	(-0.1)	0.0	(-0.5)	0.0	(0.1)
Swiss	0.2	(0.0)	0.0	(0.0)	0.0	(0.0)	0.2	(0.0)	0.0	(0.0)	0.1	(0.0)	0.1	(0.1)	0.2	(3.0)	0.2	(1.4)
Mexican	0.0	(0.0)	0.2	(0.0)	0.3	(0.3)	0.2	(0.0)	0.2	(0.2)	0.0	(0.0)	0.0	(0.0)	0.0	(0.0)	0.0	(0.0)
Central American (NS)	0.0	(0.0)	0.0	(0.0)	0.0	(0.0)	0.0	(0.0)	0.0	(0.0)	0.0	(0.0)	0.0	(0.0)	0.0	(0.0)	0.0	(0.0)
South American	0.0	(0.0)	0.0	(0.0)	0.0	(0.0)	0.0	(0.0)	0.0	(0.0)	0.0	(0.0)	0.0	(0.0)	0.0	(0.0)	0.0	(0.0)
Spanish American	0.1	(-0.1)	0.1	(-0.1)	0.2	(0.0)	0.2	(0.0)	0.1	(-0.1)	0.2	(0.1)	0.1	(0.1)	0.1	(-0.4)	0.1	(0.5)
Cuban	0.0	(-0.2)	0.0	(-0.1)	0.0	(-0.3)	0.0	(0.0)	0.0	(-0.3)	0.1	(0.0)	0.0	(-0.1)	0.0	(-0.6)	0.0	(0.2)
West Indian	0.0	(-0.2)	0.0	(-0.4)	0.0	(-0.3)	0.0	(-0.1)	0.0	(-0.3)	0.0	(-0.1)	0.0	(-0.1)	0.0	(-0.6)	0.0	(0.2)
Chinese	1.6	(2.2)	2.3	(5.7)	2.6	(3.4)	7.1	(5.6)	1.8	(3.9)	0.7	(5.0)	5.0	(4.7)	0.2	(3.0)	0.2	(1.4)
East Indian	0.0	(-0.1)	0.7	(1.6)	0.0	(-0.4)	0.0	(-0.1)	0.4	(0.8)	0.0	(-0.1)	0.0	(0.7)	0.4	(0.4)	0.4	(3.6)
Japanese	0.7	(0.8)	0.4	(0.6)	0.7	(0.9)	2.3	(2.1)	0.1	(0.0)	0.3	(0.1)	0.7	(4.7)	0.0	(-0.2)	0.0	(-0.1)
Korean	0.0	(0.0)	0.7	(1.3)	0.9	(0.9)	0.0	(-0.1)	1.2	(5.0)	4.3	(5.0)	4.5	(4.3)	0.0	(-0.6)	0.0	(-0.2)
Filipino	0.0	(-0.2)	0.0	(-0.4)	0.0	(-0.4)	0.0	(0.0)	0.0	(0.0)	0.0	(0.0)	0.0	(0.0)	0.0	(-0.4)	0.0	(0.0)
Hawaiian	0.0	(0.0)	0.0	(0.0)	0.0	(0.0)	0.0	(0.0)	0.0	(0.0)	0.0	(0.0)	0.0	(0.0)	0.0	(0.0)	0.0	(0.0)
Pacific Islander	4.0	(15.9)	0.0	(-0.4)	0.0	(0.0)	0.4	(0.2)	0.0	(-0.3)	0.3	(0.1)	0.3	(0.2)	0.0	(-0.6)	0.0	(-0.2)
African (black)	0.2	(0.0)	0.0	(0.0)	0.1	(-0.1)	0.1	(0.0)	0.2	(0.2)	0.3	(0.1)	0.3	(0.2)	0.0	(-0.2)	0.0	(0.0)
(NS)	0.0	(0.0)	0.0	(0.0)	0.0	(0.0)	0.0	(0.0)	0.0	(0.0)	0.0	(0.0)	0.0	(0.0)	0.0	(0.0)	0.0	(0.0)
Esquimauz	0.0	(0.0)	0.0	(0.0)	0.0	(0.0)	0.0	(0.0)	0.0	(0.0)	0.0	(0.0)	0.0	(0.0)	0.0	(0.0)	0.0	(0.0)
Other	0.2	(0.1)	0.4	(0.7)	0.4	(0.4)	0.9	(0.6)	0.2	(0.4)	0.8	(0.4)	0.2	(0.1)	0.1	(1.3)	0.2	(1.5)

TABLE 6-7. Deviations in Certification Rate for Trachoma, by Race, 1916–1930, Expressed as Percentage Certified and NSDMR

Race or People	1916		1917		1918		1919		1920		1921		1922		1923		1924		1925		1926		1927		1928		1929		1930	
	%	NSDMR	%	NSDMR	%	NSDMR	%	NSDMR	%	NSDMR	%	NSDMR	%	NSDMR	%	NSDMR	%	NSDMR	%	NSDMR	%	NSDMR	%	NSDMR	%	NSDMR	%	NSDMR	%	NSDMR
English	0.0	-0.2	0.0	-0.3	0.0	-0.2	0.0	-0.2	0.0	-0.2	0.0	-0.2	0.0	-0.3	0.0	-0.3	0.0	-0.2	0.0	-0.1	0.0	-0.1	0.0	0.0	0.0	(0.0)	0.0	-0.3	0.0	-0.1
Irish	0.0	-0.2	0.1	-0.2	0.0	-0.2	0.0	-0.2	0.0	-0.1	0.0	-0.1	0.0	-0.1	0.0	-0.2	0.0	-0.2	0.0	-0.1	0.0	-0.1	0.0	0.0	0.0	(0.0)	0.0	-0.3	0.0	-0.1
Scotch	0.0	-0.2	0.0	-0.2	0.0	-0.2	0.0	-0.2	0.0	-0.1	0.0	-0.2	0.0	-0.2	0.0	-0.2	0.0	-0.2	0.0	-0.1	0.0	-0.1	0.0	(0.0)	0.0	(0.0)	0.0	-0.3	0.0	-0.1
Welsh	0.0	-0.3	0.0	-0.3	0.0	-0.2	0.0	-0.2	0.0	-0.1	0.0	-0.2	0.0	-0.3	0.0	-0.3	0.0	-0.3	0.0	-0.1	0.0	-0.1	0.0	(0.0)	0.0	-0.1	0.0	-0.3	0.0	-0.1
Spanish	0.2	(0.0)	0.2	-0.1	0.2	(0.0)	0.0	-0.1	0.1	(0.1)	0.0	(0.0)	0.0	-0.3	0.1	(0.0)	0.2	(0.0)	0.2	(3.3)	0.0	(0.3)	0.0	0.0	0.0	-0.1	0.0	-0.3	0.0	-0.1
Portuguese	0.1	-0.1	0.0	-0.3	0.0	(0.0)	0.0	-0.2	0.0	-0.2	0.0	-0.1	0.0	-0.1	0.1	(0.0)	0.1	(0.1)	0.2	(0.3)	0.0	-0.1	0.1	(0.1)	0.0	-0.1	0.0	0.0	0.0	-0.1
Italian (North)	0.3	-0.1	0.0	-0.3	0.0	-0.1	0.0	-0.2	0.0	(0.1)	0.0	-0.2	0.0	-0.1	0.0	-0.3	0.1	(0.0)	0.2	(0.1)	0.0	-0.1	0.1	(0.0)	0.0	(0.0)	0.0	0.0	0.0	0.0
Italian (South)	0.2	(0.0)	0.2	(0.0)	0.5	(0.1)	0.0	(0.0)	0.0	-0.1	0.0	(0.0)	0.0	(0.1)	0.0	(0.0)	0.1	(0.1)	0.3	(0.4)	0.0	-0.1	0.0	(0.0)	0.1	(0.0)	0.0	-0.3	0.0	-0.3
Dutch and Flemish	0.1	-0.1	0.0	(0.0)	0.1	-0.1	0.0	-0.1	0.0	-0.1	0.0	(0.0)	0.0	-0.1	0.0	(0.0)	0.1	-0.1	0.1	(0.0)	0.0	-0.1	0.0	(0.0)	0.0	(0.0)	0.0	-0.3	0.0	0.0
Finnish	0.3	-0.1	0.1	-0.2	0.1	-0.1	0.2	(0.2)	0.0	(0.0)	0.0	-0.1	0.0	-0.3	0.0	(0.3)	0.0	(0.0)	0.1	(0.0)	0.2	(0.2)	0.2	(0.0)	0.0	-0.1	0.0	-0.3	0.0	-0.1
Scandinavian*	0.0	-0.2	0.2	-0.1	0.0	-0.2	0.0	-0.2	0.0	-0.2	0.0	-0.2	0.0	-0.2	0.0	-0.3	0.0	-0.2	0.3	(0.0)	0.3	(0.3)	0.0	0.0	0.3	(0.3)	0.0	-0.3	0.0	-0.1
Lithuanian	0.6	(0.4)	0.2	(0.1)	0.0	-0.1	0.0	-0.2	0.0	-0.2	0.0	-0.1	0.0	-0.3	0.0	(0.5)	0.0	(0.5)	0.3	(0.3)	0.0	-0.1	0.0	(0.0)	0.0	-0.1	0.0	-0.3	0.0	-0.1
Russian	0.2	(0.0)	0.0	(0.0)	0.2	(0.0)	0.1	(0.0)	0.1	(0.0)	0.0	-0.2	0.1	(0.3)	0.1	(0.0)	0.1	(0.2)	0.3	(0.3)	0.0	-0.1	0.0	(0.0)	0.3	(0.0)	0.0	-0.3	0.0	0.0
Polish	0.4	-0.2	0.1	-0.2	0.1	(0.0)	0.0	-0.2	0.0	(0.0)	0.0	(0.0)	0.0	-0.3	0.0	(0.0)	0.1	(0.2)	0.3	(0.3)	0.1	(0.1)	0.0	(0.0)	0.1	(0.1)	0.0	-0.3	0.0	0.0
German	0.1	-0.1	0.0	-0.3	0.0	(0.0)	0.0	-0.1	0.0	-0.1	0.0	-0.1	0.0	(0.0)	0.0	(0.0)	0.1	-0.1	0.1	-0.1	0.1	(0.1)	0.0	(0.0)	0.0	(0.0)	0.0	-0.3	0.0	-0.1
Magyar	0.3	-0.1	0.0	-0.3	0.0	-0.2	0.0	-0.2	0.3	(0.0)	0.1	(0.0)	0.0	(0.0)	0.0	(0.0)	0.0	(0.1)	0.0	-0.1	0.0	(0.1)	0.0	(0.0)	0.0	(0.1)	0.0	-0.3	0.0	-0.1
Slovak	0.2	(0.0)	0.0	-0.3	0.0	(0.0)	0.0	-0.2	0.2	(0.7)	0.1	(0.0)	0.1	(0.0)	0.1	(0.0)	0.3	(0.1)	0.1	(0.1)	0.1	(0.1)	0.0	(0.0)	0.0	-0.1	0.0	-0.3	0.0	-0.1
Croatian and Slovenian	1.3	(1.1)	0.3	(0.1)	2.6	(1.6)	2.2	(5.4)	0.2	(0.4)	0.1	(0.4)	0.1	(0.4)	0.2	(0.2)	0.3	(0.4)	0.4	(0.4)	0.1	(0.2)	0.1	(0.2)	0.1	(0.2)	0.1	(0.2)	0.1	(0.2)
Czech †	0.0	-0.2	0.0	-0.3	0.0	-0.2	0.0	-0.2	0.0	-0.2	0.0	-0.2	0.0	-0.2	0.0	(0.0)	0.0	-0.3	0.0	-0.1	0.0	-0.1	0.0	(0.0)	0.0	-0.1	0.0	-0.3	0.0	-0.1
Dalmatian, Bosnian, Herzegovinian		-0.2		-0.2		(0.0)		-0.2		(0.0)		-0.2		-0.3		-0.3		-0.3		-0.1		-0.1		0.0		-0.1		-0.3		-0.1
Ruthenian (Russniak)	0.2	(0.0)	0.0	-0.3	0.0	(0.0)	0.0	-0.2	0.1	(0.1)	0.1	(0.1)	0.0	-0.3	0.0	(0.0)	0.0	-0.3	0.1	(0.1)	0.1	(0.1)	0.0	(0.0)	0.1	(0.2)	0.0	-0.3	0.0	-0.1
Roumanian	0.3	(0.1)	0.0	-0.3	0.6	(0.2)	0.0	-0.2	0.0	-0.2	0.0	-0.1	0.0	(0.1)	0.0	(0.0)	0.1	(0.1)	0.1	(0.0)	0.1	(0.2)	0.0	(0.0)	0.0	-0.1	0.0	-0.3	0.0	-0.1
Bulgarian, Serbian, Montenegrin																														
Greek	0.5	(0.3)	0.2	(0.0)	2.1	(1.2)	0.0	-0.2	0.0	-0.1	0.0	-0.1	0.0	-0.2	0.0	-0.2	0.1	(0.1)	0.1	(0.1)	0.0	-0.1	0.0	-0.1	0.0	-0.1	0.0	0.0	0.0	(0.2)
Armenian	2.4	(2.3)	2.6	(2.6)	0.4	(0.1)	0.4	(0.5)	0.0	-0.1	0.9	(0.9)	9.8	(3.5)	1.6	(1.6)	5.1	(5.1)	3.0	(3.0)	1.0	(1.0)	0.5	(0.5)	0.3	(0.3)	0.7	(1.2)	0.7	(2.5)
Hebrew	0.4	(2.3)	0.7	-0.1	1.3	(0.1)	0.3	(0.5)	0.3	(0.9)	0.0	(0.8)	0.4	(0.0)	0.2	(0.2)	0.1	(1.1)	0.1	(1.7)	0.5	(1.0)	0.4	(0.5)	0.3	(0.3)	0.1	(0.1)	0.1	(0.1)
Syrian	5.0	(5.0)	3.5	(3.6)	7.6	(5.0)	0.6	(1.3)	0.3	(0.9)	0.3	(0.8)	0.0	(1.6)	0.4	(1.9)	1.1	(1.1)	1.1	(1.7)	0.9	(2.0)	0.6	(0.6)	0.2	(0.1)	0.3	-0.3	0.3	(1.0)
French	0.0	-0.2	0.0	-0.3	0.0	-0.2	0.0	-0.2	0.0	-0.2	0.0	(0.0)	0.0	-0.2	0.0	-0.2	0.0	-0.2	0.0	-0.1	0.0	-0.1	0.0	0.0	0.0	0.0	0.0	0.0	0.0	0.0
Turkish	2.6	(2.5)	0.6	(0.4)	3.8	(3.9)	0.0	-0.2	0.0	-0.2	1.7	(5.8)	0.9	(3.9)	0.3	(0.4)	1.0	(2.0)	0.0	(0.0)	0.8	(1.9)	0.9	(1.2)	0.3	(0.3)	0.3	(2.6)	1.5	(5.8)
Mexican	0.5	(0.3)	0.9	(0.7)	0.1	-0.1	0.0	(0.0)	0.2	(0.3)	0.4	(0.4)	0.1	(0.1)	0.4	(0.2)	0.2	(0.2)	0.1	(0.1)	0.1	(0.8)	0.1	(0.0)	0.1	(0.1)	0.2	(1.3)	0.2	(0.6)
Spanish American	0.1	-0.2	0.1	-0.2	0.0	-0.2	0.0	-0.2	0.0	-0.2	0.0	-0.2	0.0	-0.3	0.0	-0.3	0.0	-0.2	0.0	-0.1	0.0	-0.1	0.0	0.0	0.0	-0.1	0.0	-0.3	0.0	0.0
Cuban	0.0	-0.2	0.0	-0.2	0.0	-0.2	0.0	-0.2	0.0	-0.2	0.0	-0.2	0.0	-0.3	0.0	-0.2	0.0	-0.2	0.0	-0.1	0.0	-0.1	0.0	0.0	0.0	-0.1	0.0	-0.3	0.0	-0.1
West Indian	0.2	(0.0)	0.1	-0.2	0.0	-0.2	0.0	-0.2	0.0	-0.2	0.0	-0.2	0.0	-0.3	0.0	-0.3	0.0	-0.2	0.0	-0.1	0.0	-0.1	0.3	(0.2)	0.3	(0.2)	0.1	(0.0)	0.1	(0.3)
Chinese	1.8	(1.7)	2.1	(2.1)	2.4	(1.5)	0.5	(1.1)	0.4	(1.3)	0.5	(1.6)	0.6	(2.7)	0.7	(3.7)	1.4	(3.3)	3.1	(5.1)	0.7	(1.7)	4.2	(6.0)	0.2	(0.5)	4.0	-0.3	0.4	(1.5)
East Indian	1.6	(1.5)	0.0	-0.3	0.0	-0.2	0.0	-0.2	0.0	-0.2	0.5	(0.5)	0.3	(1.2)	4.2	(4.2)	0.3	(0.6)	0.0	-0.1	0.8	(1.8)	0.0	0.0	0.0	-0.1	0.0	-0.3	0.0	-0.1
Japanese	0.6	(0.4)	1.8	(1.7)	1.1	(0.6)	0.6	(1.4)	0.5	(1.4)	0.3	(0.3)	0.1	-0.2	0.0	-0.3	0.0	(0.6)	0.1	-0.1	0.0	(0.0)	0.0	0.0	0.0	-0.1	0.0	-0.3	0.0	-0.1
Korean	0.0	-0.2	3.2	(3.3)	1.8	(1.1)	1.0	(2.2)	0.0	-0.2	0.0	(0.0)	0.0	-0.3	0.0	-0.3	0.0	-0.3	0.0	-0.1	0.0	-0.1	0.0	0.0	0.0	-0.1	0.0	-0.3	0.0	-0.1
Pacific Islander	0.0	-0.2	0.0	-0.3	0.0	-0.2	0.0	-0.1	1.8	(5.8)	0.0	(0.0)	0.1	(0.1)	0.0	-0.3	0.0	-0.2	1.9	(4.5)	1.9	(4.5)	0.0	0.0	0.0	-0.1	0.0	(0.0)	0.0	-0.1
African (black)	0.1	-0.1	0.0	-0.2	0.0	-0.2	0.0	-0.1	0.0	(0.0)	0.0	-0.2	0.0	(0.1)	0.0	-0.3	0.0	-0.2	0.1	(1.0)	1.2	(2.8)	0.0	0.0	0.0	-0.1	0.0	-0.3	0.0	-0.1
Other	0.3	(0.1)	0.1	(0.5)	0.6	(0.3)	0.2	(0.3)	0.1	(0.0)	0.0	-0.1	0.1	(0.1)	0.1	(0.2)	0.5	(1.3)	0.6	(1.0)	0.2	(2.8)	0.0	0.0	0.4	(0.4)	0.2	(1.6)	0.3	(1.0)

Bold = 1 or more NSDMR below mean

= 1 to 1.99 NSDMRs above mean

= 2 or more NSDMRs above mean

* Norwegians, Danes, and Swedes.

† Bohemians and Moravians.

TABLE 6-8. Deviations in Rate of Deportation for Trachoma Certification, by Race, 1916–1930, Expressed as Percent Certified Then Deported and NSDMR

Race or People	1916 %	1916 NSDMR	1917 %	1917 NSDMR	1918 %	1918 NSDMR	1919 %	1919 NSDMR	1920 %	1920 NSDMR	1921 %	1921 NSDMR	1922 %	1922 NSDMR	1923 %	1923 NSDMR	1924 %	1924 NSDMR	1925 %	1925 NSDMR	1926 %	1926 NSDMR	1927 %	1927 NSDMR	1928 %	1928 NSDMR	1929 %	1929 NSDMR	1930 %	1930 NSDMR
English	100	(0.7)	75	(-0.3)	100	(1.0)	0	(0.0)	0	(-2.0)	100	(0.8)	100	(0.0)	71.43	(-0.6)	100	(0.2)	100	(0.0)	0	(0.0)	100	(1.1)	0	(-0.6)	0	(-0.6)	0	(0.0)
Irish	83.33	(0.2)	83.33	(0.0)	100	(1.0)	100	(0.0)	100	(1.0)	50	(-0.5)	100	(0.8)	87.5	(-0.1)	100	(0.2)	100	(0.0)	100	(0.0)	100	(-1.3)	100	(1.9)	0	(-0.6)	100	(2.0)
Scotch	100	(0.7)	100	(0.7)	100	(1.0)	100	(0.7)	100	(1.0)	0	(0.0)	75	(0.1)	100	(0.2)	100	(0.0)	100	(0.0)	100	(0.0)	0	(0.0)	100	(1.9)	0	(-0.6)	0	(0.0)
Welsh	0	(0.0)	0	(0.0)	0	(0.0)	0	(0.0)	0	(0.0)	0	(0.0)	0	(0.0)	0	(0.0)	0	(0.0)	0	(0.0)	0	(0.0)	0	(0.0)	0	(0.0)	0	(0.0)	0	(0.0)
Spanish	94.12	(0.5)	84	(0.1)	0	(0.4)	50	(-0.4)	68	(0.0)	85	(0.4)	0	(0.0)	0	(0.0)	44.44	(-1.5)	50	(-1.0)	0	(-1.9)	0	(-1.3)	0	(0.0)	0	(-0.6)	0	(-0.7)
Portuguese	100	(0.7)	100	(0.7)	100	(1.0)	0	(0.0)	0	(0.0)	85.71	(0.5)	0	(0.0)	0	(-2.6)	28.57	(-2.0)	50	(-1.0)	75	(-1.9)	50	(-1.3)	50	(0.7)	0	(-0.6)	0	(-0.7)
Italian (North)	93.33	(0.5)	100	(0.7)	100	(1.0)	0	(0.0)	50	(-0.5)	100	(0.0)	100	(0.8)	100	(0.2)	100	(0.0)	100	(0.0)	100	(0.6)	60	(0.2)	100	(1.9)	0	(0.0)	100	(2.0)
Italian (South)	68.12	(-0.3)	61.54	(-0.9)	30.77	(-1.0)	0	(-1.5)	34.78	(-0.9)	66.67	(0.0)	48.65	(-0.5)	81.48	(-0.3)	83.13	(-0.3)	100	(-1.0)	100	(0.6)	60	(0.2)	100	(1.9)	0	(0.0)	100	(2.0)
Dutch and Flemish	100	(0.7)	0	(-3.5)	100	(1.0)	0	(-1.5)	0	(0.0)	100	(0.8)	0	(0.0)	0	(0.0)	0	(0.0)	0	(0.0)	100	(0.6)	100	(0.6)	0	(0.0)	0	(0.0)	0	(0.0)
Finnish	88.24	(0.3)	91.82	(0.0)	100	(1.0)	0	(-1.5)	0	(0.0)	100	(0.8)	100	(0.8)	100	(0.2)	100	(0.2)	0	(0.0)	100	(0.6)	100	(1.1)	0	(0.0)	0	(0.0)	0	(-0.7)
Scandinavian*	100	(0.7)	98.31	(0.4)	0	(-1.9)	0	(-1.5)	0	(-0.5)	100	(0.8)	50	(0.8)	50	(0.0)	50	(-1.3)	0	(0.0)	0	(0.0)	0	(0.0)	0	(0.0)	0	(0.0)	0	(0.0)
Lithuanian	50	(-0.9)	100	(0.7)	100	(0.0)	0	(0.0)	0	(0.0)	100	(0.0)	100	(0.8)	100	(0.0)	66.67	(-0.8)	100	(0.6)	0	(0.0)	0	(0.0)	0	(-0.6)	0	(0.0)	0	(0.0)
Russian	100	(0.7)	100	(0.7)	66.67	(0.1)	100	(0.7)	50	(-0.5)	0	(-1.8)	0	(0.0)	50	(-1.2)	20	(-2.2)	25	(-1.7)	50	(-0.7)	0	(0.0)	0	(-0.6)	0	(-0.6)	0	(-0.7)
Polish	89.47	(0.4)	66.67	(-0.7)	100	(1.0)	100	(0.7)	100	(0.0)	18.18	(-1.3)	100	(0.8)	100	(0.2)	50	(-1.3)	33.33	(-1.5)	75	(-1.5)	0	(-1.3)	0	(0.0)	0	(0.0)	0	(-0.7)
German	100	(0.7)	100	(0.7)	80	(0.5)	100	(0.7)	0	(-1.0)	100	(0.8)	75	(0.1)	97.30	(0.1)	97.30	(0.1)	100	(0.6)	75	(0.2)	0	(-1.3)	100	(1.9)	90	(1.8)	66.67	(1.1)
Magyar	0	(-2.6)	0	(0.0)	0	(0.0)	0	(0.0)	0	(-2.0)	33.33	(-0.9)	33.33	(0.0)	100	(0.2)	83.33	(0.0)	0	(0.0)	75	(0.2)	100	(-1.3)	100	(1.9)	100	(2.1)	0	(0.0)
Slovak	0	(-2.6)	0	(0.0)	0	(0.0)	0	(0.0)	0	(0.0)	57.89	(-0.3)	0	(-1.8)	50	(-1.2)	100	(0.2)	0	(-2.5)	0	(0.0)	0	(0.0)	0	(0.0)	0	(0.0)	0	(0.0)
Croatian and Slovenian	100	(0.7)	100	(0.7)	0	(-1.9)	100	(0.7)	50	(-0.5)	100	(0.8)	0	(-1.8)	100	(0.2)	100	(0.2)	50	(-1.0)	0	(0.0)	0	(0.0)	0	(0.0)	0	(0.0)	0	(0.0)
Czech †	0	(0.0)	0	(0.0)	0	(0.0)	0	(0.0)	0	(0.0)	0	(0.0)	0	(0.0)	0	(0.0)	75	(-0.6)	0	(0.0)	0	(0.0)	0	(0.0)	0	(0.0)	0	(0.0)	0	(0.0)
Dalmatian, Bosnian, Herzegovinian	0	(0.0)	0	(0.7)	0	(0.0)	0	(0.0)	0	(0.0)	0	(-1.8)	0	(0.0)	0	(0.0)	0	(0.0)	0	(0.0)	0	(0.0)	0	(0.0)	0	(0.0)	0	(0.0)	0	(0.0)
Ruthenian (Russniak)	75	(-0.1)	0	(0.0)	0	(0.0)	100	(0.0)	0	(0.0)	100	(0.8)	0	(0.0)	100	(0.2)	0	(0.0)	0	(0.0)	100	(0.6)	0	(0.0)	0	(-0.6)	0	(0.0)	0	(0.0)
Roumanian	66.67	(-0.4)	0	(0.0)	0	(0.0)	66.67	(0.0)	50	(-0.5)	50	(-0.5)	50	(-0.5)	100	(0.2)	66.67	(-0.8)	50	(-1.0)	0	(-1.9)	0	(0.0)	0	(0.0)	0	(0.0)	0	(0.0)
Bulgarian, Serbian, Montenegrin	93.75	(0.5)	100	(0.7)	75	(0.3)	0	(0.0)	0	(0.0)	100	(0.8)	0	(0.0)	0	(-2.6)	100	(0.2)	0	(0.0)	100	(0.6)	0	(0.0)	0	(0.0)	0	(0.0)	0	(-0.7)
Greek	88.14	(0.3)	84.48	(0.1)	75	(0.3)	100	(0.0)	50	(-0.5)	66.67	(0.0)	100	(0.8)	100	(0.2)	100	(0.2)	80	(0.0)	100	(0.6)	75	(0.5)	33.33	(0.3)	50	(0.7)	42.86	(0.4)
Armenian	62.5	(-0.5)	78.79	(0.0)	40	(-0.7)	0	(-1.5)	60	(-0.2)	29.41	(-1.0)	38.1	(-0.8)	55.56	(-1.0)	105.4	(-1.0)	76.92	(-0.1)	60	(-0.4)	14.29	(-1.0)	25	(-0.6)	75	(1.4)	25	(1.1)
Hebrew	36.84	(-1.4)	34.29	(-2.0)	35.29	(-0.9)	100	(-1.0)	60	(-0.2)	36.36	(-0.4)	53.33	(-0.4)	88.37	(0.0)	81.82	(0.1)	83.33	(0.1)	58.33	(0.1)	0	(-1.3)	100	(0.3)	50	(0.7)	50	(0.6)
Syrian	57.5	(-0.7)	94.59	(0.5)	100	(0.7)	100	(0.7)	71.43	(0.1)	72.22	(0.1)	55.56	(-0.4)	75	(-0.5)	100	(0.2)	100	(0.1)	50	(-0.7)	50	(-0.7)	0	(-0.6)	0	(-0.6)	50	(0.6)
French	75	(-0.1)	100	(0.7)	66.67	(0.1)	100	(0.7)	0	(0.0)	0	(-1.8)	0	(0.0)	75	(0.0)	100	(0.2)	100	(0.0)	50	(-0.7)	50	(-0.1)	0	(0.0)	0	(-0.6)	0	(-0.7)
Turkish	0	(-2.6)	100	(0.7)	100	(1.0)	0	(0.0)	0	(-1.8)	0	(-1.8)	0	(-1.8)	0	(0.0)	0	(-2.8)	0	(0.0)	0	(-0.7)	0	(-1.3)	0	(0.0)	0	(-0.6)	0	(-0.2)
Mexican	100	(0.7)	100	(0.7)	100	(1.0)	16.67	(1.1)	87.16	(0.6)	98.77	(0.8)	100	(0.8)	96.77	(0.1)	97.09	(0.1)	81.08	(0.0)	68.42	(-0.2)	14.71	(-1.0)	24.07	(0.0)	16.87	(-0.1)	18.18	(-0.2)
Spanish American	100	(0.7)	50	(-1.4)	0	(0.0)	0	(0.0)	0	(0.0)	100	(0.8)	100	(0.8)	100	(0.2)	0	(-2.8)	100	(0.6)	0	(0.0)	0	(0.0)	0	(0.0)	0	(0.0)	0	(-0.7)
Cuban	100	(0.7)	100	(0.7)	0	(0.0)	0	(0.0)	0	(0.0)	0	(0.0)	0	(0.0)	100	(0.2)	0	(0.0)	100	(0.6)	0	(0.0)	0	(0.0)	0	(0.0)	0	(0.0)	0	(-0.7)
West Indian	100	(0.7)	100	(0.7)	0	(0.0)	0	(0.0)	0	(0.0)	0	(0.0)	0	(0.0)	0	(0.0)	0	(0.0)	0	(0.0)	0	(0.0)	0	(0.0)	0	(-0.6)	0	(0.0)	0	(-0.7)
Chinese	60.42	(-0.6)	67.39	(-1.7)	67.39	(0.1)	60	(-0.2)	64.52	(-0.1)	79.17	(0.2)	77.01	(0.2)	95.24	(0.1)	97.75	(0.1)	85.33	(0.1)	97.01	(0.5)	84.09	(0.7)	8	(-0.4)	0	(-0.6)	33.33	(0.2)
East Indian	42.22	(-1.7)	42.22	(-0.6)	0	(0.0)	60	(-0.2)	0	(-0.6)	23.81	(-1.2)	0	(-1.8)	0	(-1.2)	50	(-1.3)	75	(-0.2)	0	(0.0)	0	(0.0)	0	(0.0)	0	(0.0)	0	(0.0)
Japanese	74.55	(-0.1)	85.80	(0.3)	58.56	(-0.2)	0	(-1.9)	0	(0.0)	0	(0.3)	100	(0.8)	100	(0.8)	100	(0.2)	100	(0.6)	0	(-1.9)	0	(0.0)	0	(0.0)	0	(0.0)	0	(0.0)
Korean	100	(0.0)	85.71	(0.1)	0	(0.0)	0	(-1.9)	100	(1.0)	0	(0.0)	0	(0.0)	0	(0.0)	0	(0.0)	0	(-0.7)	0	(-1.9)	0	(0.0)	0	(0.0)	0	(0.0)	0	(0.0)
Pacific Islander	0	(0.0)	0	(0.0)	0	(0.0)	0	(0.0)	0	(0.0)	0	(0.0)	0	(0.0)	0	(0.0)	0	(0.0)	0	(0.0)	0	(0.0)	0	(0.0)	0	(0.0)	0	(0.0)	0	(0.0)
African (black)	100	(0.7)	100	(0.7)	100	(1.0)	100	(0.7)	0	(-2.0)	100	(0.8)	100	(0.5)	0	(-2.6)	50	(-1.3)	0	(-2.5)	0	(-1.9)	0	(0.0)	100	(2.0)	0	(0.0)	100	(2.0)
Other	120	(1.4)	80	(-0.1)	50	(-0.4)	100	(0.7)	0	(-2.0)	100	(0.8)	100	(0.5)	0	(-2.6)	42.86	(-1.5)	75	(-0.2)	6.667	(-1.8)	0	(0.0)	50	(0.7)	50	(0.7)	0	(-0.7)

Bold = 1 or more NSDMR below mean
* Norwegians, Danes, and Swedes.
† Bohemians and Moravians.
▨ = 1 to 1.99 NSDMRs above mean
▓ = 2 or more NSDMRs above mean

TABLE 6-9. Deviations in Deportation Rate for Trachoma, by Race, 1908–1930, Expressed as Percent Deported and NSDMR

Race or People	1908 %	1908 NSDMR	1909 %	1909 NSDMR	1910 %	1910 NSDMR	1911 %	1911 NSDMR	1912 %	1912 NSDMR	1913 %	1913 NSDMR	1914 %	1914 NSDMR	1915 %	1915 NSDMR	1916 %	1916 NSDMR	1917 %	1917 NSDMR	1918 %	1918 NSDMR	1919 %	1919 NSDMR	1920 %	1920 NSDMR	1921 %	1921 NSDMR	1922 %	1922 NSDMR
English	0.0	-(0.1)	0.0	-(0.1)	0.0	-(0.1)	0.0	-(0.1)	0.0	-(0.2)	0.0	-(0.2)	0.0	-(0.2)	0.1	-(0.2)	0.0	-(0.2)	0.0	-(0.2)	0.0	-(0.1)	0.0	-(0.1)	0.0	-(0.1)	0.0	-(0.5)	0.0	-(0.4)
Irish	0.0	-(0.1)	0.0	-(0.1)	0.0	-(0.1)	0.1	-(0.1)	0.0	-(0.2)	0.0	-(0.2)	0.0	-(0.2)	0.1	-(0.2)	0.0	-(0.2)	0.1	-(0.2)	0.0	-(0.1)	0.0	-(0.1)	0.0	-(0.1)	0.0	-(0.4)	0.0	-(0.1)
Scotch	0.1	-(0.1)	0.0	-(0.1)	0.0	-(0.1)	0.0	-(0.1)	0.0	-(0.2)	0.0	-(0.2)	0.0	-(0.2)	0.1	-(0.1)	0.0	-(0.2)	0.0	-(0.2)	0.0	-(0.1)	0.0	-(0.1)	0.0	-(0.1)	0.0	-(0.5)	0.0	-(0.3)
Welsh	0.0	-(0.1)	0.0	-(0.1)	0.0	-(0.2)	0.0	-(0.2)	0.0	-(0.3)	0.0	-(0.2)	0.0	-(0.2)	0.1	-(0.1)	0.0	-(0.3)	0.0	-(0.3)	0.0	-(0.1)	0.0	-(0.1)	0.0	-(0.1)	0.0	-(0.5)	0.0	-(0.4)
Australian	0.0	(0.0)	0.0	(0.0)	0.0	(0.0)	0.0	(0.0)	0.0	(0.0)	0.0	(0.0)	0.0	(0.0)	0.0	(0.0)	0.0	(0.0)	0.0	(0.0)	0.0	(0.0)	0.0	(0.0)	0.0	(0.0)	0.0	(0.0)	0.0	(0.0)
Spanish	1.0	(0.3)	0.1	-(0.1)	0.1	-(0.1)	0.0	-(0.1)	0.1	-(0.1)	0.3	(0.1)	0.4	(0.3)	0.2	-(0.1)	0.2	(0.0)	0.1	-(0.1)	0.2	(0.0)	0.0	-(0.1)	0.1	(0.1)	0.1	(0.8)	0.0	-(0.3)
Portuguese	0.3	(0.0)	0.0	-(0.1)	0.1	-(0.1)	0.0	-(0.1)	0.0	-(0.2)	0.0	-(0.2)	0.0	-(0.2)	0.1	-(0.1)	0.1	-(0.2)	0.0	-(0.3)	0.0	-(0.1)	0.0	-(0.1)	0.0	-(0.1)	0.0	-(0.1)	0.0	-(0.4)
Italian (North)	0.2	-(0.1)	0.1	-(0.1)	0.1	-(0.1)	0.1	-(0.1)	0.1	-(0.1)	0.1	-(0.1)	0.1	-(0.1)	0.3	(0.0)	0.3	(0.2)	0.0	-(0.2)	0.1	-(0.1)	0.0	-(0.1)	0.0	-(0.1)	0.0	-(0.3)	0.0	(0.1)
Italian (South)	0.3	(0.0)	0.1	(0.0)	0.2	(0.0)	0.2	(0.0)	0.1	-(0.1)	0.1	-(0.1)	0.2	(0.0)	0.4	(0.0)	0.1	(0.0)	0.1	-(0.1)	0.1	(0.0)	0.0	-(0.1)	0.0	-(0.1)	0.0	(0.0)	0.0	(0.0)
Dutch and Flemish	0.0	-(0.1)	0.1	-(0.1)	0.0	-(0.1)	0.0	-(0.1)	0.0	-(0.3)	0.0	-(0.2)	0.1	-(0.1)	0.1	-(0.2)	0.1	-(0.1)	0.0	-(0.3)	0.0	-(0.1)	0.0	-(0.1)	0.0	-(0.1)	0.0	-(0.5)	0.0	-(0.2)
Finnish	0.2	(0.0)	0.2	(0.0)	0.2	(0.0)	0.2	(0.0)	0.1	-(0.1)	0.1	-(0.1)	0.2	(0.0)	0.4	(0.0)	0.3	(0.2)	0.1	-(0.1)	0.1	-(0.1)	0.0	-(0.1)	0.0	-(0.1)	0.0	-(0.2)	0.0	-(0.4)
Scandinavian°	0.1	-(0.1)	0.0	-(0.1)	0.0	-(0.1)	0.0	-(0.1)	0.0	-(0.2)	0.0	-(0.2)	0.0	-(0.2)	0.3	-(0.1)	0.0	-(0.2)	0.1	-(0.2)	0.0	-(0.1)	0.0	-(0.1)	0.0	-(0.1)	0.0	-(0.5)	0.0	-(0.3)
Lithuanian	0.6	(0.1)	0.5	(0.1)	0.4	(0.2)	0.4	(0.2)	0.3	(0.4)	0.3	(0.1)	0.4	(0.2)	0.9	(0.3)	0.3	(0.2)	0.2	(0.0)	0.0	-(0.1)	0.0	-(0.1)	0.0	-(0.1)	0.0	-(0.5)	0.1	(0.8)
Russian	0.2	-(0.1)	0.5	(0.2)	0.3	(0.1)	0.2	(0.0)	0.2	(0.1)	0.1	-(0.1)	0.2	(0.0)	1.3	(0.5)	0.2	(0.0)	0.2	(0.0)	0.1	(0.0)	0.0	-(0.1)	0.0	(0.0)	0.0	-(0.5)	0.1	(0.3)
Polish	0.3	(0.0)	0.2	(0.0)	0.3	(0.0)	0.2	(0.0)	0.2	(0.1)	0.2	(0.0)	0.2	(0.0)	0.5	(0.1)	0.3	(0.3)	0.1	-(0.2)	0.1	(0.0)	0.0	-(0.1)	0.0	-(0.1)	0.0	-(0.4)	0.0	-(0.2)
German	0.2	-(0.1)	0.2	(0.0)	0.1	-(0.1)	0.1	-(0.1)	0.1	-(0.1)	0.1	-(0.1)	0.1	-(0.1)	0.1	-(0.1)	0.1	-(0.2)	0.0	-(0.2)	0.2	(0.0)	0.0	(0.0)	0.0	(0.0)	0.0	-(0.3)	0.0	-(0.2)
Magyar	0.2	-(0.1)	0.1	(0.0)	0.1	-(0.1)	0.2	(0.0)	0.1	(0.0)	0.1	-(0.1)	0.1	-(0.1)	0.0	-(0.2)	0.0	-(0.3)	0.0	-(0.3)	0.0	-(0.1)	0.0	-(0.1)	0.0	-(0.1)	0.0	-(0.5)	0.0	-(0.2)
Slovak	0.2	(0.0)	0.1	(0.0)	0.1	-(0.1)	0.2	(0.0)	0.2	(0.1)	0.1	-(0.1)	0.1	-(0.1)	0.1	-(0.1)	0.0	-(0.3)	0.0	-(0.3)	0.0	-(0.1)	0.0	-(0.1)	0.0	-(0.1)	0.0	-(0.1)	0.0	-(0.4)
Croatian & Slovenian	0.3	(0.0)	0.3	(0.0)	0.2	(0.0)	0.2	(0.0)	0.1	(0.0)	1.8	(1.9)	0.2	(0.1)	4.4	(2.2)	1.3	(2.0)	0.3	(0.1)	0.0	-(0.1)	2.2	(5.9)	0.1	(0.2)	0.0	-(0.3)	0.0	-(0.4)
Transylvanian (Siebenburger)	0.0	(0.0)	0.0	(0.0)	0.0	(0.0)	0.0	(0.0)	0.0	(0.0)	0.0	(0.0)	0.0	(0.0)	0.0	(0.0)	0.0	(0.0)	0.0	(0.0)	0.0	(0.0)	0.0	(0.0)	0.0	(0.0)	0.0	(0.0)	0.0	(0.0)
Czech †	0.1	-(0.1)	0.1	-(0.1)	0.2	(0.0)	0.1	-(0.1)	0.0	-(0.2)	0.1	-(0.1)	0.1	-(0.1)	0.1	-(0.2)	0.0	-(0.3)	0.0	-(0.3)	0.0	-(0.1)	0.0	-(0.1)	0.0	-(0.1)	0.0	-(0.5)	0.0	-(0.1)
Dalmatian, Bosnian, Herzegovinian	0.2	-(0.1)	0.1	-(0.1)	0.3	(0.0)	0.1	-(0.1)	0.0	-(0.2)	0.0	-(0.2)	0.1	-(0.1)	2.3	(1.0)	0.0	-(0.3)	1.0	(1.1)	0.0	-(0.1)	0.0	-(0.1)	0.0	-(0.1)	0.0	-(0.5)	0.0	-(0.4)
Ruthenian (Russniak)	0.1	-(0.1)	0.1	(0.0)	0.2	(0.0)	0.2	(0.0)	0.1	(0.0)	0.1	-(0.1)	0.1	-(0.1)	0.4	(0.0)	0.2	(0.0)	0.0	-(0.3)	0.0	-(0.1)	0.0	-(0.1)	0.0	-(0.1)	0.2	(1.7)	0.0	-(0.4)
Austro-Hungarian	0.0	(0.0)	0.0	(0.0)	0.0	(0.0)	0.0	(0.0)	0.0	(0.0)	0.0	(0.0)	0.0	(0.0)	0.0	(0.0)	0.0	(0.0)	0.0	(0.0)	0.0	(0.0)	0.0	(0.0)	0.0	(0.0)	0.0	(0.0)	0.0	(0.0)
Roumanian	0.2	-(0.1)	0.4	(0.1)	0.2	(0.0)	0.4	(0.1)	0.1	(0.0)	0.1	-(0.1)	0.1	-(0.1)	1.0	(0.4)	0.2	(0.1)	0.0	-(0.3)	0.6	(0.3)	0.0	-(0.1)	0.0	-(0.1)	0.0	-(0.3)	0.1	(0.2)
Bulgarian, Serbian, Montenegrin	0.2	-(0.1)	0.6	(0.2)	0.3	(0.1)	0.3	(0.1)	0.2	(0.1)	0.4	(0.3)	0.3	(0.1)	1.6	(0.7)	0.4	(0.5)	0.3	(0.2)	1.5	(1.0)	0.0	-(0.1)	0.0	-(0.1)	0.0	-(0.2)	0.0	-(0.4)
Greek	0.3	(0.0)	0.4	(0.1)	0.3	(0.0)	0.3	(0.1)	0.1	(0.0)	0.1	-(0.1)	0.1	(0.0)	0.4	(0.0)	0.2	(0.1)	0.2	(0.0)	0.1	(0.0)	0.0	-(0.1)	0.0	-(0.1)	0.0	-(0.3)	0.0	-(0.4)
Armenian	2.6	(1.0)	1.9	(0.8)	2.7	(2.0)	2.5	(1.8)	1.2	(2.2)	0.5	(0.4)	1.4	(1.4)	2.0	(0.9)	1.5	(2.3)	2.0	(2.5)	0.8	(0.5)	0.0	-(0.1)	0.0	-(0.1)	0.1	(0.7)	0.3	(2.9)
Hebrew	0.1	-(0.1)	0.3	(0.0)	0.2	(0.0)	0.2	(0.0)	0.2	(0.1)	0.1	-(0.1)	0.2	(0.0)	0.1	-(0.1)	0.1	-(0.1)	0.1	-(0.2)	0.2	(0.0)	0.1	(0.0)	0.0	-(0.1)	0.0	-(0.3)	0.0	-(0.1)
Arabian	0.0	(0.0)	0.0	(0.0)	0.0	(0.0)	0.0	(0.0)	0.0	(0.0)	0.0	(0.0)	0.0	(0.0)	0.0	(0.0)	0.0	(0.0)	0.0	(0.0)	0.0	(0.0)	0.0	(0.0)	0.0	(0.0)	0.0	(0.0)	0.0	(0.0)
Syrian	5.5	(2.3)	3.1	(1.4)	2.8	(2.0)	3.5	(2.7)	1.7	(3.4)	1.6	(1.6)	1.9	(2.0)	3.1	(1.5)	2.9	(4.8)	3.3	(4.1)	7.9	(5.6)	0.6	(1.5)	0.2	(0.7)	0.2	(2.3)	0.2	(1.9)
Turkish	0.5	(0.1)	0.6	(0.2)	1.0	(0.6)	1.3	(0.8)	0.6	(1.1)	0.7	(0.6)	0.9	(0.8)	1.5	(0.6)	0.0	-(0.3)	0.6	(0.6)	3.8	(2.7)	0.0	-(0.1)	0.5	(1.6)	0.0	-(0.5)	0.0	-(0.4)
French	0.2	-(0.1)	0.1	-(0.1)	0.0	-(0.1)	0.1	-(0.1)	0.1	-(0.1)	0.0	-(0.2)	0.0	-(0.2)	0.1	-(0.1)	0.0	-(0.2)	0.0	-(0.2)	0.0	-(0.1)	0.0	-(0.1)	0.0	-(0.1)	0.0	-(0.5)	0.0	-(0.3)
Swiss	0.0	(0.0)	0.0	(0.0)	0.0	(0.0)	0.0	(0.0)	0.0	(0.0)	0.0	(0.0)	0.0	(0.0)	0.0	(0.0)	0.0	(0.0)	0.0	(0.0)	0.0	(0.0)	0.0	(0.0)	0.0	(0.0)	0.0	(0.0)	0.0	(0.0)
Mexican	0.9	(0.2)	0.4	(0.1)	0.3	(0.1)	0.2	(0.0)	0.2	(0.1)	1.2	(1.2)	0.9	(0.8)	0.7	(0.2)	0.5	(0.5)	0.9	(0.9)	0.1	(0.0)	0.0	-(0.1)	0.1	(0.3)	0.2	(1.7)	0.1	(0.4)
Central American (NS)	0.0	(0.0)	0.0	(0.0)	0.0	(0.0)	0.0	(0.0)	0.0	(0.0)	0.0	(0.0)	0.0	(0.0)	0.0	(0.0)	0.0	(0.0)	0.0	(0.0)	0.0	(0.0)	0.0	(0.0)	0.0	(0.0)	0.0	(0.0)	0.0	(0.0)
South American	0.0	(0.0)	0.0	(0.0)	0.0	(0.0)	0.0	(0.0)	0.0	(0.0)	0.0	(0.0)	0.0	(0.0)	0.0	(0.0)	0.0	(0.0)	0.0	(0.0)	0.0	(0.0)	0.0	(0.0)	0.0	(0.0)	0.0	(0.0)	0.0	(0.0)
Spanish American	0.7	(0.1)	0.2	(0.0)	0.2	(0.0)	0.1	-(0.1)	0.0	-(0.3)	0.0	-(0.2)	0.1	-(0.1)	0.2	-(0.1)	0.1	-(0.2)	0.0	-(0.2)	0.0	-(0.1)	0.0	-(0.1)	0.0	-(0.1)	0.0	-(0.5)	0.0	-(0.4)
Cuban	0.1	-(0.1)	0.0	-(0.1)	0.0	-(0.2)	0.0	-(0.1)	0.0	-(0.3)	0.0	-(0.2)	0.0	-(0.2)	0.0	-(0.2)	0.0	-(0.2)	0.0	-(0.2)	0.0	-(0.1)	0.0	-(0.1)	0.0	-(0.1)	0.0	-(0.5)	0.0	-(0.4)
West Indian	0.1	-(0.1)	0.0	-(0.1)	0.0	-(0.2)	0.1	-(0.1)	0.0	-(0.3)	0.0	-(0.2)	0.0	-(0.2)	0.2	-(0.1)	0.2	(0.1)	0.1	-(0.2)	0.0	-(0.1)	0.0	-(0.1)	0.0	-(0.1)	0.0	-(0.5)	0.0	-(0.4)
Chinese	10.6	(4.5)	1.7	(0.7)	1.1	(0.7)	0.5	(0.3)	0.7	(1.1)	2.1	(2.2)	1.3	(1.3)	0.5	(0.1)	1.1	(1.6)	0.9	(0.9)	1.6	(1.1)	0.3	(0.7)	0.3	(0.9)	0.4	(4.7)	0.5	(4.8)
East Indian	8.9	(3.8)	12.0	(5.9)	7.1	(5.4)	7.3	(5.4)	2.2	(4.3)	4.2	(4.7)	5.0	(5.5)	10.9	(5.7)	1.6	(2.6)	0.0	-(0.3)	0.0	-(0.1)	0.0	-(0.1)	0.0	-(0.1)	0.0	-(0.5)	0.0	-(0.4)
Japanese	1.8	(0.7)	2.7	(1.3)	1.0	(0.6)	0.5	(0.2)	1.0	(1.8)	2.0	(2.1)	1.0	(0.9)	0.7	(0.2)	0.5	(0.5)	1.5	(1.8)	0.6	(0.3)	0.5	(1.3)	0.2	(0.6)	0.0	(0.0)	0.0	-(0.3)
Korean	0.0	-(0.1)	0.0	-(0.1)	0.0	-(0.2)	0.0	-(0.2)	0.0	-(0.3)	1.5	(1.6)	1.8	(1.8)	0.0	-(0.2)	0.6	(0.8)	2.7	(3.4)	0.0	-(0.1)	0.0	-(0.1)	0.0	-(0.1)	0.0	-(0.5)	0.0	-(0.4)
Filipino	0.0	(0.0)	0.0	(0.0)	0.4	(0.1)	0.0	(0.0)	0.0	(0.0)	1.6	(1.6)	0.5	(0.3)	0.6	(0.1)	0.1	-(0.1)	0.0	-(0.3)	0.0	-(0.1)	0.0	-(0.1)	0.0	-(0.1)	0.0	-(0.5)	0.0	-(0.4)
Hawaiian	0.0	(0.0)	0.0	(0.0)	0.0	(0.0)	0.0	(0.0)	0.0	(0.0)	0.0	(0.0)	0.0	(0.0)	0.0	(0.0)	0.0	(0.0)	0.0	(0.0)	0.0	(0.0)	0.0	(0.0)	0.0	(0.0)	0.0	(0.0)	0.0	(0.0)
Pacific Islander	0.0	-(0.1)	3.4	(1.6)	2.9	(2.1)	0.0	-(0.2)	0.0	-(0.3)	0.0	-(0.2)	0.0	-(0.2)	0.0	-(0.2)	0.0	-(0.3)	0.0	-(0.3)	0.0	-(0.1)	0.0	-(0.1)	1.5	(6.0)	0.0	-(0.5)	0.0	-(0.4)
African (black)	0.1	-(0.1)	0.1	(0.0)	0.1	-(0.1)	0.1	-(0.1)	0.0	-(0.2)	0.1	(0.0)	0.0	-(0.2)	0.2	-(0.1)	0.1	-(0.1)	0.0	-(0.2)	0.0	-(0.1)	0.0	-(0.1)	0.0	-(0.1)	0.0	-(0.5)	0.0	-(0.4)
(NS)	0.0	(0.0)	0.0	(0.0)	0.0	(0.0)	0.0	(0.0)	0.0	(0.0)	0.0	(0.0)	0.0	(0.0)	0.0	(0.0)	0.0	(0.0)	0.0	(0.0)	0.0	(0.0)	0.0	(0.0)	0.0	(0.0)	0.0	(0.0)	0.0	(0.0)
Esquimauz	0.0	(0.0)	0.0	(0.0)	0.0	(0.0)	0.0	(0.0)	0.0	(0.0)	0.0	(0.0)	0.0	(0.0)	0.0	(0.0)	0.0	(0.0)	0.0	(0.0)	0.0	(0.0)	0.0	(0.0)	0.0	(0.0)	0.0	(0.0)	0.0	(0.0)
Other	0.8	(0.2)	0.5	(0.1)	0.5	(0.2)	0.9	(0.5)	0.2	(0.2)	0.5	(0.4)	0.6	(0.5)	0.5	(0.1)	0.2	(0.1)	0.5	(0.5)	0.3	(0.1)	0.2	(0.4)	0.0	-(0.1)	0.0	-(0.1)	0.1	(0.5)

Bold = 1 or more NSDMR below mean; [light shading] = 1 to 1.99 NSDMRs above mean; [dark shading] = 2 or more NSDMRs above mean; NS = not specified

° Norwegians, Danes, and Swedes.

† Bohemians and Moravians.

TABLE 6-9. Continued

Race or People	1923 %	1923 NSDMR	1924 %	1924 NSDMR	1925 %	1925 NSDMR	1926 %	1926 NSDMR	1927 %	1927 NSDMR	1928 %	1928 NSDMR	1929 %	1929 NSDMR	1930 %	1930 NSDMR
English	0.0	-(0.3)	0.0	-(0.3)	0.0	-(0.1)	0.0	-(0.2)	0.0	(0.0)	0.0	-(0.1)	0.0	-(0.2)	0.0	-(0.1)
Irish	0.0	-(0.2)	0.0	-(0.3)	0.0	-(0.1)	0.0	-(0.1)	0.0	(0.0)	0.0	-(0.1)	0.0	-(0.2)	0.0	(0.0)
Scotch	0.0	-(0.3)	0.0	-(0.2)	0.0	-(0.1)	0.0	-(0.2)	0.0	(0.0)	0.0	-(0.1)	0.0	-(0.2)	0.0	-(0.1)
Welsh	0.0	-(0.4)	0.0	-(0.3)	0.0	-(0.1)	0.0	-(0.2)	0.0	(0.0)	0.0	-(0.1)	0.0	-(0.2)	0.0	-(0.1)
Australian	0.0	(0.0)	0.0	(0.0)	0.0	(0.0)	0.0	(0.0)	0.0	(0.0)	0.0	(0.0)	0.0	(0.0)	0.0	(0.0)
Spanish	0.0	-(0.4)	0.0	-(0.1)	0.1	(0.1)	0.0	-(0.2)	0.0	(0.0)	0.0	-(0.1)	0.0	-(0.2)	0.0	-(0.1)
Portuguese	0.0	-(0.4)	0.0	-(0.2)	0.1	(0.1)	0.0	-(0.2)	0.0	(0.0)	0.0	-(0.1)	0.0	-(0.2)	0.0	-(0.1)
Italian (North)	0.0	-(0.2)	0.1	-(0.1)	0.2	(0.3)	0.0	-(0.2)	0.0	(0.0)	0.0	(0.4)	0.0	-(0.2)	0.0	-(0.1)
Italian (South)	0.0	(0.0)	0.1	(0.1)	0.2	(0.2)	0.0	-(0.1)	0.0	(0.0)	0.0	(0.4)	0.0	-(0.1)	0.0	-(0.1)
Dutch and Flemish	0.0	-(0.4)	0.1	-(0.2)	0.1	(0.0)	0.0	-(0.2)	0.0	(0.0)	0.0	-(0.1)	0.0	-(0.2)	0.0	-(0.1)
Finnish	0.1	(0.5)	0.0	-(0.2)	0.0	-(0.1)	0.1	(0.2)	0.2	(0.3)	0.0	-(0.1)	0.0	-(0.2)	0.0	-(0.1)
Scandinavian*	0.0	-(0.4)	0.0	-(0.3)	0.0	-(0.1)	0.0	(0.2)	0.0	(0.4)	0.0	-(0.1)	0.0	-(0.2)	0.0	-(0.1)
Lithuanian	0.1	(0.7)	0.2	(0.3)	0.3	(0.4)	0.0	-(0.2)	0.0	(0.0)	0.0	-(0.1)	0.0	-(0.2)	0.0	-(0.1)
Russian	0.1	(0.1)	0.0	-(0.2)	0.1	(0.0)	0.0	-(0.1)	0.0	(0.0)	0.0	-(0.1)	0.1	(1.7)	0.0	-(0.1)
Polish	0.0	(0.0)	0.0	-(0.1)	0.0	-(0.1)	0.1	(0.1)	0.0	(0.0)	0.0	-(0.1)	0.1	(4.9)	0.0	-(0.1)
German	0.0	-(0.2)	0.0	-(0.1)	0.0	-(0.1)	0.0	-(0.1)	0.0	(0.0)	0.0	-(0.1)	0.0	-(0.1)	0.0	-(0.1)
Magyar	0.1	(0.0)	0.1	-(0.1)	0.0	-(0.1)	0.0	(0.1)	0.0	(0.0)	0.1	(1.1)	0.0	-(0.2)	0.0	-(0.1)
Slovak	0.0	-(0.3)	0.2	(0.3)	0.0	-(0.1)	0.0	-(0.2)	0.0	(0.0)	0.0	-(0.1)	0.0	-(0.2)	0.0	-(0.1)
Croatian & Slovenian	0.0	-(0.2)	0.1	(0.0)	0.2	(0.2)	0.0	-(0.2)	0.0	(0.0)	0.0	-(0.1)	0.0	-(0.2)	0.0	-(0.1)
Transylvanian (Siebenburger)	0.0	(0.0)	0.0	(0.0)	0.0	(0.0)	0.0	(0.0)	0.0	(0.0)	0.0	(0.0)	0.0	(0.0)	0.0	(0.0)
Czech †	0.0	-(0.4)	0.0	-(0.1)	0.1	(0.0)	0.0	-(0.2)	0.0	(0.0)	0.0	-(0.1)	0.0	-(0.2)	0.0	-(0.1)
Dalmatian, Bosnian, Herzegovinian	0.0	-(0.4)	0.0	-(0.3)	0.0	-(0.1)	0.4	(2.0)	0.2	(0.2)	0.0	-(0.1)	0.0	-(0.2)	0.0	-(0.1)
Ruthenian (Russniak)	0.0	-(0.4)	0.0	-(0.3)	0.2	(0.3)	0.1	(0.4)	0.2	(0.4)	0.0	-(0.1)	0.0	(0.0)	0.0	-(0.1)
Austro-Hungarian	0.0	(0.0)	0.0	(0.0)	0.0	(0.0)	0.0	(0.0)	0.0	(0.0)	0.0	(0.0)	0.0	(0.0)	0.0	(0.0)
Roumanian	0.1	(0.1)	0.1	(0.1)	0.0	-(0.1)	0.0	-(0.2)	0.0	(0.0)	0.0	-(0.1)	0.0	-(0.2)	0.0	-(0.1)
Bulgarian, Serbian, Montenegrin	0.0	(0.0)	0.0	(0.0)	0.0	(0.0)	0.0	(0.0)	0.0	(0.0)	0.0	(0.0)	0.0	(0.0)	0.0	(0.0)
Greek	0.0	-(0.4)	0.1	(0.1)	0.2	(0.3)	0.0	-(0.2)	0.0	(0.0)	0.0	-(0.1)	0.0	-(0.2)	0.0	-(0.3)
Armenian	0.2	(1.1)	1.1	(3.8)	1.4	(2.9)	0.3	(1.5)	0.3	(0.2)	0.1	(1.1)	0.1	(2.5)	0.3	(3.1)
Hebrew	0.1	(0.2)	0.1	(0.1)	0.1	(0.2)	0.1	(0.3)	0.0	(0.4)	0.0	(0.0)	0.0	(0.4)	0.0	(0.1)
Arabian	0.0	(0.0)	0.0	(0.0)	0.0	(0.0)	0.0	(0.0)	0.0	(0.0)	0.0	(0.0)	0.0	(0.0)	0.0	(0.0)
Syrian	0.3	(1.9)	0.4	(1.0)	0.9	(1.7)	0.5	(2.9)	0.0	(0.0)	0.0	-(0.1)	0.0	-(0.2)	0.1	(1.5)
Turkish	0.0	-(0.4)	0.0	-(0.3)	0.0	-(0.1)	0.4	(2.3)	0.0	(0.0)	0.0	-(0.1)	0.0	-(0.2)	0.3	(3.2)
French	0.0	-(0.3)	0.0	-(0.2)	0.0	-(0.1)	0.0	-(0.1)	0.0	(0.0)	0.0	-(0.1)	0.0	-(0.2)	0.0	-(0.1)
Swiss	0.0	(0.0)	0.0	(0.0)	0.0	(0.3)	0.0	(0.0)	0.0	(0.0)	0.0	(0.0)	0.0	(0.0)	0.0	(0.0)
Mexican	0.1	(0.6)	0.2	(0.3)	0.1	(0.1)	0.0	(0.1)	0.0	(0.0)	0.0	(0.2)	0.0	(0.8)	0.0	(0.3)
Central American (NS)	0.0	(0.0)	0.0	(0.0)	0.0	(0.0)	0.0	(0.0)	0.0	(0.0)	0.0	(0.0)	0.0	(0.0)	0.0	(0.0)
South American	0.0	(0.0)	0.0	(0.0)	0.0	(0.0)	0.0	(0.0)	0.0	(0.0)	0.0	(0.0)	0.0	(0.0)	0.0	(0.0)
Spanish American	0.0	-(0.4)	0.0	-(0.2)	0.0	-(0.1)	0.0	-(0.2)	0.0	(0.0)	0.0	-(0.1)	0.0	-(0.2)	0.0	-(0.1)
Cuban	0.0	-(0.2)	0.0	-(0.3)	0.0	-(0.1)	0.0	-(0.2)	0.0	(0.0)	0.0	-(0.1)	0.0	-(0.2)	0.0	-(0.1)
West Indian	0.0	-(0.4)	0.0	-(0.3)	0.0	-(0.1)	0.0	-(0.2)	0.0	(0.0)	0.0	-(0.1)	0.0	-(0.2)	0.0	-(0.1)
Chinese	0.6	(4.9)	1.4	(4.6)	2.7	(5.4)	0.7	(4.2)	3.5	(6.3)	0.4	(5.5)	0.1	(0.4)	0.0	(0.3)
East Indian	0.4	(2.7)	0.3	-(0.1)	0.3	-(0.1)	0.4	(2.2)	0.4	(0.0)	0.0	-(0.1)	0.0	-(0.2)	0.4	(4.4)
Japanese	0.1	(0.4)	0.0	-(0.3)	0.1	(0.1)	0.0	-(0.2)	0.0	(0.0)	0.0	-(0.1)	0.0	-(0.2)	0.0	-(0.1)
Korean	0.0	-(0.4)	0.0	-(0.3)	0.0	-(0.1)	0.0	-(0.2)	0.0	(0.0)	0.0	-(0.1)	0.0	-(0.2)	0.0	-(0.1)
Filipino	0.0	-(0.4)	0.0	-(0.3)	0.0	-(0.1)	0.0	-(0.2)	0.0	(0.0)	0.0	-(0.1)	0.0	-(0.2)	0.0	(0.0)
Hawaiian	0.0	(0.0)	0.0	(0.0)	0.0	(0.0)	0.0	(0.0)	0.0	(0.0)	0.0	(0.0)	0.0	(0.0)	0.0	(0.0)
Pacific Islander	0.0	-(0.4)	0.0	-(0.3)	0.0	-(0.1)	0.0	-(0.2)	0.0	(0.0)	0.0	-(0.1)	0.0	-(0.2)	0.0	-(0.1)
African (black)	0.0	-(0.4)	0.0	-(0.3)	0.0	-(0.1)	0.0	-(0.1)	0.0	(0.0)	0.0	-(0.1)	0.0	-(0.2)	0.0	-(0.1)
(NS)	0.0	(0.0)	0.0	(0.0)	0.0	(0.0)	0.0	(0.0)	0.0	(0.0)	0.0	(0.0)	0.0	(0.0)	0.0	(0.0)
Esquimaux	0.0	(0.0)	0.0	(0.0)	0.0	(0.0)	0.0	(0.0)	0.0	(0.0)	0.0	(0.0)	0.0	(0.0)	0.0	(0.0)
Other	0.0	-(0.4)	0.2	(0.5)	0.5	(0.9)	0.1	(0.3)	0.3	(0.4)	0.2	(2.6)	0.1	(3.3)	0.0	-(0.1)

TABLE 6-10. Deviations in Certification Rate for Parasitic Infection, by Race, 1916–1930, Expressed as Percent Certified and NSDMR

Race or People	1916 %	1916 NSDMR	1917 %	1917 NSDMR	1918 %	1918 NSDMR	1919 %	1919 NSDMR	1920 %	1920 NSDMR	1921 %	1921 NSDMR	1922 %	1922 NSDMR	1923 %	1923 NSDMR	1924 %	1924 NSDMR	1925 %	1925 NSDMR	1926 %	1926 NSDMR	1927 %	1927 NSDMR	1928 %	1928 NSDMR	1929 %	1929 NSDMR	1930 %	1930 NSDMR
English	0.0	(0.0)	0.0	-0.1	0.0	-0.3	0.0	-0.2	0.0	-0.2	0.0	-0.2	0.0	-0.1	0.0	-0.1	0.0	-0.1	0.0	-0.1	0.0	0.0	0.0	0.0	0.0	0.0	0.0	-0.1	0.0	-0.1
Irish	0.0	(0.0)	0.0	-0.1	0.0	-0.3	0.0	-0.2	0.0	-0.2	0.0	-0.2	0.0	-0.1	0.0	-0.1	0.0	-0.1	0.0	-0.1	0.0	0.0	0.0	0.0	0.0	0.0	0.0	-0.1	0.0	-0.1
Scotch	0.0	(0.0)	0.0	-0.1	0.0	-0.3	0.0	-0.2	0.0	-0.2	0.0	-0.2	0.0	-0.1	0.0	-0.1	0.0	-0.1	0.0	-0.1	0.0	0.0	0.0	0.0	0.0	0.0	0.0	-0.1	0.0	-0.1
Welsh	0.0	(0.0)	0.0	-0.1	0.0	-0.3	0.1	-0.1	0.0	-0.2	0.0	-0.2	0.0	-0.1	0.0	-0.1	0.0	-0.1	0.0	-0.1	0.0	0.0	0.0	0.0	0.0	0.0	0.0	-0.1	0.0	-0.1
Spanish	0.0	(0.0)	0.0	-0.1	0.0	-0.3	0.0	-0.2	0.0	-0.2	0.0	-0.2	0.0	0.0	0.0	-0.1	0.0	-0.1	0.0	-0.1	0.0	0.0	0.0	0.0	0.0	0.0	0.0	-0.1	0.0	-0.1
Portuguese	0.0	(0.0)	0.0	-0.1	0.0	-0.3	0.0	-0.2	0.0	-0.2	0.0	-0.2	0.0	0.0	0.0	-0.1	0.0	0.0	0.0	-0.1	0.0	0.0	0.0	0.0	0.0	0.0	0.0	-0.1	0.0	-0.1
Italian (North)	0.0	(0.0)	0.0	-0.1	0.0	-0.3	0.0	-0.2	0.0	-0.2	0.0	-0.2	0.0	-0.1	0.0	-0.1	0.0	-0.1	0.0	-0.1	0.0	0.0	0.0	0.0	0.0	0.0	0.0	-0.1	0.0	-0.1
Italian (South)	0.0	(0.0)	0.0	-0.1	0.0	-0.3	0.0	-0.2	0.0	-0.2	0.0	-0.1	0.5	-0.1	0.0	-0.1	0.0	-0.1	0.0	-0.1	0.0	0.0	0.0	0.0	0.0	0.0	0.0	-0.1	0.0	-0.1
Dutch and Flemish	0.0	(0.0)	0.0	-0.1	0.0	-0.3	0.0	-0.2	0.0	-0.2	0.0	-0.2	0.0	-0.1	0.0	-0.1	0.0	-0.1	0.0	-0.1	0.0	0.0	0.0	0.0	0.0	0.0	0.0	-0.1	0.0	-0.1
Finnish	0.0	(0.0)	0.0	-0.1	0.0	-0.3	0.1	-0.1	0.0	-0.2	0.0	-0.2	0.1	0.0	0.0	-0.1	0.0	-0.1	0.0	-0.1	0.0	0.0	0.2	0.0	0.0	0.0	0.0	-0.1	0.0	-0.1
Scandinavian°	0.0	(0.0)	0.0	-0.1	0.0	-0.3	0.0	-0.2	0.0	-0.2	0.0	-0.2	0.0	0.0	0.0	-0.1	0.0	-0.1	0.0	-0.1	0.0	0.0	0.0	0.0	0.0	0.0	0.0	-0.1	0.0	-0.1
Lithuanian	0.0	(0.0)	0.0	-0.1	0.0	-0.3	0.0	-0.2	0.0	-0.2	0.0	-0.2	0.1	0.0	0.0	-0.1	0.0	-0.1	0.0	0.3	0.0	0.1	0.0	0.0	0.0	0.0	0.0	0.0	0.0	-0.1
Russian	0.1	(0.0)	0.0	-0.1	0.0	-0.3	0.0	-0.2	0.0	-0.2	0.0	-0.2	0.1	0.0	0.0	-0.1	0.0	-0.1	0.0	0.1	0.0	0.0	0.0	0.0	0.0	0.0	0.1	0.0	0.0	-0.1
Polish	0.0	(0.0)	0.0	-0.1	0.0	-0.3	0.0	-0.2	0.0	-0.2	0.0	-0.2	0.0	0.0	0.0	-0.1	0.0	-0.1	0.0	-0.1	0.0	0.0	0.0	0.0	0.0	0.0	0.0	0.0	0.0	-0.1
German	0.0	(0.0)	0.0	-0.1	0.0	-0.3	0.0	-0.2	0.0	-0.2	0.0	-0.2	0.0	0.0	0.0	-0.1	0.0	-0.1	0.0	0.0	0.0	0.0	0.0	0.0	0.0	0.0	0.0	0.1	0.0	0.0
Magyar	0.0	(0.0)	0.0	-0.1	0.0	0.0	0.0	-0.2	0.8	(1.0)	0.0	-0.2	0.1	0.0	0.0	-0.1	0.0	-0.1	0.0	-0.1	0.0	0.0	0.0	-0.1	0.0	0.0	0.0	0.0	0.0	-0.1
Slovak	0.0	(0.0)	0.0	-0.1	0.0	0.0	0.0	-0.2	0.0	-0.2	0.0	-0.2	0.0	0.0	0.0	-0.1	0.0	-0.1	0.0	-0.1	0.0	0.0	0.1	0.0	0.0	0.0	0.0	0.0	0.0	-0.1
Croatian and Slovenian	0.0	(0.0)	0.0	-0.1	0.0	0.0	0.0	-0.2	0.0	-0.2	0.0	-0.2	0.1	-0.1	0.0	-0.1	0.0	-0.1	0.0	-0.1	0.0	0.0	0.2	0.0	0.0	0.0	0.0	-0.1	0.0	-0.1
Czech †	0.0	(0.0)	0.0	-0.1	0.0	-0.3	0.0	-0.2	0.0	-0.2	0.0	-0.2	0.0	0.0	0.0	-0.1	0.0	-0.1	0.0	-0.1	0.0	0.1	0.1	0.0	0.0	0.0	0.0	-0.1	0.0	-0.1
Dalmatian, Bosnian, Herzegovinian	0.0	(0.0)	0.0	-0.1	0.0	0.0	0.0	-0.2	0.0	-0.2	0.0	-0.2	0.8	-0.1	0.0	-0.1	0.0	-0.1	0.0	-0.1	0.0	0.0	0.0	0.0	0.2	0.0	0.0	0.0	0.0	-0.1
Ruthenian (Russniak)	0.1	(0.0)	0.0	-0.1	0.0	0.0	0.0	-0.2	0.0	-0.2	0.0	-0.1	0.8	-0.1	0.0	-0.1	0.0	-0.1	0.0	-0.1	0.0	0.1	0.0	0.0	0.0	0.0	0.1	0.0	0.0	-0.1
Rounmanian	0.0	(0.0)	0.0	-0.1	0.0	-0.3	0.0	-0.2	0.0	-0.2	0.0	-0.1	0.0	0.0	0.0	-0.1	0.0	0.0	0.0	-0.1	0.0	0.1	0.0	0.0	0.0	0.0	0.0	0.0	0.0	-0.1
Bulgarian, Serbian, Montenegrin	0.0	(0.0)	0.2	(0.0)	0.0	-0.3	0.0	-0.2	0.0	-0.2	0.0	-0.2	24.8	(6.2)	0.0	-0.1	0.0	-0.1	0.0	-0.1	0.0	0.0	0.0	0.0	0.2	0.0	0.1	0.0	0.0	-0.1
Greek	0.0	(0.0)	0.0	-0.1	0.0	-0.2	0.0	-0.2	0.0	-0.2	0.0	-0.1	0.2	0.0	0.0	0.0	0.0	0.0	0.0	-0.1	0.0	0.0	0.0	0.0	0.0	0.0	0.0	0.0	0.0	-0.1
Armenian	0.0	(0.0)	0.0	-0.1	0.3	-0.2	0.0	-0.2	0.0	-0.2	0.0	-0.1	0.2	0.0	0.0	0.0	0.1	0.0	0.0	-0.1	0.0	0.0	0.1	0.0	0.0	0.0	0.2	0.0	0.0	-0.1
Hebrew	0.0	(0.0)	0.0	-0.1	0.0	-0.3	0.0	-0.2	0.0	-0.2	0.0	-0.1	0.0	0.0	0.0	0.0	0.0	0.0	0.0	-0.1	0.0	0.0	0.0	0.0	0.0	0.0	0.0	0.0	0.0	-0.1
Syrian	0.0	(0.0)	0.0	-0.1	0.0	-0.3	0.0	-0.2	0.0	-0.2	0.0	-0.1	0.0	0.0	0.0	-0.1	0.0	0.0	0.0	-0.1	0.0	0.0	0.0	0.0	0.0	0.0	0.0	0.1	0.0	-0.1
French	0.0	(0.0)	0.0	-0.1	0.0	-0.3	0.0	-0.2	0.1	-0.2	0.0	0.0	0.0	0.0	0.0	0.0	0.0	0.0	0.0	0.0	0.0	0.0	0.0	0.0	0.0	0.0	0.0	0.1	0.0	-0.1
Turkish	0.4	(0.0)	0.0	-0.1	0.0	-0.3	0.0	-0.2	0.0	-0.2	0.0	-0.2	0.0	0.0	0.0	0.2	0.0	0.0	0.0	-0.1	0.0	0.0	0.0	0.0	0.0	0.0	0.0	0.0	0.0	-0.1
Mexican	0.1	(0.0)	0.1	-0.1	0.0	-0.3	0.2	-0.1	0.0	-0.1	0.1	-0.1	0.0	0.0	0.0	0.0	0.0	0.1	0.0	0.0	0.0	0.0	0.0	0.0	0.0	0.0	0.0	0.0	0.0	-0.1
Spanish American	0.3	(0.0)	0.0	-0.1	0.0	-0.3	0.0	-0.2	0.0	-0.2	0.0	-0.2	0.0	0.0	0.0	0.0	0.0	0.1	0.0	-0.1	0.0	0.0	0.0	0.0	0.0	0.0	0.0	0.0	0.0	-0.1
Cuban	0.0	(0.0)	0.0	-0.1	0.0	-0.3	0.0	-0.2	0.0	-0.2	0.0	-0.2	0.0	0.0	0.0	0.0	0.0	0.1	0.0	0.0	0.0	0.0	0.0	0.0	0.0	0.0	0.0	0.0	0.0	-0.1
West Indian	0.2	(0.0)	0.1	-0.1	0.0	-0.3	0.0	-0.2	0.0	-0.2	0.0	-0.2	0.0	0.0	0.0	-0.1	0.0	0.0	0.0	-0.1	0.0	0.0	0.0	0.0	0.0	0.0	0.0	-0.1	0.0	-0.1
Chinese	12.1	(1.3)	19.9	(6.6)	12.7	(5.2)	2.9	(1.5)	1.7	(2.3)	2.0	(2.8)	1.4	(0.3)	4.0	(4.6)	2.9	(2.8)	9.0	(5.2)	1.4	(1.4)	13.3	(4.8)	43.4	(5.7)	6.8	(5.7)	3.5	(5.9)
East Indian	0.0	(0.0)	3.2	(0.8)	0.0	0.0	2.9	(1.7)	3.6	(0.8)	0.8	(1.1)	0.6	(0.1)	1.0	(0.1)	3.1	(1.0)	3.1	(1.7)	0.3	(0.3)	3.5	(1.2)	17.8	(2.3)	0.8	(0.6)	0.4	(0.6)
Japanese	3.5	(0.4)	5.6	(1.5)	5.4	(2.0)	3.2	(1.7)	4.0	(5.5)	3.6	(5.3)	0.0	0.1	1.0	(1.0)	1.1	(1.0)	5.1	(2.9)	0.3	(0.3)	3.5	(1.2)	2.3	(2.3)	0.8	(0.8)	1.0	(0.6)
Korean	4.9	(0.5)	8.6	(2.4)	5.4	(0.7)	1.9	(0.9)	1.9	(0.9)	0.0	0.2	0.7	(0.1)	3.6	(3.6)	5.6	(5.6)	0.0	(0.0)	5.8	(6.1)	10.6	(6.1)	9.1	(1.2)	2.7	(2.2)	1.0	(1.7)
Pacific Islander	55.6	(6.2)	0.0	-0.1	4.2	(1.5)	10.0	(5.7)	0.0	-0.2	0.0	(0.0)	4.0	(1.0)	0.0	0.0	0.0	0.0	0.0	0.0	0.0	0.0	0.0	0.0	0.0	0.0	0.0	0.0	0.0	0.0
African (black)	0.8	(0.1)	0.1	-0.1	0.1	0.0	0.0	0.0	0.0	-0.2	0.0	-0.2	0.0	0.0	0.0	0.0	0.0	0.0	0.2	0.0	0.1	0.0	0.0	0.0	0.0	0.0	0.0	0.0	0.0	-0.1
Other	0.1	(0.0)	0.0	-0.1	0.0	-0.3	0.0	-0.2	0.0	-0.2	0.0	0.0	0.1	0.0	0.0	0.0	0.0	0.0	0.0	0.0	0.0	0.0	0.0	0.0	0.0	0.0	0.0	0.0	0.0	-0.1

Bold = 1 or more NSDMR below mean   ▨ = 1 to 1.99 NSDMRs above mean   ▩ = 2 or more NSDMRs above mean

° Norwegians, Danes, and Swedes.   † Bohemians and Moravians.

TABLE 6-11. Deviations in Certification Rate for TB, by Race, 1916–1930, Expressed as Percent Certified and NSDMR

Race or People	1916 %	NSDMR	1917 %	NSDMR	1918 %	NSDMR	1919 %	NSDMR	1920 %	NSDMR	1921 %	NSDMR	1922 %	NSDMR	1923 %	NSDMR	1924 %	NSDMR	1925 %	NSDMR	1926 %	NSDMR	1927 %	NSDMR	1928 %	NSDMR	1929 %	NSDMR	1930 %	NSDMR
English	0.0	(0.1)	0.0	(0.0)	0.1	(0.1)	0.0	(0.0)	0.0	(0.0)	0.0	(0.6)	0.0	(0.1)	0.0	(0.2)	0.0	(0.1)	0.0	(0.0)	0.0	(0.0)	0.1	(0.1)	0.0	(0.0)	0.0	(0.3)	0.0	(0.1)
Irish	0.0	(0.1)	0.0	(0.0)	0.2	(0.6)	0.1	(0.2)	0.0	(0.3)	0.0	(0.9)	0.0	(0.1)	0.0	(0.2)	0.0	(0.1)	0.0	(0.1)	0.0	(0.2)	0.0	(0.0)	0.0	(0.0)	0.0	(0.2)	0.0	(0.2)
Scotch	0.1	(0.5)	0.0	(0.1)	0.2	(0.8)	0.0	(0.1)	0.0	(0.4)	0.0	(0.2)	0.0	(0.2)	0.0	(0.3)	0.0	(0.4)	0.0	(0.1)	0.0	(0.2)	0.0	(0.0)	0.0	(0.0)	0.0	(0.3)	0.0	(0.2)
Welsh	0.0	(0.9)	0.0	(0.2)	0.7	(2.8)	0.0	(0.2)	0.0	(0.3)	0.0	(0.1)	0.0	(0.5)	0.0	(0.7)	0.0	(0.7)	0.0	(0.3)	0.0	(0.7)	0.0	(0.3)	0.0	(0.0)	0.0	(0.5)	0.0	(0.6)
Spanish	0.0	(0.1)	0.0	(0.0)	0.0	(0.2)	0.0	(0.0)	0.0	(0.3)	0.0	(0.1)	0.0	(0.2)	0.0	(0.7)	0.0	(0.3)	0.0	(0.2)	0.0	(0.2)	0.1	(0.3)	0.1	(0.0)	0.0	(0.1)	0.0	(0.8)
Portuguese	0.0	(0.7)	0.0	(0.0)	0.0	(0.4)	0.0	(0.2)	0.0	(0.4)	0.0	(0.8)	0.0	(0.5)	0.0	(0.7)	0.0	(0.1)	0.1	(0.7)	0.0	(0.7)	0.1	(0.5)	0.1	(0.1)	0.0	(0.2)	0.0	(0.3)
Italian (North)	0.0	(0.4)	0.1	(0.1)	0.2	(0.4)	0.0	(0.0)	0.0	(0.4)	0.0	(0.4)	0.0	(0.5)	0.0	(0.4)	0.0	(0.4)	0.1	(0.1)	0.0	(0.7)	0.0	(0.0)	0.1	(0.1)	0.0	(0.2)	0.0	(0.1)
Italian (South)	0.0	(0.3)	0.0	(0.1)	0.0	(0.3)	0.0	(0.0)	0.0	(0.1)	0.0	(0.3)	0.0	(0.0)	0.0	(0.0)	0.0	(0.5)	0.2	(0.1)	0.0	(0.0)	0.0	(0.0)	0.0	(0.0)	0.0	(0.5)	0.0	(0.2)
Dutch and Flemish	0.0	(0.9)	0.0	(0.0)	0.0	(0.3)	0.0	(0.0)	0.0	(0.4)	0.0	(0.8)	0.0	(0.2)	0.0	(0.0)	0.0	(0.7)	0.0	(0.3)	0.0	(0.3)	0.0	(0.3)	0.0	(0.0)	0.0	(0.1)	0.0	(0.2)
Finnish	0.1	(0.7)	0.0	(0.1)	0.1	(0.1)	0.0	(0.2)	0.0	(0.4)	0.0	(0.8)	0.0	(0.3)	0.0	(0.3)	0.0	(1.3)	0.0	(0.7)	0.0	(0.7)	0.5	(0.2)	0.4	(0.0)	0.2	(0.4)	0.0	(0.3)
Scandinavian*	0.0	(0.2)	0.0	(0.1)	0.0	(0.0)	0.0	(0.0)	0.0	(0.2)	0.0	(0.2)	0.0	(0.0)	0.0	(0.5)	0.0	(0.5)	0.0	(0.0)	0.0	(0.0)	0.0	(0.0)	0.0	(0.0)	0.0	(0.4)	0.0	(0.1)
Lithuanian	0.0	(0.9)	0.4	(1.9)	0.0	(0.4)	0.0	(0.2)	0.0	(0.4)	0.0	(0.5)	0.0	(0.5)	0.0	(0.7)	0.0	(0.7)	0.0	(0.3)	0.1	(3.2)	0.4	(2.0)	0.0	(0.0)	0.1	(1.1)	0.0	(0.3)
Russian	0.0	(0.5)	0.0	(0.1)	0.0	(0.2)	0.0	(0.2)	0.0	(0.4)	0.0	(0.8)	0.0	(0.3)	0.0	(2.0)	0.0	(0.2)	0.2	(1.4)	0.0	(0.7)	0.2	(0.7)	0.0	(0.0)	0.1	(1.1)	0.0	(0.3)
Polish	0.0	(0.4)	0.0	(0.1)	0.0	(0.3)	0.1	(0.3)	0.0	(0.4)	0.0	(0.4)	0.1	(1.3)	0.1	(1.3)	0.0	(1.0)	0.0	(0.1)	0.0	(0.0)	0.2	(0.7)	0.0	(0.0)	0.1	(0.4)	0.0	(0.0)
German	0.1	(0.9)	0.0	(0.1)	0.1	(0.2)	0.0	(0.0)	0.0	(0.1)	0.0	(0.9)	0.0	(0.3)	0.0	(0.0)	0.0	(0.3)	0.0	(0.2)	0.0	(0.2)	0.0	(0.2)	0.0	(0.0)	0.0	(0.4)	0.0	(0.4)
Magyar	0.0	(1.5)	0.0	(0.0)	0.0	(0.4)	0.0	(0.2)	0.0	(0.4)	0.0	(0.2)	0.0	(0.9)	0.0	(0.7)	0.0	(0.5)	0.2	(1.3)	0.0	(2.9)	0.3	(1.5)	0.0	(0.1)	0.0	(0.3)	0.1	(1.2)
Slovak	0.2	(3.3)	0.4	(1.9)	0.0	(0.4)	0.0	(0.2)	0.0	(0.4)	0.0	(0.0)	0.0	(0.0)	0.0	(0.7)	0.0	(0.2)	0.2	(1.8)	0.0	(0.7)	0.2	(0.9)	0.2	(0.0)	0.0	(0.5)	0.0	(0.5)
Croatian and Slovenian	0.0	(0.9)	0.3	(1.5)	0.0	(0.4)	0.0	(0.2)	0.0	(0.4)	0.0	(0.0)	0.0	(0.5)	0.0	(0.1)	0.0	(1.5)	0.2	(1.0)	0.0	(0.7)	0.2	(1.2)	0.2	(0.0)	0.1	(2.5)	0.1	(2.4)
Czech †	0.0	(0.9)	0.0	(0.0)	1.1	(5.0)	0.4	(1.7)	0.0	(0.4)	0.1	(4.3)	0.0	(0.5)	0.0	(0.1)	0.0	(0.7)	0.0	(0.3)	0.0	(0.7)	0.0	(0.3)	0.0	(0.0)	0.0	(0.5)	0.0	(0.3)
Dalmatian, Bosnian, Herzegovinian																														
Ruthenian (Russniak)	0.1	(2.1)	0.0	(0.2)	0.0	(0.4)	1.2	(6.0)	0.0	(0.4)	0.0	(0.8)	0.0	(0.5)	0.0	(0.7)	0.0	(0.7)	0.1	(0.5)	0.0	(0.7)	0.2	(1.1)	0.0	(0.0)	0.0	(0.0)	0.0	(0.3)
Roumanian	0.1	(1.4)	0.2	(0.7)	0.0	(0.4)	0.0	(0.0)	0.1	(3.3)	0.0	(2.1)	0.0	(0.5)	0.1	(3.7)	0.2	(0.7)	0.2	(1.3)	0.0	(0.7)	0.0	(0.3)	0.0	(0.0)	0.0	(0.5)	0.0	(0.3)
Bulgarian, Serbian, Montenegrin																														
Greek	0.1	(0.5)	0.0	(0.2)	0.0	(0.0)	0.0	(0.2)	0.1	(3.2)	0.0	(0.4)	0.0	(0.1)	0.0	(0.9)	0.0	(1.0)	0.0	(0.3)	0.0	(0.7)	0.2	(0.8)	0.2	(0.2)	0.1	(1.3)	0.1	(3.0)
Armenian	0.0	(0.9)	0.1	(0.2)	0.1	(0.1)	0.0	(0.0)	0.0	(0.4)	0.0	(0.8)	0.2	(4.3)	0.1	(2.6)	0.1	(4.9)	0.1	(0.3)	0.1	(1.2)	0.1	(0.4)	0.0	(0.0)	0.0	(0.5)	0.1	(2.2)
Hebrew	0.0	(0.0)	0.1	(0.2)	0.1	(0.1)	0.0	(0.1)	0.0	(0.0)	0.0	(0.8)	0.0	(0.2)	0.0	(0.2)	0.0	(0.3)	0.0	(0.9)	0.1	(0.7)	0.1	(0.4)	0.0	(0.0)	0.0	(0.5)	0.0	(0.1)
Syrian	0.1	(2.1)	0.1	(0.1)	0.1	(0.2)	0.0	(0.0)	0.0	(0.0)	0.0	(0.0)	0.0	(0.5)	0.0	(0.7)	0.0	(1.3)	0.0	(0.3)	0.1	(1.9)	0.1	(0.5)	0.3	(0.4)	0.1	(0.5)	0.1	(3.6)
French	0.1	(0.4)	0.1	(0.1)	0.2	(0.4)	0.0	(0.1)	0.0	(0.4)	0.0	(0.1)	0.0	(0.4)	0.0	(0.1)	0.0	(0.0)	0.0	(0.0)	0.1	(0.0)	0.1	(0.5)	0.1	(0.0)	0.0	(0.1)	0.0	(0.1)
Turkish	0.0	(0.9)	0.0	(0.3)	0.0	(0.4)	0.0	(0.2)	0.0	(0.4)	0.0	(0.8)	0.0	(0.5)	0.0	(0.7)	0.0	(0.7)	0.0	(0.3)	0.0	(0.7)	0.0	(0.2)	0.0	(0.0)	0.0	(0.5)	0.0	(0.3)
Mexican	0.1	(0.5)	0.1	(0.3)	0.0	(0.2)	0.0	(0.1)	0.0	(0.4)	0.0	(0.0)	0.0	(0.5)	0.0	(0.3)	0.0	(0.2)	0.0	(0.3)	0.1	(0.2)	0.0	(0.2)	0.1	(0.0)	0.1	(1.3)	0.0	(0.8)
Spanish American	0.1	(0.4)	0.0	(0.2)	0.0	(0.4)	0.0	(0.2)	0.0	(0.3)	0.0	(0.0)	0.0	(0.5)	0.0	(0.7)	0.0	(0.7)	0.0	(0.6)	0.0	(0.3)	0.0	(0.3)	0.0	(0.0)	0.0	(0.5)	0.0	(0.3)
Cuban	0.1	(0.5)	0.0	(0.2)	0.0	(0.3)	0.0	(0.2)	0.0	(0.0)	0.0	(1.3)	0.0	(0.5)	0.0	(0.6)	0.0	(0.1)	0.0	(0.4)	0.0	(0.1)	0.1	(0.1)	0.1	(0.1)	0.0	(0.2)	0.0	(0.3)
West Indian	0.0	(0.9)	0.0	(0.2)	0.1	(0.3)	0.0	(0.2)	0.0	(0.4)	0.0	(0.1)	0.0	(0.8)	0.0	(0.2)	0.0	(0.4)	0.1	(0.6)	0.0	(0.3)	0.1	(0.1)	0.0	(0.0)	0.0	(0.5)	0.0	(0.3)
Chinese	0.0	(0.9)	0.0	(0.0)	0.2	(0.4)	0.0	(0.1)	0.0	(0.2)	0.0	(0.1)	0.0	(0.5)	0.0	(0.2)	0.0	(0.3)	0.1	(0.4)	0.0	(0.7)	0.4	(0.3)	0.3	(0.4)	0.1	(1.7)	0.0	(0.6)
East Indian	0.0	(0.4)	1.1	(5.5)	0.0	(0.4)	0.0	(0.2)	0.0	(0.4)	0.0	(0.7)	0.0	(0.3)	0.0	(0.3)	0.0	(0.7)	0.0	(0.3)	0.0	(0.3)	0.4	(0.3)	2.6	(3.2)	0.1	(0.5)	0.0	(0.3)
Japanese	0.0	(0.9)	0.0	(0.2)	0.0	(0.4)	0.0	(0.2)	0.0	(0.4)	0.0	(0.8)	0.0	(0.0)	0.0	(0.3)	0.0	(0.7)	0.2	(1.8)	0.0	(0.7)	0.0	(0.3)	0.0	(0.0)	0.0	(0.1)	0.0	(0.3)
Korean	0.0	(0.9)	0.0	(0.2)	0.0	(0.4)	0.0	(0.2)	0.0	(0.4)	0.0	(0.8)	0.0	(0.0)	0.0	(0.7)	0.0	(0.7)	0.0	(0.0)	0.0	(0.7)	0.0	(0.3)	4.5	(5.5)	0.0	(0.5)	0.0	(0.3)
Pacific Islander	0.0	(0.9)	0.0	(0.1)	0.0	(0.4)	0.0	(0.2)	0.0	(0.4)	0.0	(0.0)	0.0	(0.0)	0.0	(0.7)	0.0	(0.7)	0.0	(0.4)	0.0	(0.7)	0.0	(0.0)	0.0	(0.0)	0.0	(0.5)	0.0	(0.3)
African (black)	0.1	(0.6)	0.0	(0.1)	0.0	(0.2)	0.0	(0.1)	0.0	(0.4)	0.0	(0.1)	0.0	(0.0)	0.0	(0.2)	0.0	(0.1)	0.1	(0.4)	0.0	(0.7)	0.0	(0.0)	0.0	(0.0)	0.0	(0.5)	0.0	(0.3)
Other	0.0	(0.2)	0.1	(0.5)	0.0	(0.0)	0.0	(0.2)	0.1	(3.4)	0.0	(0.8)	0.0	(0.5)	0.1	(2.4)	0.0	(0.7)	0.5	(3.7)	0.1	(1.7)	0.8	(4.4)	0.0	(0.0)	0.0	(0.5)	0.1	(2.2)

Bold = 1 or more NSDMR below mean   ▨ = 1 to 1.99 NSDMRs above mean   ▨ = 2 or more NSDMRs above mean.

* Norwegians, Danes, and Swedes.

† Bohemians and Moravians.

TABLE 6-12. Deviations in Rate of Deportation for Those Certified with TB, by Race, 1916–1930, Expressed as Percent Certified then Deported and NSDMR

Race or People	1916 %	1916 NSDMR	1917 %	1917 NSDMR	1918 %	1918 NSDMR	1919 %	1919 NSDMR	1920 %	1920 NSDMR	1921 %	1921 NSDMR	1922 %	1922 NSDMR	1923 %	1923 NSDMR	1924 %	1924 NSDMR	1925 %	1925 NSDMR	1926 %	1926 NSDMR	1927 %	1927 NSDMR	1928 %	1928 NSDMR	1929 %	1929 NSDMR	1930 %	1930 NSDMR
English	81.25	(-0.4)	92.31	(-0.1)	100	(0.3)	92.86	(0.2)	72.73	(-0.2)	100	(0.4)	75	(-0.8)	72.73	(0.7)	73.33	(-0.7)	52.38	(-0.7)	50	(-0.7)	30.43	(-0.4)	0	(-0.6)	0	(-0.7)	0	(-0.6)
Irish	85.71	(-0.2)	100	(0.1)	90	(0.1)	62.5	(0.1)	60	(-0.6)	87.5	(0.1)	100	(0.1)	100	(0.7)	50	(-1.5)	92.86	(0.6)	75	(0.2)	30.77	(-0.4)	0	(-0.6)	0	(-0.7)	25	(0.0)
Scotch	87.5	(-0.1)	71.43	(-0.8)	84.62	(-0.1)	85.71	(0.0)	100	(0.4)	100	(0.4)	80	(-0.6)	58.33	(-0.6)	100	(0.3)	80	(0.2)	83.33	(0.5)	25	(-0.5)	0	(-0.6)	33.33	(0.2)	0	(-0.6)
Welsh	0	(0.0)	0	(0.0)	50	(-1.1)	0	(0.0)	0	(0.0)	0	(0.0)	0	(0.0)	0	(0.0)	0	(0.0)	0	(0.0)	0	(0.0)	0	(0.0)	0	(0.0)	0	(0.0)	0	(-0.6)
Spanish	100	(0.5)	100	(0.1)	66.67	(-0.6)	100	(0.4)	100	(0.6)	66.67	(-0.4)	0	(0.0)	0	(0.0)	100	(0.3)	75	(0.1)	50	(-0.7)	0	(0.0)	0	(0.0)	0	(-0.7)	33.33	(0.2)
Portuguese	100	(0.5)	33.33	(-2.0)	0	(-1.1)	0	(0.0)	0	(0.0)	0	(0.0)	0	(0.0)	0	(0.0)	100	(0.3)	0	(-1.8)	0	(0.0)	0	(-1.2)	0	(0.0)	0	(-0.7)	0	(0.0)
Italian (North)	100	(0.5)	100	(0.1)	0	(0.0)	100	(0.4)	50	(-0.9)	0	(-2.1)	0	(0.0)	100	(0.7)	100	(0.3)	100	(0.3)	0	(0.0)	0	(0.0)	100	(1.7)	0	(-0.7)	100	(1.8)
Italian (South)	66.67	(-1.1)	90.91	(-0.2)	100	(0.3)	50	(-0.9)	50	(0.0)	100	(0.4)	80	(-0.6)	40	(-1.1)	68.75	(-0.8)	42.86	(-0.7)	50	(-0.7)	71.43	(0.7)	66.67	(-1.0)	0	(0.0)	0	(-0.6)
Dutch and Flemish	0	(0.0)	0	(0.0)	100	(0.3)	0	(-2.2)	0	(0.0)	100	(0.4)	0	(0.0)	0	(0.0)	100	(0.3)	0	(0.0)	0	(0.0)	0	(0.0)	0	(0.0)	0	(-0.7)	0	(0.0)
Finnish	100	(0.5)	0	(0.0)	100	(0.3)	0	(0.0)	0	(0.0)	0	(0.0)	100	(0.1)	100	(0.7)	100	(0.3)	100	(0.3)	0	(0.0)	100	(1.4)	0	(0.0)	50	(0.6)	0	(0.0)
Scandinavian*	83.33	(-0.3)	100	(0.1)	85.71	(-0.1)	75	(-0.3)	100	(0.6)	100	(0.4)	85.71	(-0.4)	100	(0.7)	100	(0.3)	66.67	(-0.1)	100	(1.0)	100	(1.4)	0	(-0.6)	0	(0.0)	0	(-0.6)
Lithuanian	0	(0.0)	50	(-1.5)	0	(0.0)	0	(0.0)	0	(0.0)	0	(0.0)	0	(0.0)	0	(0.0)	0	(0.0)	0	(0.0)	0	(-2.4)	0	(-1.2)	0	(0.0)	0	(0.0)	0	(0.0)
Russian	100	(0.5)	100	(0.1)	100	(0.3)	100	(0.4)	100	(0.6)	0	(0.4)	75	(-0.8)	50	(-0.8)	0	(-3.2)	33.33	(-1.0)	0	(0.0)	100	(1.4)	0	(-0.6)	100	(2.0)	0	(-0.6)
Polish	100	(0.5)	100	(0.1)	100	(0.3)	0	(-2.2)	0	(0.0)	100	(0.4)	0	(0.1)	66.67	(-0.3)	57.14	(-1.2)	50	(-0.5)	50	(0.0)	100	(1.2)	0	(-0.6)	100	(2.0)	100	(1.8)
German	88.89	(0.0)	66.67	(-0.9)	100	(0.3)	100	(0.4)	100	(0.6)	100	(0.4)	100	(0.1)	92.96	(0.5)	88.89	(-0.1)	77.78	(0.2)	92.31	(0.8)	33.33	(-0.3)	0	(-0.6)	0	(-0.7)	0	(0.0)
Magyar	100	(0.5)	0	(0.0)	0	(0.0)	0	(0.0)	0	(0.0)	100	(0.4)	66.67	(-1.1)	0	(0.0)	100	(0.3)	100	(0.3)	100	(1.0)	66.67	(-0.6)	100	(1.7)	100	(2.0)	0	(0.0)
Slovak	0	(-4.1)	100	(0.1)	0	(0.0)	0	(0.0)	0	(0.0)	100	(0.4)	0	(0.0)	100	(0.7)	0	(0.0)	0	(-1.8)	0	(0.0)	50	(0.1)	0	(0.0)	0	(0.0)	0	(0.0)
Croatian and Slovenian	0	(0.0)	0	(-3.1)	0	(0.0)	0	(-2.2)	0	(0.0)	100	(0.4)	0	(0.0)	0	(-2.3)	50	(-1.5)	0	(-1.8)	0	(0.0)	50	(0.1)	100	(1.7)	100	(2.0)	100	(1.8)
Czech †	0	(0.0)	0	(0.0)	100	(0.3)	0	(0.0)	0	(0.0)	0	(-2.1)	0	(0.0)	0	(-2.3)	0	(0.0)	0	(0.0)	0	(0.0)	0	(0.0)	0	(0.0)	0	(0.0)	0	(0.0)
Dalmatian, Bosnian, Herzegovinian	0	(0.0)	0	(0.0)	0	(0.0)	0	(0.0)	0	(0.0)	0	(0.0)	0	(0.0)	0	(0.0)	0	(0.0)	0	(0.0)	0	(0.0)	0	(0.0)	0	(0.0)	0	(0.0)	0	(0.0)
Ruthenian (Russniak)	100	(0.5)	100	(0.1)	0	(0.0)	100	(0.4)	100	(0.6)	100	(0.4)	0	(0.0)	0	(0.0)	0	(0.0)	100	(0.3)	0	(0.0)	0	(0.0)	0	(0.0)	0	(0.0)	0	(0.0)
Roumanian	100	(0.5)	100	(0.1)	0	(0.0)	0	(0.0)	100	(0.6)	100	(0.4)	0	(0.0)	50	(-0.8)	0	(0.0)	100	(0.8)	0	(0.0)	0	(0.0)	0	(0.0)	0	(0.0)	0	(0.0)
Bulgarian, Serbian, Montenegrin	100	(0.5)	100	(0.1)	0	(0.0)	100	(0.4)	100	(0.6)	100	(0.4)	0	(0.0)	0	(0.0)	100	(0.3)	100	(0.5)	0	(0.0)	0	(0.0)	0	(0.0)	0	(0.0)	0	(0.0)
Greek	100	(0.5)	71.43	(-0.8)	0	(0.0)	0	(0.0)	66.67	(-0.4)	100	(0.4)	0	(0.0)	0	(-2.3)	100	(0.3)	0	(-1.8)	0	(0.0)	0	(-1.2)	100	(1.7)	0	(0.0)	0	(-0.6)
Armenian	0	(0.0)	0	(-3.1)	0	(0.0)	100	(0.4)	0	(-2.4)	75	(-0.2)	50	(-1.7)	50	(-0.8)	100	(0.3)	0	(-1.8)	50	(-0.7)	100	(1.7)	100	(1.7)	0	(0.0)	0	(-0.6)
Hebrew	83.33	(-0.3)	55.56	(-1.3)	50	(-1.1)	100	(0.4)	100	(0.6)	42.86	(-1.0)	50	(0.1)	50	(0.7)	100	(0.3)	100	(0.3)	55.56	(-0.5)	58.33	(0.3)	0	(-0.6)	66.67	(1.1)	100	(1.8)
Syrian	100	(0.5)	100	(0.1)	0	(0.0)	0	(0.0)	100	(0.6)	100	(0.0)	100	(0.0)	100	(0.7)	100	(0.3)	100	(0.0)	0	(0.0)	0	(0.0)	0	(-0.6)	0	(-0.7)	0	(0.0)
French	100	(0.5)	92.31	(0.1)	100	(0.3)	100	(0.4)	100	(0.6)	100	(0.4)	100	(0.1)	70	(-0.2)	87.5	(0.0)	63.64	(-0.2)	64.29	(-0.2)	17.39	(-0.7)	0	(-0.6)	0	(-0.7)	50	(0.6)
Turkish	0	(0.0)	0	(0.0)	0	(0.0)	0	(0.0)	0	(0.0)	0	(0.0)	0	(0.1)	0	(-0.2)	0	(0.0)	0	(0.0)	0	(0.0)	0	(0.0)	0	(0.0)	0	(0.0)	0	(0.0)
Mexican	100	(0.5)	100	(0.1)	87.5	(0.0)	100	(0.4)	100	(0.6)	75	(-0.2)	90	(-0.3)	77.78	(0.0)	90.91	(-0.1)	66.67	(-0.1)	50	(-0.7)	37.5	(-0.2)	43.33	(0.4)	45.16	(0.5)	25	(0.6)
Spanish American	100	(0.5)	100	(0.0)	0	(0.0)	0	(0.0)	0	(-2.4)	0	(-2.1)	0	(-3.5)	66.67	(-0.3)	0	(-3.2)	0	(-1.8)	33.33	(-1.3)	100	(1.4)	0	(-0.6)	0	(-0.7)	50	(0.6)
Cuban	50	(-1.8)	0	(0.0)	0	(-2.5)	0	(0.0)	100	(0.6)	100	(0.4)	0	(0.0)	66.67	(-0.3)	100	(0.3)	0	(-1.8)	100	(1.0)	100	(1.4)	0	(-0.6)	0	(-0.7)	0	(0.0)
West Indian	0	(0.0)	0	(0.0)	0	(-2.5)	0	(-2.2)	0	(0.0)	0	(-2.1)	0	(0.0)	66.67	(-0.3)	0	(0.0)	50	(-0.5)	0	(0.0)	50	(0.1)	0	(-0.6)	0	(-0.4)	0	(-0.6)
Chinese	0	(0.0)	100	(0.1)	100	(0.3)	100	(0.4)	100	(0.6)	0	(-2.1)	0	(0.0)	100	(0.7)	100	(0.3)	50	(0.0)	100	(1.0)	50	(0.1)	0	(-0.6)	14.29	(-0.4)	0	(-0.6)
East Indian	100	(0.5)	100	(0.1)	0	(0.0)	0	(0.0)	0	(0.0)	0	(0.0)	0	(0.0)	0	(0.0)	0	(0.0)	0	(0.0)	0	(0.0)	0	(0.0)	0	(0.0)	0	(0.0)	0	(0.0)
Japanese	100	(0.5)	100	(0.1)	100	(0.3)	100	(0.4)	80	(0.0)	0	(-2.1)	33.33	(-2.3)	66.67	(-0.3)	100	(0.3)	50	(0.0)	0	(0.0)	0	(0.0)	0	(0.0)	0	(-0.7)	50	(0.6)
Korean	0	(0.0)	0	(0.0)	0	(0.0)	0	(0.0)	0	(0.0)	0	(0.0)	0	(0.0)	0	(0.0)	0	(0.0)	0	(0.0)	0	(0.0)	0	(0.0)	0	(0.0)	0	(0.0)	0	(0.0)
Pacific Islander	0	(0.0)	0	(0.0)	0	(0.0)	0	(0.0)	0	(0.0)	0	(0.0)	0	(0.0)	0	(0.0)	0	(0.0)	0	(0.0)	0	(0.0)	0	(0.0)	0	(0.0)	0	(0.0)	0	(0.0)
African (black)	66.67	(-1.1)	50	(-1.5)	100	(0.3)	100	(0.4)	0	(0.0)	100	(0.4)	50	(-1.7)	50	(-0.8)	0	(0.0)	100	(0.8)	0	(0.0)	0	(0.0)	0	(0.0)	0	(0.0)	0	(0.0)
Other	100	(0.5)	100	(0.1)	0	(0.0)	0	(0.0)	50	(-0.9)	0	(0.0)	0	(0.0)	100	(0.7)	0	(0.0)	100	(0.5)	100	(1.0)	66.67	(0.6)	0	(0.0)	0	(0.0)	100	(1.8)

Bold = 1 or more NSDMR below mean
[shaded] = 1 to 1.99 NSDMRs above mean
[shaded] = 2 or more NSDMRs above mean
* Norwegians, Danes, and Swedes.
† Bohemians and Moravians.

TABLE 6-13. Deviations in Rate of Deportation for TB, by Race, 1908–1930, Expressed as Percent Deported and NSDMR

Race or People	1908 %	1908 NSDMR	1909 %	1909 NSDMR	1910 %	1910 NSDMR	1911 %	1911 NSDMR	1912 %	1912 NSDMR	1913 %	1913 NSDMR	1914 %	1914 NSDMR	1915 %	1915 NSDMR	1916 %	1916 NSDMR	1917 %	1917 NSDMR	1918 %	1918 NSDMR	1919 %	1919 NSDMR
English	0.0	(0.3)	0.0	(1.1)	0.0	(0.7)	0.0	(0.6)	0.0	(0.3)	0.0	(0.1)	0.0	(1.2)	0.0	(0.6)	0.0	(0.0)	0.0	(0.0)	0.1	(0.2)	0.0	(0.0)
Irish	0.0	(0.3)	0.0	(0.3)	0.0	(0.8)	0.0	(0.7)	0.0	(0.3)	0.0	(0.2)	0.0	(0.6)	0.0	(0.9)	0.0	(0.0)	0.0	(0.0)	0.2	(0.6)	0.0	(0.1)
Scotch	0.0	(0.4)	0.0	(1.0)	0.0	(1.6)	0.0	(0.7)	0.0	(0.6)	0.0	(0.3)	0.0	(1.0)	0.1	(1.3)	0.0	(0.1)	0.0	(0.2)	0.2	(0.7)	0.0	(0.1)
Welsh	0.0	(0.3)	0.0	(0.7)	0.0	(2.8)	0.0	(0.7)	0.0	(0.1)	0.0	(0.0)	0.0	(0.6)	0.0	(1.2)	0.0	(0.3)	0.0	(0.2)	0.3	(1.5)	0.0	(0.1)
Australian	0.0	(0.0)	0.0	(0.0)	0.0	(0.0)	0.0	(0.0)	0.0	(0.0)	0.0	(0.0)	0.0	(0.0)	0.0	(0.0)	0.0	(0.0)	0.1	(0.0)	0.0	(0.0)	0.0	(0.0)
Spanish	0.0	(0.9)	0.0	(0.0)	0.0	(0.8)	0.0	(0.2)	0.0	(0.1)	0.0	(0.0)	0.0	(0.2)	0.0	(0.4)	0.0	(0.0)	0.1	(0.5)	0.1	(0.1)	0.0	(0.0)
Portuguese	0.0	(0.3)	0.0	(0.7)	0.0	(0.8)	0.0	(0.1)	0.0	(0.1)	0.0	(0.8)	0.0	(0.8)	0.0	(0.1)	0.0	(0.2)	0.0	(0.0)	0.1	(0.1)	0.0	(0.0)
Italian (North)	0.0	(0.0)	0.0	(0.7)	0.0	(0.6)	0.0	(0.2)	0.0	(0.1)	0.0	(0.0)	0.0	(0.3)	0.0	(0.1)	0.0	(0.1)	0.1	(0.1)	0.0	(0.4)	0.0	(0.0)
Italian (South)	0.0	(0.0)	0.0	(0.4)	0.0	(0.2)	0.0	(0.2)	0.0	(0.1)	0.0	(0.0)	0.0	(0.2)	0.0	(0.3)	0.0	(0.1)	0.0	(0.1)	0.0	(0.3)	0.0	(0.0)
Dutch and Flemish	0.0	(0.3)	0.0	(0.1)	0.0	(0.1)	0.0	(0.3)	0.0	(0.0)	0.0	(0.0)	0.0	(0.2)	0.0	(0.3)	0.0	(0.3)	0.0	(0.2)	0.1	(0.1)	0.0	(0.0)
Finnish	0.0	(0.9)	0.0	(0.7)	0.0	(0.8)	0.0	(1.1)	0.0	(0.4)	0.0	(0.1)	0.0	(0.1)	0.0	(1.2)	0.0	(0.3)	0.0	(0.1)	0.1	(0.2)	0.0	(0.1)
Scandinavian*	0.0	(0.2)	0.0	(0.5)	0.0	(0.3)	0.0	(0.0)	0.0	(0.0)	0.0	(0.1)	0.0	(0.3)	0.0	(0.4)	0.0	(0.1)	0.0	(0.1)	0.0	(0.4)	0.0	(0.0)
Lithuanian	0.0	(0.3)	0.0	(0.7)	0.0	(0.8)	0.0	(0.3)	0.0	(0.0)	0.0	(0.1)	0.0	(0.6)	0.0	(1.2)	0.2	(0.9)	0.0	(0.9)	0.0	(0.4)	0.0	(0.0)
Russian	0.0	(0.1)	0.0	(0.7)	0.0	(0.8)	0.0	(0.1)	0.0	(0.1)	0.0	(0.0)	0.0	(0.5)	0.0	(0.4)	0.0	(0.0)	0.2	(0.0)	0.0	(0.0)	0.0	(0.1)
Polish	0.0	(0.1)	0.0	(0.6)	0.0	(0.4)	0.0	(0.7)	0.0	(0.1)	0.0	(0.1)	0.0	(0.3)	0.0	(0.7)	0.0	(0.1)	0.0	(0.0)	0.0	(0.3)	0.0	(0.1)
German	0.0	(0.0)	0.0	(0.3)	0.0	(0.2)	0.0	(0.1)	0.0	(0.1)	0.0	(0.1)	0.0	(0.4)	0.0	(1.2)	0.0	(0.3)	0.0	(0.1)	0.1	(0.4)	0.0	(0.1)
Magyar	0.0	(0.3)	0.0	(0.4)	0.0	(0.1)	0.0	(0.7)	0.0	(0.1)	0.0	(0.1)	0.0	(0.6)	0.1	(1.2)	0.1	(0.5)	0.0	(0.2)	0.0	(0.4)	0.0	(0.1)
Slovak	0.0	(0.3)	0.0	(0.7)	0.0	(0.8)	0.0	(0.7)	0.0	(0.1)	0.0	(0.1)	0.0	(0.3)	0.0	(1.2)	0.0	(0.3)	0.0	(0.2)	0.0	(0.4)	0.0	(0.1)
Croatian & Slovenian	0.0	(0.3)	0.0	(0.3)	0.0	(0.1)	0.0	(0.7)	0.0	(0.0)	0.0	(0.3)	0.0	(0.2)	0.0	(1.2)	0.4	(0.3)	0.4	(2.1)	0.0	(0.4)	0.0	(0.1)
Transylvanian (Siebenburger)	0.0	(0.0)	0.0	(0.0)	0.0	(0.0)	0.0	(0.0)	0.0	(0.0)	0.0	(0.0)	0.0	(0.0)	0.0	(0.0)	0.0	(0.0)	0.0	(0.2)	0.0	(0.0)	0.0	(0.0)
Czech †	0.0	(0.3)	0.0	(0.3)	0.0	(0.3)	0.0	(0.7)	0.0	(0.1)	0.0	(0.1)	0.0	(0.6)	0.0	(0.0)	0.0	(0.3)	0.0	(0.2)	1.1	(5.7)	0.4	(1.7)
Dalmatian, Bosnian, Herzegovinian	0.0	(0.3)	0.0	(0.7)	0.0	(0.8)	0.0	(0.1)	0.0	(0.1)	0.0	(0.1)	0.0	(0.6)	0.0	(1.2)	0.8	(6.2)	0.0	(0.2)	0.0	(0.4)	0.0	(0.1)
Ruthenian (Russniak)	0.0	(0.9)	0.0	(0.2)	0.0	(0.8)	0.0	(0.7)	0.0	(0.1)	0.0	(0.1)	0.0	(0.6)	0.0	(1.2)	0.1	(0.7)	0.0	(0.2)	0.0	(0.4)	1.2	(6.1)
Austro-Hungarian	0.0	(0.1)	0.0	(0.2)	0.0	(0.0)	0.0	(0.0)	0.0	(0.0)	0.0	(0.0)	0.0	(0.4)	0.0	(0.0)	0.0	(0.0)	0.0	(0.0)	0.0	(0.2)	0.0	(0.0)
Roumanian	0.0	(0.3)	0.0	(0.2)	0.0	(0.8)	0.0	(0.4)	0.0	(0.1)	0.0	(0.0)	0.0	(0.3)	0.0	(0.0)	0.0	(0.0)	0.2	(0.8)	0.0	(0.4)	0.0	(0.1)
Bulgarian, Serbian, Montenegrin	0.0	(0.1)	0.0	(0.3)	0.0	(0.2)	0.0	(0.1)	0.0	(0.1)	0.0	(0.1)	0.0	(0.2)	0.0	(0.9)	0.1	(0.4)	0.0	(0.0)	0.0	(0.4)	0.0	(0.1)
Greek	0.0	(0.0)	0.0	(0.2)	0.0	(0.1)	0.0	(0.2)	0.0	(0.0)	0.0	(0.1)	0.0	(0.3)	0.0	(0.0)	0.0	(0.1)	0.0	(0.1)	0.0	(0.4)	0.0	(0.2)
Armenian	0.0	(0.9)	0.0	(1.5)	0.0	(1.0)	0.0	(0.7)	0.0	(0.0)	0.0	(0.1)	0.0	(0.6)	0.0	(0.0)	0.0	(0.0)	0.0	(0.2)	0.0	(0.2)	0.1	(0.1)
Hebrew	0.0	(0.1)	0.0	(0.5)	0.0	(0.7)	0.0	(0.2)	0.0	(0.1)	0.0	(0.1)	0.0	(0.4)	0.0	(0.2)	0.0	(0.0)	0.0	(0.1)	0.0	(0.2)	0.1	(0.1)
Arabian	0.0	(0.0)	0.0	(0.0)	0.0	(0.0)	0.0	(0.0)	0.0	(0.0)	0.0	(0.0)	0.0	(0.0)	0.0	(0.0)	0.0	(0.0)	0.0	(0.0)	0.0	(0.0)	0.0	(0.0)
Syrian	0.0	(0.3)	0.0	(1.0)	0.0	(0.6)	0.0	(0.0)	0.0	(0.5)	0.0	(0.1)	0.0	(0.1)	0.0	(0.2)	0.0	(0.0)	0.0	(0.2)	0.0	(0.4)	0.0	(0.1)
Turkish	0.0	(0.0)	0.0	(0.7)	0.0	(0.8)	0.0	(0.7)	0.0	(0.1)	0.0	(0.3)	0.0	(0.6)	0.0	(1.2)	0.0	(0.3)	0.0	(0.2)	0.0	(0.0)	0.0	(0.1)
French	0.0	(0.0)	0.0	(2.7)	0.0	(1.2)	0.0	(1.5)	0.0	(0.3)	0.0	(0.3)	0.1	(3.4)	0.1	(2.6)	0.1	(0.1)	0.0	(0.1)	0.2	(0.6)	0.0	(0.1)
Swiss	0.0	(0.0)	0.0	(0.0)	0.0	(0.0)	0.0	(0.0)	0.0	(0.0)	0.0	(0.0)	0.0	(0.0)	0.0	(0.0)	0.2	(0.0)	0.0	(0.3)	0.0	(0.0)	0.0	(0.1)
Mexican	0.1	(2.3)	0.1	(3.6)	0.0	(2.8)	0.0	(0.7)	0.0	(0.2)	0.0	(0.1)	0.0	(2.6)	0.1	(1.2)	0.0	(0.2)	0.0	(0.3)	0.0	(0.2)	0.0	(0.1)
Central American (NS)	0.0	(0.0)	0.0	(0.0)	0.0	(0.0)	0.0	(0.0)	0.0	(0.0)	0.0	(0.0)	0.0	(0.0)	0.0	(0.0)	0.0	(0.0)	0.0	(0.0)	0.0	(0.0)	0.0	(0.0)
South American	0.0	(0.3)	0.0	(0.7)	0.0	(0.8)	0.0	(0.0)	0.0	(0.0)	0.0	(0.0)	0.0	(1.6)	0.0	(0.0)	0.0	(0.0)	0.0	(0.0)	0.0	(0.0)	0.0	(0.0)
Spanish American	0.0	(0.3)	0.0	(0.7)	0.0	(0.8)	0.0	(0.7)	0.0	(0.0)	0.0	(0.1)	0.0	(0.6)	0.0	(1.2)	0.1	(0.1)	0.0	(0.2)	0.0	(0.4)	0.0	(0.1)
Cuban	0.0	(0.3)	0.0	(0.7)	0.0	(0.8)	0.0	(0.7)	0.0	(0.0)	0.0	(0.2)	0.0	(0.6)	0.0	(0.2)	0.0	(0.0)	0.0	(0.2)	0.0	(0.4)	0.0	(0.1)
West Indian	0.0	(0.3)	0.0	(0.7)	0.0	(0.8)	0.0	(0.7)	0.0	(0.1)	0.1	(0.2)	0.0	(0.6)	0.0	(0.0)	0.0	(0.3)	0.0	(0.2)	0.0	(0.4)	0.0	(0.1)
Chinese	0.1	(4.7)	0.0	(1.9)	0.0	(0.8)	0.0	(1.3)	0.0	(0.2)	0.0	(0.3)	0.0	(1.3)	0.1	(2.3)	0.0	(0.3)	0.1	(0.3)	0.0	(0.4)	0.0	(0.1)
East Indian	0.0	(1.6)	0.0	(0.0)	0.0	(0.8)	0.1	(4.0)	0.0	(6.2)	0.0	(6.4)	0.0	(0.6)	0.0	(1.2)	0.0	(0.3)	1.1	(5.8)	0.0	(0.3)	0.0	(0.1)
Japanese	0.0	(0.0)	0.0	(0.7)	0.0	(0.5)	0.0	(0.7)	0.0	(0.7)	0.0	(0.1)	0.0	(0.6)	0.0	(0.1)	0.0	(0.1)	0.0	(0.2)	0.0	(0.3)	0.0	(0.1)
Korean	0.0	(0.3)	0.0	(0.7)	0.0	(0.8)	0.0	(0.7)	0.0	(0.1)	0.0	(0.1)	0.0	(0.6)	0.0	(1.2)	0.0	(0.3)	0.0	(0.2)	0.0	(0.4)	0.0	(0.1)
Filipino	0.0	(0.0)	0.0	(0.0)	0.0	(0.0)	0.0	(0.0)	0.0	(0.0)	0.0	(0.0)	0.0	(0.0)	0.0	(0.0)	0.0	(0.0)	0.0	(0.0)	0.0	(0.0)	0.0	(0.0)
Hawaiian	0.0	(0.0)	0.0	(0.0)	0.0	(0.0)	0.0	(0.0)	0.0	(0.0)	0.0	(0.0)	0.0	(0.0)	0.0	(0.0)	0.0	(0.0)	0.0	(0.0)	0.0	(0.0)	0.0	(0.0)
Pacific Islander	0.0	(0.3)	0.0	(0.7)	0.0	(0.8)	0.0	(0.7)	0.0	(0.0)	0.0	(0.1)	0.0	(0.6)	0.0	(0.4)	0.0	(0.0)	0.0	(0.2)	0.0	(0.4)	0.0	(0.1)
African (black) (NS)	0.0	(1.4)	0.0	(0.0)	0.0	(0.7)	0.0	(0.0)	0.0	(0.1)	0.0	(0.0)	0.0	(0.0)	0.0	(0.0)	0.0	(0.1)	0.0	(0.0)	0.0	(0.2)	0.0	(0.0)
Esquimauz	0.0	(0.0)	0.0	(0.0)	0.0	(0.0)	0.0	(0.0)	0.0	(0.0)	0.0	(0.0)	0.0	(0.0)	0.0	(0.0)	0.0	(0.0)	0.0	(0.0)	0.0	(0.0)	0.0	(0.0)
Other	0.1	(2.1)	0.0	(0.7)	0.0	(1.8)	0.0	(2.6)	0.0	(0.1)	0.1	(0.5)	0.0	(2.6)	0.0	(1.2)	0.0	(0.0)	0.1	(0.6)	0.0	(0.4)	0.0	(0.1)

Bold = 1 or more NSDMR below mean   = 1 to 1.99 NSDMRs above mean   = 2 or more NSDMRs above mean   NS = not specified

* Norwegians, Danes, and Swedes.   † Bohemians and Moravians.

TABLE 6-13. Continued

Race or People	1920 %	1920 NSDMR	1921 %	1921 NSDMR	1922 %	1922 NSDMR	1923 %	1923 NSDMR	1924 %	1924 NSDMR	1925 %	1925 NSDMR	1926 %	1926 NSDMR	1927 %	1927 NSDMR	1928 %	1928 NSDMR	1929 %	1929 NSDMR	1930 %	1930 NSDMR
English	0.0	(0.0)	0.0	(1.1)	0.0	(-0.2)	0.0	(0.2)	0.0	(-0.1)	0.0	(0.0)	0.0	(-0.2)	0.0	(0.0)	0.0	(-0.2)	0.0	(-0.2)	0.0	(-0.1)
Irish	0.0	(0.1)	0.0	(1.2)	0.0	(-0.1)	0.0	(0.7)	0.0	(-0.2)	0.0	(0.0)	0.0	(-0.2)	0.0	(-0.1)	0.0	(-0.2)	0.0	(-0.2)	0.0	(0.0)
Scotch	0.0	(0.7)	0.0	(0.0)	0.0	(0.1)	0.0	(0.1)	0.0	(-0.2)	0.0	(0.1)	0.0	(-0.1)	0.0	(-0.1)	0.0	(-0.2)	0.0	(-0.1)	0.0	(-0.1)
Welsh	0.0	(-0.4)	0.0	(-0.9)	0.0	(-0.7)	0.0	(-0.8)	0.0	(-0.5)	0.0	(-0.1)	0.0	(-0.6)	0.0	(-0.2)	0.0	(-0.2)	0.0	(-0.2)	0.0	(-0.1)
Australian	0.0	(0.0)	0.0	(0.0)	0.0	(0.0)	0.0	(0.0)	0.0	(0.0)	0.0	(0.0)	0.0	(0.0)	0.0	(0.0)	0.0	(0.0)	0.0	(0.0)	0.0	(0.0)
Spanish	0.0	(-0.3)	0.0	(-0.1)	0.0	(0.0)	0.0	(0.0)	0.0	(0.3)	0.3	(1.6)	0.0	(-0.1)	0.0	(0.0)	0.0	(-0.2)	0.0	(-0.2)	0.0	(0.3)
Portuguese	0.0	(-0.4)	0.0	(-0.9)	0.0	(-0.7)	0.0	(-0.8)	0.0	(0.1)	0.0	(-0.1)	0.0	(-0.6)	0.0	(-0.2)	0.0	(0.0)	0.0	(-0.2)	0.0	(-0.1)
Italian (North)	0.0	(-0.1)	0.0	(-0.9)	0.0	(-0.7)	0.0	(0.0)	0.0	(0.2)	0.1	(0.6)	0.0	(0.0)	0.1	(0.8)	0.1	(1.4)	0.0	(-0.2)	0.0	(0.3)
Italian (South)	0.0	(-0.2)	0.0	(-0.3)	0.0	(-0.3)	0.0	(-0.4)	0.0	(0.1)	0.1	(0.3)	0.0	(-0.4)	0.0	(0.1)	0.0	(0.4)	0.0	(-0.2)	0.0	(-0.1)
Dutch and Flemish	0.0	(-0.4)	0.0	(-0.9)	0.0	(-0.7)	0.0	(-0.8)	0.0	(-0.1)	0.0	(0.0)	0.0	(-0.6)	0.0	(0.0)	0.0	(-0.2)	0.0	(-0.2)	0.0	(-0.1)
Finnish	0.0	(-0.4)	0.0	(-0.9)	0.0	(0.8)	0.0	(0.7)	0.0	(1.1)	0.0	(0.4)	0.0	(-0.6)	0.5	(3.8)	0.0	(-0.2)	0.1	(3.8)	0.0	(-0.1)
Scandinavian*	0.0	(-0.1)	0.0	(-0.1)	0.0	(0.4)	0.0	(-0.3)	0.0	(-0.3)	0.0	(-0.1)	0.0	(0.2)	0.0	(0.1)	0.0	(-0.2)	0.0	(-0.2)	0.0	(-0.1)
Lithuanian	0.0	(-0.4)	0.0	(-0.9)	0.0	(-0.7)	0.0	(-0.8)	0.0	(-0.5)	0.0	(-0.1)	0.0	(-0.6)	0.0	(-0.2)	0.0	(-0.2)	0.0	(-0.2)	0.0	(-0.1)
Russian	0.0	(1.0)	0.0	(-0.9)	0.1	(3.5)	0.0	(1.2)	0.0	(0.0)	0.0	(0.2)	0.0	(-0.6)	0.2	(1.2)	0.0	(-0.2)	0.0	(0.4)	0.0	(-0.1)
Polish	0.0	(-0.4)	0.0	(-0.4)	0.0	(-0.2)	0.0	(1.2)	0.0	(0.2)	0.0	(0.0)	0.0	(-0.6)	0.0	(-0.2)	0.0	(-0.2)	0.0	(-0.2)	0.0	(-0.1)
German	0.0	(0.0)	0.0	(1.5)	0.0	(0.6)	0.0	(0.2)	0.0	(0.4)	0.0	(0.9)	0.1	(4.0)	0.2	(1.4)	0.1	(1.7)	0.0	(1.2)	0.0	(-0.1)
Magyar	0.0	(-0.4)	0.0	(0.5)	0.0	(-0.2)	0.0	(-0.8)	0.2	(0.4)	0.2	(0.9)	0.0	(0.0)	0.0	(-0.1)	0.0	(0.0)	0.0	(0.0)	0.0	(-0.1)
Slovak	0.0	(-0.4)	0.0	(0.2)	0.0	(-0.7)	0.0	(0.0)	0.0	(0.2)	0.0	(0.0)	0.0	(0.0)	0.0	(0.0)	0.0	(0.0)	0.0	(-0.2)	0.0	(-0.1)
Croatian & Slovenian	0.0	(-0.4)	0.0	(0.3)	0.0	(0.0)	0.0	(0.0)	0.0	(0.4)	0.0	(-0.1)	0.0	(0.0)	0.1	(0.9)	0.2	(-4.3)	0.1	(-4.5)	0.1	(2.9)
Transylvanian (Siebenburger)	0.0	(0.0)	0.0	(0.0)	0.0	(0.0)	0.0	(0.0)	0.0	(0.0)	0.0	(0.0)	0.0	(0.0)	0.0	(0.0)	0.0	(0.0)	0.0	(0.0)	0.0	(0.0)
Czech †	0.0	(-0.4)	0.0	(-0.9)	0.0	(-0.7)	0.0	(-0.8)	0.0	(-0.5)	0.0	(0.4)	0.0	(-0.6)	0.0	(0.5)	0.0	(-0.2)	0.0	(-0.2)	0.0	(-0.1)
Dalmatian, Bosnian, Herzegovinian	0.0	(-0.4)	0.0	(-0.9)	0.0	(-0.7)	0.0	(-0.8)	0.0	(-0.5)	0.0	(-0.1)	0.0	(-0.6)	0.0	(-0.2)	0.0	(-0.2)	0.0	(-0.2)	0.0	(-0.1)
Ruthenian (Russniak)	0.0	(-0.4)	0.0	(-0.9)	0.0	(-0.7)	0.0	(-0.8)	0.0	(0.9)	0.1	(0.3)	0.0	(-0.6)	0.2	(1.7)	0.0	(-0.2)	0.0	(-0.2)	0.0	(-0.1)
Austro-Hungarian	0.0	(0.0)	0.0	(0.0)	0.0	(0.0)	0.0	(0.0)	0.0	(0.0)	0.0	(0.0)	0.0	(0.0)	0.0	(0.0)	0.0	(0.0)	0.0	(0.0)	0.0	(0.0)
Roumanian	0.1	(4.5)	0.0	(3.1)	0.0	(-0.7)	0.1	(2.5)	0.1	(2.9)	0.2	(0.9)	0.0	(-0.6)	0.0	(0.0)	0.0	(-0.2)	0.0	(-0.2)	0.0	(-0.1)
Bulgarian, Serbian, Montenegrin	0.1	(2.8)	0.0	(2.2)	0.0	(-0.7)	0.0	(-0.8)	0.1	(3.3)	0.2	(0.8)	0.0	(-0.6)	0.2	(1.8)	0.2	(3.5)	0.0	(-0.2)	0.0	(-0.1)
Greek	0.0	(-0.4)	0.0	(0.3)	0.0	(0.2)	0.0	(-0.8)	0.0	(0.6)	0.0	(-0.1)	0.0	(0.3)	0.0	(-0.2)	0.1	(-1.3)	0.0	(-0.2)	0.0	(-0.1)
Armenian	0.0	(0.4)	0.0	(-0.9)	0.1	(2.7)	0.0	(1.2)	0.1	(3.8)	0.0	(0.0)	0.0	(-0.6)	0.0	(0.7)	0.0	(-0.2)	0.0	(-0.2)	0.1	(2.8)
Hebrew	0.0	(0.4)	0.0	(-0.5)	0.0	(0.3)	0.0	(0.2)	0.0	(0.6)	0.0	(0.0)	0.0	(0.7)	0.1	(0.3)	0.0	(-0.2)	0.0	(0.8)	0.0	(-0.1)
Arabian	0.0	(-0.4)	0.0	(0.0)	0.0	(0.0)	0.0	(0.0)	0.0	(0.0)	0.0	(0.0)	0.0	(0.0)	0.1	(0.0)	0.0	(0.0)	0.0	(0.0)	0.0	(0.0)
Syrian	0.0	(-0.4)	0.0	(-0.9)	0.0	(-0.7)	0.0	(0.5)	0.0	(-0.5)	0.0	(-0.1)	0.0	(-0.6)	0.0	(-0.3)	0.0	(-0.2)	0.0	(-0.2)	0.0	(-0.1)
Turkish	0.0	(-0.4)	0.0	(-0.9)	0.0	(-0.7)	0.0	(-0.8)	0.0	(1.0)	1.1	(5.6)	0.0	(-0.6)	0.0	(-0.2)	0.0	(-0.2)	0.0	(-0.2)	0.1	(4.2)
French	0.0	(0.6)	0.0	(0.4)	0.0	(-0.5)	0.0	(0.1)	0.0	(-0.5)	0.0	(0.0)	0.1	(1.6)	0.0	(0.0)	0.0	(-0.2)	0.0	(-0.2)	0.0	(-0.1)
Swiss	0.0	(0.0)	0.0	(0.4)	0.0	(0.0)	0.0	(0.0)	0.0	(0.0)	0.0	(0.0)	0.0	(0.0)	0.0	(0.0)	0.0	(0.0)	0.0	(0.0)	0.0	(0.0)
Mexican	0.0	(0.2)	0.0	(0.0)	0.0	(0.6)	0.0	(-0.3)	0.0	(-0.1)	0.0	(-0.1)	0.0	(-0.3)	0.0	(-0.1)	0.0	(0.0)	0.0	(0.3)	0.0	(0.2)
Central American (NS)	0.0	(0.0)	0.0	(0.0)	0.0	(0.0)	0.0	(0.0)	0.0	(0.0)	0.0	(0.0)	0.0	(0.0)	0.0	(0.0)	0.0	(0.0)	0.0	(0.0)	0.0	(0.0)
South American	0.0	(-0.4)	0.0	(-0.9)	0.0	(-0.7)	0.0	(-0.8)	0.0	(-0.5)	0.0	(-0.1)	0.0	(-0.6)	0.0	(-0.1)	0.0	(-0.2)	0.0	(0.0)	0.0	(0.0)
Spanish American	0.0	(0.0)	0.0	(0.0)	0.0	(0.0)	0.0	(0.0)	0.0	(0.0)	0.0	(0.0)	0.0	(0.0)	0.0	(0.0)	0.0	(0.0)	0.0	(0.0)	0.0	(0.3)
Cuban	0.0	(0.1)	0.0	(3.0)	0.0	(-0.7)	0.0	(0.5)	0.0	(-0.5)	0.0	(-0.1)	0.0	(-0.2)	0.0	(0.3)	0.0	(-0.2)	0.0	(-0.2)	0.0	(-0.1)
West Indian	0.0	(-0.4)	0.0	(-0.9)	0.0	(-0.7)	0.0	(-0.8)	0.0	(1.2)	0.0	(1.2)	0.0	(-0.6)	0.0	(-0.2)	0.0	(-0.2)	0.0	(-0.2)	0.0	(-0.1)
Chinese	0.0	(-0.1)	0.0	(-0.9)	0.1	(2.8)	0.0	(0.5)	0.0	(0.5)	0.0	(-0.1)	0.1	(1.6)	0.2	(1.4)	0.0	(-0.2)	0.0	(0.3)	0.0	(-0.1)
East Indian	0.0	(0.7)	0.0	(-0.9)	0.0	(-0.4)	0.0	(0.2)	0.0	(0.2)	0.0	(0.5)	0.0	(-0.6)	0.0	(-0.2)	0.0	(-0.2)	0.0	(-0.2)	0.0	(-0.1)
Japanese	0.0	(-0.4)	0.0	(-0.9)	0.0	(-0.4)	0.0	(-0.8)	0.0	(-0.5)	0.0	(-0.1)	0.0	(-0.6)	0.0	(-0.2)	0.0	(-0.2)	0.0	(-0.2)	0.0	(-0.1)
Korean	0.0	(-0.4)	0.0	(-0.9)	0.0	(-0.7)	0.0	(-0.8)	0.0	(-0.5)	0.0	(-0.1)	0.0	(-0.6)	0.0	(-0.2)	0.0	(-0.2)	0.0	(-0.2)	0.0	(-0.1)
Filipino	0.0	(-0.4)	0.0	(-0.9)	0.0	(-0.7)	0.0	(-0.8)	0.0	(-0.5)	0.0	(-0.1)	0.0	(-0.6)	0.0	(-0.2)	0.0	(-0.2)	0.0	(-0.2)	0.0	(-0.1)
Hawaiian	0.0	(0.0)	0.0	(0.0)	0.0	(0.0)	0.0	(0.0)	0.0	(0.0)	0.0	(0.0)	0.0	(0.0)	0.0	(0.0)	0.0	(0.0)	0.0	(0.0)	0.0	(0.0)
Pacific Islander	0.0	(-0.4)	0.0	(-0.9)	0.0	(-0.2)	0.0	(-0.4)	0.0	(0.1)	0.0	(0.1)	0.0	(-0.6)	0.0	(-0.2)	0.0	(0.0)	0.0	(1.1)	0.0	(-0.1)
African (black)	0.0	(-0.4)	0.0	(0.0)	0.0	(-0.7)	0.0	(-0.8)	0.0	(0.1)	0.1	(0.3)	0.0	(0.0)	0.0	(-0.2)	0.0	(-0.2)	0.0	(-0.2)	0.0	(-0.1)
(NS)	0.0	(0.0)	0.0	(0.0)	0.0	(0.0)	0.0	(0.0)	0.0	(0.0)	0.0	(0.0)	0.0	(0.0)	0.0	(0.0)	0.0	(0.0)	0.0	(0.0)	0.0	(0.0)
Esquimauz	0.0	(0.0)	0.0	(0.0)	0.0	(0.0)	0.0	(0.0)	0.0	(0.0)	0.0	(0.0)	0.0	(0.0)	0.0	(0.0)	0.0	(0.0)	0.0	(0.0)	0.0	(0.0)
Other	0.1	(2.1)	0.0	(0.0)	0.0	(-0.7)	0.1	(3.9)	0.0	(-0.5)	0.5	(2.4)	0.1	(2.6)	0.5	(4.0)	0.0	(-0.2)	0.1	(-0.2)	0.1	(2.7)

TABLE 6-14. Deviations in Certifications for Mental Conditions, by Race, 1916–1930, Expressed as Percent Certified and NSDMR

Race or People	1916 %	1916 NSDMR	1917 %	1917 NSDMR	1918 %	1918 NSDMR	1919 %	1919 NSDMR	1920 %	1920 NSDMR	1921 %	1921 NSDMR	1922 %	1922 NSDMR	1923 %	1923 NSDMR	1924 %	1924 NSDMR	1925 %	1925 NSDMR	1926 %	1926 NSDMR	1927 %	1927 NSDMR	1928 %	1928 NSDMR	1929 %	1929 NSDMR	1930 %	1930 NSDMR
English	0.2	(0.6)	0.2	(0.6)	0.2	(0.5)	0.1	(-0.1)	0.0	(0.4)	0.0	(0.7)	0.0	(-0.3)	0.1	(0.2)	0.1	(0.2)	0.1	(0.0)	0.1	(0.2)	0.1	(0.1)	0.1	(0.3)	0.0	(0.2)	0.0	(-0.1)
Irish	0.3	(1.3)	0.2	(0.3)	0.2	(0.6)	0.1	(0.6)	0.0	(0.3)	0.0	(1.2)	0.0	(0.6)	0.1	(1.8)	0.1	(1.1)	0.1	(0.1)	0.1	(0.1)	0.1	(0.1)	0.1	(0.0)	0.0	(0.4)	0.0	(0.3)
Scotch	0.1	(-0.4)	0.2	(0.6)	0.2	(0.4)	0.1	(0.4)	0.0	(0.4)	0.1	(1.2)	0.1	(0.6)	0.0	(-0.1)	0.0	(-0.1)	0.1	(0.2)	0.1	(0.6)	0.1	(0.1)	0.1	(0.2)	0.0	(-0.2)	0.0	(0.0)
Welsh	0.0	(-0.5)	0.0	(-1.4)	0.7	(3.6)	0.2	(1.2)	0.0	(-0.6)	0.0	(0.4)	0.0	(0.1)	0.0	(-1.3)	0.0	(-0.1)	0.0	(-0.6)	0.0	(-0.5)	0.0	(0.6)	0.0	(-0.6)	0.0	(-0.9)	0.0	(0.0)
Spanish	0.1	(-0.9)	0.1	(-0.8)	0.0	(-0.6)	0.0	(-0.4)	0.0	(-0.5)	0.0	(-0.9)	0.0	(-0.2)	0.0	(-0.9)	0.0	(-0.8)	0.0	(-0.6)	0.2	(0.6)	0.0	(-0.2)	0.0	(-0.6)	0.0	(-0.5)	0.0	(-0.5)
Portuguese	0.1	(-0.3)	0.0	(0.0)	0.0	(0.0)	0.0	(-0.5)	0.0	(-0.6)	0.0	(-1.0)	0.0	(-0.6)	0.0	(-0.4)	0.0	(-0.6)	0.0	(0.6)	0.0	(-0.7)	0.0	(-0.2)	0.0	(-0.6)	0.0	(-0.9)	0.0	(-0.5)
Italian (North)	0.1	(-0.3)	0.1	(-1.2)	0.0	(-0.8)	0.1	(0.6)	0.0	(-0.3)	0.0	(-0.6)	0.0	(-0.6)	0.0	(-1.0)	0.0	(-0.6)	0.0	(0.6)	0.0	(-0.7)	0.0	(-0.1)	0.0	(-0.3)	0.0	(-0.9)	0.0	(-0.2)
Italian (South)	0.3	(1.2)	0.4	(2.0)	0.3	(0.9)	0.2	(1.4)	0.0	(0.0)	0.0	(-0.5)	0.1	(0.0)	0.0	(-0.5)	0.0	(-0.5)	0.0	(0.9)	0.0	(0.1)	0.0	(-0.1)	0.0	(-0.6)	0.0	(-0.6)	0.0	(-0.1)
Dutch and Flemish	0.1	(-0.3)	0.1	(-0.6)	0.1	(0.1)	0.0	(-0.5)	0.0	(0.0)	0.0	(-0.3)	0.0	(-0.3)	0.0	(-0.8)	0.0	(-0.5)	0.0	(0.1)	0.0	(-0.2)	0.0	(-0.2)	0.0	(0.0)	0.0	(0.3)	0.0	(-0.2)
Finnish	0.1	(-0.8)	0.0	(-1.4)	0.0	(-0.1)	0.0	(-0.2)	0.0	(-0.6)	0.0	(-0.4)	0.0	(0.3)	0.0	(-1.3)	0.0	(-1.1)	0.0	(-0.6)	0.0	(-0.2)	0.0	(-0.2)	0.4	(3.1)	0.0	(-0.9)	0.0	(-0.5)
Scandinavian*	0.1	(-0.2)	0.1	(-0.1)	0.1	(0.0)	0.0	(-0.2)	0.0	(0.0)	0.0	(-0.3)	0.1	(-0.4)	0.0	(-0.7)	0.0	(-0.2)	0.0	(-0.4)	0.0	(-0.1)	0.0	(-0.2)	0.1	(0.0)	0.0	(-0.5)	0.0	(-0.2)
Lithuanian	0.0	(-1.5)	0.0	(-1.4)	0.0	(-0.8)	0.0	(-0.5)	0.0	(-0.6)	0.1	(3.0)	0.0	(0.0)	0.0	(0.5)	0.0	(-0.2)	0.0	(-0.6)	0.0	(-0.7)	0.0	(-0.2)	0.1	(0.2)	0.1	(2.3)	0.0	(-0.5)
Russian	0.0	(-1.1)	0.0	(-0.7)	0.0	(-0.8)	0.0	(0.7)	0.1	(1.9)	0.1	(1.0)	0.0	(-0.7)	0.0	(-0.6)	0.0	(2.9)	0.1	(1.0)	0.1	(0.6)	0.1	(0.1)	0.1	(0.2)	0.1	(2.6)	0.0	(0.0)
Polish	0.1	(-0.5)	0.1	(-0.3)	0.1	(0.0)	0.1	(0.2)	0.0	(0.6)	0.0	(-0.1)	0.1	(0.7)	0.0	(-0.1)	0.0	(-0.2)	0.0	(-0.3)	0.1	(0.5)	0.2	(0.5)	0.0	(-0.2)	0.1	(1.7)	0.0	(0.0)
German	0.1	(-0.7)	0.1	(-0.3)	0.6	(3.0)	0.2	(0.9)	0.0	(0.6)	0.0	(0.1)	0.0	(0.7)	0.0	(-0.6)	0.0	(-0.5)	0.0	(-0.5)	0.0	(-0.3)	0.0	(-0.7)	0.0	(-0.5)	0.0	(-0.4)	0.0	(-0.3)
Magyar	0.1	(-0.5)	0.4	(2.6)	0.0	(-0.8)	0.0	(-0.5)	0.3	(5.0)	0.0	(0.1)	0.0	(-1.1)	0.0	(-0.8)	0.0	(-0.8)	0.0	(-0.6)	0.0	(-0.7)	0.5	(2.1)	0.1	(0.3)	0.1	(0.9)	0.0	(-0.5)
Slovak	0.0	(-1.5)	0.0	(-1.4)	0.0	(-0.8)	0.0	(-0.5)	0.0	(-0.6)	0.0	(-0.1)	0.0	(-0.9)	0.0	(-0.7)	0.0	(-0.8)	0.2	(2.4)	0.3	(4.8)	0.0	(-0.2)	0.2	(1.5)	0.0	(0.4)	0.0	(0.0)
Croatian and Slovenian	0.0	(-1.5)	0.0	(-1.4)	0.0	(-0.8)	0.0	(-0.5)	0.0	(-0.6)	0.0	(-0.8)	0.0	(-1.3)	0.0	(-1.3)	0.0	(-1.3)	0.2	(1.4)	0.0	(-0.7)	0.0	(-0.2)	0.0	(-0.6)	0.0	(-0.9)	0.1	(0.6)
Czech †	0.2	(0.0)	0.0	(-1.4)	0.0	(-0.8)	0.0	(-0.5)	0.0	(-0.6)	0.0	(-1.2)	0.0	(-1.6)	0.0	(-1.3)	0.1	(-0.8)	0.1	(0.0)	0.0	(-0.7)	0.0	(-0.2)	0.0	(-0.6)	0.0	(-0.9)	0.0	(-0.5)
Dalmatian, Bosnian, Herzegovinian	0.0	(-1.5)	0.0	(-1.4)	0.0	(0.0)	0.0	(-0.5)	0.0	(0.0)	0.0	(-1.2)	0.0	(-1.6)	0.0	(-1.3)	0.0	(-0.8)	0.0	(-0.8)	0.0	(0.8)	0.0	(0.8)	0.0	(0.0)	0.0	(0.0)	0.0	(-0.5)
Ruthenian (Russniak)	0.1	(-0.9)	0.0	(-0.1)	0.0	(-0.8)	0.0	(-0.5)	0.0	(0.0)	0.1	(2.1)	0.0	(-1.6)	0.1	(-1.3)	0.1	(3.7)	0.1	(0.0)	0.0	(0.9)	0.0	(4.0)	0.2	(1.9)	0.1	(1.8)	0.3	(5.6)
Roumanian	0.2	(0.4)	0.2	(0.2)	0.0	(0.0)	0.6	(5.0)	0.0	(-0.6)	0.0	(-1.2)	0.1	(-0.1)	0.0	(1.1)	0.0	(0.6)	0.0	(-0.6)	0.0	(-0.7)	0.0	(-0.2)	0.0	(-0.6)	0.0	(-0.9)	0.0	(-0.5)
Bulgarian, Serbian, Montenegrin	0.2	(0.3)	0.2	(0.8)	0.0	(-0.8)	0.0	(-0.5)	0.1	(0.8)	0.0	(-0.7)	0.1	(0.0)	0.0	(-1.3)	0.0	(-0.8)	0.2	(1.6)	0.0	(-0.7)	0.0	(-0.2)	0.4	(3.2)	0.0	(-0.9)	0.0	(-0.5)
Greek	0.1	(-0.4)	0.1	(-0.6)	0.1	(-0.1)	0.1	(0.6)	0.0	(-0.1)	0.0	(-0.6)	0.0	(-0.5)	0.0	(0.2)	0.0	(-0.3)	0.1	(0.4)	0.0	(0.9)	0.0	(0.3)	0.0	(-0.6)	0.0	(-0.4)	0.0	(0.2)
Armenian	0.0	(-1.5)	0.0	(0.0)	0.0	(-0.1)	0.0	(-0.5)	0.0	(-0.6)	0.0	(-0.5)	0.0	(-1.6)	0.0	(-1.3)	0.0	(-0.8)	0.0	(-0.6)	0.1	(0.1)	0.0	(-0.2)	0.0	(-0.6)	0.0	(-0.9)	0.0	(-0.5)
Hebrew	0.1	(-0.3)	0.2	(-0.6)	0.2	(0.4)	0.0	(0.6)	0.0	(0.2)	0.1	(0.8)	0.1	(1.4)	0.0	(1.1)	0.0	(0.6)	0.0	(-0.1)	0.1	(0.3)	0.1	(0.0)	0.0	(-0.4)	0.0	(-0.2)	0.0	(0.1)
Syrian	0.2	(1.0)	0.3	(1.2)	0.3	(1.4)	0.2	(1.2)	0.0	(-0.1)	0.1	(-0.5)	0.1	(-0.4)	0.0	(0.0)	0.0	(0.4)	0.0	(-0.6)	0.1	(0.5)	0.2	(0.6)	0.0	(-0.6)	0.0	(-0.9)	0.0	(0.0)
French	0.1	(-0.4)	0.1	(-0.1)	0.2	(0.2)	0.1	(0.1)	0.1	(0.5)	0.1	(1.9)	0.1	(1.3)	0.1	(3.4)	0.1	(1.0)	0.1	(0.7)	0.1	(1.7)	0.2	(0.6)	0.2	(1.2)	0.0	(0.9)	0.0	(0.7)
Turkish	0.4	(2.9)	0.2	(0.5)	0.0	(-0.8)	0.0	(-0.5)	0.0	(-0.6)	0.0	(-1.2)	0.0	(-1.6)	0.0	(-1.3)	0.0	(-0.8)	0.0	(-0.6)	0.1	(-0.7)	0.9	(4.0)	0.0	(-0.6)	0.0	(-0.9)	0.0	(-0.5)
Mexican	0.2	(0.3)	0.2	(0.2)	0.1	(-0.4)	0.1	(0.1)	0.0	(-0.2)	0.0	(-0.7)	0.0	(-0.9)	0.0	(-1.0)	0.0	(-0.8)	0.0	(-0.3)	0.0	(-0.4)	0.1	(0.1)	0.1	(0.0)	0.0	(0.4)	0.1	(1.1)
Spanish American	0.0	(-1.5)	0.0	(-1.1)	0.0	(-0.3)	0.0	(-0.5)	0.0	(-0.1)	0.0	(-0.2)	0.0	(-1.6)	0.0	(-0.6)	0.0	(-0.8)	0.0	(-0.1)	0.0	(-0.5)	0.0	(0.1)	0.0	(-0.1)	0.0	(-0.6)	0.0	(-0.5)
Cuban	0.0	(-1.2)	0.0	(-1.4)	0.0	(0.0)	0.0	(-0.5)	0.0	(-0.1)	0.0	(-0.3)	0.0	(-1.6)	0.0	(-1.3)	0.0	(-0.8)	0.0	(0.7)	0.1	(0.1)	0.1	(0.1)	0.1	(0.4)	0.0	(-0.6)	0.0	(0.0)
West Indian	0.3	(1.6)	0.1	(-0.1)	0.1	(0.1)	0.0	(-0.5)	0.0	(-0.6)	0.0	(-1.2)	0.0	(-1.6)	0.0	(0.0)	0.0	(-0.8)	0.0	(-0.6)	0.0	(-0.7)	0.0	(-0.2)	0.0	(-0.6)	0.0	(-0.6)	0.0	(-0.5)
Chinese	0.0	(-1.5)	0.1	(-0.1)	0.0	(-0.5)	0.0	(-0.4)	0.0	(-0.2)	0.0	(-0.2)	0.0	(-0.6)	0.0	(-0.3)	0.0	(-0.8)	0.0	(-0.4)	0.0	(-0.3)	0.0	(-0.2)	0.0	(-0.6)	0.0	(-0.1)	0.0	(0.0)
East Indian	0.0	(-1.5)	0.0	(-1.4)	0.0	(-0.6)	0.0	(-0.5)	0.0	(-0.6)	0.0	(0.0)	0.0	(-1.6)	0.0	(-1.3)	0.0	(-0.8)	0.1	(0.9)	0.0	(-0.4)	0.0	(-0.2)	0.0	(-0.6)	0.0	(-0.9)	0.0	(-0.5)
Japanese	0.0	(-1.4)	0.0	(-0.4)	0.0	(-0.6)	0.0	(-0.4)	0.0	(-0.3)	0.0	(-1.2)	0.0	(-1.1)	0.0	(-0.6)	0.0	(-0.8)	0.0	(0.0)	0.0	(-0.7)	0.2	(-0.2)	0.2	(1.3)	0.0	(-0.9)	0.0	(0.0)
Korean	0.0	(-1.5)	0.0	(-1.4)	0.0	(-0.8)	0.0	(-0.5)	0.0	(-0.6)	0.0	(-1.2)	0.0	(-1.6)	0.0	(-1.3)	0.0	(-0.8)	0.0	(0.0)	0.0	(-0.7)	0.0	(-0.2)	0.0	(-0.6)	0.0	(-0.9)	0.0	(-0.5)
Pacific Islander	0.0	(-1.5)	0.0	(-1.4)	0.0	(-0.6)	0.0	(-0.4)	0.0	(-0.2)	0.1	(0.0)	0.1	(0.0)	0.0	(-0.7)	0.0	(-0.8)	0.0	(-0.6)	0.0	(-0.7)	0.0	(0.0)	0.0	(-0.6)	0.0	(0.0)	0.0	(-0.5)
African (black)	0.1	(-0.5)	0.2	(0.0)	0.0	(-0.8)	0.0	(-0.4)	0.0	(-0.6)	0.0	(0.1)	0.0	(0.1)	0.0	(-0.2)	0.0	(-0.8)	0.2	(2.4)	0.0	(-0.7)	0.1	(0.2)	0.1	(0.4)	0.1	(0.1)	0.0	(-0.5)
Other	0.1	(-0.4)	0.0	(0.0)	0.0	(-0.8)	0.0	(-0.5)	0.1	(0.5)	0.1	(0.9)	0.0	(0.0)	0.0	(-1.3)	0.0	(-0.8)	0.3	(3.4)	0.0	(-0.7)	0.0	(0.1)	0.0	(-0.6)	0.0	(-0.9)	0.0	(-0.5)

Bold = 1 or more NSDMR below mean   [shaded light] = 1 to 1.99 NSDMRs above mean   [shaded dark] = 2 or more NSDMRs above mean

* Norwegians, Danes, and Swedes.   † Bohemians and Moravians.

TABLE 6-15. Deviations in Rate of Deportations for those Certified for Mental Conditions, by Race, 1916–1930, Expressed as Percent Deported and NSDMR

Race or People	1916 %	1916 NSDMR	1917 %	1917 NSDMR	1918 %	1918 NSDMR	1919 %	1919 NSDMR	1920 %	1920 NSDMR	1921 %	1921 NSDMR	1922 %	1922 NSDMR	1923 %	1923 NSDMR	1924 %	1924 NSDMR	1925 %	1925 NSDMR	1926 %	1926 NSDMR	1927 %	1927 NSDMR	1928 %	1928 NSDMR	1929 %	1929 NSDMR	1930 %	1930 NSDMR
English	80.0	(0.1)	76.3	(0.0)	77.8	(0.0)	100.0	(0.1)	80.4	(-0.5)	95.2	(0.6)	92.6	(0.1)	91.4	(-0.2)	100.0	(0.0)	65.5	(-0.3)	88.9	(-0.1)	34.4	(-0.5)	3.1	(-0.4)	0.0	(-0.4)	33.3	(0.6)
Irish	90.2	(0.5)	74.3	(0.5)	81.8	(0.2)	100.0	(0.1)	81.8	(-0.4)	85.7	(0.3)	100.0	(0.4)	96.4	(-0.1)	100.0	(0.0)	66.7	(0.2)	100.0	(0.3)	68.4	(0.4)	16.7	(0.0)	11.1	(-0.1)	12.5	(-0.4)
Scotch	62.5	(-0.6)	74.2	(-0.1)	90.0	(0.4)	100.0	(0.1)	100.0	(0.0)	100.0	(0.8)	82.4	(0.2)	100.0	(0.0)	100.0	(0.0)	85.7	(0.2)	90.0	(0.0)	44.4	(-0.2)	22.2	(0.2)	0.0	(-0.4)	33.3	(0.6)
Welsh	100.0	(0.9)	0.0	(0.0)	100.0	(0.7)	100.0	(0.1)	0.0	(0.0)	100.0	(0.8)	100.0	(0.4)	0.0	(0.0)	0.0	(0.0)	0.0	(0.0)	100.0	(0.3)	0.0	(-1.4)	0.0	(-0.5)	0.0	(0.0)	0.0	(-1.0)
Spanish	100.0	(0.9)	90.0	(0.5)	100.0	(0.7)	0.0	(-3.1)	100.0	(0.3)	66.7	(-0.4)	100.0	(0.0)	100.0	(0.0)	0.0	(-2.9)	0.0	(0.0)	100.0	(0.3)	100.0	(1.3)	0.0	(0.0)	0.0	(0.0)	0.0	(0.0)
Portuguese	86.7	(0.3)	100.0	(0.8)	0.0	(0.0)	100.0	(0.1)	100.0	(0.3)	100.0	(0.8)	100.0	(0.4)	100.0	(0.0)	0.0	(0.0)	50.0	(0.0)	100.0	(0.3)	0.0	(0.0)	100.0	(1.3)	66.7	(1.6)	0.0	(0.0)
Italian (North)	33.3	(-1.8)	100.0	(0.0)	50.0	(-0.8)	0.0	(-0.1)	0.0	(0.3)	100.0	(0.8)	33.3	(-2.1)	100.0	(0.0)	0.0	(-2.9)	50.0	(-0.1)	0.0	(-1.2)	0.0	(-1.4)	100.0	(-2.5)	66.7	(1.6)	60.7	(-2.3)
Italian (South)	76.6	(-0.1)	70.1	(-0.2)	21.4	(-1.7)	92.9	(-0.2)	100.0	(0.3)	85.4	(0.3)	87.5	(-0.1)	100.0	(0.0)	85.7	(-0.4)	75.0	(-0.1)	55.6	(-1.2)	0.0	(-1.4)	100.0	(1.3)	0.0	(-0.4)	0.0	(-1.0)
Dutch and Flemish	62.5	(-0.6)	100.0	(0.8)	100.0	(0.7)	0.0	(0.0)	80.0	(-0.5)	100.0	(0.3)	33.3	(-2.1)	100.0	(0.0)	80.0	(-0.6)	50.0	(-0.8)	0.0	(0.3)	0.0	(0.0)	0.0	(-0.5)	0.0	(0.0)	0.0	(0.0)
Finnish	50.0	(-1.1)	0.0	(0.0)	100.0	(0.7)	0.0	(0.0)	0.0	(0.0)	100.0	(0.8)	100.0	(0.0)	0.0	(0.0)	100.0	(0.0)	0.0	(-0.8)	100.0	(0.3)	0.0	(0.0)	50.0	(1.0)	0.0	(-0.4)	0.0	(0.0)
Scandinavian*	76.0	(-0.1)	100.0	(0.8)	72.7	(-0.1)	75.0	(-0.7)	100.0	(0.3)	87.5	(0.3)	100.0	(0.0)	100.0	(0.0)	100.0	(0.0)	0.0	(0.7)	72.7	(0.0)	0.0	(1.3)	18.2	(0.1)	33.3	(0.0)	33.3	(0.6)
Lithuanian	0.0	(0.0)	0.0	(0.0)	0.0	(0.0)	0.0	(0.0)	0.0	(0.0)	0.0	(0.0)	0.0	(0.0)	100.0	(0.0)	100.0	(0.0)	100.0	(0.7)	0.0	(0.0)	0.0	(0.0)	0.0	(0.0)	0.0	(-0.4)	0.0	(0.0)
Russian	50.0	(-1.1)	66.7	(-0.3)	0.0	(0.0)	66.7	(-1.0)	75.0	(-0.7)	50.0	(-1.0)	100.0	(0.4)	100.0	(0.0)	100.0	(0.0)	100.0	(0.7)	100.0	(0.3)	0.0	(-1.4)	0.0	(0.0)	16.7	(0.1)	0.0	(-1.0)
Polish	80.0	(0.1)	100.0	(0.8)	100.0	(0.7)	100.0	(0.1)	0.0	(0.0)	100.0	(0.8)	85.7	(-0.1)	80.0	(0.0)	100.0	(0.0)	100.0	(0.0)	0.0	(0.0)	28.6	(-0.6)	0.0	(-0.5)	0.0	(0.0)	0.0	(-1.0)
German	80.0	(0.1)	75.0	(-0.1)	100.0	(0.7)	100.0	(0.1)	100.0	(0.3)	88.9	(0.3)	100.0	(0.4)	83.3	(-0.3)	88.9	(-0.3)	100.0	(0.7)	87.5	(-0.1)	42.9	(-0.2)	22.2	(0.2)	43.8	(0.9)	0.0	(-1.0)
Magyar	100.0	(0.9)	100.0	(0.8)	0.0	(0.0)	0.0	(0.0)	100.0	(0.3)	75.0	(0.4)	100.0	(0.4)	0.0	(0.0)	0.0	(0.0)	0.0	(-0.8)	33.3	(-2.0)	60.0	(0.2)	100.0	(2.5)	50.0	(1.1)	0.0	(0.0)
Slovak	0.0	(0.0)	100.0	(0.8)	0.0	(0.0)	0.0	(0.0)	100.0	(0.3)	60.0	(-0.1)	50.0	(-1.5)	0.0	(-2.8)	0.0	(0.0)	50.0	(0.0)	0.0	(0.0)	0.0	(0.0)	100.0	(-0.5)	0.0	(-0.4)	0.0	(-1.0)
Croatian and Slovenian	0.0	(0.0)	0.0	(0.0)	0.0	(0.0)	100.0	(0.0)	0.0	(0.0)	100.0	(0.0)	100.0	(0.0)	0.0	(0.0)	100.0	(0.0)	50.0	(-0.8)	0.0	(0.0)	60.0	(0.2)	100.0	(-2.5)	0.0	(0.0)	0.0	(-1.0)
Czech†	100.0	(0.9)	0.0	(0.0)	0.0	(0.0)	0.0	(0.1)	0.0	(0.0)	100.0	(0.0)	0.0	(-2.8)	0.0	(-2.8)	100.0	(0.0)	0.0	(-2.3)	33.3	(0.0)	0.0	(0.0)	0.0	(0.0)	0.0	(0.0)	0.0	(0.0)
Dalmatian, Bosnian, Herzegovinian	0.0	(0.0)	0.0	(0.0)	0.0	(0.0)	0.0	(0.0)	0.0	(0.0)	0.0	(0.0)	0.0	(0.0)	0.0	(0.0)	0.0	(0.0)	0.0	(0.0)	0.0	(0.0)	0.0	(0.0)	0.0	(0.0)	0.0	(0.0)	0.0	(0.0)
Ruthenian (Russniak)	100.0	(0.9)	100.0	(0.8)	0.0	(0.0)	100.0	(0.0)	100.0	(0.0)	100.0	(0.0)	100.0	(0.0)	100.0	(0.0)	100.0	(0.0)	0.0	(0.0)	100.0	(0.3)	25.0	(-0.7)	0.0	(0.0)	0.0	(0.0)	0.0	(0.0)
Roumanian	50.0	(-1.1)	100.0	(0.8)	0.0	(0.0)	100.0	(0.1)	0.0	(0.0)	0.0	(0.0)	0.0	(-2.8)	0.0	(-2.8)	0.0	(0.0)	0.0	(0.0)	0.0	(0.0)	0.0	(0.0)	0.0	(0.0)	0.0	(0.0)	0.0	(-1.0)
Bulgarian, Serbian, Montenegrin	83.3	(0.2)	100.0	(0.8)	0.0	(0.0)	0.0	(0.0)	100.0	(0.3)	0.0	(-2.7)	100.0	(0.4)	0.0	(0.0)	0.0	(0.0)	100.0	(0.7)	0.0	(0.0)	0.0	(0.0)	50.0	(1.0)	0.0	(0.0)	0.0	(0.0)
Greek	93.1	(0.6)	69.6	(-0.2)	100.0	(0.7)	100.0	(0.1)	100.0	(0.3)	0.0	(0.8)	100.0	(0.4)	100.0	(0.0)	0.0	(-2.9)	100.0	(0.7)	0.0	(0.0)	0.0	(-1.4)	0.0	(0.0)	100.0	(2.6)	0.0	(0.0)
Armenian	0.0	(0.0)	100.0	(0.8)	0.0	(0.0)	0.0	(0.0)	0.0	(-0.4)	50.0	(-1.0)	100.0	(0.0)	100.0	(0.0)	0.0	(0.0)	66.7	(0.1)	100.0	(0.3)	0.0	(0.0)	0.0	(0.0)	0.0	(0.0)	0.0	(-1.0)
Hebrew	57.9	(-0.8)	50.0	(-0.9)	71.4	(-0.2)	60.0	(-1.2)	85.7	(-0.3)	34.4	(-1.5)	80.0	(-0.4)	100.0	(0.0)	80.0	(-0.6)	80.0	(0.1)	88.9	(-0.1)	100.0	(1.3)	33.3	(0.5)	0.0	(-0.4)	25.0	(0.2)
Syrian	100.0	(-1.5)	100.0	(0.8)	100.0	(0.7)	100.0	(0.1)	100.0	(0.3)	100.0	(0.8)	82.6	(0.4)	100.0	(0.0)	100.0	(0.0)	0.0	(0.0)	100.0	(0.3)	0.0	(0.0)	0.0	(0.0)	0.0	(0.0)	0.0	(-1.0)
French	52.2	(-1.1)	69.4	(-0.2)	91.7	(0.5)	100.0	(0.1)	100.0	(0.3)	92.3	(0.5)	82.6	(-0.2)	98.1	(0.0)	100.0	(0.0)	81.5	(0.1)	100.0	(0.3)	61.8	(0.3)	3.0	(-0.4)	28.6	(-0.4)	28.6	(0.4)
Turkish	100.0	(0.9)	0.0	(0.0)	0.0	(0.0)	0.0	(-1.2)	100.0	(0.3)	100.0	(0.8)	100.0	(0.0)	0.0	(0.0)	82.6	(-0.5)	100.0	(0.7)	0.0	(0.0)	0.0	(-1.4)	0.0	(0.0)	0.0	(0.0)	0.0	(0.0)
Mexican	95.0	(0.7)	100.0	(0.8)	84.6	(0.2)	60.0	(-1.2)	100.0	(0.3)	100.0	(0.8)	100.0	(0.4)	100.0	(0.0)	0.0	(-0.5)	0.0	(-2.3)	25.0	(-2.3)	11.1	(-1.1)	23.7	(0.2)	4.8	(-0.3)	7.1	(-0.7)
Spanish American	0.0	(0.0)	0.0	(-2.7)	0.0	(-1.6)	0.0	(0.0)	100.0	(0.3)	0.0	(-2.7)	0.0	(0.0)	100.0	(0.0)	0.0	(0.0)	0.0	(-2.3)	100.0	(0.3)	50.0	(0.0)	0.0	(-0.5)	100.0	(2.6)	0.0	(0.0)
Cuban	100.0	(0.9)	0.0	(0.0)	100.0	(0.7)	0.0	(0.0)	50.0	(0.0)	50.0	(-1.0)	0.0	(0.0)	0.0	(-2.6)	0.0	(0.0)	0.0	(-2.3)	0.0	(-3.1)	0.0	(-1.4)	0.0	(-0.5)	0.0	(-0.4)	0.0	(-1.0)
West Indian	0.0	(-3.2)	100.0	(0.8)	0.0	(-2.3)	0.0	(0.0)	0.0	(-3.6)	0.0	(-1.0)	0.0	(0.0)	0.0	(-2.6)	0.0	(-2.6)	0.0	(0.0)	0.0	(0.0)	0.0	(-1.4)	0.0	(-0.5)	0.0	(0.0)	0.0	(-1.0)
Chinese	0.0	(0.0)	66.7	(-0.3)	0.0	(-2.3)	100.0	(0.1)	33.3	(-2.3)	50.0	(0.0)	60.0	(0.0)	100.0	(0.0)	0.0	(0.0)	50.0	(-0.8)	50.0	(-1.4)	0.0	(0.0)	0.0	(0.0)	66.7	(1.6)	0.0	(-1.0)
East Indian	0.0	(0.0)	0.0	(0.0)	0.0	(0.0)	0.0	(-3.1)	50.0	(-1.6)	0.0	(-1.0)	0.0	(-1.1)	0.0	(0.0)	0.0	(0.0)	0.0	(0.7)	0.0	(0.0)	0.0	(0.0)	0.0	(0.0)	0.0	(0.0)	0.0	(0.0)
Japanese	100.0	(0.9)	100.0	(0.8)	66.7	(-0.3)	0.0	(0.0)	100.0	(0.3)	0.0	(0.0)	0.0	(0.0)	0.0	(0.0)	100.0	(0.0)	100.0	(0.7)	100.0	(0.3)	0.0	(0.0)	0.0	(0.0)	33.3	(0.6)	50.0	(1.5)
Korean	0.0	(0.0)	0.0	(0.0)	0.0	(0.0)	0.0	(0.0)	0.0	(0.0)	0.0	(0.0)	0.0	(0.0)	0.0	(0.0)	0.0	(0.0)	0.0	(0.0)	0.0	(0.0)	0.0	(0.0)	0.0	(0.0)	0.0	(0.0)	0.0	(0.0)
Pacific Islander	0.0	(0.0)	0.0	(0.0)	0.0	(0.0)	0.0	(0.0)	0.0	(0.3)	100.0	(0.1)	100.0	(0.0)	100.0	(0.0)	0.0	(0.0)	0.0	(-1.3)	0.0	(0.0)	0.0	(0.0)	0.0	(0.0)	25.0	(0.3)	0.0	(0.0)
African (black)	80.0	(0.1)	84.6	(0.3)	50.0	(-0.8)	100.0	(0.1)	100.0	(0.3)	80.0	(0.1)	100.0	(0.4)	0.0	(0.0)	0.0	(0.0)	0.0	(0.0)	0.0	(0.0)	0.0	(-1.4)	0.0	(0.0)	0.0	(0.3)	0.0	(0.0)
Other	50.0	(-1.1)	100.0	(-1.1)	50.0	(0.0)	100.0	(0.0)	0.0	(0.0)	0.0	(0.0)	0.0	(0.0)	0.0	(0.0)	0.0	(0.0)	100.0	(0.7)	0.0	(0.0)	0.0	(0.0)	0.0	(0.0)	0.0	(0.3)	0.0	(0.0)

Bold = 1 or more NSDMR below mean.   [shaded] = 1 to 1.99 NSDMRs above mean   [dark shaded] = 2 or more NSDMRs above mean

* Norwegians, Danes, and Swedes.
† Bohemians and Moravians.

TABLE 6-16. Deviations in Deportation Rate for Mental Conditions, by Race, 1899–1930, Expressed as Percent Deported and NSDMR

Race or People	1899 %	NSDMR	1900 %	NSDMR	1901 %	NSDMR	1902 %	NSDMR	1903 %	NSDMR	1904 %	NSDMR	1905 %	NSDMR	1906 %	NSDMR	1907 %	NSDMR	1908 %	NSDMR	1909 %	NSDMR	1910 %	NSDMR
English	0.0	(1.0)	0.0	(0.1)	0.0	(0.8)	0.0	(0.8)	0.0	(1.1)	0.0	-0.2	0.0	(1.0)	0.1	(0.4)	0.1	(4.4)	0.1	(2.3)	0.1	(2.8)	0.0	(0.9)
Irish	0.0	(0.5)	0.0	(0.5)	0.0	(0.8)	0.0	(0.8)	0.0	(1.6)	0.0	(0.6)	0.0	(1.0)	0.1	(0.4)	0.1	(1.2)	0.1	(2.3)	0.1	(2.8)	0.1	(1.7)
Scotch	0.0	-0.5	0.0	-0.3	0.0	-0.4	0.0	-0.2	0.0	-0.4	0.0	-0.2	0.0	-0.7	0.0	-0.1	0.0	0.0	0.0	-0.3	0.0	-0.1	0.1	(1.3)
Welsh	0.0	-0.5	0.0	-0.3	0.0	-0.4	0.0	-0.2	0.0	-0.4	0.0	-0.2	0.0	(1.4)	0.0	-0.1	0.0	-0.7	0.0	(-1.3)	0.1	(1.7)	0.0	(0.2)
Australian	0.0	-0.5	0.0	-0.3	0.0	0.0	0.0	0.0	0.0	0.0	0.0	0.0	0.0	0.0	0.0	0.0	0.0	0.0	0.0	0.0	0.0	0.0	0.0	0.0
Spanish	0.0	-0.5	0.0	0.0	0.0	-0.4	0.1	0.0	0.0	-0.4	0.0	0.4	0.0	0.3	0.0	0.0	0.0	-0.7	0.0	-0.4	0.0	(-1.1)	0.0	-0.4
Portuguese	0.0	-0.5	0.0	-0.3	0.0	-0.4	0.0	-0.2	0.0	-0.4	0.0	-0.2	0.0	-0.7	0.0	0.1	0.0	-0.3	0.1	(-1.3)	0.0	0.0	0.0	0.0
Italian (North)	0.0	-0.5	0.0	-0.1	0.0	-0.4	0.0	-0.2	0.0	-0.4	0.0	-0.2	0.0	-0.4	0.0	-0.1	0.0	-0.2	0.1	0.4	0.0	-0.4	0.0	-0.3
Italian (South)	0.0	-0.1	0.0	0.0	0.0	0.1	0.0	-0.2	0.0	0.2	0.0	0.0	0.0	0.0	0.0	0.0	0.1	0.2	0.1	0.2	0.0	-0.1	0.0	0.0
Dutch and Flemish	0.0	-0.5	0.0	-0.3	0.0	-0.4	0.0	-0.2	0.0	(1.8)	0.0	-0.2	0.0	(1.2)	0.0	-0.1	0.0	0.7	0.0	-0.7	0.0	0.0	0.0	-0.1
Finnish	0.0	-0.5	0.0	0.4	0.0	0.3	0.0	0.1	0.0	-0.2	0.0	-0.1	0.0	0.3	0.0	0.0	0.0	0.7	0.0	-0.4	0.0	-0.2	0.0	(-1.2)
Scandinavian*	0.0	-0.2	0.0	-0.3	0.0	-0.4	0.0	0.2	0.0	-0.2	0.0	-0.2	0.0	0.3	0.0	0.0	0.0	-0.2	0.0	0.7	0.0	-0.3	0.0	-0.5
Lithuanian	0.0	-0.5	0.0	-0.3	0.0	-0.4	0.0	-0.2	0.0	-0.4	0.0	0.0	0.0	0.2	0.0	-0.1	0.0	-0.2	0.0	-0.9	0.0	-0.7	0.0	-0.6
Russian	0.0	-0.5	0.0	-0.3	0.0	-0.4	0.0	-0.2	0.0	(3.5)	0.0	-0.2	0.0	(3.6)	0.0	-0.2	0.0	-0.7	0.0	(-1.1)	0.0	-0.8	0.0	-0.3
Polish	0.0	-0.2	0.0	0.1	0.0	-0.4	0.0	0.4	0.0	-0.2	0.1	-0.1	0.0	-0.5	0.0	-0.1	0.0	-0.5	0.0	-0.8	0.0	0.1	0.0	0.2
German	0.0	0.4	0.0	0.1	0.0	-0.4	0.0	0.1	0.0	-0.2	0.0	0.2	0.0	0.2	0.0	0.0	0.0	0.3	0.0	-0.3	0.0	0.2	0.0	0.2
Magyar	0.0	-0.5	0.0	-0.2	0.0	-0.4	0.0	-0.2	0.0	-0.4	0.0	0.1	0.0	-0.7	0.0	-0.1	0.0	-0.5	0.0	-0.7	0.0	-0.6	0.0	-0.9
Slovak	0.0	-0.5	0.0	-0.2	0.0	-0.4	0.0	-0.2	0.0	-0.4	0.0	-0.1	0.0	-0.7	0.0	-0.1	0.0	-0.4	0.0	-0.7	0.0	0.1	0.0	-0.7
Croatian & Slovenian	0.0	-0.5	0.0	-0.1	0.0	-0.4	0.0	-0.1	0.0	-0.4	0.0	0.1	0.0	-0.7	0.0	0.0	0.0	-0.4	0.0	(-1.0)	0.0	-0.6	0.0	-0.9
Transylvanian (Siebenburger)	0.0	-0.5	0.0	0.0	0.0	-0.4	0.0	0.4	0.0	0.0	0.0	0.0	0.0	0.1	0.0	0.0	0.0	0.3	0.0	0.4	0.0	0.0	0.0	0.0
Czech †	0.0	-0.5	0.0	(1.1)	0.0	-0.4	0.0	0.0	0.0	-0.4	0.0	-0.2	0.0	-0.2	0.1	0.3	0.0	-0.1	0.0	0.5	0.0	0.2	0.0	(-1.2)
Dalmatian, Bosnian, Herzegovinian	0.0	-0.5	0.0	-0.3	0.0	-0.4	0.0	-0.2	0.0	-0.4	0.0	-0.1	0.0	-0.7	0.0	-0.2	0.0	-0.2	0.1	(-1.3)	0.1	(1.8)	0.0	0.0
Ruthenian (Russniak)	0.0	-0.5	0.0	-0.3	0.0	-0.4	0.0	-0.2	0.0	-0.4	0.0	0.1	0.0	0.1	0.0	0.0	0.0	-0.2	0.0	0.4	0.0	-0.6	0.0	-0.2
Austro-Hungarian	0.0	-0.5	0.0	-0.3	0.0	0.0	0.0	0.0	0.0	0.0	0.0	0.0	0.0	0.0	0.0	0.0	0.1	0.1	0.0	0.1	0.0	0.0	0.0	0.0
Roumanian	0.0	-0.5	0.0	-0.3	0.0	-0.4	0.0	-0.2	0.0	-0.4	0.0	-0.2	0.0	-0.7	0.0	-0.1	0.0	-0.5	0.0	(-1.3)	0.0	0.0	0.0	(-1.2)
Bulgarian, Serbian, Montenegrin	0.0	-0.5	0.0	-0.3	0.0	-0.4	0.0	-0.2	0.0	-0.4	0.0	-0.2	0.0	-0.7	0.0	0.0	0.0	-0.4	0.0	(-1.1)	0.0	-0.7	0.0	-0.9
Greek	0.0	-0.5	0.0	-0.3	0.0	-0.4	0.0	-0.2	0.0	-0.4	0.0	-0.2	0.0	-0.7	0.0	0.0	0.0	-0.6	0.0	-0.9	0.0	(-1.1)	0.0	-0.5
Armenian	0.0	-0.1	0.0	-0.1	0.0	0.2	0.1	(3.4)	0.0	0.0	0.2	0.0	0.0	0.0	0.1	0.0	0.0	0.0	0.0	0.3	0.1	0.8	0.1	(1.0)
Hebrew	0.0	0.0	0.0	0.0	0.0	0.0	0.0	-0.1	0.0	0.0	0.0	0.0	0.0	0.1	0.0	0.1	0.0	0.0	0.1	0.3	0.1	0.1	0.0	0.0
Arabian	0.0	0.0	0.0	0.0	0.0	-0.4	0.0	-0.2	0.0	-0.4	0.0	0.0	0.0	0.0	0.0	0.0	0.0	0.0	0.0	0.0	0.0	0.0	0.0	0.0
Syrian	0.0	-0.5	0.0	-0.3	0.0	-0.4	0.0	-0.2	0.0	-0.4	0.0	-0.2	0.0	-0.7	0.0	-0.2	0.0	-0.7	0.1	-0.3	0.0	0.3	0.0	(-1.2)
Turkish	0.0	-0.5	0.0	-0.3	0.0	-0.4	0.0	-0.2	0.0	-0.4	0.0	-0.2	0.0	-0.7	0.0	-0.2	0.0	-0.2	0.0	(-1.3)	0.0	(-1.1)	0.0	(-1.2)
French	0.0	(2.9)	0.0	-0.3	0.1	-0.4	0.0	-0.2	0.0	-0.4	0.0	0.3	0.0	0.3	0.1	0.5	0.0	0.2	0.1	0.7	0.1	0.8	0.1	(2.3)
Swiss	0.0	0.0	0.0	0.0	0.0	0.0	0.0	0.0	0.0	0.0	0.0	0.0	0.0	0.0	0.0	0.0	0.0	-0.5	0.0	0.0	0.0	0.0	0.0	0.0
Mexican	0.0	-0.5	0.0	-0.5	0.0	-0.4	0.0	-0.2	0.0	-0.4	0.2	(6.1)	0.1	0.0	0.0	0.0	0.0	-0.7	0.1	(1.1)	0.0	(2.6)	0.1	(1.9)
Central American (NS)	0.0	-0.5	0.0	0.0	0.0	0.0	0.0	0.0	0.0	0.0	0.0	0.0	0.0	0.0	0.0	0.0	0.0	0.0	0.0	0.0	0.0	0.0	0.0	0.0
South American	0.0	0.0	0.0	0.0	0.0	-0.4	0.0	0.0	0.0	-0.4	0.0	0.0	0.0	0.0	0.0	0.1	0.0	0.0	0.0	0.0	0.0	0.0	0.0	0.0
Spanish American	0.0	0.0	0.0	-0.3	0.0	-0.4	0.0	-0.2	0.0	-0.4	0.0	-0.2	0.0	-0.7	0.0	-0.2	0.0	-0.7	0.1	(1.5)	0.0	(-1.1)	0.0	(-1.2)
Cuban	0.1	(5.2)	0.0	(1.3)	0.0	(5.5)	0.0	-0.2	0.0	-0.4	0.0	(1.0)	0.0	0.1	0.0	-0.2	0.1	(2.4)	0.0	-0.4	0.0	-0.6	0.0	(-1.2)
West Indian	0.0	-0.5	0.0	-0.3	0.0	-0.3	0.0	0.0	0.0	-0.4	0.0	-0.2	0.1	(2.8)	0.1	(0.4)	0.1	(2.4)	0.0	(-1.3)	0.0	(-1.1)	0.0	(-1.2)
Chinese	0.0	-0.5	0.0	-0.3	0.0	-0.4	0.0	-0.2	0.0	-0.4	0.0	-0.2	0.0	(2.0)	0.0	-0.2	0.0	-0.7	0.0	(-1.3)	0.0	(-1.1)	0.0	(-1.2)
East Indian	0.0	-0.5	0.0	0.4	0.0	-0.4	0.0	-0.2	0.0	-0.4	0.0	-0.2	0.0	-0.7	0.7	(6.2)	0.0	-0.7	0.0	(-1.3)	0.0	(-1.1)	0.1	(2.4)
Japanese	0.0	-0.5	0.0	-0.3	0.0	-0.4	0.0	-0.2	0.0	-0.4	0.0	-0.2	0.0	-0.2	0.0	-0.2	0.0	-0.7	0.0	(-1.3)	0.0	-0.4	0.0	(-1.2)
Korean	0.0	0.0	0.0	0.0	0.0	0.0	0.0	0.0	0.0	0.0	0.0	0.0	0.0	0.0	0.0	0.0	0.0	-0.7	0.0	(-1.3)	0.0	(-1.1)	0.0	(-1.2)
Filipino	0.0	0.0	0.0	0.0	0.0	0.0	0.0	0.0	0.0	0.0	0.0	0.0	0.0	0.0	0.0	0.0	0.0	0.0	0.0	0.0	0.0	0.0	0.0	(-1.2)
Hawaiian	0.0	0.0	0.0	0.0	0.0	0.0	0.0	0.0	0.0	0.0	0.0	0.0	0.0	0.0	0.0	0.0	0.0	0.0	0.0	0.0	0.0	0.0	0.0	0.0
Pacific Islander	0.0	-0.5	0.0	-0.3	0.0	-0.4	0.0	-0.2	0.0	-0.4	0.0	-0.2	0.0	-0.7	0.0	-0.2	0.0	-0.7	0.0	(-1.3)	0.0	(-1.1)	0.0	(-1.2)
African (black)	0.0	-0.5	0.1	(5.9)	0.0	-0.4	0.1	(4.7)	0.0	-0.4	0.0	-0.2	0.0	-0.7	0.0	-0.2	0.0	-0.7	0.0	-0.7	0.0	-0.5	0.0	(-1.2)
(NS)	0.0	0.0	0.0	0.0	0.0	0.0	0.0	0.0	0.0	0.0	0.0	0.0	0.0	0.0	0.0	0.1	0.0	0.1	0.0	0.0	0.0	0.0	0.0	0.0
Esquimaux	0.0	0.0	0.0	0.0	0.0	0.0	0.0	0.0	0.0	0.0	0.0	0.0	0.0	0.0	0.0	0.6	0.0	0.0	0.0	0.0	0.0	0.0	0.0	0.0
Other	0.0	0.0	0.0	0.0	0.0	0.0	0.0	-0.2	0.0	0.0	0.0	0.0	0.0	-0.7	0.1	-0.7	0.1	-0.7	0.1	(2.2)	0.1	(0.5)	0.0	-0.2

Bold = 1 or more NSDMR below mean   = 1 to 1.99 NSDMRs above mean   = 2 or more NSDMRs above mean   NS = not specified

* Norwegians, Danes, and Swedes.

† Bohemians and Moravians.

TABLE 6-16. Continued

Race or People	1911 %	1911 NSDMR	1912 %	1912 NSDMR	1913 %	1913 NSDMR	1914 %	1914 NSDMR	1915 %	1915 NSDMR	1916 %	1916 NSDMR	1917 %	1917 NSDMR	1918 %	1918 NSDMR	1919 %	1919 NSDMR	1920 %	1920 NSDMR	1921 %	1921 NSDMR	1922 %	1922 NSDMR
English	0.0	(0.3)	0.0	(0.9)	0.1	(0.4)	0.1	(0.8)	0.1	(0.2)	0.2	(0.6)	0.2	(0.6)	0.2	(0.4)	0.1	(0.9)	0.0	(0.3)	0.0	(0.9)	0.0	-(0.2)
Irish	0.1	(1.3)	0.1	(0.9)	0.1	(1.0)	0.1	(1.0)	0.1	(0.1)	0.3	(1.6)	0.1	(0.3)	0.2	(0.6)	0.2	(0.9)	0.0	(0.2)	0.0	(0.7)	0.2	(3.2)
Scotch	0.0	(0.3)	0.0	(0.9)	0.0	(0.5)	0.1	(0.8)	0.0	(0.3)	0.1	(0.5)	0.1	(0.5)	0.1	(0.5)	0.1	(0.5)	0.0	(0.4)	0.1	(1.8)	0.1	(0.5)
Welsh	0.2	(4.6)	0.0	(0.3)	0.0	(0.5)	0.1	-0.7	0.1	-0.3	0.1	-0.2	0.0	-(1.3)	0.1	(5.9)	0.2	(1.3)	0.0	-0.6	0.0	(0.7)	0.1	(0.5)
Australian	0.0	(0.0)	0.0	(0.3)	0.0	(0.0)	0.0	0.0	0.0	0.0	0.0	0.0	0.0	0.0	0.0	0.0	0.0	0.0	0.1	(1.4)	0.0	(0.7)	0.0	0.0
Spanish	0.0	-(0.9)	0.0	-0.7	0.1	-0.2	0.0	-(1.5)	0.0	-0.5	0.1	-0.7	0.1	-0.6	0.0	-0.4	0.0	-0.5	0.0	0.0	0.0	-0.7	0.0	-(1.5)
Portuguese	0.0	-0.5	0.0	-0.9	0.0	-0.6	0.1	-0.5	0.0	-0.7	0.1	-0.1	0.0	-0.9	0.0	-0.4	0.1	(0.7)	0.0	-0.4	0.0	-0.7	0.0	-0.4
Italian (North)	0.0	-(0.1)	0.0	(0.3)	0.0	-0.2	0.0	-0.7	0.0	-0.4	0.1	-0.8	0.1	-0.7	0.1	-0.7	0.1	(0.7)	0.0	-0.3	0.0	0.0	0.0	-(1.1)
Italian (South)	0.0	(0.0)	0.1	(0.3)	0.1	(0.6)	0.2	(2.6)	0.4	(1.5)	0.2	(1.0)	0.3	(1.6)	0.1	-0.2	0.2	(1.3)	0.0	-0.1	0.0	-0.3	0.0	(0.0)
Dutch and Flemish	0.0	-0.7	0.0	-(1.1)	0.0	-0.7	0.0	-(1.8)	0.1	-0.4	0.1	-0.4	0.1	-0.3	0.1	(0.0)	0.0	-0.5	0.0	-0.1	0.0	0.0	0.0	-(1.0)
Finnish	0.0	-(0.9)	0.1	(1.4)	0.0	-0.5	0.0	-(1.3)	0.1	-0.3	0.0	-0.9	0.1	-(1.3)	0.2	(0.5)	0.0	-0.5	0.0	-0.6	0.1	-0.1	0.1	(0.5)
Scandinavian*	0.0	-(0.1)	0.0	(0.1)	0.0	-0.4	0.0	-(1.0)	0.0	-0.5	0.0	-0.2	0.1	-0.2	0.1	(0.0)	0.0	-0.3	0.0	(0.1)	0.0	-0.1	0.0	0.0
Lithuanian	0.0	-(0.1)	0.0	-0.7	0.0	-0.2	0.1	(0.1)	0.2	(0.4)	0.0	(0.0)	0.0	-(1.3)	0.0	-0.6	0.0	-0.5	0.0	-0.6	0.1	(3.3)	0.0	-(1.5)
Russian	0.0	-(0.9)	0.0	-(1.2)	0.0	-0.6	0.1	-0.9	0.1	-0.2	0.0	-(1.1)	0.1	-0.7	0.0	-0.6	0.0	-0.5	0.1	(1.4)	0.0	(0.2)	0.0	-0.5
Polish	0.0	-0.4	0.0	(0.0)	0.0	-0.3	0.1	-0.2	0.1	-0.2	0.0	-0.4	0.0	(0.0)	0.1	(0.2)	0.1	(0.3)	0.0	(0.1)	0.0	(0.1)	0.0	(0.6)
German	0.0	(0.0)	0.0	(0.4)	0.0	-0.4	0.0	-(1.0)	0.1	-0.4	0.1	-0.5	0.1	-(0.3)	0.1	(3.6)	0.2	(1.3)	0.3	(1.4)	0.0	(0.2)	0.0	-0.5
Magyar	0.0	-(0.2)	0.0	-0.8	0.0	-0.7	0.0	-(1.5)	0.1	-0.4	0.1	-0.2	0.4	(3.4)	0.0	-0.6	0.0	-0.5	0.3	(5.3)	0.0	(0.3)	0.0	-(1.5)
Slovak	0.0	-0.1	0.0	-(1.2)	0.0	-0.4	0.1	-0.5	0.1	-0.4	0.0	-(1.3)	0.0	-(1.3)	0.0	(0.0)	0.0	-0.2	0.0	-0.2	0.0	-0.3	0.0	-0.5
Croatian & Slovenian	0.0	-(0.2)	0.0	-0.6	0.3	(4.3)	0.1	-0.2	0.1	-0.2	0.0	-(1.3)	0.0	-(1.3)	0.1	(0.0)	0.0	-0.5	0.0	-0.6	0.0	-0.6	0.0	-0.8
Transylvanian (Siebenburger)	0.0	-(0.9)	0.0	(0.0)	0.0	(0.0)	0.0	(0.0)	0.0	(0.0)	0.0	(0.4)	0.0	-(1.3)	0.0	(0.0)	0.0	(0.0)	0.0	(0.0)	0.0	(0.0)	0.0	(0.0)
Czech †	0.0	-(0.9)	0.0	-0.8	0.1	(0.0)	0.0	-0.8	0.0	-0.7	0.2	(0.4)	0.0	-0.8	0.1	(0.3)	0.1	(0.2)	0.0	(0.2)	0.0	-0.2	0.0	-0.6
Dalmatian, Bosnian, Herzegovinian	0.0	-(0.2)	0.0	-(0.1)	0.1	(0.8)	0.1	(0.1)	0.0	(0.0)	0.0	-(1.3)	0.0	-(1.3)	0.0	(0.0)	0.0	-0.5	0.0	(0.0)	0.0	-0.9	0.0	-(1.5)
Ruthenian (Russniak)	0.0	-0.5	0.0	(0.0)	0.1	(1.3)	0.1	(0.7)	0.1	-0.1	0.1	-0.6	0.1	(0.0)	0.0	-0.6	0.0	-0.5	0.0	-0.6	0.1	-(1.3)	0.0	-(1.5)
Austro-Hungarian	0.0	(0.0)	0.0	(0.0)	0.0	(0.0)	0.0	(0.0)	0.0	(0.0)	0.0	(0.0)	0.0	(0.0)	0.0	(0.0)	0.0	(0.0)	0.0	(0.0)	0.0	(0.0)	0.0	(0.0)
Roumanian	0.0	-(0.9)	0.0	-(1.5)	0.0	-0.8	0.0	-(1.8)	0.0	-0.7	0.0	-0.2	0.2	(0.6)	0.0	-0.6	0.1	(0.9)	0.0	(0.0)	0.1	(0.1)	0.1	(0.1)
Bulgarian, Serbian, Montenegrin	0.0	-(0.9)	0.0	-(1.0)	0.0	-0.3	0.0	-(1.4)	0.1	-0.3	0.1	(0.4)	0.2	-(1.3)	0.2	(0.0)	0.1	(0.0)	0.0	(0.0)	0.1	(0.2)	0.1	(0.2)
Greek	0.0	-(0.6)	0.0	-(1.3)	0.0	-0.4	0.1	-0.1	0.1	-0.1	0.1	-0.2	0.1	-0.6	0.1	(0.0)	0.0	(0.9)	0.0	(0.0)	0.0	-0.2	0.0	-0.3
Armenian	0.1	(1.0)	0.0	-(1.5)	0.0	-0.4	0.0	-0.8	0.0	-0.7	0.0	-(1.3)	0.0	(0.4)	0.0	-0.2	0.2	(1.2)	0.0	-0.6	0.0	-0.6	0.1	-(1.5)
Hebrew	0.0	(0.4)	0.0	(0.5)	0.1	-0.2	0.0	-0.5	0.0	(0.0)	0.0	-0.5	0.0	-0.8	0.1	(0.3)	0.1	(0.2)	0.0	(0.2)	0.1	-0.2	0.1	(1.1)
Arabian	0.0	(0.0)	0.0	(0.0)	0.0	(0.0)	0.0	(0.0)	0.0	(0.0)	0.0	(0.0)	0.0	(0.0)	0.0	(0.0)	0.0	(0.0)	0.0	(0.0)	0.0	(0.0)	0.0	(0.0)
Syrian	0.0	(0.6)	0.0	-(1.5)	0.1	(0.2)	0.1	(0.0)	0.1	(0.0)	0.0	(1.5)	0.1	-0.2	0.3	(1.6)	0.2	(1.2)	0.0	(0.0)	0.0	-0.3	0.0	-0.1
Turkish	0.0	-0.9	0.1	(2.7)	0.1	(0.6)	0.0	-(1.9)	0.0	-0.7	0.2	(3.6)	0.0	-(1.3)	0.1	-0.6	0.1	-0.5	0.0	-0.6	0.0	-0.9	0.0	-(1.5)
French	0.0	(0.6)	0.0	(0.0)	0.1	(0.6)	0.0	-(1.1)	0.0	-0.6	0.0	-0.1	0.1	-0.2	0.1	(0.4)	0.1	-0.1	0.1	(0.6)	0.1	(2.0)	0.1	(1.1)
Swiss	0.0	-(0.9)	0.0	(0.0)	0.1	(0.0)	0.0	(0.0)	0.0	(0.0)	0.0	(0.0)	0.2	(0.7)	0.0	-0.2	0.1	(0.0)	0.0	(0.0)	0.0	-0.4	0.0	-0.6
Mexican	0.1	(1.4)	0.0	(1.0)	0.1	(1.2)	0.1	(0.6)	0.2	(0.5)	0.2	(0.6)	0.1	(0.0)	0.1	-0.2	0.0	-0.1	0.0	-0.2	0.0	-0.4	0.0	-0.6
Central American (NS)	0.0	(0.0)	0.0	(0.0)	0.0	(0.0)	0.0	(0.0)	0.0	(0.0)	0.0	(0.0)	0.0	(0.0)	0.0	(0.0)	0.0	(0.0)	0.0	(0.0)	0.0	(0.0)	0.0	(0.0)
South American	0.0	-(0.9)	0.0	(0.0)	0.0	(0.0)	0.0	(0.0)	0.0	(0.0)	0.0	(0.0)	0.0	(0.0)	0.0	(0.0)	0.0	(0.0)	0.0	(0.0)	0.0	(0.0)	0.0	(0.0)
Spanish American	0.0	-(0.9)	0.0	-(1.5)	0.1	(1.4)	0.0	-0.7	0.0	-0.6	0.0	-(1.3)	0.0	-(1.3)	0.0	-0.3	0.0	-0.5	0.0	-0.3	0.0	-0.9	0.0	-(1.5)
Cuban	0.0	-0.4	0.0	-(1.5)	0.0	-0.5	0.0	-(1.9)	0.0	-0.6	0.0	-(1.0)	0.1	-(1.3)	0.1	-0.6	0.0	-0.5	0.0	-0.6	0.0	-0.5	0.0	-(1.5)
West Indian	0.0	-(0.9)	0.2	(1.8)	0.0	(0.0)	0.0	-(1.9)	0.0	-0.7	0.0	-(1.3)	0.1	(0.3)	0.1	-0.6	0.0	-0.5	0.0	-0.6	0.0	-0.9	0.0	-0.8
Chinese	0.0	-(0.9)	0.0	-(1.5)	0.0	-(1.1)	0.0	-(1.9)	0.0	-0.7	0.0	-(1.3)	0.0	-0.3	0.0	-0.6	0.0	-0.3	0.0	-0.4	0.0	-0.4	0.0	-(1.5)
East Indian	0.0	-(0.9)	0.0	-(1.5)	0.0	-(1.1)	0.0	-(1.9)	0.1	(5.4)	0.0	-(1.3)	0.0	-(1.3)	0.0	-0.6	0.0	-0.6	0.0	-0.6	0.0	-0.9	0.0	-(1.5)
Japanese	0.0	-(0.9)	0.0	-(1.5)	0.0	-0.9	0.0	-(1.7)	0.0	-0.7	0.0	-(1.3)	0.0	-(1.3)	0.0	-0.5	0.0	-0.5	0.0	-0.4	0.0	-0.9	0.0	-(1.5)
Korean	0.0	-(0.9)	0.0	-(1.5)	0.0	-(1.1)	0.0	-(1.9)	0.0	-0.7	0.0	-(1.1)	0.0	-(1.3)	0.0	-0.6	0.0	-0.5	0.0	-0.4	0.0	-0.9	0.0	-(1.0)
Filipino	0.0	-(0.9)	0.0	-(1.5)	0.0	-0.7	0.0	-(1.9)	0.0	-0.7	0.0	-(1.3)	0.0	-(1.1)	0.0	-0.6	0.0	-0.5	0.0	-0.6	0.0	-0.9	0.0	-(1.5)
Hawaiian	0.0	(0.0)	0.0	(0.0)	0.0	(0.0)	0.0	(0.0)	0.0	(0.0)	0.0	(0.0)	0.0	(0.0)	0.0	(0.0)	0.0	(0.0)	0.0	(0.0)	0.0	(0.0)	0.0	(0.0)
Pacific Islander	0.0	-(0.9)	0.0	-(1.5)	0.0	-(1.1)	0.0	-(1.9)	0.0	-0.7	0.0	-0.4	0.1	(0.1)	0.0	-0.5	0.0	-0.4	0.0	-0.9	0.0	-0.9	0.0	-(1.5)
African (black)	0.0	-0.6	0.0	(0.0)	0.1	-(0.1)	0.1	(0.0)	0.1	(0.2)	0.1	-0.4	0.1	(0.1)	0.0	(0.0)	0.0	-0.1	0.0	-0.6	0.1	(0.1)	0.1	(0.1)
(NS)	0.0	(0.0)	0.0	(0.0)	0.0	(0.0)	0.0	(0.0)	0.0	(0.0)	0.0	(0.0)	0.0	(0.0)	0.0	(0.0)	0.0	(0.0)	0.0	(0.0)	0.0	(0.0)	0.0	(0.0)
Esquimaux	0.0	(0.0)	0.0	(0.0)	0.0	(0.0)	0.0	(0.0)	0.0	(0.0)	0.0	(0.0)	0.0	(0.0)	0.0	(0.0)	0.0	(0.0)	0.0	(0.0)	0.0	(0.0)	0.0	(0.0)
Other	0.0	-(0.9)	0.0	-(1.5)	0.0	(0.0)	0.0	(0.4)	0.0	-0.4	0.1	-0.6	0.0	-(1.3)	0.0	-0.6	0.0	-0.5	0.1	(0.6)	0.1	(1.2)	0.0	-(1.5)

TABLE 6-16. Continued

Race or People	1923 %	1923 NSDMR	1924 %	1924 NSDMR	1925 %	1925 NSDMR	1926 %	1926 NSDMR	1927 %	1927 NSDMR	1928 %	1928 NSDMR	1929 %	1929 NSDMR	1930 %	1930 NSDMR
English	0.0	(0.1)	0.0	(0.3)	0.0	(0.1)	0.0	(0.2)	0.0	(0.0)	0.0	(0.1)	0.0	-0.3	0.0	(0.4)
Irish	0.1	(1.6)	0.1	1.0	0.0	(0.0)	0.0	-0.1	0.0	-0.1	0.0	(0.0)	0.0	-0.3	0.0	-0.1
Scotch	0.0	-0.1	0.0	(0.2)	0.1	(0.2)	0.0	(0.7)	0.0	(0.1)	0.0	(0.1)	0.0	-0.3	0.0	(0.6)
Welsh	0.0	(1.2)	0.1	(1.7)	0.0	-0.5	0.0	(0.1)	0.0	-0.4	0.0	-0.2	0.0	-0.3	0.0	(1.2)
Australian	0.0	(0.0)	0.0	(0.0)	0.0	(0.0)	0.0	(0.1)	0.0	(0.0)	0.0	(0.0)	0.0	(0.0)	0.0	(0.0)
Spanish	0.0	-0.9	0.0	-0.7	0.0	-0.5	0.0	-0.5	0.2	2.9	0.0	-0.2	0.0	-0.3	0.0	(1.2)
Portuguese	0.0	-0.4	0.0	-0.7	0.1	(0.9)	0.0	-0.2	0.0	-0.4	0.0	(0.0)	0.0	-0.3	0.0	(1.2)
Italian (North)	0.0	-0.7	0.0	-0.7	0.0	(0.1)	0.0	-0.1	0.0	-0.4	0.0	-0.2	0.0	-0.3	0.0	(1.2)
Italian (South)	0.0	-0.4	0.0	-0.4	0.1	(0.6)	0.0	-0.5	0.0	-0.1	0.0	-0.1	0.0	(0.1)	0.0	(1.6)
Dutch and Flemish	0.0	(0.5)	0.0	(0.2)	0.0	-0.1	0.0	-0.5	0.0	(0.5)	0.0	-0.2	0.0	-0.3	0.0	(1.2)
Finnish	0.0	-0.2	0.1	(0.9)	0.0	-0.5	0.0	(0.4)	0.0	-0.4	0.2	(3.3)	0.0	-0.3	0.0	(1.2)
Scandinavian*	0.0	-0.6	0.0	-0.2	0.0	-0.2	0.1	(0.4)	0.1	(0.1)	0.0	-0.2	0.0	-0.3	0.0	(0.0)
Lithuanian	0.0	(0.5)	0.1	2.6	0.0	-0.5	0.0	-0.2	0.0	-0.4	0.0	-0.2	0.0	-0.3	0.0	(1.2)
Russian	0.0	-0.5	0.0	-0.2	0.2	(1.9)	0.1	(1.8)	0.0	(0.3)	0.0	-0.2	0.0	(0.7)	0.0	(1.2)
Polish	0.0	-0.3	0.0	-0.6	0.0	-0.1	0.0	-0.9	0.0	-0.4	0.0	-0.2	0.0	(0.2)	0.0	(1.2)
German	0.0	-0.7	0.0	-0.3	0.0	-0.3	0.0	-0.4	0.0	-0.2	0.0	-0.2	0.0	(0.0)	0.0	(1.2)
Magyar	0.0	-0.3	0.0	-0.7	0.0	-0.5	0.0	-0.9	0.0	-0.4	0.1	(1.5)	0.0	(1.7)	0.0	(1.2)
Slovak	0.0	(1.2)	0.0	-0.7	0.1	(1.0)	0.1	(1.6)	0.3	(5.9)	0.0	-0.2	0.0	-0.3	0.0	(1.2)
Croatian & Slovenian	0.0	(0.5)	0.0	(0.5)	0.2	(1.5)	0.0	-0.9	0.1	(1.4)	0.2	(3.8)	0.1	(3.1)	0.0	(1.2)
Transylvanian (Siebenburger)	0.0	(0.0)	0.0	(0.0)	0.0	(0.0)	0.0	(0.0)	0.0	(0.0)	0.0	(0.0)	0.0	(0.0)	0.0	(0.0)
Czech †	0.0	-0.6	0.0	-0.7	0.0	-0.5	0.0	(0.2)	0.0	-0.4	0.0	-0.2	0.0	-0.3	0.0	(1.2)
Dalmatian, Bosnian, Herzegovinian	0.0	(1.2)	0.0	-0.7	0.0	-0.5	0.0	(0.0)	0.0	-0.4	0.0	(0.0)	0.0	-0.3	0.0	(1.2)
Ruthenian (Russniak)	0.0	(1.2)	0.2	4.3	0.1	(0.6)	0.2	3.1	0.2	2.9	0.0	-0.2	0.0	(1.2)	0.0	(1.2)
Austro-Hungarian	0.0	(0.0)	0.0	(0.0)	0.0	(0.0)	0.0	(0.0)	0.0	(0.0)	0.0	(0.0)	0.0	(0.0)	0.0	(0.0)
Roumanian	0.0	(1.2)	0.0	(0.5)	0.0	-0.5	0.0	-0.9	0.0	-0.4	0.0	-0.2	0.1	-4.6	0.0	(1.2)
Bulgarian, Serbian, Montenegrin	0.0	(1.2)	0.0	-0.7	0.3	3.8	0.1	(0.7)	0.0	-0.4	0.2	3.4	0.1	(0.9)	0.0	(1.2)
Greek	0.0	(0.2)	0.0	-0.7	0.1	(0.5)	0.0	-0.9	0.0	-0.4	0.0	-0.2	0.0	(0.9)	0.0	(1.2)
Armenian	0.0	(1.2)	0.0	(0.1)	0.0	-0.5	0.1	(1.2)	0.0	-0.4	0.0	-0.2	0.0	-0.3	0.0	(1.2)
Hebrew	0.1	(1.1)	0.0	(0.3)	0.0	(0.0)	0.1	(0.4)	0.0	(0.5)	0.0	(0.0)	0.0	-0.3	0.0	(0.4)
Arabian	0.0	(0.0)	0.0	(0.0)	0.0	(0.1)	0.1	(0.8)	0.0	(0.1)	0.0	(0.0)	0.0	(0.0)	0.0	(0.0)
Syrian	0.0	(0.6)	0.0	(0.4)	0.0	-0.5	0.1	(0.1)	0.0	(0.1)	0.0	-0.2	0.0	(0.5)	0.0	(1.2)
Turkish	0.0	(0.0)	0.0	-0.7	0.0	-0.5	0.0	-0.9	0.1	(1.7)	0.0	(0.0)	0.0	-0.3	0.0	(1.2)
French	0.1	3.2	0.1	(1.0)	0.1	(0.6)	0.1	2.5	0.1	(1.2)	0.0	-0.1	0.0	-0.3	0.0	52.9
Swiss	0.0	(0.0)	0.0	(0.0)	0.0	(0.0)	0.0	(0.0)	0.0	(0.0)	0.0	(0.1)	0.0	(0.0)	0.0	(0.0)
Mexican	0.0	-0.6	0.0	-0.2	0.0	-0.1	0.0	-0.7	0.0	-0.3	0.0	(0.1)	0.0	-0.1	0.0	(0.2)
Central American (NS)	0.0	(0.0)	0.0	(0.0)	0.0	(0.0)	0.0	(0.0)	0.0	(0.0)	0.0	(0.0)	0.0	(0.0)	0.0	(0.0)
South American	0.0	-0.6	0.0	-0.3	0.0	-0.5	0.0	(0.0)	0.0	(0.1)	0.0	(0.0)	0.0	(0.5)	0.0	(0.0)
Spanish American	0.0	(0.0)	0.0	(0.0)	0.0	(0.0)	0.0	(0.0)	0.0	(0.0)	0.0	(0.0)	0.0	(0.0)	0.0	(0.0)
Cuban	0.0	-0.6	0.0	-0.7	0.0	-0.5	0.0	-0.9	0.0	-0.4	0.0	-0.2	0.0	-0.3	0.0	(1.2)
West Indian	0.0	(1.2)	0.0	-0.7	0.0	-0.5	0.0	-0.9	0.0	-0.4	0.0	-0.2	0.0	-0.3	0.0	(1.2)
Chinese	0.0	-0.3	0.0	-0.7	0.0	(0.0)	0.0	-0.6	0.0	-0.4	0.0	(0.0)	0.0	(1.2)	0.0	(1.2)
East Indian	0.0	(1.2)	0.0	-0.7	0.0	-0.5	0.0	-0.9	0.0	-0.4	0.0	-0.2	0.0	-0.3	0.0	(1.2)
Japanese	0.0	-0.6	0.0	-0.7	0.1	(1.0)	0.0	-0.5	0.0	-0.4	0.0	-0.2	0.0	(0.5)	0.0	(1.5)
Korean	0.0	(1.2)	0.0	-0.7	0.0	-0.5	0.0	-0.9	0.0	-0.4	0.0	-0.2	0.0	-0.3	0.0	(1.2)
Filipino	0.0	(0.0)	0.0	(0.0)	0.0	(0.0)	0.0	(0.0)	0.0	(0.0)	0.0	(0.0)	0.0	(0.0)	0.0	(0.0)
Hawaiian	0.0	(0.0)	0.0	(0.0)	0.0	(0.0)	0.0	(0.0)	0.0	(0.0)	0.0	(0.0)	0.0	(0.0)	0.0	(0.0)
Pacific Islander	0.0	(1.2)	0.0	-0.7	0.1	(0.5)	0.0	-0.9	0.0	-0.4	0.0	-0.2	0.0	-0.3	0.0	(1.2)
African (black)	0.0	(0.0)	0.0	-0.2	0.0	(0.0)	0.0	(0.0)	0.0	(0.0)	0.0	-0.2	0.0	(1.1)	0.0	(0.0)
(NS)	0.0	(0.0)	0.0	(0.0)	0.0	(0.0)	0.0	(0.0)	0.0	(0.0)	0.0	(0.0)	0.0	(0.0)	0.0	(0.0)
Esquimauz	0.0	(0.0)	0.0	(0.0)	0.0	(0.0)	0.0	(0.0)	0.0	(0.0)	0.0	(0.0)	0.0	(0.0)	0.0	(0.0)
Other	0.1	2.0	0.0	-0.7	0.3	5.4	0.1	(0.9)	0.0	-0.4	0.0	-0.2	0.0	-0.3	0.0	(1.2)

TABLE 6-17. Deviations in Certifications for Senility, by Race, 1916–1930, Expressed as Percent Certified and NSDMR

Race or People	1916 %	1916 NSDMR	1917 %	1917 NSDMR	1918 %	1918 NSDMR	1919 %	1919 NSDMR	1920 %	1920 NSDMR	1921 %	1921 NSDMR	1922 %	1922 NSDMR	1923 %	1923 NSDMR	1924 %	1924 NSDMR	1925 %	1925 NSDMR	1926 %	1926 NSDMR	1927 %	1927 NSDMR	1928 %	1928 NSDMR	1929 %	1929 NSDMR	1930 %	1930 NSDMR
English	0.8	(0.0)	0.9	(-0.3)	1.0	(0.0)	0.3	(-0.3)	0.7	(-0.3)	1.0	(-0.6)	0.9	(-0.8)	0.4	(-0.6)	0.3	(-0.5)	1.0	(-0.2)	0.6	(-0.2)	1.4	(-0.1)	1.7	(0.0)	0.7	(-0.2)	0.7	(-0.2)
Irish	1.2	(0.0)	1.2	(-0.3)	1.4	(0.5)	0.5	(0.5)	0.8	(-0.1)	0.9	(-0.6)	0.9	(-0.7)	0.5	(-0.4)	0.4	(-0.4)	0.7	(-0.3)	0.6	(-0.2)	0.8	(-0.3)	1.0	(-0.2)	0.7	(-0.3)	0.5	(-0.3)
Scotch	0.7	(0.0)	0.9	(-0.3)	0.9	(0.1)	0.5	(0.1)	0.8	(-0.1)	1.1	(-0.4)	0.9	(-0.7)	0.3	(-0.7)	0.6	(-0.4)	0.8	(-0.2)	0.5	(-0.2)	1.1	(-0.2)	1.0	(-0.2)	0.6	(-0.4)	0.3	(-0.4)
Welsh	0.6	(-0.1)	0.8	(-0.4)	1.0	(0.0)	0.5	(0.4)	1.5	(1.0)	1.4	(-0.1)	1.1	(-1.1)	0.5	(-0.5)	0.6	(-0.1)	1.3	(-0.1)	0.8	(-0.3)	1.8	(0.1)	1.3	(-0.1)	0.4	(-0.5)	0.3	(-0.4)
Spanish	0.3	(-0.1)	**0.2**	(0.2)	**0.3**	(-0.8)	**0.1**	(-1.0)	**0.2**	(-1.2)	**0.3**	(-1.3)	**0.5**	(-1.1)	**0.5**	(-0.5)	0.6	(-0.6)	1.3	(-0.1)	0.3	(-0.3)	1.5	(0.1)	2.5	(-0.3)	0.2	(-0.6)	0.3	(-0.4)
Portuguese	1.3	(0.1)	**2.3**	(1.8)	**1.3**	(-1.3)	0.4	(-1.0)	0.7	(-1.2)	0.8	(-0.8)	1.2	(-0.5)	0.3	(-0.9)	0.2	(-0.6)	1.6	(0.0)	0.2	(-0.3)	1.4	(0.0)	0.6	(-0.4)	0.2	(-0.6)	0.2	(-0.5)
Italian (North)	0.8	(0.0)	1.1	(0.0)	0.8	(-0.1)	0.4	(-0.1)	0.9	(0.0)	**0.5**	(-1.1)	**0.4**	(-1.4)	**0.1**	(-1.0)	0.1	(-0.7)	1.6	(0.0)	0.4	(0.0)	0.8	(-0.1)	1.5	(-0.1)	1.0	(0.0)	0.2	(-0.3)
Italian (South)	2.5	(0.3)	**2.8**	(2.7)	1.3	(0.3)	0.5	(0.6)	2.1	(1.9)	2.1	(0.7)	3.0	(1.3)	1.8	(0.7)	0.7	(0.1)	7.0	(1.8)	3.5	(1.1)	6.9	(1.8)	5.9	(1.5)	4.2	(2.0)	4.3	(1.6)
Finnish	1.0	(0.0)	0.9	(-0.2)	1.4	(0.5)	0.4	(0.1)	1.2	(0.5)	0.9	(-0.6)	0.9	(-0.7)	0.5	(-0.4)	0.2	(-0.5)	1.5	(0.0)	0.9	(-0.1)	2.4	(0.3)	2.4	(0.2)	0.9	(-0.2)	0.9	(-0.1)
Scandinavian*	0.3	(-0.1)	0.6	(-0.7)	0.5	(-0.5)	0.4	(-1.2)	1.1	(0.2)	1.6	(0.2)	0.8	(-0.6)	0.5	(-0.3)	0.2	(-0.5)	0.8	(0.2)	0.3	(-0.3)	1.7	(0.2)	2.2	(0.2)	0.6	(-0.6)	0.2	(-0.5)
Lithuanian	1.0	(0.0)	1.4	(0.4)	1.8	(1.0)	**0.9**	(0.9)	1.0	(0.2)	1.4	(-0.2)	1.1	(-0.5)	1.1	(-0.4)	0.6	(-0.1)	1.0	(-0.2)	0.7	(-0.1)	1.1	(-0.1)	1.3	(-0.1)	0.6	(-0.3)	0.7	(-0.3)
Russian	0.3	(-0.1)	0.6	(-0.7)	**0.0**	(-1.1)	**0.0**	(-1.5)	**0.6**	(-0.5)	3.2	(1.9)	0.9	(0.9)	1.4	(0.7)	**2.5**	(2.4)	9.2	(3.5)	4.4	(1.5)	8.6	(2.3)	13.5	(4.2)	2.4	(4.1)	5.2	(2.0)
Polish	0.1	(-0.2)	**0.1**	(-1.5)	0.1	(-1.0)	**0.0**	(-1.1)	**0.5**	(-0.9)	**0.5**	(-1.1)	0.7	(-0.1)	0.7	(-0.4)	0.7	(-0.1)	1.1	(-0.1)	0.4	(-0.2)	2.3	(0.3)	3.0	(0.5)	0.0	(0.0)	1.1	(0.0)
German	0.3	(-0.1)	0.4	(-1.0)	**0.1**	(-0.7)	0.3	(-0.5)	0.8	(-1.2)	0.8	(-0.7)	0.5	(-0.5)	1.0	(0.2)	1.2	(0.7)	4.6	(1.0)	2.6	(0.7)	3.9	(0.8)	5.6	(1.4)	3.3	(3.3)	3.1	(1.0)
Magyar	1.5	(0.1)	**0.0**	(-1.6)	1.5	(0.6)	1.4	(1.6)	2.1	(0.8)	2.1	(0.7)	1.8	(0.1)	1.1	(0.4)	2.2	(1.0)	3.7	(0.7)	3.3	(1.0)	6.3	(2.3)	4.7	(1.1)	0.8	(0.8)	2.4	(0.6)
Slovak	0.9	(0.0)	0.4	(-0.5)	2.6	(1.9)	**0.0**	(-1.5)	1.0	(0.1)	2.1	(0.6)	1.8	(0.2)	1.4	(0.7)	1.3	(0.9)	13.7	(4.0)	12.4	(4.8)	12.7	(3.7)	5.9	(1.5)	1.3	(1.3)	0.7	(0.7)
Croatian and Slovenian	0.7	(0.0)	1.2	(0.3)	2.6	(2.0)	**0.0**	(-1.5)	1.3	(-1.2)	0.4	(-1.3)	0.6	(-1.1)	0.2	(-0.2)	0.8	(-0.1)	3.3	(0.6)	3.5	(1.1)	6.1	(1.6)	6.1	(1.2)	2.6	(2.6)	0.2	(-0.5)
Czech †	1.4	(0.1)	1.8	(1.1)	**0.0**	(-1.1)	**0.0**	(-1.5)	0.3	(-0.9)	0.4	(-1.2)	0.4	(-1.3)	0.4	(-0.5)	0.8	(0.2)	0.4	(-0.4)	0.0	(-0.4)	0.0	(-0.5)	0.4	(-0.4)	0.0	(-0.8)	3.6	(3.7)
Dalmatian, Bosnian, Herzegovinian	0.8	(0.0)	**0.0**	(-1.6)	0.0	(0.0)	**0.0**	(-1.5)	0.0	(0.0)	0.1	(-1.6)	1.0	(-0.6)	1.0	(-0.9)	1.0	(0.5)	0.0	(0.0)	0.1	(-0.4)	0.0	(0.0)	0.0	(0.0)	0.0	(0.0)	3.6	(1.2)
Ruthenian (Russniak)	0.2	(-0.1)	**0.3**	(0.3)	**0.0**	(-1.2)	**0.0**	(-1.5)	**0.0**	(-1.4)	0.6	(0.6)	**0.0**	(-1.3)	0.0	(0.0)	0.0	(-0.8)	0.2	(-0.4)	0.1	(-0.4)	1.1	(-0.1)	0.7	(-0.3)	0.1	(-0.7)	0.1	(-0.5)
Roumanian	1.0	(0.0)	1.4	(0.4)	0.6	(0.0)	**0.0**	(-1.5)	0.3	(-0.9)	1.0	(-0.6)	1.1	(-0.6)	0.9	(0.3)	0.9	(0.3)	4.4	(1.0)	2.5	(0.6)	7.3	(2.1)	9.3	(2.7)	1.3	(0.2)	2.2	(0.5)
Bulgarian, Serbian, Montenegrin	0.2	(-0.1)	**0.4**	(0.4)	1.0	(0.1)	0.6	(0.9)	**0.1**	(-1.2)	**0.3**	(-1.3)	**0.7**	(-1.0)	0.6	(-0.2)	0.3	(-0.4)	1.9	(0.1)	0.4	(-0.2)	1.3	(-0.1)	0.9	(-0.3)	0.3	(-0.5)	5.4	(2.1)
Greek	1.1	(0.0)	0.9	(-0.3)	1.0	(0.7)	0.2	(-0.8)	1.1	(1.1)	1.4	(-0.2)	1.9	(0.2)	1.3	(0.7)	1.9	(1.7)	3.9	(0.8)	0.8	(-0.2)	2.5	(0.3)	1.9	(0.1)	1.1	(0.0)	1.5	(0.2)
Armenian	0.8	(0.0)	0.9	(-0.3)	1.3	(0.3)	1.5	(1.0)	**1.5**	(1.0)	1.7	(-0.2)	0.6	(-0.6)	2.0	(1.8)	1.6	(0.0)	1.6	(0.0)	1.4	(0.2)	1.3	(0.3)	1.1	(-0.2)	0.8	(-0.1)	0.7	(-0.2)
Hebrew	0.9	(0.0)	1.7	(0.9)	1.6	(0.7)	0.6	(0.9)	2.1	(2.0)	3.5	(2.0)	5.6	(3.3)	3.6	(3.8)	3.3	(3.5)	10.3	(2.9)	7.2	(3.0)	5.8	(1.4)	5.5	(1.4)	2.1	(2.1)	5.3	(2.1)
Syrian	0.9	(0.0)	0.5	(-0.8)	1.9	(1.9)	0.7	(1.2)	1.4	(0.8)	1.8	(0.3)	0.4	(-0.4)	1.2	(0.1)	1.2	(0.7)	3.7	(0.7)	3.4	(0.6)	5.5	(1.4)	1.8	(0.2)	0.4	(0.4)	1.6	(0.2)
French	0.5	(-0.1)	0.6	(-0.8)	**0.0**	(-0.7)	**0.0**	(-1.5)	1.0	(0.7)	1.0	(0.3)	0.9	(-0.8)	0.4	(-0.6)	0.4	(-0.3)	1.1	(-0.1)	0.8	(0.1)	1.4	(0.6)	1.1	(-0.2)	0.7	(-0.2)	0.4	(-0.4)
Turkish	0.4	(-0.1)	**0.0**	(-1.6)	1.0	(1.0)	0.0	(-1.5)	2.0	(1.8)	2.0	(0.6)	0.9	(-0.8)	0.3	(-0.7)	1.2	(0.8)	0.5	(-0.5)	0.4	(-0.3)	0.7	(-0.3)	0.7	(-0.3)	0.0	(-0.8)	1.2	(0.0)
Mexican	0.4	(-0.1)	**0.4**	(0.4)	**0.0**	(-0.9)	0.5	(0.3)	**0.1**	(-1.4)	0.6	(0.0)	**0.1**	(-1.6)	**0.1**	(-1.0)	**0.2**	(-0.6)	0.5	(-0.5)	0.4	(-0.3)	0.5	(-0.4)	0.7	(-0.3)	0.4	(-0.7)	0.1	(-0.6)
Spanish American	0.2	(-0.1)	**0.1**	(0.1)	0.3	(-0.8)	0.1	(-0.9)	**0.2**	(-1.2)	**0.2**	(-1.5)	**0.1**	(-1.3)	**0.1**	(-1.0)	0.2	(-0.6)	0.6	(-0.3)	0.4	(-0.4)	0.5	(-0.4)	0.5	(-0.4)	0.0	(-0.7)	0.0	(-0.6)
Cuban	0.1	(-0.2)	**0.1**	(0.1)	0.3	(-0.9)	0.1	(-1.1)	**0.2**	(-1.2)	**0.2**	(-1.5)	**0.1**	(-1.3)	**0.3**	(-1.0)	0.1	(-0.8)	1.3	(-0.1)	0.1	(-0.4)	0.4	(-0.4)	0.5	(-0.4)	0.0	(-0.7)	0.0	(-0.6)
West Indian	0.0	(-0.2)	**0.3**	(0.3)	0.9	(0.0)	0.4	(-0.1)	**0.1**	(-1.3)	**0.0**	(-1.6)	**0.2**	(-1.4)	0.1	(-0.9)	0.0	(-0.8)	0.3	(-0.5)	0.0	(-0.4)	0.8	(-0.3)	0.8	(-0.3)	0.0	(-0.7)	0.0	(-0.6)
Chinese	0.0	(-0.2)	**0.0**	(-1.6)	0.1	(-1.1)	**0.0**	(-1.5)	**0.0**	(-1.4)	**0.1**	(-1.7)	**0.0**	(-1.6)	0.0	(-1.1)	0.3	(-0.5)	0.3	(-0.4)	0.1	(-0.4)	0.8	(-0.5)	0.8	(-0.3)	0.0	(-0.8)	0.0	(-0.6)
East Indian	0.0	(-0.2)	**0.0**	(-1.6)	1.3	(0.3)	**0.0**	(-1.5)	**0.0**	(-1.4)	**0.0**	(-1.7)	**0.3**	(-1.3)	**0.0**	(-1.1)	0.0	(-0.5)	0.2	(-0.4)	0.0	(-0.5)	0.0	(-0.5)	0.0	(-0.5)	0.0	(-0.8)	0.0	(-0.6)
Japanese	0.2	(-0.2)	**0.2**	(0.2)	**0.1**	(-0.8)	0.1	(-1.0)	**0.1**	(-1.2)	**0.1**	(-1.6)	**0.1**	(-1.5)	0.2	(-1.0)	0.2	(-0.6)	0.2	(-0.4)	0.0	(-0.4)	1.1	(-0.2)	1.1	(-0.2)	0.0	(-0.8)	0.1	(-0.6)
Korean	0.0	(-0.2)	**0.0**	(-1.6)	1.2	(0.3)	**0.0**	(-1.5)	0.0	(-1.4)	**0.0**	(-1.7)	**0.0**	(-1.6)	0.7	(-0.2)	0.0	(-0.9)	0.0	(-0.5)	0.0	(-0.4)	0.0	(-0.5)	0.0	(-0.5)	0.0	(-0.8)	0.0	(-0.6)
Pacific Islander	33.3	(6.2)	**1.6**	(1.6)	**0.0**	(-1.1)	**0.0**	(-1.5)	0.4	(-1.4)	**0.0**	(0.0)	**1.6**	(-1.6)	3.3	(3.3)	0.0	(0.0)	0.0	(0.0)	12.4	(4.8)	0.0	(0.0)	0.0	(-0.6)	0.0	(0.0)	3.6	(3.7)
African (black)	0.4	(-0.1)	**0.2**	(0.2)	0.3	(-0.7)	0.2	(-0.8)	**0.3**	(-1.1)	**0.3**	(-1.3)	**0.4**	(-1.4)	0.1	(-0.9)	0.1	(-0.7)	1.4	(0.0)	0.2	(-0.3)	1.4	(0.1)	0.4	(-0.4)	0.1	(-0.7)	0.3	(-0.4)
Other	0.4	(-0.1)	0.7	(-0.5)	0.6	(0.7)	0.9	(0.7)	0.9	(0.0)	0.5	(-1.3)	1.0	(-1.0)	1.1	(0.0)	0.5	(0.3)	0.5	(-0.3)	0.5	(-0.2)	2.8	(0.4)	4.5	(1.0)	1.8	(0.6)	2.8	(0.8)

Bold = 1 or more NSDMR below mean   = 1 to 1.99 NSDMRs above mean   = 2 or more NSDMRs above mean

* Norwegians, Danes, and Swedes.   † Bohemians and Moravians.

TABLE 6-18. Deviations in Certifications for Poor Physical Development, by Race, 1916–1930, Expressed as Percent Certified and NSDMR

Race or People	1916 %	1916 NSDMR	1917 %	1917 NSDMR	1918 %	1918 NSDMR	1919 %	1919 NSDMR	1920 %	1920 NSDMR	1921 %	1921 NSDMR	1922 %	1922 NSDMR	1923 %	1923 NSDMR	1924 %	1924 NSDMR	1925 %	1925 NSDMR	1926 %	1926 NSDMR	1927 %	1927 NSDMR	1928 %	1928 NSDMR	1929 %	1929 NSDMR	1930 %	1930 NSDMR
English	0.1	(-0.7)	0.1	(-0.1)	0.1	(-0.1)	0.0	(-0.1)	0.0	(-0.4)	0.0	(-0.8)	0.0	(-0.5)	0.0	(-0.4)	0.0	(-0.3)	0.1	(-0.1)	0.0	(-0.1)	0.0	(-0.1)	0.0	(-0.2)	0.0	(-0.2)	0.0	(-0.1)
Irish	0.2	(-0.5)	0.1	(-0.1)	0.0	(-0.2)	**0.0**	**(-1.0)**	0.0	(-0.3)	0.0	(-0.5)	0.1	(-0.1)	0.0	(-0.5)	0.0	(0.4)	0.0	(-0.1)	0.1	(0.1)	0.0	(0.0)	0.0	(-0.4)	0.0	(0.0)	0.0	(-0.2)
Scotch	0.0	(-0.8)	0.1	(-0.1)	0.0	(-0.1)	**0.0**	**(1.3)**	0.0	(-0.3)	0.0	(-0.7)	0.0	(-0.7)	0.0	(-0.2)	0.0	(0.4)	0.0	(0.0)	0.0	(-0.1)	0.0	(0.1)	0.0	(0.5)	0.0	(-0.3)	0.0	(-0.5)
Welsh	0.1	(-0.7)	0.0	(0.0)	0.0	(-0.3)	**0.0**	**(-1.0)**	0.0	(-0.6)	0.0	(-0.4)	0.0	(-0.9)	0.0	(-0.7)	0.1	(-0.2)	0.0	(-0.4)	0.1	(4.3)	0.1	(0.3)	0.1	(0.4)	0.0	(-0.5)	0.0	(-0.7)
Spanish	0.2	(-0.4)	0.3	(0.0)	0.1	(0.0)	**0.0**	**(-1.0)**	0.1	(0.6)	0.1	(0.3)	0.0	(-0.9)	0.0	(-0.5)	0.1	(-0.4)	0.1	(0.6)	0.0	(-0.3)	0.0	(-0.2)	0.1	(-0.8)	0.0	(-0.3)	0.0	(-0.4)
Portuguese	0.5	(0.8)	1.0	(0.5)	0.2	(0.5)	**0.0**	**(-1.0)**	0.1	(0.7)	0.1	(-0.1)	0.1	(0.8)	0.0	(-0.1)	0.2	(-1.1)	0.2	(1.8)	0.0	(-0.5)	0.0	(-0.2)	0.1	(1.7)	0.0	(-0.5)	0.0	(-0.7)
Italian (North)	0.1	(-0.6)	0.2	(0.0)	0.3	(0.1)	**0.0**	**(1.8)**	0.1	(0.7)	0.1	(-0.1)	0.1	(0.8)	0.1	(1.0)	0.1	(-0.7)	0.1	(-0.4)	0.0	(-0.3)	0.0	(-0.2)	0.1	(0.0)	0.1	(1.0)	0.1	(0.9)
Italian (South)	**1.0**	**(2.6)**	0.7	(0.3)	0.0	(0.3)	**0.0**	**(0.3)**	0.1	(0.5)	0.2	(0.7)	0.2	(1.5)	0.1	(1.0)	0.1	(1.4)	0.1	(0.1)	0.0	(-0.3)	0.1	(-0.1)	0.1	(0.0)	0.1	(0.9)	0.1	(0.4)
Dutch and Flemish	0.0	(-0.9)	0.0	(-0.1)	0.0	(-0.3)	**0.0**	**(-1.0)**	0.0	(-0.5)	0.0	(-0.9)	0.0	(-0.6)	0.0	(-0.6)	**0.0**	**(-0.8)**	0.0	(-0.4)	0.0	(-0.5)	0.0	(0.0)	0.1	(1.4)	0.0	(-0.3)	0.0	(-0.2)
Finnish	0.0	(-0.9)	0.0	(-0.1)	0.0	(-0.1)	**0.0**	**(-1.0)**	0.0	(-0.6)	0.0	(-0.8)	0.0	(-0.9)	0.0	(-0.3)	**0.0**	**(-0.8)**	0.0	(-0.4)	0.0	(-0.5)	0.2	(0.8)	0.0	(-0.8)	0.0	(-0.5)	0.0	(-0.7)
Scandinavian*	0.1	(-0.7)	0.0	(-0.1)	0.1	(0.0)	**0.0**	**(-1.0)**	0.0	(-0.1)	0.0	(-0.9)	0.0	(-0.8)	0.0	(-0.7)	0.0	(-0.6)	0.0	(-0.3)	0.0	(-0.5)	0.0	(0.0)	0.0	(-0.8)	0.0	(-0.2)	0.0	(-0.2)
Lithuanian	**0.0**	**(-1.0)**	0.1	(-0.1)	0.6	(1.9)	**0.0**	**(-1.0)**	0.0	(-0.6)	**0.0**	**(-1.2)**	0.0	(-0.9)	0.0	(-0.7)	**0.0**	**(-1.1)**	0.0	(-0.4)	0.0	(-0.5)	0.0	(-0.2)	0.0	(-0.8)	0.0	(-0.5)	0.0	(-0.7)
Russian	**0.0**	**(-1.0)**	0.1	(-0.1)	0.1	(0.6)	**0.0**	**(-1.0)**	0.0	(-0.6)	**0.0**	**(-1.2)**	0.0	(-0.9)	0.0	(-0.7)	**0.0**	**(-1.1)**	0.5	(4.7)	0.0	(-0.5)	0.0	(-0.2)	0.0	(-0.8)	0.0	(0.0)	0.0	(0.1)
Polish	0.1	(-0.8)	0.0	(-0.1)	0.1	(0.2)	**0.0**	**(-1.0)**	0.0	(-0.4)	0.0	(-0.8)	0.1	(-0.1)	0.0	(-0.1)	0.1	(-0.2)	0.1	(0.3)	0.1	(0.2)	0.0	(-0.1)	0.1	(1.2)	0.1	(0.5)	0.1	(0.7)
German	**0.0**	**(-1.0)**	0.0	(-0.1)	0.0	(-0.1)	**0.0**	**(-1.0)**	0.0	(-0.6)	0.0	(-0.2)	0.0	(-0.1)	0.0	(-0.1)	0.0	(0.6)	0.0	(-0.2)	0.0	(0.2)	0.1	(-0.1)	0.1	(0.6)	0.0	(0.1)	0.1	(0.7)
Magyar	0.1	(-0.7)	0.0	(-0.1)	0.0	(-0.3)	**0.0**	**(-1.0)**	0.0	(-0.6)	0.0	(-0.4)	0.1	(0.2)	0.1	(0.4)	0.2	(1.4)	0.2	(1.4)	0.1	(0.5)	0.3	(1.6)	0.1	(0.6)	0.2	(2.1)	0.1	(0.4)
Slovak	0.3	(0.2)	0.0	(-0.1)	0.0	(0.0)	**0.0**	**(-1.0)**	0.0	(-0.2)	0.1	(0.0)	0.1	(0.8)	0.1	(0.4)	0.1	(1.4)	0.2	(-0.4)	0.3	(3.4)	0.1	(0.4)	0.1	(1.1)	0.1	(1.8)	0.0	(-0.3)
Croatian and Slovenian	0.1	(-0.6)	0.0	(-0.1)	0.0	(-0.3)	**0.0**	**(-1.0)**	0.0	(-0.6)	**0.0**	**(-1.0)**	0.0	(-0.4)	0.0	(-0.4)	0.2	(-0.8)	0.2	(1.1)	0.0	(-0.5)	0.0	(-0.2)	0.0	(-0.8)	0.1	(1.3)	0.0	(-0.7)
Czech †	0.3	(0.1)	0.0	(-0.1)	1.1	(3.7)	**0.0**	**(-1.0)**	0.0	(-0.6)	**0.0**	**(-1.2)**	0.0	(-0.9)	0.0	(-0.3)	0.1	(0.9)	0.1	(0.1)	0.0	(-0.5)	0.0	(-0.2)	0.0	(-0.8)	0.2	(3.3)	0.2	(0.4)
Dalmatian, Bosnian, Herzegovinian	**0.0**	**(-1.0)**	0.0	(-0.1)	0.0	(0.0)	**0.0**	**(-1.0)**	0.0	(0.0)	**0.0**	**(-1.2)**	0.3	(2.2)	0.0	(-0.7)	**0.0**	**(-1.1)**	0.0	(0.0)	0.0	(-0.5)	0.0	(0.0)	0.0	(0.0)	0.0	(0.0)	0.0	(-0.7)
Ruthenian (Russniak)	0.1	(-0.6)	0.1	(-0.1)	0.0	(-0.3)	**0.0**	**(-1.0)**	0.0	(-0.6)	**0.0**	**(-1.2)**	0.1	(0.4)	0.0	(-0.7)	**0.0**	**(-1.1)**	0.0	(-0.4)	0.0	(-0.5)	0.2	(1.2)	0.0	(-0.1)	0.0	(-0.5)	0.0	(-0.7)
Roumanian	0.3	(0.0)	0.0	(-0.1)	0.0	(-0.3)	**0.0**	**(-1.0)**	0.0	(-0.6)	**0.0**	**(-1.2)**	0.0	(-0.3)	0.1	(1.0)	0.0	(-0.4)	0.0	(0.0)	0.0	(-0.5)	0.0	(-0.2)	0.0	(-0.8)	0.0	(-0.5)	0.0	(-0.7)
Bulgarian, Serbian, Montenegrin	0.5	(0.9)	0.3	(0.1)	0.0	(-0.3)	**0.0**	**(-1.0)**	0.0	(0.1)	0.0	(-0.8)	0.1	(-0.3)	0.1	(0.5)	0.1	(1.4)	0.0	(-0.4)	0.0	(-0.5)	0.0	(-0.2)	0.0	(-0.8)	0.0	(-0.5)	0.0	(-0.8)
Greek	0.7	(1.6)	0.5	(0.2)	0.4	(1.2)	**0.0**	**(-1.0)**	0.1	(1.5)	0.1	(1.2)	0.1	(0.7)	0.1	(0.1)	0.1	(1.4)	0.0	(0.0)	0.0	(0.0)	0.0	(0.0)	0.0	(-0.1)	0.0	(-0.5)	0.3	(0.5)
Armenian	0.2	(-0.3)	0.1	(-0.1)	0.1	(0.1)	**0.0**	**(-1.0)**	0.1	(0.8)	0.0	(-0.5)	0.2	(1.4)	0.4	(4.4)	0.1	(2.1)	0.1	(1.0)	0.1	(0.5)	0.1	(0.4)	0.1	(1.2)	0.0	(-0.5)	0.0	(-0.4)
Hebrew	0.2	(-0.1)	0.1	(-0.1)	0.1	(0.1)	0.1	(3.3)	**0.1**	**(1.3)**	0.2	(1.8)	0.2	(1.9)	0.2	(0.2)	0.2	(2.1)	0.1	(0.4)	0.1	(0.6)	0.1	(0.3)	0.1	(0.3)	0.1	(0.1)	0.1	(0.8)
Syrian	0.1	(-0.6)	0.1	(-0.1)	0.0	(-0.3)	**0.0**	**(-1.0)**	0.1	(0.9)	0.1	(0.4)	0.0	(-0.9)	0.0	(-0.1)	**0.0**	**(-1.1)**	0.0	(-0.4)	0.0	(-0.5)	0.0	(-0.2)	0.0	(-0.8)	0.0	(-0.5)	0.0	(-0.7)
French	0.1	(-0.6)	0.1	(-0.1)	0.1	(0.0)	**0.0**	**(-1.0)**	0.0	(-0.6)	**0.0**	**(-1.0)**	0.0	(-0.7)	0.0	(-0.3)	0.0	(-0.4)	0.1	(0.2)	0.1	(1.0)	0.1	(0.2)	0.0	(-0.5)	0.0	(-0.5)	0.0	(-0.5)
Turkish	**0.0**	**(-1.0)**	0.2	(0.0)	0.0	(-0.3)	**0.0**	**(-1.0)**	0.0	(-0.6)	**0.2**	**(0.7)**	0.0	(-0.9)	0.0	(-0.7)	**0.0**	**(-1.0)**	0.0	(-0.4)	0.1	(1.0)	0.1	(5.6)	0.0	(-0.8)	0.0	(-0.5)	0.0	(-0.7)
Mexican	0.0	(-0.9)	0.0	(-0.1)	0.0	(-0.2)	**0.0**	**(1.3)**	0.0	(-0.6)	**0.0**	**(-1.2)**	0.0	(-0.8)	0.0	(-0.6)	**0.0**	**(-1.0)**	0.0	(-0.3)	0.0	(-0.4)	0.0	(-0.1)	0.0	(-0.4)	0.0	(-0.1)	0.0	(-0.2)
Spanish American	0.0	(-0.8)	0.0	(-0.1)	0.1	(0.1)	**0.0**	**(0.8)**	0.0	(-0.4)	**0.0**	**(-1.2)**	0.0	(-0.9)	0.0	(-0.5)	**0.0**	**(-1.0)**	0.0	(0.0)	0.0	(-0.5)	0.0	(0.0)	0.0	(-0.8)	0.0	(-0.3)	0.0	(-0.7)
Cuban	0.0	(-0.9)	0.0	(-0.1)	0.0	(-0.3)	**0.0**	**(0.8)**	0.0	(-0.6)	**0.0**	**(-1.2)**	0.0	(-0.9)	0.0	(-0.7)	0.0	(-0.3)	0.0	(-0.4)	0.0	(-0.5)	0.0	(-0.2)	0.0	(-0.8)	0.0	(-0.3)	0.0	(-0.7)
West Indian	0.1	(-0.6)	0.3	(0.0)	0.0	(-0.3)	**0.0**	**(-1.0)**	0.1	(0.6)	**0.0**	**(-1.2)**	0.0	(-0.9)	0.0	(-0.3)	**0.0**	**(-1.1)**	0.0	(-0.4)	0.0	(-0.5)	0.0	(-0.2)	0.0	(-0.8)	0.0	(-0.5)	0.0	(-0.7)
Chinese	0.2	(-0.3)	0.0	(0.0)	0.1	(-0.1)	**0.0**	**(-1.0)**	0.0	(-0.3)	**0.0**	**(-1.1)**	0.0	(-0.8)	0.0	(-0.7)	0.0	(-0.3)	0.2	(1.6)	0.0	(-0.3)	0.2	(1.0)	0.2	(3.7)	0.0	(-0.3)	0.0	(0.0)
East Indian	**0.0**	**(-1.0)**	0.0	(-0.1)	1.8	(4.2)	**0.0**	**(-1.0)**	0.0	(-0.6)	**0.0**	**(-1.2)**	0.3	(2.9)	0.0	(-0.7)	**0.0**	**(-1.1)**	0.0	(-0.4)	0.0	(-0.5)	0.0	(-0.2)	0.0	(-0.8)	0.0	(-0.5)	0.0	(-0.7)
Japanese	**0.0**	**(-1.0)**	0.0	(-0.1)	0.1	(-0.1)	**0.0**	**(-1.0)**	0.0	(-0.6)	**0.0**	**(-1.2)**	0.0	(-0.9)	0.0	(-0.6)	**0.0**	**(-1.1)**	0.0	(0.0)	0.0	(-0.5)	0.0	(-0.2)	0.0	(-0.8)	0.0	(-0.5)	0.0	(-0.7)
Korean	**0.0**	**(-1.0)**	0.0	(-0.1)	0.0	(-0.3)	**0.0**	**(-1.0)**	0.0	(-0.6)	**0.0**	**(-1.2)**	0.0	(-0.9)	0.0	(-0.7)	**0.0**	**(-1.1)**	0.0	(-0.4)	0.0	(-0.5)	0.0	(-0.2)	0.0	(-0.8)	0.0	(-0.5)	0.0	(-0.7)
Pacific Islander	**0.0**	**(-1.0)**	10.0	(6.2)	0.0	(-0.3)	**0.0**	**(0.3)**	0.0	(-0.4)	0.0	(-0.7)	0.0	(-0.8)	0.0	(-0.4)	0.0	(-0.9)	0.2	(1.1)	0.0	(-0.5)	0.0	(0.0)	0.0	(-0.8)	0.1	(0.6)	0.0	(-0.7)
African (black)	0.1	(-0.7)	0.1	(-0.1)	0.1	(0.1)	**0.0**	**(0.3)**	0.0	(-0.4)	0.0	(-0.7)	0.0	(-0.8)	0.0	(-0.4)	0.0	(0.0)	0.0	(-0.4)	0.0	(-0.5)	0.0	(-0.2)	0.0	(-0.8)	0.1	(0.1)	0.0	(-0.7)
Other	1.3	(3.7)	0.8	(0.4)	0.0	(-0.3)	**0.0**	**(-1.0)**	0.3	(4.9)	0.1	(0.4)	0.0	(-0.9)	0.0	(-0.7)	**0.0**	**(-1.1)**	0.0	(-0.4)	0.1	(0.5)	0.0	(-0.2)	0.0	(-0.8)	0.2	(3.1)	0.1	(1.0)

Bold = 1 or more NSDMR below mean   [shaded] = 1 to 1.99 NSDMRs above mean   [dark shaded] = 2 or more NSDMRs above mean

* Norwegians, Danes, and Swedes.

† Bohemians and Moravians.

TABLE 6-19. Deviations in Certification Rate for Venereal Diseases, by Race, 1916–1930, Expressed as Percent Certified and NSDMR

Race or People	1916 %	1916 NSDMR	1917 %	1917 NSDMR	1918 %	1918 NSDMR	1919 %	1919 NSDMR	1920 %	1920 NSDMR	1921 %	1921 NSDMR	1922 %	1922 NSDMR	1923 %	1923 NSDMR	1924 %	1924 NSDMR	1925 %	1925 NSDMR	1926 %	1926 NSDMR	1927 %	1927 NSDMR	1928 %	1928 NSDMR	1929 %	1929 NSDMR	1930 %	1930 NSDMR
English	0.0	-0.1	0.1	-0.1	0.0	-0.2	0.0	-0.1	0.0	-0.3	0.0	-0.4	0.0	(0.0)	0.0	-0.2	0.0	-0.6	0.0	-0.1	0.0	-0.4	0.0	-0.1	0.0	-0.1	0.0	-0.4	0.0	-0.3
Irish	0.1	(0.0)	0.1	-0.1	0.2	(0.0)	0.0	-0.1	0.0	-0.3	0.0	-0.5	0.0	(0.0)	0.0	-0.1	0.0	-0.5	0.0	-0.1	0.0	-0.3	0.0	-0.2	0.0	-0.1	0.0	-0.4	0.0	-0.3
Scotch	0.0	-0.2	0.1	-0.1	0.0	(0.0)	0.0	-0.2	0.0	-0.3	0.0	-0.5	0.0	(0.0)	0.0	-0.2	0.0	-0.7	0.0	-0.2	0.0	-0.3	0.0	-0.1	0.0	-0.1	0.0	-0.4	0.0	-0.3
Welsh	0.0	-0.4	0.1	(0.0)	0.0	-0.2	0.0	-0.2	0.0	-0.4	0.0	(0.1)	0.1	(0.0)	0.0	-0.3	0.0	-0.2	0.0	-0.2	0.0	-0.5	0.1	(0.0)	0.0	-0.1	0.1	-0.1	0.0	-0.3
Spanish	0.1	(0.2)	0.1	-0.1	0.0	(0.3)	0.0	-0.1	0.0	(0.1)	0.0	(0.1)	0.2	(0.1)	0.2	(0.0)	0.2	(0.3)	0.1	(0.3)	0.1	(0.2)	0.4	(0.6)	0.5	(0.5)	0.1	-0.3	0.1	(0.4)
Portuguese	0.0	-0.1	0.1	-0.1	0.0	(0.4)	0.0	(0.8)	0.1	(0.4)	0.1	(1.2)	0.0	(0.1)	0.0	(0.9)	0.2	(1.7)	0.1	(0.3)	0.1	(0.5)	0.4	(0.6)	0.5	(0.8)	0.1	(0.1)	0.1	(0.9)
Italian (North)	0.0	-0.1	0.1	(0.0)	0.0	(0.4)	0.0	(0.2)	0.0	(0.0)	0.0	-0.3	0.0	-0.1	0.0	-0.2	0.0	-0.3	0.1	(0.0)	0.0	-0.2	0.0	-0.2	0.1	-0.1	0.1	-0.4	0.2	-0.2
Italian (South)	0.0	-0.1	0.1	-0.1	0.1	-0.2	0.1	(0.1)	0.0	(0.0)	0.0	(0.1)	0.2	(0.1)	0.1	(0.2)	0.1	(0.5)	0.4	(0.9)	0.1	-0.2	0.1	-0.2	0.0	-0.1	0.0	-0.4	0.0	-0.3
Dutch and Flemish	0.0	-0.1	0.1	(0.0)	0.0	-0.2	0.0	(0.1)	0.0	(0.0)	0.0	-0.4	0.1	-0.1	0.1	-0.1	0.1	-0.4	0.0	-0.1	0.1	-0.4	0.1	(0.1)	0.0	-0.1	0.0	-0.4	0.0	-0.3
Finnish	0.0	-0.1	0.1	-0.1	0.0	(0.7)	0.0	(0.0)	0.1	(0.9)	0.0	(0.9)	0.0	(0.1)	0.0	(0.1)	0.0	(0.3)	0.0	(0.6)	0.1	(0.0)	0.6	(1.1)	0.2	(0.1)	0.1	(0.2)	0.0	-0.3
Scandinavian*	0.0	-0.2	0.1	-0.1	0.0	-0.7	0.1	(0.0)	0.0	(0.9)	0.0	(0.6)	0.1	(0.1)	0.1	-0.1	0.1	-0.4	0.3	(0.6)	0.1	-0.1	0.1	-0.2	0.1	(0.1)	0.1	(0.2)	0.0	-0.3
Lithuanian	0.0	-0.4	0.0	-0.2	0.1	(1.4)	0.1	(2.1)	0.2	(0.2)	0.0	-0.7	0.0	(0.0)	0.0	(0.0)	0.0	-0.7	0.0	-0.2	0.0	-0.5	0.0	-0.2	0.0	-0.1	0.0	-0.5	0.1	(0.5)
Russian	0.1	(0.1)	0.2	(0.1)	0.1	(0.2)	0.1	(0.4)	0.1	(0.3)	0.0	-0.7	0.1	(0.0)	0.1	(0.1)	0.1	-0.1	0.1	(0.3)	0.0	-0.2	0.2	(0.2)	0.0	-0.1	0.0	-0.3	0.0	-0.2
Polish	0.0	-0.4	0.2	(0.1)	0.0	-0.1	0.0	(0.2)	0.0	(0.1)	0.0	-0.4	0.1	(0.0)	0.1	(0.1)	0.0	(0.0)	0.1	(0.0)	0.0	(0.0)	0.0	-0.1	0.1	(0.0)	0.0	-0.4	0.0	-0.2
German	0.0	-0.4	0.0	(0.0)	0.0	-0.3	0.0	-0.2	0.0	(0.0)	0.0	-0.2	0.0	(0.0)	0.0	(0.1)	0.0	(0.0)	0.0	(0.3)	0.0	-0.1	0.0	-0.2	0.0	(0.0)	0.0	-0.4	0.0	-0.2
Magyar	0.0	-0.4	0.0	-0.2	0.0	-0.3	0.0	-0.2	0.0	-0.4	0.0	-0.1	0.0	(0.0)	0.1	(0.4)	0.1	(0.2)	0.2	(0.3)	0.0	-0.1	0.0	-0.2	0.1	(0.0)	0.1	-0.5	0.1	(0.1)
Slovak	0.0	-0.4	0.0	-0.2	0.0	(0.0)	0.0	(0.7)	0.0	(0.0)	0.0	-0.5	0.2	(0.1)	0.1	(0.1)	0.1	(0.1)	0.0	-0.2	0.1	(0.5)	0.2	(0.1)	0.2	(0.1)	0.1	(0.1)	0.0	-0.3
Croatian and Slovenian	0.0	-0.4	0.0	-0.2	0.0	-0.3	0.1	(2.2)	0.1	(0.6)	0.0	-0.2	0.1	(0.1)	0.0	(0.1)	0.0	(0.4)	0.0	-0.2	0.0	-0.5	0.1	(0.1)	0.0	-0.1	0.0	-0.1	0.0	-0.3
Czech †	0.0	-0.4	0.0	-0.2	0.0	-0.3	0.0	-0.2	0.2	(1.7)	0.0	-0.7	0.0	-0.1	0.0	-0.3	0.0	-0.5	0.0	-0.2	0.0	-0.5	0.0	-0.1	0.1	-0.1	0.0	-0.5	0.0	-0.3
Dalmatian, Bosnian, Herzegovinian	0.0	-0.4	0.0	(1.0)	0.0	(0.0)	0.0	-0.2	0.0	(0.0)	0.0	-0.7	0.0	-0.1	0.0	-0.3	0.0	-0.7	0.0	-0.7	0.0	-0.5	0.0	(0.0)	0.0	(0.0)	0.0	-0.5	0.0	-0.3
Ruthenian (Russniak)	0.0	-0.4	0.0	(1.7)	0.0	-0.3	0.0	-0.2	0.0	(5.0)	0.0	-0.7	0.0	(0.0)	0.1	(0.3)	0.0	-0.7	0.0	-0.2	0.0	(0.3)	0.0	(0.3)	0.0	-0.1	0.0	-0.5	0.3	(1.7)
Roumanian	0.0	-0.4	0.0	-0.2	0.0	-0.3	0.1	(1.5)	0.0	(0.2)	0.0	(0.7)	0.1	(0.1)	0.2	(0.9)	0.1	(1.4)	0.8	(1.7)	0.1	(0.5)	0.0	(0.3)	0.0	(0.7)	0.0	-0.5	0.5	(2.9)
Bulgarian, Serbian, Montenegrin	0.0	-0.2	0.0	(0.1)	0.0	-0.3	0.0	(0.8)	0.0	(0.8)	0.0	-0.4	0.0	(0.0)	0.1	(0.2)	0.0	-0.2	0.0	(0.7)	0.0	-0.5	0.0	-0.2	0.0	(0.3)	0.0	-0.5	0.2	(1.0)
Greek	0.0	-0.1	0.2	(0.1)	0.0	(0.3)	0.0	(0.9)	0.0	(0.1)	0.0	(0.0)	0.1	(0.0)	0.1	(0.3)	0.0	(0.2)	0.0	(0.2)	0.0	-0.5	0.0	(0.0)	0.0	(0.0)	0.1	-0.2	0.1	(0.3)
Armenian	0.7	(4.6)	0.3	(0.4)	0.0	(0.3)	0.1	(0.7)	0.1	(0.1)	0.1	(0.0)	0.0	(0.0)	0.0	(0.3)	0.0	(0.2)	0.1	(0.1)	0.0	-0.4	0.1	(0.0)	0.1	(0.0)	0.0	(0.0)	0.1	(0.3)
Hebrew	0.1	(0.0)	0.0	(0.0)	0.0	-0.1	0.0	(0.0)	0.0	(0.1)	0.0	(0.7)	0.0	(0.0)	0.0	-0.1	0.0	-0.3	0.0	(0.1)	0.0	-0.4	0.0	-0.1	0.0	(0.0)	0.0	-0.4	0.0	-0.2
Syrian	0.2	(1.4)	0.1	(0.0)	0.0	-0.1	0.0	-0.2	0.0	(0.2)	0.0	-0.3	0.1	(0.0)	0.0	(0.4)	0.0	-0.1	0.0	(0.2)	0.0	-0.5	0.0	-0.2	0.2	(0.1)	0.0	-0.5	0.0	-0.3
French	0.1	(0.0)	0.0	(0.0)	0.0	-0.2	0.0	(0.0)	0.0	-0.2	0.0	-0.3	0.0	(0.0)	0.0	-0.2	0.0	-0.4	0.1	-0.1	0.1	-0.1	0.0	-0.1	0.0	(0.0)	0.0	-0.3	0.0	-0.1
Turkish	0.4	(2.8)	0.2	(0.2)	0.0	-0.3	0.0	(0.0)	0.0	(0.2)	0.0	-0.7	0.9	(0.6)	0.0	-0.3	0.0	-0.7	1.1	(2.4)	0.4	(3.6)	0.0	-0.2	0.7	(0.7)	0.0	-0.5	0.0	(7.8)
Mexican	0.0	(0.0)	0.0	-0.2	0.0	-0.1	0.0	(0.5)	0.0	(0.2)	0.2	(2.5)	0.0	(0.0)	0.0	-0.1	0.0	(0.6)	0.0	(0.1)	0.1	(0.9)	0.2	(0.3)	0.3	(0.3)	0.2	(3.1)	0.6	(0.6)
Spanish American	0.2	(0.7)	0.0	-0.2	0.0	-0.1	0.0	(0.0)	0.0	(0.1)	0.0	-0.7	0.0	(0.0)	0.0	(0.8)	0.0	(0.8)	0.4	(0.7)	0.1	(0.1)	0.2	(0.1)	0.1	(0.0)	0.1	-0.1	0.1	(0.3)
Cuban	0.1	(0.0)	0.0	-0.2	0.0	-0.3	0.0	(0.0)	0.0	-0.4	0.0	(0.9)	0.0	(0.0)	0.0	(0.1)	0.0	-0.4	0.0	-0.2	0.0	-0.5	0.0	-0.2	0.0	-0.1	0.0	-0.4	0.0	(0.0)
West Indian	0.1	(0.4)	0.0	-0.1	0.0	-0.1	0.0	-0.2	0.0	-0.3	0.1	(0.0)	0.2	(0.1)	0.1	(0.4)	0.1	(0.3)	0.0	-0.2	0.1	(0.5)	0.0	-0.2	0.0	-0.1	0.0	-0.5	0.0	(0.0)
Chinese	0.2	(1.2)	0.3	(0.3)	0.1	(0.1)	0.0	-0.2	0.0	-0.3	0.0	(0.0)	0.0	(0.0)	0.0	(0.4)	0.0	(0.4)	0.1	(1.1)	0.4	(2.8)	0.0	(3.0)	0.0	(1.8)	0.2	(0.5)	0.2	(1.0)
East Indian	0.0	-0.4	3.2	(5.9)	1.3	(1.5)	0.0	-0.2	0.0	-0.4	0.2	(3.1)	0.0	-0.1	1.1	(5.5)	0.3	(3.6)	0.4	(3.3)	0.5	(2.8)	2.0	(1.6)	1.5	-0.1	0.4	(1.7)	0.4	(2.5)
Japanese	0.2	(0.7)	0.0	(0.0)	0.0	-0.2	0.0	-0.1	0.0	-0.3	0.0	-0.7	0.0	(0.5)	0.0	-0.3	0.0	(0.8)	0.0	-0.2	0.0	-0.5	0.0	-0.2	0.0	(1.1)	0.0	(0.0)	0.0	-0.2
Korean	0.0	-0.4	0.0	-0.2	0.0	(4.2)	0.0	(0.0)	0.0	-0.4	0.0	(0.0)	0.7	(6.3)	0.0	-0.3	0.0	-0.7	0.0	(0.0)	0.0	-0.5	0.0	(0.0)	4.5	(5.6)	0.9	(4.3)	0.0	-0.3
Pacific Islander	0.0	-0.4	0.0	(0.0)	0.0	(5.6)	0.0	-0.2	0.0	-0.4	0.0	(3.2)	8.0	(6.3)	0.0	(1.5)	0.0	(2.6)	0.0	(2.0)	0.0	(3.7)	0.0	(3.7)	0.0	(2.0)	0.0	(0.0)	0.0	(0.7)
African (black)	0.3	(2.0)	0.8	(1.3)	0.6	(0.6)	0.1	(0.1)	0.2	(1.7)	0.2	(3.2)	0.4	(0.3)	0.4	(1.5)	0.2	(2.6)	0.9	(2.0)	0.4	(3.7)	2.6	(3.7)	1.7	(2.0)	0.2	(3.1)	0.3	(1.8)
Other	0.1	(0.2)	0.0	(0.0)	0.3	(0.1)	0.2	(0.3)	0.2	-0.4	0.1	(1.3)	0.1	(0.1)	0.3	(2.3)	0.1	(1.5)	0.2	(0.2)	0.0	-0.5	0.0	-0.2	0.0	-0.1	0.1	(0.1)	0.3	(1.6)

Bold = 1 or more NSDMR below mean = 1 to 1.99 NSDMRs above mean = 2 or more NSDMRs above mean

* Norwegians, Danes, and Swedes.
† Bohemians and Moravians.

TABLE 6-20. Deviations in Certification Rate for Class B and C Conditions, by Race, 1916–1930, Expressed as Percent Certified and NSDMR

Race or People	1916 %	1916 NSDMR	1917 %	1917 NSDMR	1918 %	1918 NSDMR	1919 %	1919 NSDMR	1920 %	1920 NSDMR	1921 %	1921 NSDMR	1922 %	1922 NSDMR	1923 %	1923 NSDMR	1924 %	1924 NSDMR	1925 %	1925 NSDMR	1926 %	1926 NSDMR	1927 %	1927 NSDMR	1928 %	1928 NSDMR	1929 %	1929 NSDMR	1930 %	1930 NSDMR
English	3.5	(-0.1)	4.5	(0.0)	5.7	(0.4)	2.4	(-0.1)	1.6	(-0.3)	2.0	(-0.6)	2.5	(-0.4)	2.1	(-0.4)	1.6	(-0.6)	4.2	(-0.2)	3.1	(-0.2)	4.9	(-0.1)	5.3	(-0.1)	2.4	(-0.3)	2.7	(-0.2)
Irish	4.5	(0.1)	4.8	(0.0)	7.3	(0.9)	4.1	(0.7)	2.0	(0.1)	2.3	(-0.5)	3.5	(-0.2)	2.8	(-0.3)	2.0	(-0.4)	5.5	(-0.1)	7.8	(0.5)	5.1	(0.0)	3.7	(-0.2)	2.5	(-0.3)	2.5	(-0.2)
Scotch	3.9	(0.0)	4.4	(0.0)	5.6	(0.4)	3.3	(0.4)	1.9	(0.0)	2.3	(-0.4)	2.7	(-0.4)	2.1	(-0.4)	1.8	(-0.5)	3.6	(-0.3)	2.7	(-0.2)	4.2	(-0.2)	4.2	(-0.2)	2.3	(-0.3)	1.7	(-0.4)
Welsh	3.2	(-0.1)	6.3	(0.9)	8.3	(1.2)	2.1	(-0.1)	2.9	(0.9)	2.1	(-0.6)	2.1	(-0.5)	2.5	(-0.4)	2.2	(-0.3)	4.8	(-0.2)	5.2	(0.1)	6.8	(0.2)	5.1	(0.0)	2.3	(-0.4)	1.1	(-0.5)
Spanish	2.0	(-0.3)	2.1	(-0.7)	2.1	(-0.7)	0.6	(-0.8)	0.9	(-0.9)	1.3	(-1.0)	1.5	(-0.6)	1.3	(-0.9)	1.9	(-0.5)	11.4	(0.7)	1.3	(-0.4)	8.8	(0.4)	10.3	(0.5)	1.4	(-0.7)	2.1	(-0.3)
Portuguese	4.2	(0.0)	6.5	(1.0)	3.2	(-0.4)	1.3	(-0.5)	1.3	(-0.6)	1.6	(-0.9)	2.2	(-0.5)	2.2	(-0.5)	1.8	(-0.5)	10.1	(0.5)	2.0	(-0.3)	10.1	(0.6)	6.4	(0.1)	1.8	(-0.5)	1.9	(-0.3)
Italian (North)	3.2	(-0.1)	3.7	(-0.4)	5.5	(0.3)	2.0	(0.0)	1.5	(-0.4)	1.1	(-1.1)	1.7	(-0.7)	0.6	(-1.3)	0.7	(-1.0)	4.2	(-0.2)	1.4	(-0.4)	2.2	(-0.4)	3.6	(-0.2)	3.5	(-0.1)	9.0	(1.1)
Italian (South)	7.0	(0.4)	7.2	(1.3)	7.3	(0.9)	1.9	(-0.2)	3.1	(1.1)	4.1	(0.5)	6.7	(0.5)	5.3	(0.9)	4.4	(0.6)	22.7	(2.1)	6.5	(0.3)	13.4	(1.0)	10.6	(0.6)	6.6	(1.1)	8.1	(0.9)
Dutch and Flemish	3.2	(-0.1)	3.4	(-0.6)	4.3	(0.0)	1.6	(-0.4)	1.8	(-0.1)	1.8	(-0.8)	2.4	(-0.5)	2.1	(-0.4)	1.6	(-0.6)	4.9	(-0.2)	3.0	(-0.2)	7.9	(0.3)	9.4	(0.4)	3.5	(-0.1)	3.2	(-0.1)
Finnish	2.4	(-0.3)	2.7	(-0.9)	3.4	(-0.3)	2.1	(-0.1)	2.5	(0.6)	3.3	(0.1)	4.6	(0.0)	3.1	(-0.1)	3.1	(0.0)	6.3	(0.0)	2.4	(-0.2)	12.7	(0.7)	11.6	(0.7)	2.8	(-0.2)	3.2	(-0.3)
Scandinavian*	2.9	(-0.2)	3.6	(-0.5)	5.0	(0.2)	2.2	(-0.1)	1.8	(-0.1)	2.5	(-0.4)	3.6	(-0.2)	2.2	(-0.5)	2.6	(-0.2)	5.0	(-0.1)	3.2	(-0.1)	5.7	(0.0)	7.0	(0.2)	3.7	(0.1)	3.6	(0.0)
Lithuanian	3.6	(-0.1)	6.5	(1.0)	6.7	(0.7)	4.3	(0.8)	2.6	(0.7)	5.6	(1.3)	7.0	(0.5)	4.7	(0.6)	6.6	(1.6)	22.5	(2.1)	10.7	(0.9)	18.0	(1.5)	25.2	(2.2)	8.7	(1.8)	9.6	(1.2)
Russian	2.9	(-0.2)	3.0	(-0.7)	4.0	(-0.1)	2.4	(0.0)	2.0	(0.1)	2.7	(-0.3)	2.7	(-0.4)	3.3	(-0.1)	3.3	(0.1)	5.7	(-0.1)	2.7	(-0.2)	7.7	(0.3)	6.3	(0.1)	2.7	(-0.2)	2.5	(-0.2)
Polish	4.0	(0.0)	4.7	(0.1)	5.2	(0.2)	3.7	(0.6)	0.6	(-1.3)	2.1	(-0.6)	5.9	(0.3)	5.6	(1.0)	6.0	(1.3)	13.1	(0.9)	9.3	(0.7)	12.4	(0.8)	12.5	(0.8)	11.0	(2.7)	6.0	(0.5)
German	3.9	(0.0)	4.5	(0.0)	6.9	(0.8)	5.9	(1.5)	2.2	(0.3)	3.9	(0.4)	6.1	(0.3)	5.0	(0.8)	4.9	(0.8)	7.4	(0.3)	5.6	(0.2)	4.5	(-0.1)	5.1	(0.0)	2.6	(-0.3)	3.0	(-0.1)
Magyar	5.2	(0.1)	3.9	(-0.1)	7.9	(1.1)	6.9	(2.0)	5.1	(3.0)	4.9	(0.9)	6.7	(0.5)	4.7	(0.6)	7.1	(1.8)	12.7	(0.8)	8.6	(0.6)	17.7	(1.5)	12.6	(0.8)	6.1	(1.0)	7.1	(0.7)
Slovak	3.3	(-0.1)	3.6	(-0.5)	0.0	(0.0)	4.1	(0.7)	1.9	(1.5)	5.2	(1.1)	10.4	(1.3)	2.6	(0.6)	10.1	(3.1)	42.2	(4.5)	46.3	(5.5)	46.0	(4.9)	16.3	(1.2)	6.9	(1.3)	6.0	(0.3)
Croatian and Slovenian	3.8	(0.0)	4.3	(-0.1)	13.2	(2.7)	10.9	(3.7)	2.0	(0.1)	1.1	(-1.2)	4.5	(0.0)	6.1	(1.3)	6.8	(1.7)	33.3	(3.3)	12.9	(1.1)	36.7	(3.6)	24.9	(2.2)	12.4	(3.1)	0.4	(-0.6)
Czech †	5.9	(0.3)	5.3	(0.4)	7.7	(1.0)	3.6	(0.5)	1.7	(-0.2)	0.8	(-1.3)	1.1	(-0.7)	1.7	(-0.8)	4.1	(0.5)	0.9	(-0.7)	0.1	(-0.5)	0.4	(-0.6)	1.6	(-0.4)	0.5	(-1.0)	26.7	(3.4)
Dalmatian, Bosnian, Herzegovinian	3.3	(-0.1)	3.0	(-0.3)	0.0	(0.0)	9.1	(2.9)	0.0	(0.0)	0.4	(-1.5)	4.9	(0.1)	1.4	(-0.9)	5.2	(0.9)	0.0	(0.0)	1.9	(-0.3)								
Ruthenian (Russniak)	4.4	(0.0)	6.3	(0.9)	10.1	(1.8)	2.4	(0.0)	3.3	(1.3)	2.3	(-0.5)	2.6	(-0.4)	3.8	(0.2)	1.6	(-0.6)	3.0	(-0.4)	1.7	(0.0)	8.5	(0.4)	8.8	(0.4)	2.6	(-0.2)	1.2	(-0.5)
Bulgarian, Serbian, Montenegrin	5.0	(0.1)	6.7	(1.1)	4.1	(-0.1)	3.2	(0.3)	1.2	(-0.7)	2.5	(-0.4)	5.1	(0.1)	5.7	(1.1)	5.9	(1.2)	16.6	(1.3)	7.9	(0.5)	18.5	(1.6)	18.3	(1.4)	5.5	(0.7)	9.5	(1.2)
Greek	4.3	(0.0)	3.9	(-0.3)	7.2	(0.9)	2.7	(0.1)	1.4	(-0.5)	1.2	(-1.1)	4.5	(0.1)	3.4	(0.0)	1.8	(-0.5)	5.8	(0.0)	1.9	(-0.3)	6.8	(0.2)	6.2	(0.1)	1.5	(-0.6)	20.6	(3.4)
Armenian	3.9	(0.0)	3.5	(-0.5)	5.7	(0.4)	1.8	(-0.3)	2.3	(0.4)	3.1	(-0.1)	4.5	(0.3)	3.8	(0.2)	4.1	(0.5)	12.1	(0.8)	3.2	(-0.1)	8.4	(0.4)	6.1	(0.1)	3.2	(-0.1)	3.7	(0.0)
Hebrew	6.9	(0.4)	6.9	(1.2)	5.0	(0.2)	4.4	(0.9)	3.6	(1.6)	4.4	(0.6)	5.7	(0.3)	6.4	(1.4)	6.2	(1.4)	9.2	(0.4)	5.7	(1.3)	7.3	(0.2)	5.9	(0.1)	4.3	(0.3)	4.4	(0.2)
Syrian	10.3	(0.9)	7.2	(1.4)	13.9	(3.0)	2.6	(0.1)	3.3	(1.3)	6.9	(2.0)	10.8	(1.3)	9.1	(2.7)	7.4	(1.9)	17.7	(1.5)	14.2	(1.3)	12.7	(0.9)	11.9	(0.7)	8.4	(1.7)	10.0	(1.3)
French	3.3	(-0.1)	3.3	(-0.6)	6.4	(0.6)	3.2	(0.3)	1.9	(0.0)	4.8	(0.8)	2.7	(-0.1)	2.4	(-0.5)	4.3	(0.6)	10.9	(0.6)	6.1	(0.2)	8.2	(0.3)	4.0	(-0.2)	2.4	(-0.3)	1.7	(-0.4)
Turkish	7.0	(0.4)	6.1	(0.8)	3.8	(-0.2)	1.9	(0.2)	3.0	(3.0)	2.3	(0.4)	10.6	(1.3)	4.0	(0.3)	6.1	(1.3)	5.3	(-0.1)	2.9	(-0.2)	8.9	(0.4)	7.0	(0.2)	1.3	(-0.7)	4.0	(0.1)
Mexican	3.9	(0.0)	3.0	(-0.8)	2.2	(-0.7)	4.8	(1.0)	1.5	(-0.3)	7.7	(2.4)	1.2	(-0.7)	1.1	(-1.0)	1.3	(-0.7)	2.9	(-0.1)	1.9	(-0.3)	2.5	(-0.4)	3.1	(-0.3)	4.4	(0.4)	4.5	(0.2)
Spanish American	1.4	(-0.4)	0.8	(-1.9)	1.1	(-1.0)	0.6	(-0.8)	0.7	(-1.1)	1.4	(-1.0)	0.9	(-0.8)	0.9	(-1.1)	1.2	(-0.8)	2.7	(-0.4)	0.9	(-0.4)	2.5	(-0.4)	2.3	(-0.4)	0.6	(-0.9)	0.6	(-0.6)
Cuban	0.7	(-0.5)	0.5	(-1.9)	1.0	(-1.1)	0.2	(-1.0)	0.2	(-1.6)	0.2	(-1.7)	0.5	(-0.9)	0.7	(-1.2)	0.7	(-1.0)	4.2	(-0.2)	0.5	(-0.5)	2.2	(-0.4)	1.7	(-0.4)	0.6	(-1.0)	0.3	(-0.6)
West Indian	2.7	(-0.2)	1.7	(-1.4)	1.9	(-0.8)	0.7	(-0.7)	0.4	(-1.5)	0.2	(-1.6)	0.8	(-0.8)	0.4	(-1.3)	0.4	(-1.1)	1.1	(-0.6)	0.6	(-0.5)	1.6	(-0.5)	2.3	(-0.4)	0.5	(-1.0)	0.6	(-0.6)
Chinese	7.4	(0.5)	6.0	(0.8)	5.7	(0.4)	1.2	(-1.0)	0.8	(-1.1)	0.8	(-1.3)	4.4	(0.0)	4.1	(0.4)	3.1	(0.1)	10.1	(0.5)	2.0	(-0.3)	13.0	(0.5)	5.3	(0.0)	1.7	(-0.5)	0.6	(-0.5)
East Indian	8.1	(0.6)	8.4	(2.0)	11.3	(2.1)	0.2	(-1.0)	0.8	(-1.5)	0.6	(-1.4)	1.9	(-0.6)	1.7	(-0.7)	3.3	(0.1)	4.6	(-0.2)	2.4	(-0.3)	5.9	(0.1)	16.8	(1.3)	2.1	(-0.4)	2.1	(-0.3)
Japanese	3.7	(-0.1)	4.7	(0.1)	3.9	(-0.3)	1.6	(-0.4)	1.7	(-0.4)	1.5	(-1.0)	2.9	(-0.3)	1.7	(-0.7)	2.0	(-0.4)	9.5	(0.4)	0.5	(-0.5)	3.3	(-0.3)	11.7	(0.7)	0.8	(-0.9)	3.1	(-0.1)
Korean	1.8	(-0.3)	3.6	(-0.4)	5.5	(0.3)	1.0	(-0.7)	2.4	(0.5)	0.0	(-1.8)	2.1	(-0.5)	1.4	(-0.9)	0.6	(-1.0)	0.0	(0.0)	1.2	(-0.4)	2.1	(-0.4)	18.2	(1.4)	5.4	(0.7)	3.1	(-0.1)
Pacific Islander	-44.4	(-6.1)	10.0	(2.8)	12.5	(2.5)	0.0	(0.0)	3.6	(1.6)	0.0	(-1.8)	1.6	(-0.6)	2.3	(-0.5)	1.9	(-0.5)	10.0	(0.5)	1.9	(-0.3)	6.8	(0.2)	50.0	(5.6)	1.0	(-0.8)	14.3	(2.1)
African (black)	2.9	(-0.2)	2.4	(-1.1)	2.3	(-0.7)	0.9	(-0.7)	0.6	(-1.2)	1.0	(-1.2)	1.6	(-0.6)	1.3	(-0.8)	1.3	(-0.7)	7.9	(0.2)	1.7	(-0.2)	6.8	(0.2)	5.1	(0.0)	1.0	(-0.8)	1.1	(-0.5)
Other	4.3	(0.0)	4.4	(0.0)	2.3	(-0.7)	3.1	(0.3)	3.1	(1.1)	1.9	(-0.7)	2.5	(-0.4)	5.3	(0.9)	4.2	(0.5)	10.2	(0.5)	4.5	(0.0)	11.4	(0.7)	11.2	(0.6)	6.5	(1.2)	8.2	(0.9)

Bold = 1 or more NSDMR below mean   = 1 to 1.99 NSDMRs above mean   = 2 or more NSDMRs above mean

* Norwegians, Danes, and Swedes.
† Bohemians and Moravians.

TABLE 6-21. Deviations in Rate of Deportation for Those Certified with Class B and C Conditions, by Race, 1916–1930, Expressed as Percent Deported and NSDMR

Race or People	1916 %	1916 NSDMR	1917 %	1917 NSDMR	1918 %	1918 NSDMR	1919 %	1919 NSDMR	1920 %	1920 NSDMR	1921 %	1921 NSDMR	1922 %	1922 NSDMR	1923 %	1923 NSDMR	1924 %	1924 NSDMR	1925 %	1925 NSDMR	1926 %	1926 NSDMR	1927 %	1927 NSDMR	1928 %	1928 NSDMR	1929 %	1929 NSDMR	1930 %	1930 NSDMR
English	12.9	(0.0)	14.5	(0.2)	4.5	(-0.2)	3.6	(-0.2)	3.6	(0.3)	2.3	(0.1)	3.5	(0.2)	3.1	(0.0)	3.5	(0.2)	2.0	(0.0)	1.4	(0.0)	0.9	(0.0)	0.1	(-0.3)	0.2	(-0.2)	0.0	(-0.2)
Irish	15.3	(0.2)	17.7	(0.5)	5.4	(-0.1)	5.4	(-0.1)	2.6	(-0.1)	3.0	(0.4)	4.8	(0.4)	2.8	(0.2)	3.4	(0.2)	2.1	(0.0)	0.5	(-0.1)	0.9	(0.0)	0.0	(-0.3)	0.1	(-0.3)	0.0	(-0.1)
Scotch	12.1	(-0.1)	17.5	(0.5)	5.8	(0.0)	5.2	(-0.1)	2.1	(-0.1)	3.3	(0.0)	2.7	(-0.1)	3.3	(0.1)	2.7	(0.0)	2.0	(-0.1)	2.0	(0.0)	0.9	(0.0)	0.2	(-0.2)	0.0	(-0.3)	0.1	(-0.2)
Welsh	18.2	(0.4)	9.4	(-0.3)	4.0	(-0.2)	4.8	(-0.1)	3.1	(0.0)	1.9	(0.0)	6.3	(0.8)	3.4	(0.1)	4.1	(0.4)	1.5	(-0.1)	1.7	(0.1)	2.2	(0.0)	0.0	(-0.3)	0.0	(-0.3)	3.2	(1.8)
Spanish	26.7	(1.0)	16.1	(0.4)	21.9	(1.7)	13.7	(0.4)	8.4	(0.9)	10.1	(2.9)	7.3	(1.0)	4.8	(0.3)	5.9	(0.9)	1.5	(-0.2)	1.0	(0.2)	1.1	(0.0)	1.1	(0.0)	0.0	(-0.3)	0.0	(-0.3)
Portuguese	9.7	(-0.3)	6.2	(-0.6)	2.7	(-0.4)	3.6	(-0.2)	0.9	(-0.3)	1.9	(0.0)	1.1	(-0.4)	4.8	(0.3)	3.4	(0.2)	4.5	(0.5)	4.1	(0.5)	4.7	(0.5)	7.1	(3.7)	0.0	(-0.3)	0.6	(0.6)
Italian (North)	10.6	(-0.2)	9.6	(-0.3)	8.2	(0.2)	4.9	(-0.1)	1.6	(-0.2)	1.8	(-0.1)	12.1	(2.1)	16.3	(2.2)	0.2	(2.1)	3.5	(0.0)	3.5	(0.3)	1.8	(-0.1)	2.1	(0.8)	0.0	(-0.3)	0.0	(-0.2)
Italian (South)	11.1	(-0.2)	10.8	(-0.2)	2.0	(-0.4)	3.4	(-0.2)	0.6	(-0.4)	1.8	(-0.1)	3.2	(-0.1)	2.4	(-0.1)	2.1	(-0.2)	1.6	(0.0)	3.5	(-0.1)	1.8	(-0.1)	2.1	(-0.1)	0.1	(-0.3)	0.1	(-0.2)
Dutch and Flemish	8.6	(-0.3)	6.9	(-0.5)	1.0	(-0.5)	2.2	(-0.3)	1.9	(-0.2)	1.0	(-0.2)	2.6	(-0.1)	2.1	(-0.1)	3.3	(0.2)	2.3	(0.1)	0.5	(-0.1)	0.8	(0.0)	0.7	(-0.1)	0.4	(-0.1)	0.0	(-0.2)
Finnish	10.7	(-0.3)	10.0	(-0.2)	6.2	(0.0)	0.0	(-0.4)	0.0	(-0.5)	0.0	(-0.5)	0.0	(-0.5)	2.1	(-0.1)	2.0	(-0.2)	5.3	(0.6)	4.4	(0.5)	0.8	(0.1)	0.5	(0.1)	0.0	(-0.3)	0.0	(-0.2)
Scandinavian*	6.4	(-0.5)	6.1	(-0.7)	1.6	(-0.5)	1.7	(-0.3)	0.9	(-0.3)	1.0	(-0.4)	1.9	(0.0)	0.9	(-0.3)	1.1	(-0.2)	1.5	(-0.1)	0.5	(-0.1)	0.6	(0.0)	0.2	(-0.3)	0.2	(-0.3)	0.0	(-0.2)
Lithuanian	4.2	(-0.7)	5.0	(-0.5)	4.5	(-0.2)	0.0	(-0.4)	0.0	(-0.5)	0.0	(-0.5)	0.0	(-0.7)	0.0	(-0.3)	0.0	(-0.6)	0.0	(-0.1)	0.0	(-0.1)	0.2	(0.0)	0.5	(0.1)	1.6	(1.0)	0.0	(-0.2)
Russian	12.5	(-0.5)	16.7	(0.4)	4.5	(-0.1)	1.9	(-0.3)	4.7	(0.3)	5.0	(1.1)	2.3	(0.0)	0.9	(-0.3)	1.1	(-0.5)	2.3	(0.0)	1.2	(0.0)	3.1	(0.0)	0.2	(-0.3)	1.3	(0.4)	0.1	(-0.2)
Polish	8.1	(-0.4)	16.9	(0.4)	4.5	(0.0)	4.5	(-0.1)	7.7	(0.3)	0.6	(-0.5)	1.2	(-0.4)	1.9	(-0.2)	1.8	(-0.3)	1.8	(0.0)	0.4	(-0.2)	0.3	(-0.1)	0.0	(-0.3)	0.0	(-0.3)	0.0	(-0.2)
German	7.5	(-0.4)	11.6	(-0.1)	6.5	(0.0)	4.8	(-0.1)	3.7	(0.1)	3.7	(0.1)	2.0	(-0.2)	1.5	(-0.3)	1.5	(-0.2)	0.3	(-0.3)	0.2	(-0.1)	0.3	(-0.1)	0.1	(-0.3)	0.0	(-0.3)	0.0	(-0.2)
Magyar	0.0	(-1.0)	5.6	(-0.7)	0.0	(-0.6)	0.0	(-0.4)	0.0	(-0.5)	1.0	(-0.1)	3.2	(-0.1)	0.5	(-0.4)	2.4	(-0.1)	1.5	(-0.1)	0.4	(-0.2)	0.0	(-0.1)	0.0	(-0.3)	0.0	(-0.3)	0.0	(-0.2)
Slovak	0.0	(-1.0)	0.0	(0.0)	0.0	(-0.6)	0.0	(-0.5)	0.0	(-0.5)	0.0	(-0.2)	0.2	(-0.2)	1.5	(-0.3)	3.3	(-0.1)	0.3	(-0.3)	0.2	(-0.1)	0.0	(-0.1)	0.1	(-0.3)	0.1	(-0.3)	0.0	(-0.2)
Croatian and Slovenian	3.0	(-0.8)	21.4	(0.9)	0.0	(-0.6)	20.0	(0.8)	4.0	(0.2)	4.1	(0.8)	4.6	(0.4)	1.1	(-0.3)	3.3	(-0.1)	0.7	(-0.2)	0.0	(0.0)	0.0	(-0.1)	0.4	(-0.1)	0.0	(-0.3)	1.3	(0.4)
Czech†	2.6	(-0.8)	5.6	(-0.7)	29.6	(2.4)	10.0	(1.2)	10.0	(1.2)			15.0	(2.8)	20.0	(2.8)	1.0	(-0.5)	11.8	(1.7)	40.0	(6.0)	25.0	(2.3)	0.0	(-0.3)	8.3	(4.4)	0.0	(-0.2)
Dalmatian, Bosnian, Herzegovinian	0.0	(-1.0)	0.0	(0.0)	0.0	(0.0)	100.0	(5.6)	13.3	(1.7)	0.0	(-0.7)	0.0	(-0.7)	20.0	(2.8)	0.0	(-0.8)	0.0	(-0.8)	0.0	(0.0)	0.0	(0.0)	0.0	(0.0)	0.0	(0.0)	0.0	(-0.2)
Ruthenian (Russniak)	6.9	(-0.5)	20.2	(0.8)	0.0	(-0.6)	25.0	(1.1)	0.0	(-0.7)	0.0	(-0.7)	0.0	(-0.7)	30.4	(4.5)	4.5	(0.5)	3.1	(0.2)	0.0	(-0.2)	2.6	(0.2)	0.0	(0.0)	0.0	(-0.3)	0.0	(-0.2)
Roumanian	21.2	(0.6)	30.8	(1.8)	14.3	(0.9)	0.0	(-0.5)	1.2	(-0.5)	1.2	(-0.3)	1.0	(-0.4)	2.0	(-0.1)	2.3	(-0.1)	0.0	(-0.3)	0.0	(-0.2)	0.0	(-0.1)	0.0	(-0.3)	0.0	(-0.3)	0.0	(-0.2)
Bulgarian, Serbian, Montenegrin	36.5	(1.8)	27.1	(1.5)	28.6	(2.4)	7.7	(0.1)	4.9	(0.3)	6.9	(1.8)	0.5	(-0.8)	2.5	(-0.1)	5.6	(0.8)	0.0	(-0.3)	3.6	(0.4)	0.0	(0.0)	6.1	(3.0)	0.0	(-0.3)	0.0	(-0.2)
Greek	10.0	(-0.2)	15.5	(0.1)	7.8	(0.2)	3.6	(-0.2)	3.4	(0.1)	5.9	(1.4)	1.8	(0.3)	2.5	(0.1)	1.4	(-0.4)	1.3	(-0.1)	0.0	(-0.2)	0.5	(0.0)	0.6	(0.0)	0.0	(2.3)	0.2	(-0.1)
Armenian	7.6	(-0.4)	5.4	(-0.7)	4.7	(-0.1)	0.0	(-0.5)	0.0	(-0.5)	1.1	(-0.3)	1.3	(-0.4)	0.6	(-0.4)	0.9	(-0.5)	0.9	(-0.2)	0.6	(-0.2)	0.4	(0.0)	0.1	(-0.3)	0.0	(-0.3)	0.2	(-0.1)
Hebrew	7.2	(-0.4)	10.4	(-0.4)	2.4	(-0.4)	3.5	(-0.4)	1.2	(-0.4)	0.6	(-0.5)	1.6	(-0.3)	1.8	(-0.4)	1.6	(-0.3)	4.8	(0.0)	0.6	(-0.4)	1.9	(0.0)	2.4	(1.0)	1.5	(0.5)	0.0	(-0.2)
Syrian	20.0	(0.5)	17.4	(0.5)	5.1	(-0.1)	8.9	(0.1)	6.7	(0.6)	1.7	(-0.1)	13.4	(2.4)	5.2	(0.4)	7.6	(1.4)	0.7	(-0.2)	4.6	(0.5)	1.0	(-0.1)	0.0	(-0.3)	0.5	(0.0)	0.0	(-0.2)
French	20.0	(0.5)	12.5	(0.0)	0.0	(0.0)	0.0	(0.0)	0.0	(-0.5)	0.0	(-0.7)	0.0	(-0.7)	0.0	(-0.5)	0.0	(-0.8)	5.3	(0.0)	0.0	(-0.2)	0.0	(-0.1)	0.0	(-0.3)	0.0	(0.0)	0.0	(-0.2)
Turkish	56.3	(3.2)	50.0	(3.7)	0.0	(-0.6)	20.9	(0.9)	33.3	(5.1)	3.2	(0.4)	8.6	(1.3)	7.7	(0.8)	8.1	(1.6)	20.0	(3.2)	2.0	(0.2)	1.3	(0.1)	3.8	(1.8)	4.7	(2.7)	0.0	(-0.2)
Mexican	3.7	(-0.7)	15.2	(0.3)	17.0	(1.2)	3.1	(-0.2)	1.5	(-0.2)	4.3	(0.8)	2.6	(-0.1)	6.0	(0.5)	2.4	(-0.1)	4.2	(0.4)	2.0	(0.2)	2.5	(0.2)	1.3	(0.4)	3.2	(1.5)	1.9	(0.9)
Spanish American	0.0	(-1.0)	4.8	(-0.8)	0.0	(-0.6)	0.0	(-0.4)	0.0	(-0.5)	0.0	(-0.7)	4.0	(0.2)	5.1	(0.4)	3.2	(0.1)	1.5	(-0.1)	7.9	(1.1)	5.9	(0.1)	3.9	(2.9)	1.5	(3.2)	2.0	(4.0)
Cuban	16.7	(0.3)	7.7	(-0.5)	8.3	(0.1)	0.0	(-0.4)	0.0	(-0.5)	0.0	(-0.7)	4.0	(0.5)	5.1	(0.5)	3.2	(0.5)	5.0	(0.5)	0.0	(-0.2)	2.5	(0.2)	0.0	(-0.3)	2.0	(2.2)	4.0	(4.5)
West Indian	11.5	(-0.1)	4.2	(-1.1)	21.4	(1.6)	0.0	(-0.4)	0.0	(-0.5)	0.0	(-0.7)	0.0	(-0.5)	7.1	(0.7)	5.6	(0.8)	14.0	(3.0)	0.0	(-0.2)	50.0	(5.8)	0.0	(-0.3)	4.2	(4.5)	7.7	(4.5)
Chinese	0.0	(-1.0)	1.5	(-1.1)	0.0	(-0.6)	0.0	(-0.4)	7.3	(0.7)	1.0	(-0.3)	0.0	(-0.6)	0.7	(-0.4)	0.6	(-0.6)	0.4	(-0.3)	2.2	(0.2)	0.0	(-0.1)	0.0	(-0.3)	0.0	(-0.3)	0.0	(-0.2)
East Indian	30.0	(1.3)	12.5	(0.0)	0.0	(-0.6)	0.0	(-0.4)	0.0	(-0.5)	0.0	(-0.7)	0.0	(-0.7)	0.0	(-0.5)	0.9	(-0.5)	5.3	(0.0)	0.0	(-0.2)	0.0	(-0.1)	0.0	(-0.3)	0.0	(0.0)	0.0	(-0.2)
Japanese	2.1	(-0.8)	0.7	(-1.2)	0.5	(-0.6)	0.0	(-0.4)	0.7	(-0.5)	0.0	(-0.7)	0.0	(-0.7)	0.0	(-0.5)	0.0	(-0.8)	0.0	(-0.3)	0.0	(-0.2)	0.0	(-0.1)	0.0	(-0.3)	0.0	(0.0)	0.0	(-0.2)
Korean	0.0	(-1.0)	0.0	(0.0)	0.0	(-0.6)	0.0	(0.0)	0.0	(-0.5)	0.0	(0.0)	0.0	(-0.7)	0.0	(-0.5)	0.0	(-0.5)	0.0	(-0.3)	0.0	(-0.2)	0.0	(0.0)	0.0	(-0.3)	0.0	(0.0)	0.0	(-0.2)
Pacific Islander	0.0	(-1.0)	0.0	(0.0)	0.0	(-0.6)	0.0	(0.0)	0.0	(-0.5)	0.0	(0.0)	0.0	(-0.7)	0.0	(-0.5)	0.0	(-0.5)	0.0	(-0.3)	0.0	(-0.2)	0.0	(0.0)	0.0	(-0.3)	0.0	(0.0)	0.0	(-0.2)
African (black)	29.0	(1.2)	27.1	(1.5)	33.3	(2.9)	1.7	(-0.3)	8.4	(0.1)	6.9	(1.8)	9.9	(1.6)	14.4	(1.9)	18.9	(4.6)	19.4	(3.0)	7.7	(1.0)	4.6	(0.5)	0.0	(-0.3)	0.0	(0.0)	1.9	(1.0)
Other	59.7	(2.8)	19.4	(0.7)	25.0	(1.1)	0.0	(-0.4)	3.4	(0.1)	0.0	(1.8)	6.5	(0.8)	3.1	(0.0)	3.4	(0.0)	3.1	(0.2)	5.3	(0.6)	2.2	(0.2)	1.9	(0.7)	0.0	(0.0)	0.0	(-0.2)

Bold = 1 or more NSDMR below mean
= 1 to 1.99 NSDMRs above mean
= 2 or more NSDMRs above mean
* Norwegians, Danes, and Swedes.
† Bohemians and Moravians.

TABLE 6-22. Deviations in Deportation Rate for Class B and C Conditions, by Race, 1908–1930, Expressed as Percent Deported and NSDMR

Race or People	1908 %	1908 NSDMR	1909 %	1909 NSDMR	1910 %	1910 NSDMR	1911 %	1911 NSDMR	1912 %	1912 NSDMR	1913 %	1913 NSDMR	1914 %	1914 NSDMR	1915 %	1915 NSDMR	1916 %	1916 NSDMR	1917 %	1917 NSDMR	1918 %	1918 NSDMR	1919 %	1919 NSDMR	1920 %	1920 NSDMR
English	0.1	(-0.1)	0.1	(0.0)	0.0	(-0.1)	0.1	(-0.4)	0.0	(-0.4)	0.1	(-0.4)	0.1	(-1.0)	0.1	(-0.3)	0.5	(-0.1)	0.7	(0.2)	0.3	(0.0)	0.1	(0.0)	0.1	(0.0)
Irish	0.1	(0.0)	0.1	(0.1)	0.0	(0.0)	0.1	(-0.3)	0.1	(-0.3)	0.1	(-0.4)	0.1	(-0.8)	0.2	(-0.2)	0.7	(0.2)	0.9	(0.7)	0.4	(0.2)	0.2	(0.0)	0.1	(0.0)
Scotch	0.1	(0.0)	0.0	(0.0)	0.1	(0.3)	0.1	(-0.4)	0.1	(-0.4)	0.1	(-0.4)	0.1	(-1.0)	0.1	(-0.2)	0.5	(-0.1)	0.8	(0.5)	0.3	(0.1)	0.2	(0.0)	0.0	(-0.1)
Welsh	0.0	(-0.1)	0.0	(0.0)	0.1	(1.0)	0.3	(-0.1)	0.1	(-0.4)	0.2	(-0.2)	0.1	(-1.1)	0.2	(-0.1)	0.6	(0.1)	0.6	(0.1)	0.3	(0.1)	0.1	(0.0)	0.0	(0.2)
Australian	0.0	(0.0)	0.0	(0.0)	0.0	(0.0)	0.0	(0.0)	0.0	(0.0)	0.0	(0.0)	0.0	(0.0)	0.0	(0.0)	0.0	(0.0)	0.3	(0.0)	0.4	(0.3)	0.1	(0.0)	0.0	(0.0)
Spanish	0.1	(-0.1)	0.0	(0.0)	0.0	(-0.1)	0.0	(-0.5)	0.0	(-0.4)	0.0	(-0.2)	0.3	(-0.5)	0.4	(0.1)	0.5	(0.0)	0.3	(-0.5)	0.4	(-0.3)	0.1	(0.0)	0.1	(0.1)
Portuguese	0.0	(0.0)	0.0	(0.0)	0.0	(-0.2)	0.1	(-0.4)	0.1	(-0.3)	0.1	(-0.4)	0.3	(-0.5)	0.4	(0.1)	0.4	(-0.2)	0.4	(-0.3)	0.4	(-0.3)	0.1	(0.0)	0.0	(-0.3)
Italian (North)	0.1	(0.0)	0.0	(0.0)	0.0	(-0.1)	0.1	(-0.1)	0.1	(-0.3)	0.1	(-0.4)	0.4	(-0.3)	0.3	(0.1)	0.4	(-0.3)	0.4	(-0.4)	0.4	(-0.3)	0.0	(-0.1)	0.0	(-0.2)
Italian (South)	0.1	(0.0)	0.0	(0.0)	0.0	(-0.2)	0.5	(0.4)	0.2	(0.4)	0.3	(0.2)	0.7	(0.6)	0.4	(0.2)	0.8	(0.3)	0.5	(0.5)	0.1	(0.1)	0.0	(0.0)	0.0	(-0.2)
Dutch and Flemish	0.1	(0.0)	0.0	(0.0)	0.0	(-0.2)	0.0	(-0.4)	0.0	(-0.4)	0.1	(-0.4)	0.4	(-0.9)	0.4	(-0.4)	0.3	(-0.3)	0.2	(-0.7)	0.2	(-0.3)	0.0	(0.0)	0.0	(-0.1)
Finnish	0.0	(-0.1)	0.0	(0.0)	0.0	(0.0)	0.1	(-0.5)	0.0	(-0.3)	0.1	(-0.4)	0.1	(-0.8)	0.4	(-0.4)	0.3	(-0.4)	0.3	(-0.6)	0.2	(-0.1)	0.0	(0.0)	0.0	(-0.3)
Scandinavian*	0.0	(-0.1)	0.0	(0.0)	0.0	(-0.1)	0.0	(-0.5)	0.0	(-0.4)	0.0	(-0.5)	0.1	(-0.9)	0.3	(-0.3)	0.3	(-0.2)	0.3	(-0.6)	0.2	(-0.1)	0.0	(0.0)	0.0	(-0.2)
Lithuanian	0.1	(0.0)	0.0	(0.0)	0.0	(-0.1)	0.2	(-0.3)	0.1	(-0.3)	0.1	(-0.3)	0.2	(-0.7)	0.3	(-0.3)	0.4	(-0.2)	0.4	(-0.3)	0.6	(0.5)	0.0	(0.0)	0.0	(-0.3)
Russian	0.1	(0.0)	0.0	(0.0)	0.0	(-0.2)	0.3	(-0.3)	0.2	(-0.2)	0.2	(-0.3)	0.4	(-0.7)	0.2	(-0.2)	0.4	(-0.2)	0.5	(-0.1)	0.2	(0.0)	0.1	(0.0)	0.1	(0.2)
Polish	0.1	(0.0)	0.0	(0.0)	0.0	(-0.2)	0.3	(-0.1)	0.2	(-0.1)	0.4	(-0.2)	0.4	(-0.2)	0.3	(-0.3)	0.3	(-0.5)	0.8	(0.6)	0.2	(0.0)	0.2	(0.0)	0.0	(-0.1)
German	0.1	(-0.1)	0.0	(0.0)	0.0	(-0.1)	0.1	(-0.3)	0.1	(-0.3)	0.1	(-0.6)	0.1	(-0.6)	0.1	(-0.3)	0.3	(-0.3)	0.5	(0.0)	0.4	(-0.3)	0.3	(0.1)	0.1	(0.2)
Magyar	0.1	(0.0)	0.0	(0.0)	0.0	(-0.2)	0.3	(0.1)	0.3	(0.0)	0.3	(-0.1)	0.6	(0.0)	0.4	(0.1)	0.2	(-0.7)	0.2	(-0.7)	0.0	(-0.3)	0.0	(-0.1)	0.0	(-0.3)
Slovak	0.2	(0.1)	0.0	(0.0)	0.0	(-0.2)	0.3	(0.1)	0.3	(0.3)	0.4	(-0.2)	0.4	(-0.2)	0.7	(0.6)	0.3	(-0.7)	0.0	(-1.2)	0.0	(-0.3)	0.0	(0.4)	0.0	(-0.3)
Croatian & Slovenian	0.1	(0.0)	0.0	(0.0)	0.0	(-0.2)	0.3	(0.1)	0.3	(0.2)	3.7	(4.9)	0.5	(0.2)	0.3	(0.0)	0.2	(-0.5)	0.3	(0.6)	0.2	(-0.3)	2.2	(1.4)	0.1	(0.2)
Transylvanian (Siebenburger)	0.0	(0.0)	0.0	(0.0)	0.0	(0.0)	0.0	(-0.3)	0.1	(0.0)	0.0	(0.0)	0.2	(-0.5)	0.1	(-0.3)	0.1	(-0.5)	0.3	(0.8)	0.0	(-0.3)	0.3	(0.1)	0.2	(0.2)
Czech †	0.1	(0.0)	0.0	(0.0)	0.1	(0.1)	0.2	(0.0)	0.1	(-0.3)	0.2	(0.0)	0.2	(0.0)	0.0	(0.0)	0.0	(-0.5)	0.0	(0.0)	0.0	(0.0)	0.1	(0.1)	0.2	(0.7)
Dalmatian, Bosnian, Herzegovinian	0.1	(0.0)	0.0	(0.0)	0.0	(-0.2)	0.3	(0.0)	0.4	(0.4)	0.3	(-0.1)	0.5	(0.1)	0.6	(0.5)	0.0	(-0.7)	0.0	(-1.2)	2.2	(2.6)	9.1	(6.2)	0.4	(2.33)
Ruthenian (Russniak)	0.1	(0.0)	0.0	(0.0)	0.0	(-0.2)	0.3	(0.0)	0.3	(0.4)	0.4	(0.3)	0.6	(0.5)	0.4	(0.2)	0.3	(-0.3)	1.3	(1.6)	0.0	(-0.3)	0.6	(0.3)	0.0	(-0.3)
Austro-Hungarian	0.0	(0.0)	0.0	(0.0)	0.0	(0.0)	0.0	(0.0)	0.0	(0.0)	0.0	(0.0)	0.0	(0.0)	0.0	(0.0)	0.0	(0.0)	0.0	(0.0)	0.0	(0.0)	0.0	(0.0)	0.0	(0.0)
Roumanian	0.1	(0.0)	0.0	(0.0)	0.0	(-0.1)	0.6	(0.5)	0.3	(0.3)	0.9	(0.7)	1.1	(1.6)	0.9	(1.0)	1.0	(0.7)	2.1	(3.3)	0.6	(0.4)	0.0	(-0.1)	0.0	(-0.3)
Bulgarian, Serbian, Montenegrin	0.2	(0.1)	0.0	(0.1)	0.1	(0.1)	0.6	(0.7)	0.5	(0.6)	0.7	(0.5)	0.8	(0.8)	0.5	(0.5)	1.6	(1.4)	1.0	(1.1)	2.1	(2.4)	0.2	(0.1)	0.0	(0.1)
Greek	0.2	(0.1)	0.0	(0.1)	0.0	(0.0)	1.4	(2.3)	1.0	(1.6)	1.9	(2.3)	0.8	(0.8)	1.1	(1.4)	0.8	(0.8)	0.5	(0.0)	0.8	(0.8)	0.1	(0.0)	0.1	(0.1)
Armenian	0.1	(0.0)	0.0	(0.1)	0.1	(0.1)	0.5	(0.5)	0.5	(0.6)	0.5	(0.4)	0.4	(0.4)	0.4	(0.4)	0.7	(0.2)	0.1	(-1.0)	0.8	(0.8)	0.0	(-0.1)	0.0	(-0.3)
Hebrew	0.1	(0.0)	0.0	(0.1)	0.0	(0.0)	0.4	(0.0)	0.3	(0.2)	0.6	(0.1)	0.6	(0.3)	0.3	(0.0)	0.4	(0.3)	0.3	(-0.6)	0.3	(0.3)	0.2	(0.0)	0.0	(-0.1)
Arabian	0.0	(0.0)	0.0	(0.0)	0.0	(0.0)	0.0	(0.0)	0.0	(0.0)	0.0	(0.0)	0.0	(0.0)	0.0	(0.0)	0.0	(0.0)	0.0	(0.0)	0.0	(0.0)	0.0	(0.0)	0.0	(0.0)
Syrian	0.0	(-0.1)	0.0	(0.0)	0.0	(-0.1)	0.0	(-0.6)	0.0	(-0.5)	0.2	(-0.2)	0.0	(-1.1)	0.0	(-0.4)	0.7	(0.3)	0.7	(0.4)	0.3	(0.1)	0.0	(-0.1)	0.0	(-0.3)
Turkish	0.2	(0.2)	0.0	(0.0)	0.0	(-0.2)	0.8	(1.2)	0.6	(0.8)	0.0	(0.0)	0.0	(0.0)	0.3	(0.1)	3.0	(4.4)	1.5	(2.1)	0.0	(-0.3)	1.0	(0.3)	7.5	(5.6)
French	0.1	(0.0)	0.0	(0.0)	0.0	(-0.1)	0.2	(-0.3)	0.1	(-0.3)	0.1	(-0.3)	0.1	(-0.8)	0.3	(-0.3)	0.7	(0.2)	0.6	(0.1)	0.3	(0.1)	0.3	(0.1)	0.0	(0.0)
Swiss	0.0	(0.0)	0.0	(0.0)	0.0	(0.0)	0.0	(0.0)	0.0	(0.0)	0.0	(0.0)	0.0	(0.0)	0.0	(0.0)	0.0	(0.0)	0.0	(0.0)	0.0	(0.0)	0.0	(0.0)	0.0	(0.0)
Mexican	0.2	(0.1)	0.0	(0.0)	0.0	(0.2)	0.1	(-0.5)	0.0	(-0.5)	0.0	(-0.5)	0.1	(-1.0)	0.0	(-0.3)	0.1	(-0.5)	0.0	(0.8)	0.4	(0.2)	1.0	(0.6)	0.0	(0.5)
Central American (NS)	0.0	(0.0)	0.0	(0.0)	0.0	(0.0)	0.0	(-0.6)	2.1	(-4.5)	0.0	(-0.5)	0.0	(-1.1)	0.0	(0.0)	0.0	(-0.7)	0.9	(-1.0)	0.8	(0.8)	0.1	(0.0)	0.0	(0.0)
South American	0.0	(0.0)	0.0	(0.0)	0.0	(0.0)	0.0	(0.0)	0.0	(0.0)	0.0	(0.0)	0.0	(0.0)	0.0	(0.0)	0.0	(0.0)	0.0	(0.0)	0.0	(0.0)	0.0	(0.0)	0.0	(0.0)
Spanish American	0.0	(-0.1)	0.0	(0.0)	0.0	(-0.1)	0.0	(-0.6)	0.0	(-0.5)	0.0	(-0.4)	0.0	(-1.1)	0.0	(-0.4)	0.7	(0.3)	0.0	(-1.1)	0.0	(-0.3)	0.0	(-0.1)	0.0	(-0.3)
Cuban	0.0	(-0.1)	0.0	(0.0)	0.0	(-0.1)	0.0	(-0.4)	0.2	(0.2)	0.1	(-0.2)	0.0	(-1.1)	0.0	(-0.4)	0.1	(-0.5)	0.1	(-1.1)	0.1	(-0.4)	0.1	(0.1)	0.0	(-0.3)
Chinese	0.0	(-0.1)	0.0	(0.0)	0.0	(-0.2)	0.1	(-0.6)	0.2	(0.0)	0.1	(-0.4)	0.1	(-0.9)	0.1	(0.0)	0.3	(-0.3)	0.7	(0.4)	0.3	(0.1)	0.0	(-0.1)	0.0	(-0.3)
East Indian	5.0	(5.0)	6.9	(6.3)	6.2	(6.2)	2.4	(4.6)	1.5	(3.1)	1.9	(2.2)	1.6	(1.5)	3.6	(5.5)	2.4	(2.5)	1.1	(1.1)	1.1	(1.1)	1.0	(1.0)	7.5	(7.5)
Japanese	0.1	(0.1)	0.1	(0.1)	0.0	(-0.2)	0.0	(-0.6)	0.0	(-0.5)	0.0	(-0.5)	0.0	(-1.0)	0.0	(-0.4)	0.1	(-0.6)	0.0	(-1.1)	0.0	(-0.3)	0.1	(0.6)	0.1	(0.0)
Korean	0.0	(-0.1)	0.0	(0.0)	0.0	(-0.2)	0.0	(-0.6)	0.0	(-0.5)	0.0	(-0.5)	0.0	(-1.1)	0.0	(-0.4)	0.0	(-0.7)	0.0	(-1.1)	0.0	(-0.3)	0.0	(-0.1)	0.0	(-0.3)
Filipino	0.0	(0.0)	0.0	(0.0)	0.0	(-0.2)	0.0	(0.0)	0.0	(0.0)	0.0	(-0.5)	0.0	(-1.1)	0.0	(0.0)	0.0	(-0.7)	0.0	(-1.2)	0.0	(-0.3)	0.0	(-0.1)	0.0	(-0.3)
Hawaiian	0.0	(0.0)	0.0	(0.0)	0.0	(0.0)	0.0	(0.0)	0.0	(0.0)	0.0	(0.0)	0.0	(0.0)	0.0	(0.0)	0.0	(0.0)	0.0	(0.0)	0.0	(0.0)	0.0	(0.0)	0.0	(0.0)
Pacific Islander	0.0	(-0.1)	0.0	(0.0)	0.0	(-0.2)	0.0	(-0.6)	0.0	(-0.5)	0.0	(-0.5)	0.0	(-1.1)	0.0	(-0.4)	0.0	(-0.7)	0.0	(-1.1)	5.2	(5.2)	0.0	(-0.1)	0.0	(0.0)
African (black)	0.0	(0.0)	0.0	(0.0)	0.0	(0.2)	0.1	(-0.4)	0.2	(0.2)	0.1	(-0.4)	0.1	(-0.9)	0.3	(0.0)	0.8	(0.4)	0.6	(0.2)	0.4	(0.2)	0.0	(-0.1)	0.1	(0.0)
(NS)	0.0	(-0.1)	0.0	(0.0)	0.0	(-0.2)	0.0	(-0.6)	0.0	(-0.5)	0.1	(-0.3)	0.1	(-1.0)	0.1	(0.0)	0.7	(0.2)	0.3	(-1.0)	0.3	(0.1)	0.3	(0.1)	0.0	(0.0)
Esquimaux	0.0	(0.0)	0.2	(0.1)	0.0	(0.0)	0.0	(0.0)	0.0	(0.0)	0.0	(0.0)	0.0	(0.0)	0.0	(0.0)	0.1	(0.0)	0.0	(0.0)	0.2	(0.0)	0.0	(0.0)	0.0	(0.0)
Other	0.5	(0.5)	0.2	(0.0)	0.0	(0.0)	0.8	(2.3)	0.8	(1.4)	1.2	(1.3)	1.1	(1.5)	1.6	(2.2)	2.2	(2.1)	0.9	(0.7)	0.6	(0.4)	0.1	(-0.1)	0.1	(0.3)

Bold = 1 or more NSDMR below mean ▨ = 1 to 1.99 NSDMRs above mean ▨ = 2 or more NSDMRs above mean

* Norwegians, Danes, and Swedes. † Bohemians and Moravians. NS = not specified

TABLE 6-22. Continued

Race or People	1921 %	1921 NSDMR	1922 %	1922 NSDMR	1923 %	1923 NSDMR	1924 %	1924 NSDMR	1925 %	1925 NSDMR	1926 %	1926 NSDMR	1927 %	1927 NSDMR	1928 %	1928 NSDMR	1929 %	1929 NSDMR	1930 %	1930 NSDMR
English	0.0	(-0.3)	0.0	(-0.4)	0.1	(-0.3)	0.1	(-0.3)	0.1	(-0.1)	0.0	(-0.1)	0.0	(0.0)	0.0	(-0.3)	0.0	(-0.4)	0.0	(-0.3)
Irish	0.1	(-0.1)	0.2	(-0.4)	0.1	(-0.2)	0.1	(-0.2)	0.1	(0.0)	0.0	(-0.1)	0.0	(0.0)	0.0	(-0.3)	0.0	(-0.4)	0.0	(-0.3)
Scotch	0.0	(-0.3)	0.1	(-0.5)	0.1	(-0.3)	0.0	(-0.4)	0.0	(-0.2)	0.0	(-0.2)	0.0	(-0.2)	0.0	(-0.2)	0.0	(-0.5)	0.0	(-0.5)
Welsh	0.0	(-0.4)	0.1	(0.0)	0.1	(-0.1)	0.1	(0.1)	0.1	(-0.1)	0.1	(0.8)	0.2	(0.7)	0.0	(-0.3)	0.0	(-0.5)	0.0	(-0.5)
Australian	0.0	(0.0)	0.1	(0.0)	0.1	(-0.1)	0.1	(0.1)	0.1	(-0.1)	0.1	(0.8)	0.2	(0.7)	0.0	(0.0)	0.0	(0.0)	0.0	(0.6)
Spanish	0.1	(1.2)	0.1	(-0.2)	0.1	(-0.3)	0.1	(0.0)	0.1	(0.0)	0.1	(0.0)	0.1	(0.3)	0.0	(-0.3)	0.0	(0.0)	0.0	(-0.4)
Portuguese	0.0	(-0.5)	0.0	(-0.9)	0.0	(-0.3)	0.0	(-0.3)	0.5	(1.0)	0.1	(0.7)	0.5	(2.8)	0.5	(-1.3)	0.0	(-0.5)	0.0	(-0.4)
Italian (North)	0.0	(-0.3)	0.1	(0.1)	0.1	(0.1)	0.1	(-0.2)	0.1	(0.1)	0.1	(0.0)	0.0	(0.1)	0.1	(0.4)	0.0	(-0.3)	0.0	(-0.4)
Italian (South)	0.1	(0.2)	0.2	(0.9)	0.1	(0.3)	0.1	(0.2)	0.4	(0.7)	0.0	(-0.4)	0.0	(-0.1)	0.1	(0.3)	0.0	(-0.2)	0.0	(-0.1)
Dutch and Flemish	0.0	(-0.7)	0.1	(-0.7)	0.0	(-0.5)	0.0	(-0.4)	0.0	(0.0)	0.0	(0.0)	0.0	(-0.1)	0.1	(0.3)	0.0	(-0.5)	0.0	(-0.4)
Finnish	0.0	(-0.7)	0.0	(-0.9)	0.0	(-0.6)	0.0	(-0.3)	0.3	(0.6)	0.1	(1.2)	0.1	(0.8)	0.2	(1.5)	0.0	(-0.5)	0.0	(-0.4)
Scandinavian*	0.0	(-0.6)	0.1	(-0.2)	0.1	(-0.6)	0.0	(-0.7)	0.3	(-0.1)	0.0	(-0.5)	0.0	(-0.3)	0.0	(-0.2)	0.0	(-0.3)	0.0	(-0.3)
Lithuanian	0.0	(-1.0)	0.1	(0.0)	0.0	(0.0)	0.0	(-0.6)	0.1	(0.4)	0.1	(1.5)	0.2	(1.3)	0.1	(0.7)	0.1	(2.2)	0.0	(-0.4)
Russian	0.1	(0.4)	0.1	(-0.6)	0.1	(0.1)	0.1	(0.0)	0.1	(0.0)	0.0	(0.0)	0.0	(-0.1)	0.0	(-0.3)	0.0	(0.2)	0.0	(-0.4)
Polish	0.0	(-0.8)	0.1	(-0.6)	0.1	(0.1)	0.1	(0.4)	0.0	(0.0)	0.0	(-0.2)	0.0	(-0.1)	0.0	(-0.3)	0.0	(-0.1)	0.0	(-0.4)
German	0.1	(0.0)	0.1	(-0.1)	0.1	(-0.3)	0.1	(-0.3)	0.1	(0.1)	0.0	(-0.3)	0.0	(-0.2)	0.0	(-0.3)	0.0	(-0.4)	0.0	(-0.3)
Magyar	0.1	(-0.2)	0.1	(-0.6)	0.1	(-0.4)	0.1	(0.0)	0.2	(0.2)	0.0	(-0.5)	0.0	(-0.3)	0.0	(-0.3)	0.0	(-0.4)	0.0	(-0.4)
Slovak	0.0	(-0.4)	0.2	(0.8)	0.2	(0.4)	0.2	(1.2)	0.1	(0.0)	0.1	(1.2)	0.1	(1.0)	0.0	(-0.3)	0.0	(-0.5)	0.0	(-0.4)
Croatian & Slovenian	0.0	(-0.3)	0.2	(3.1)	0.1	(0.4)	0.2	(3.8)	0.0	(0.0)	0.0	(0.0)	0.0	(0.0)	0.0	(0.0)	0.0	(0.4)	0.0	(0.0)
Transylvanian (Siebenburger)	0.0	(0.0)	0.0	(0.0)	0.0	(-0.5)	0.0	(-0.6)	0.0	(0.0)	0.0	(0.1)	0.0	(0.3)	0.0	(0.0)	0.0	(0.0)	0.0	(0.0)
Czech †	0.0	(-1.0)	0.2	(0.4)	0.1	(0.0)	0.1	(0.0)	0.1	(0.0)	0.1	(0.0)	0.1	(0.0)	0.1	(0.0)	0.0	(0.0)	0.0	(0.0)
Dalmatian, Bosnian, Herzegovinian	0.0	(-1.0)	0.0	(-1.3)	0.3	(1.9)	0.0	(-1.1)	1.3	(3.4)	0.0	(-0.8)	0.0	(1.2)	0.0	(0.0)	0.0	(-0.5)	0.0	(0.0)
Ruthenian (Russniak)	0.0	(-1.0)	0.0	(-1.3)	0.5	(3.8)	0.1	(-1.1)	0.1	(-3.1)	0.2	(1.2)	0.0	(0.0)	0.0	(0.0)	0.0	(0.4)	0.0	(-0.4)
Austro-Hungarian	0.0	(0.0)	0.0	(0.0)	0.0	(0.0)	0.0	(0.0)	0.0	(0.0)	0.0	(0.1)	0.0	(0.0)	0.0	(0.0)	0.0	(0.0)	0.0	(0.0)
Roumanian	0.0	(-0.5)	0.1	(-0.7)	0.1	(0.2)	0.1	(0.0)	0.0	(0.0)	0.0	(-0.8)	0.0	(-0.3)	0.0	(-0.3)	0.0	(-0.5)	0.0	(-0.4)
Bulgarian, Serbian, Montenegrin	0.1	(0.3)	0.1	(-0.2)	0.1	(-0.1)	0.1	(0.3)	0.0	(-0.3)	0.1	(0.5)	0.0	(-0.3)	0.4	(3.3)	0.1	(1.2)	0.0	(-0.4)
Greek	0.2	(2.0)	0.1	(-0.5)	0.1	(0.0)	0.1	(-0.3)	0.1	(0.1)	0.0	(-0.8)	0.0	(0.0)	0.0	(0.0)	0.1	(1.6)	0.1	(1.5)
Armenian	0.0	(-0.3)	0.1	(-0.5)	0.0	(-0.6)	0.0	(-0.3)	0.0	(-0.3)	0.0	(-0.8)	0.0	(0.0)	0.0	(-0.2)	0.0	(-0.2)	0.0	(0.2)
Arabian	0.0	(0.0)	0.2	(0.4)	0.2	(0.7)	0.1	(0.6)	0.2	(0.2)	0.1	(0.8)	0.1	(0.1)	0.0	(0.0)	0.0	(0.0)	0.0	(0.2)
Syrian	0.1	(0.3)	0.0	(-0.8)	0.2	(1.4)	0.0	(-1.1)	0.5	(1.2)	0.0	(-0.8)	0.0	(-0.3)	0.2	(1.3)	0.1	(1.3)	0.0	(-0.4)
Turkish	0.2	(3.0)	0.0	(-1.3)	0.3	(2.2)	0.0	(-1.1)	1.1	(2.8)	0.0	(-0.8)	0.3	(4.8)	0.0	(-0.3)	0.0	(-0.5)	0.0	(-0.4)
French	0.2	(2.9)	0.0	(2.3)	0.3	(2.1)	0.1	(0.9)	0.1	(-0.3)	0.1	(1.5)	0.1	(0.2)	0.0	(-0.3)	0.0	(-0.5)	0.0	(-0.4)
Swiss	0.0	(0.0)	0.0	(0.0)	0.0	(0.0)	0.0	(0.0)	0.0	(0.0)	0.0	(0.0)	0.0	(0.0)	0.0	(0.0)	0.0	(0.0)	0.0	(0.0)
Mexican	0.1	(-0.1)	0.1	(-0.3)	0.1	(-0.1)	0.1	(0.4)	0.1	(0.0)	0.0	(-0.1)	0.0	(-0.1)	0.1	(0.8)	0.2	(3.3)	0.2	(5.6)
Central American (NS)	0.0	(0.0)	0.0	(0.0)	0.0	(0.0)	0.0	(0.0)	0.0	(0.0)	0.0	(0.0)	0.0	(0.0)	0.0	(0.0)	0.0	(0.0)	0.0	(0.0)
South American	0.0	(0.0)	0.0	(0.0)	0.0	(-0.4)	0.0	(0.0)	0.0	(-0.3)	0.0	(0.0)	0.0	(-0.3)	0.0	(-0.1)	0.0	(0.0)	0.0	(0.0)
Spanish American	0.0	(-1.0)	0.0	(-1.0)	0.1	(-0.4)	0.0	(-0.7)	0.0	(-0.2)	0.1	(0.5)	0.1	(0.1)	0.1	(0.6)	0.1	(0.1)	0.0	(-0.1)
Cuban	0.0	(-1.0)	0.0	(-1.1)	0.0	(-0.6)	0.0	(-0.8)	0.2	(0.3)	0.0	(0.0)	0.0	(-0.3)	0.1	(0.6)	0.1	(0.1)	0.0	(0.0)
West Indian	0.0	(-1.0)	0.0	(-0.9)	0.0	(-0.6)	0.3	(-0.8)	0.3	(0.5)	0.0	(0.0)	0.3	(4.8)	0.0	(-0.3)	0.0	(-0.5)	0.0	(0.9)
Chinese	0.0	(-0.9)	0.0	(-1.2)	0.0	(-0.8)	0.0	(-0.9)	0.0	(-0.2)	0.0	(0.0)	0.0	(-0.3)	0.0	(-0.3)	0.0	(-0.5)	0.0	(-0.4)
East Indian	0.0	(-1.0)	0.3	(1.9)	0.0	(-1.0)	0.0	(-1.1)	0.0	(-0.3)	0.0	(0.0)	0.0	(-0.3)	0.0	(-0.3)	0.0	(-0.5)	0.0	(-0.4)
Japanese	0.0	(-1.0)	0.0	(-1.3)	0.0	(-1.0)	0.0	(-0.9)	0.5	(1.1)	0.0	(-0.8)	0.0	(-0.3)	0.0	(-0.3)	0.0	(-0.5)	0.0	(-0.4)
Korean	0.0	(-1.0)	0.0	(-1.3)	0.0	(-1.0)	0.0	(-1.1)	0.0	(-0.3)	0.0	(-0.8)	0.0	(-0.3)	0.0	(-0.3)	0.0	(-0.5)	0.0	(-0.4)
Filipino	0.0	(-1.0)	0.0	(-1.3)	0.0	(-1.0)	0.0	(-1.1)	0.0	(-0.3)	0.0	(0.0)	0.0	(-0.3)	0.0	(-0.3)	0.0	(-0.5)	0.0	(-0.4)
Hawaiian	0.0	(0.0)	0.0	(0.0)	0.0	(0.0)	0.0	(0.0)	0.0	(0.0)	0.0	(0.0)	0.0	(0.0)	0.0	(0.0)	0.0	(0.0)	0.0	(0.0)
Pacific Islander	0.0	(-1.0)	0.0	(-1.3)	0.0	(-1.0)	0.0	(-1.1)	0.0	(0.0)	0.0	(0.0)	0.0	(-0.3)	0.0	(-0.3)	0.0	(-0.5)	0.0	(-0.4)
African (black) (NS)	0.1	(0.1)	0.2	(0.3)	0.2	(0.3)	0.3	(2.4)	1.5	(4.2)	0.3	(1.6)	0.3	(1.8)	0.2	(0.8)	0.2	(3.3)	0.2	(0.2)
Esquimauz	0.0	(0.0)	0.0	(0.0)	0.0	(0.0)	0.0	(0.0)	0.0	(0.0)	0.0	(0.0)	0.0	(0.0)	0.0	(0.0)	0.0	(0.0)	0.0	(0.0)
Other	0.1	(1.1)	0.2	(0.4)	0.2	(0.7)	0.1	(0.9)	0.3	(0.6)	0.2	(3.6)	0.3	(1.4)	0.2	(1.7)	0.0	(-0.5)	0.0	(-0.4)

TABLE 6-23. Deviations in Certification Rate for Pregnancy, by Race, 1916–1930, Expressed as Percent Certified and NSDMR

Race or People	1916 %	1916 NSDMR	1917 %	1917 NSDMR	1918 %	1918 NSDMR	1919 %	1919 NSDMR	1920 %	1920 NSDMR	1921 %	1921 NSDMR	1922 %	1922 NSDMR	1923 %	1923 NSDMR	1924 %	1924 NSDMR	1925 %	1925 NSDMR	1926 %	1926 NSDMR	1927 %	1927 NSDMR	1928 %	1928 NSDMR	1929 %	1929 NSDMR	1930 %	1930 NSDMR
English	0.2	(0.0)	0.2	(-0.1)	0.2	(-0.1)	0.1	(0.0)	0.0	(-0.5)	0.1	(-0.4)	0.0	(-0.7)	0.0	(-0.6)	0.0	(-0.8)	0.1	(-0.1)	0.0	(-0.2)	0.1	(-0.2)	0.1	(0.0)	0.0	(-0.2)	0.0	(-0.1)
Irish	0.1	(-0.2)	0.1	(-0.1)	0.2	(0.1)	0.1	(-0.1)	0.0	(-0.3)	0.1	(-0.4)	0.1	(-0.6)	0.1	(-0.2)	0.0	(-0.6)	0.1	(-0.1)	0.1	(-0.2)	0.1	(-0.2)	0.0	(-0.1)	0.0	(-0.2)	0.0	(-0.1)
Scotch	0.2	(-0.1)	0.3	(0.2)	0.2	(-0.1)	0.1	(0.1)	0.0	(-0.4)	0.1	(-0.4)	0.1	(-0.6)	0.0	(-0.6)	0.0	(-0.7)	0.1	(-0.1)	0.1	(-0.2)	0.0	(-0.2)	0.0	(-0.1)	0.0	(-0.2)	0.0	(-0.1)
Welsh	0.1	(-0.3)	0.1	(-0.2)	0.2	(0.2)	0.1	(0.0)	0.0	(-0.3)	0.1	(-0.3)	0.1	(-0.3)	0.1	(-0.3)	0.0	(-0.2)	0.1	(-0.1)	0.1	(0.0)	0.0	(0.0)	0.0	(-0.1)	0.0	(-0.2)	0.0	(-0.2)
Spanish	0.0	(-0.5)	0.1	(-0.3)	0.0	(0.0)	0.0	(0.0)	0.0	(-0.5)	0.0	(-0.6)	0.0	(-0.9)	0.1	(-0.5)	0.1	(-0.2)	0.4	(0.1)	0.1	(0.1)	0.5	(0.4)	0.4	(0.1)	0.1	(0.1)	0.2	(-0.2)
Portuguese	0.1	(-0.2)	0.0	(-0.5)	0.0	(-0.4)	0.0	(-0.3)	0.1	(-0.7)	0.0	(-0.5)	0.0	(-0.8)	0.1	(-0.1)	0.1	(-0.1)	0.5	(0.2)	0.1	(0.4)	0.6	(0.6)	0.2	(0.1)	0.1	(0.2)	0.2	(0.5)
Italian (North)	0.1	(-0.2)	0.0	(-0.5)	0.1	(-0.2)	0.1	(-0.2)	0.1	(-0.2)	0.1	(-0.4)	0.1	(-0.6)	0.3	(-0.7)	0.0	(-1.0)	0.0	(0.0)	0.1	(-0.2)	0.2	(-0.2)	0.0	(0.0)	0.0	(0.0)	0.3	(0.5)
Italian (South)	0.1	(-0.3)	0.1	(-0.3)	0.1	(-0.2)	0.1	(-0.1)	0.2	(0.9)	0.3	(0.5)	0.3	(0.8)	0.3	(0.5)	0.3	(0.3)	1.5	(0.8)	0.3	(0.5)	0.9	(1.0)	0.2	(0.0)	0.0	(0.0)	0.0	(-0.1)
Dutch and Flemish	0.1	(-0.2)	0.2	(0.0)	0.1	(-0.3)	0.1	(0.1)	0.1	(-0.4)	0.0	(-0.3)	0.3	(-0.3)	0.3	(-0.3)	0.2	(-0.2)	0.3	(0.1)	0.3	(0.1)	0.9	(0.0)	0.1	(0.0)	0.1	(0.0)	0.0	(-0.1)
Finnish	0.1	(-0.2)	0.2	(0.1)	0.1	(0.1)	0.1	(0.1)	0.1	(-0.8)	0.1	(-0.3)	0.3	(-0.5)	0.3	(-0.3)	0.3	(0.2)	0.2	(0.0)	0.3	(-0.3)	0.8	(0.9)	0.9	(0.5)	0.1	(0.2)	0.0	(-0.1)
Scandinavian*	0.1	(-0.4)	0.1	(0.1)	0.1	(0.1)	0.0	(-0.1)	0.0	(-0.8)	0.1	(-0.3)	0.3	(-0.3)	0.3	(-0.6)	0.2	(-0.4)	0.0	(0.0)	0.3	(0.9)	0.8	(0.8)	0.9	(0.5)	0.1	(0.2)	0.1	(0.1)
Lithuanian	0.4	(0.8)	0.6	(1.7)	0.6	(0.7)	0.4	(1.1)	0.0	(-0.8)	0.3	(0.3)	0.4	(0.6)	0.4	(0.8)	0.2	(0.7)	1.3	(0.7)	0.5	(1.0)	0.9	(1.1)	0.6	(0.3)	2.9	(1.1)	0.6	(1.1)
Russian	0.3	(0.5)	0.2	(0.1)	0.2	(0.1)	0.3	(0.5)	0.2	(1.2)	0.3	(0.2)	0.3	(0.6)	0.3	(0.5)	0.3	(1.2)	0.2	(0.2)	0.4	(0.3)	0.2	(0.1)	0.2	(0.0)	0.0	(-0.2)	0.1	(0.0)
Polish	1.1	(2.9)	0.9	(1.7)	0.5	(0.5)	0.3	(0.5)	0.2	(1.2)	0.2	(-0.4)	0.5	(1.7)	0.5	(2.0)	0.3	(1.2)	0.5	(0.2)	0.1	(0.0)	0.2	(0.1)	0.1	(0.3)	1.1	(2.9)	0.1	(0.0)
German	0.3	(0.3)	0.3	(0.4)	0.3	(0.2)	0.4	(0.8)	0.1	(0.1)	0.1	(0.1)	0.2	(-0.3)	0.2	(0.0)	0.2	(0.2)	0.1	(0.0)	0.1	(0.0)	0.1	(-0.2)	0.1	(0.0)	0.1	(-0.1)	0.0	(0.0)
Magyar	0.0	(2.1)	0.6	(1.3)	0.4	(4.4)	1.4	(4.1)	0.3	(1.5)	0.3	(0.3)	0.5	(1.7)	0.5	(2.0)	0.3	(2.1)	0.5	(0.2)	0.1	(0.1)	0.9	(1.0)	0.2	(0.2)	0.1	(0.0)	0.2	(0.2)
Slovak	0.2	(-0.1)	0.4	(0.6)	0.0	(0.0)	0.4	(1.4)	0.4	(1.5)	0.4	(0.4)	0.4	(1.4)	0.4	(0.9)	0.3	(1.3)	1.8	(1.0)	1.6	(4.0)	2.4	(3.3)	0.5	(0.2)	0.1	(0.0)	0.1	(0.2)
Croatian and Slovenian	0.5	(0.8)	0.6	(1.2)	0.0	(-0.4)	0.0	(-0.3)	0.2	(0.5)	0.2	(-0.6)	0.2	(-0.1)	0.1	(-0.2)	0.3	(0.8)	2.5	(1.4)	1.1	(2.7)	2.3	(3.2)	2.5	(1.4)	1.3	(3.5)	0.0	(-0.2)
Czech †	0.2	(0.9)	0.9	(2.0)	0.0	(-0.4)	0.0	(-0.3)	0.2	(0.6)	0.0	(-0.6)	0.0	(-0.9)	0.1	(-0.6)	0.1	(-0.1)	0.2	(0.0)	0.0	(-0.3)	0.0	(-0.3)	0.0	(-0.1)	0.0	(-0.1)	0.0	(-0.1)
Dalmatian, Bosnian, Herzegovinian	0.0	(-0.6)	0.0	(-0.5)	0.0	(0.0)	0.0	(-0.3)	0.0	(0.0)	0.0	(-0.6)	0.0	(-1.0)	0.3	(-0.8)	0.3	(0.8)	0.0	(0.0)	0.0	(0.0)	0.0	(0.0)	0.0	(0.0)	0.0	(0.0)	1.0	(2.0)
Ruthenian (Russniak)	1.0	(2.5)	1.5	(3.7)	0.0	(-0.4)	0.0	(-0.3)	0.2	(1.0)	0.5	(1.3)	0.2	(0.2)	0.7	(0.7)	0.3	(1.8)	0.4	(0.1)	0.3	(0.4)	0.7	(0.7)	0.5	(0.2)	0.4	(0.8)	0.0	(-0.2)
Roumanian	0.9	(2.0)	0.3	(0.4)	0.6	(0.7)	0.0	(-0.3)	0.2	(0.5)	0.2	(0.2)	0.6	(2.3)	0.3	(1.6)	0.3	(0.9)	2.3	(1.3)	1.3	(3.1)	2.4	(3.3)	1.4	(0.7)	0.5	(1.4)	1.2	(2.5)
Bulgarian, Serbian, Montenegrin	0.1	(-0.3)	0.1	(-0.3)	0.0	(0.0)	0.0	(-0.3)	0.1	(0.1)	0.1	(-0.3)	0.2	(-0.1)	0.2	(-0.1)	0.3	(0.8)	0.0	(-0.1)	0.1	(0.0)	0.7	(0.7)	0.9	(0.5)	0.0	(-0.2)	1.3	(2.6)
Greek	0.0	(-0.3)	0.0	(-0.4)	1.0	(1.5)	0.0	(-0.1)	0.1	(-0.1)	0.2	(0.1)	0.2	(-0.2)	0.2	(0.3)	0.2	(0.3)	1.1	(0.5)	0.4	(0.6)	1.4	(1.9)	0.5	(0.5)	0.6	(1.5)	0.5	(0.8)
Armenian	0.1	(-0.3)	0.5	(0.8)	0.8	(1.1)	0.1	(0.1)	0.1	(0.3)	0.5	(0.5)	0.3	(0.5)	0.6	(1.9)	0.2	(0.3)	1.0	(0.5)	0.6	(1.3)	1.0	(1.3)	0.2	(0.6)	0.6	(1.5)	0.2	(0.3)
Hebrew	0.5	(0.9)	0.3	(0.4)	0.1	(0.1)	0.4	(1.0)	0.5	(0.0)	0.2	(1.2)	0.3	(0.7)	0.6	(0.6)	0.2	(0.3)	0.2	(0.0)	0.1	(0.0)	0.0	(-0.1)	0.0	(0.0)	0.1	(0.1)	0.1	(0.0)
Syrian	0.7	(1.7)	0.0	(-0.3)	0.0	(-0.4)	0.0	(0.9)	0.5	(3.0)	0.5	(1.2)	0.3	(0.7)	0.5	(1.5)	0.4	(2.0)	1.2	(0.6)	0.9	(2.0)	1.0	(1.3)	1.3	(0.7)	0.7	(1.9)	0.8	(1.6)
French	0.2	(-0.1)	0.2	(-0.2)	0.2	(0.0)	0.1	(-0.1)	0.0	(-0.5)	0.2	(-0.3)	0.1	(-0.5)	0.1	(-0.1)	0.1	(-0.4)	0.1	(0.0)	0.1	(-0.1)	0.1	(-0.1)	0.1	(0.0)	0.0	(0.0)	0.1	(0.0)
Turkish	0.0	(-0.6)	0.0	(0.1)	0.0	(0.0)	0.0	(0.1)	0.0	(-0.8)	0.0	(0.0)	0.0	(0.0)	0.0	(-0.8)	0.2	(0.2)	1.1	(1.1)	0.1	(-0.1)	1.0	(1.1)	0.7	(0.3)	0.0	(0.0)	0.1	(0.0)
Mexican	0.0	(-0.5)	0.0	(-0.5)	0.0	(-0.4)	0.1	(0.1)	0.1	(-0.8)	0.0	(-0.5)	0.0	(-0.4)	0.1	(-0.3)	0.2	(-0.2)	1.1	(1.1)	0.2	(0.0)	0.9	(1.1)	0.3	(0.3)	0.3	(0.5)	0.3	(0.5)
Spanish American	0.0	(-0.6)	0.0	(0.0)	0.0	(-0.4)	0.1	(0.1)	0.0	(-0.7)	0.1	(-0.6)	0.1	(-0.6)	0.1	(-0.3)	0.2	(-0.2)	0.2	(0.2)	0.0	(0.0)	0.1	(0.0)	0.1	(-0.1)	0.1	(-0.1)	0.0	(-0.2)
Cuban	0.0	(-0.6)	0.0	(-0.5)	0.0	(-0.4)	0.0	(-0.3)	0.0	(-0.7)	0.0	(-0.6)	0.0	(-1.0)	0.0	(-0.8)	0.0	(-0.9)	0.0	(-0.1)	0.0	(-0.3)	0.0	(-0.3)	0.0	(-0.1)	0.0	(-0.2)	0.0	(-0.2)
West Indian	0.0	(-0.6)	0.0	(-0.5)	0.0	(-0.4)	0.0	(-0.3)	0.0	(-0.8)	0.0	(-0.6)	0.0	(-1.0)	0.0	(-0.6)	0.1	(-1.0)	0.0	(0.0)	0.0	(-0.3)	0.0	(-0.3)	0.0	(-0.1)	0.0	(-0.1)	0.0	(-0.2)
Chinese	0.3	(-0.3)	0.3	(0.4)	0.0	(-0.4)	0.0	(-0.3)	0.0	(-0.8)	0.3	(0.0)	0.3	(-1.0)	0.0	(-0.8)	0.0	(-1.0)	0.2	(0.9)	0.0	(-0.3)	0.8	(0.9)	1.1	(0.6)	0.2	(0.4)	0.1	(-0.2)
East Indian	0.0	(-0.6)	0.0	(-0.5)	0.0	(-0.4)	0.0	(-0.3)	0.0	(-0.6)	0.0	(0.0)	0.0	(-1.0)	0.0	(-0.8)	0.0	(-1.0)	0.0	(-0.1)	0.0	(-0.3)	0.0	(-0.3)	0.0	(-0.1)	0.0	(-0.2)	0.0	(-0.1)
Japanese	1.1	(2.9)	1.2	(2.9)	1.2	(1.7)	0.4	(0.8)	0.6	(3.7)	0.8	(2.5)	0.7	(3.1)	0.9	(3.4)	0.5	(2.5)	1.2	(0.6)	1.1	(2.1)	0.9	(1.1)	5.0	(2.9)	0.2	(0.4)	1.3	(2.6)
Korean	0.0	(-0.6)	0.0	(-0.5)	1.8	(2.9)	0.0	(-0.3)	1.2	(-0.8)	1.2	(3.9)	0.0	(-1.0)	0.9	(-0.8)	0.5	(2.5)	10.0	(5.9)	0.0	(0.0)	0.3	(-0.3)	9.1	(5.5)	2.1	(-0.2)	2.1	(4.3)
Pacific Islander	0.0	(-0.6)	0.0	(-0.5)	0.0	(-0.4)	0.0	(-0.3)	0.0	(-0.8)	0.0	(0.0)	0.0	(-1.0)	0.0	(-0.8)	0.0	(-1.0)	0.0	(0.0)	0.0	(-0.3)	0.0	(-0.3)	0.0	(-0.1)	0.0	(0.0)	0.0	(0.0)
African (black)	0.1	(-0.4)	0.1	(-0.4)	0.0	(-0.3)	0.0	(-0.3)	0.0	(-0.6)	0.1	(-0.5)	0.1	(-0.6)	0.1	(-0.6)	0.1	(-0.6)	0.1	(-0.1)	0.1	(0.1)	0.2	(0.2)	0.1	(0.0)	0.1	(0.0)	0.0	(-0.2)
Other	0.1	(-0.2)	0.2	(-0.3)	0.0	(-0.4)	0.0	(-0.3)	0.1	(0.1)	0.1	(-0.4)	0.1	(-0.6)	0.2	(0.3)	0.4	(1.9)	0.6	(0.3)	0.1	(-0.1)	0.5	(0.5)	1.2	(0.7)	1.2	(3.2)	0.9	(1.9)

Bold = 1 or more NSDMR below mean

[shaded] = 1 to 1.99 NSDMRs above mean

[darker shaded] = 2 or more NSDMRs above mean

* Norwegians, Danes, and Swedes.

† Bohemians and Moravians.

THE END OF THE LINE

Immigrant Medical Inspection after 1924

This year, for the first time in the history of Ellis Island, it has been possible to abandon the old "routine" examination entirely. It is doubtful whether a medical officer ever served at Ellis Island who did not appreciate the weakness of the only method of examination possible when numbers far beyond the capabilities of the station . . . were presented for examination, and it is probable that none ever served here who did not realize that it was impossible to carry out the examination of aliens in the manner expected and which the law contemplated.

—Surgeon General, 1925

I wish to firmly enter a violent protest against such action. . . . America could never afford to relinquish her right of final examination of aliens on our own shores by Public Health Doctors and Immigration Inspectors. . . . There is no more important function of the American Government than the selection of their future citizens. Let us always have final inspection and selection made by American Doctors and Inspectors on American shores and surrounded by American influence. . . . Ellis Island stands as a silent guard to future America.

—U.S. Rep. John L. Cable, 1925

THE STORY of immigrant medical inspection began in 1891 with the "line." It was a world in which disease, social class, race, and therefore industrial fitness were written clearly on the immigrant body, a world that attempted to rigidly, efficiently, scientifically manage the factory worker in the industrial plant as well as in the community. By the 1920s, however, the world was changing. Immigration restriction, consumerism and mass culture, growing social and racial cohesion, and critical shifts in America's industrial economy and corporate managerial philosophy combined to alter the national imperative to discipline the nation's work force.[1] Henry Yu argues that the very act of migration gave a certain dynamism to the concept of race, which began to stagnate in a nation "that had clear boundaries."[2]

A fundamental reorganization of industrial labor and, critically, management philosophy was under way by the 1920s. It was fueled in part by wartime labor militancy and postwar strikes, as well as a spate of corporate mergers and consolidations in industries from mining and manufacturing to steel, utilities, and merchandising.[3] Scientific management and Fordism did not disappear, but with increasing mechanization and employers' desires to stabilize both production and their work force, large corporations augmented scientific management by segmenting the labor force.[4] This helped reduce labor turnover rates to half the prewar levels.[5] Corporations endured a 10 percent decline in productivity during the war years and became increasingly concerned with how to control absenteeism, turnover, and, above all, devastating strikes. David Brody argues, "The practice of labor management was thus spurred forward and beyond the simplistic economic psychology of Frederick W. Taylor."[6] In Lizabeth Cohen's view, "there is no denying that managerial ideology almost everywhere underwent a sea change during the 1920s."[7] Several factors combined to alter corporate ideology, including immigration restriction, technological advances, and the labor strikes of World War I.

Corporate America's efforts to develop new means of organizing the work force in the face of immigration restriction reflected a shift in philosophy after 1921—a shift that would have an impact on medical inspection at the nation's borders.[8] Immigration restriction reduced the number of bodies entering the American industrial labor force, limiting the employer's ability to reserve the newest immigrants for the most backbreaking and least desirable work, thus contributing to a cycle of mobility upward within the work force. Economist and labor-relations expert Sumner Slichter explains that managers increasingly found it impossible to follow the dictum of scientific management "to adapt jobs to men rather than men to jobs"—suggesting that the worker was being viewed as less of a cog suited to one task only or a senseless gorilla, and was now seen as more flexible and adaptable to many different work environments or tasks.[9]

At the same time, industry began to view machinery itself and not the individual laborer as the interchangeable cog in production. Irving Bernstein attributes "the quickening pace of technological change" to the fact that "machinery was cheaper than labor," making management eager to "replace workers with machines, to scrap old machines for new ones."[10] While Frederick W. Taylor, in the decades before the war, had almost casually dismissed the worker as an ignorant ox, the labor turmoil accompanying World War I forced corporate America to regard the worker's thinking and organizational capabilities as a potential threat. Labor would now have to be managed strategically. As Slichter wrote in 1929, management techniques in the 1920s were geared toward "counteract[ing] the effect of modern technique upon the *mind* of the worker, to prevent him from becoming class conscious and from organizing trade unions."[11] Indeed, Gerard Swope stressed to General Electric that the company's workers "must be dealt with as *thinking men.*"[12]

Employers sought—though largely without effect—to institute a system of benefits and incentives that would maintain worker loyalty and eliminate the high turnover rates characterizing production before and during World War I, when labor reached the pre–Depression era zenith of its political influence over working conditions. Employers in a variety of industries and businesses sought to counter the increasing power that unions demonstrated during World War I by establishing "personal" relationships with individual workers. They competed for worker loyalty with the various ethnic mutual aid, fraternal, banking, and charity societies and agencies. In addition, employers attempted to make the workplace a focus for social and leisure activity. At the same time, mass producers competed for the patronage of consumers, and newspapers and radios "centralized information gathering and processing."[13] Hand in hand with this shift came an increase in white-collar employment— the number of nonmanual workers increasing over 38 percent between 1920 and 1930 compared to under 10 percent for manual workers—as large corporations adopted new policies of personnel management.[14]

Of far more significance than welfare capitalism, however, was the reorganization of labor within the factory.[15] Under prewar systems of scientific management (the "drive system") corporations frequently moved workers around, from one department to another, from one place on the assembly line to another. Many companies observed during the wave of strikes gripping the nation following World War I that ethnic solidarity promoted worker unity and discovered that by exploiting interracial tension and hatred a company might break a strike.[16] In the 1920s, employers began to separate ethnic or racial groups within the workplace as part of the new strategy of segmenting labor. Corporations like Ford, General Electric, Westinghouse, Procter and Gamble, U.S. Rubber, and International Harvester experimented with "the systematic allocation of different groups of workers to different plant divisions in order

to embed job segregation into the job structure itself."[17] For example, Lizabeth Cohen observes that by 1920 International Harvester's McCormick
Works in Chicago "felt that a socially atomized work force drawn from numerous ethnic communities would minimize militance at the company's most
radical plant, still notorious for the 1886 strike that precipitated the 'Haymar-
ket riot.'"[18] A superintendent at Wisconsin Steel explained, "We try never to
allow two of a nationality to work together if we can help it. Nationalities tend
to be clannish and naturally it interferes with the work and the morale of the
place. You see here in this loading department, for instance, we have an Italian, an Irishman, a Pole, a Dane and two Mexicans working."[19] In a context
where the aim was to mix races rather than target racial groups to specific types
of industrial tasks, the racial profile of workers mattered less. As Henry Yu explains, "the social landscape of racial consciousness changed, and theories
about culture, assimilation, and racial consciousness lost their emphasis on migration and movement. In a world where new immigrants were few, ethnicity
became a matter of eradication."[20]

Ethnic or racial differences in the United States did not simply disappear
in the 1920s, but as ethnic blocs in the workplace and urban neighborhoods
began to break apart, so too did some of the barriers dividing the nation.[21] The
National Origins Act, in addition to drastically cutting the flow of people to the
United States, also expressed a federal perception of the transformation of
America into a nation of "whites" and "nonwhites," increasing a sense of internal unity and homogeneity.[22] This is not to say that this perception was
shared, for certainly the importance of drawing distinctions among European
immigrant groups in the United States persisted, but it did represent the perspective of the federal government when it came to assessing the industrial citizen.[23]

In addition to changes in corporate labor-management strategy and the reorganization of labor came improvements in standard of living for the employed sector of the economy. Such improvements were measured in increases
in earnings between 1922 and 1929, and also in living conditions. A 1929 Bureau of Labor Statistics study of Ford Motor Company employees found, for
example, that industrial workers lived in far more salubrious conditions than
they had at the turn of the century. Employed workers lived in houses that provided, on average, one room per person. They enjoyed electricity, central heat-
ing, inside running water, and toilets. That the achievement of this new standard of living came at the cost of deficit spending in the form of credit and

payment in installment plans did not strike a chord of alarm as it had at the dawn of the twentieth century.[24] In the gloss of the productive economy of the 1920s, the worker ceased to be viewed as a potential dependent, as a drain on precious charity resources. The notion of abundance and consumerism became a means of establishing American unity.[25]

Thus, the 1920s were widely regarded as an era of prosperity.[26] Unemployment among urban workers remained, on average, under 7 percent. Per-capita income grew by one-third during a decade of economic expansion that remained relatively unmarred by inflation and recession. But while the nation gloried in the bloom of apparent prosperity, little changed for the industrial worker. Unemployment in this period was lower than it had been in previous decades, but significant levels of unemployment and job turnover characterized the industrial working experience. Studies in Philadelphia and Buffalo in 1929 revealed that over 10 percent of wage earners had been unemployed for more than ten weeks; another 6.5 percent in Buffalo were employed only part time.[27] In manufacturing, with the exception of depression years, unemployment in the period 1923–27 reached its highest point since 1900.[28] A continued labor surplus was fueled not by immigration but by African American migration and migration from the farm to the city, along with the displacement of both skilled and unskilled workers with machines. This ensured continued levels of high unemployment and job insecurity along with limited improvements in wages and working conditions.[29] Brody notes that "from 1910 to 1930, the percentage of unskilled workers in the industrial labor force fell from 36 to 30.5, the semiskilled rose from 36 to 39, the skilled from 28 to 30.5." While production increased 5 percent per year in the 1920s, wages advanced at a rate of only 2.5 percent per year.[30]

Only a small segment of social workers perceived unemployment to be a significant social problem during the prosperity decade.[31] Yet even those concerned with unemployment underscored an important reversal in thinking about the relationship of the worker to the economy. While settlement house workers, for example, acknowledged that "sickness, bad habits, insanity, irresponsibility, incapacity, accidents, old age, and death put families on the rocks," they were beginning to cast problems of unemployment and dependency as stemming primarily from the economy rather than from the immigrant worker. The Unemployment Committee of the National Federation of Settlements undertook a national study of unemployment in the prosperous years before the crash of 1929 to help isolate the role that "industrial causes outside" the control of the worker played in dependency.[32] A similar shift in thinking was evident in the Children's Bureau. In 1918 it had concluded, based on a study in Gary, Indiana, that "illness . . . was the major cause of nonemployment."[33] By the early 1920s, however, the bureau saw more complex causal pathways linking illness, unemployment, poverty, and ultimately dependency. Still, it reported that among 366 families the bureau studied in Racine, Wisconsin, and Springfield, Massachusetts, 63 percent reported illness of the chief breadwinner or other family member during periods of extended unemployment during the depression of 1921 and 1922.[34]

Yet despite the continuing poor working conditions and consequent labor unrest, industrial, ideological, and cultural forces combined in the postwar period to deemphasize the need to discipline the immigrant labor force. The context and priorities of the immigrant medical examination accordingly changed. During this critical period, the law also changed. A key provision of the Immigration Restriction Act of 1924 (National Origins Act) moved immigrant medical inspection abroad and established the visa system: intending immigrants now could not depart for the United States until a U.S. Consular Office abroad had issued them visas. The legal precedent for the visa system was established during World War I as part of the Passport Control Act of 22 May 1918, which stipulated that no individual could depart from or enter the United States without approval from the State Department.[35] After 1924, consulates took on the responsibility of enforcing not only the quota law, but *all* U.S. immigration law. As quickly as they were able to establish the facilities, medical inspection became a prerequisite for consular approval for visas. The Rogers Act, passed after the Immigration Restriction Act of 1924, established the U.S. Foreign Service abroad. The Rogers Act gave the United States consulates the staff needed to inspect all immigrants abroad.[36]

Overseas, a new power to bar immigrants from the United States and a new philosophy and method of medical inspection contributed to the resounding success of immigrant medical inspection abroad. As early as 1927, IS Second Assistant Secretary W. W. Husband wrote to the surgeon general that the inspection of immigrants abroad "has . . . proven successful beyond any of our fondest dreams, and I certainly hope . . . that it has come to stay."[37] But it was ultimately a hollow success for the PHS medical officers, one that reflected larger cultural, social, economic, and industrial change in the United States.

The "British Plan"

With the passage of the Immigration Restriction Act of 1924, the goal of conducting and enforcing effective medical inspection abroad was realized.[38] It was most often billed as a "humanitarian" measure, intended to benefit immigrants and prevent the separation of families.[39] The PHS argued in addition that the medical examination, by uncovering disease, "exercise[s] a wholesome influence for the betterment of personal and community hygiene in the districts in which such cases reside."[40] The U.S. government felt that the system of inspections abroad promoted better international relations.[41] It was popular because, in practice, consular officers rejected diseased immigrants at much higher rates than had PHS and IS officers stationed at domestic immigration posts. The quota law, perhaps, had its most profound impact on the philosophy of immigrant medical inspection both at home and abroad. Rather than seeking to manage, discipline, and define the immigrant labor force, consular officers sought to fill the quota for each country with only the most desirable immigrants. Selection with attention to "quality" became, at least in theory, the basis for inspection.[42]

On 1 August 1925 consular officers at seven ports in Great Britain and Ireland began the first test of the visa system, which came to be known as the "British Plan."[43] British ports were selected for the test of the visa system because officials felt that they were more likely to get satisfactory cooperation from the British Government. If it reduced the number of immigrants examined, sent to Ellis Island, and deported, they felt that other European governments would quickly provide similar facilities.[44] From 1 August to 31 October 1925, 19,435 prospective immigrants were medically examined. Of these, 202 (1%) were "notified" (medical certification was now called "notification") for Class A conditions, and 1,797 (9%) were notified for Class B conditions. All of the candidates with Class A conditions and 41 percent of the candidates with Class B conditions were refused visas.

In 1926 the United States opened inspection stations in two more European cities: Rotterdam, Holland, and Antwerp, Belgium. As shown in table E-1, which presents the first published results of the British Plan for the nine cities issuing visas, notification rates tended to increase slightly. A total of 12.14 percent of prospective immigrants were notified for medical conditions (1.41% were notified for Class A conditions and 10.72% for Class B conditions), and 44.3 percent of those notified were refused a visa. The Secretary of

TABLE E-1. Results of British Plan, 1926

| | Class A | | | | | Class B | | | | | Total Class A and B | | | | |
| | | Certified | | Refused Visa | | Certified | | Refused Visa | | | Certified | | Refused Visa | | |
City	Total Examined	No.	%	No.	% Total Examined	No.	%	No.	% Total Examined	% No. Certified	No.	% Examined	No.	% Examined	% Certified Refused
Antwerp	266	4	1.50	4	1.50	18	6.77	7	2.63	38.89	22	8.27	11	4.14	50.00
Belfast	3,217	10	0.31	10	0.31	701	21.79	242	7.52	34.52	711	22.10	252	7.83	35.44
Cobh	9,721	277	2.85	277	2.85	1,515	15.58	548	5.64	36.17	1,792	18.43	825	8.49	46.04
Dublin	15,092	416	2.76	416	2.76	1,571	10.41	424	2.81	26.99	1,987	13.17	840	5.57	42.27
Glasgow	14,742	32	0.22	32	0.22	674	4.57	208	1.41	30.86	706	4.79	240	1.63	33.99
Liverpool	5,927	65	1.10	65	1.10	980	16.53	560	9.45	57.14	1,045	17.63	625	10.54	59.81
London	6,375	16	0.25	16	0.25	473	7.42	244	3.83	51.59	489	7.67	260	4.08	53.17
Rotterdam	601	6	1.00	6	1.00	75	12.48	40	6.66	53.33	81	13.48	46	7.65	56.79
Southampton	3,111	8	0.26	8	0.26	326	10.48	68	2.19	20.86	334	10.74	76	2.44	22.75
Total	59,052	834	1.41	834	1.41	6,333	10.72	2,341	3.96	36.95	7,167	12.14	3,175	5.37	44.30

SOURCE: SGAR, 1926.
NOTE: All immigrants certified for Class A conditions were refused visas without exception.

the Treasury concluded that "The number of physical and mental disabilities and defects certified and the high percentage of refusal of visas for medical reasons are evidences of the value of these medical examinations," which "provide a very practical means of elimination of undesirable elements to our population."[45]

Consular inspection rapidly expanded. The German cities of Berlin, Bremen, Cologne, Hamburg, and Stuttgart opened inspection stations in 1927.[46] At the same time inspection stations were established in Warsaw, Poland; Bergen and Oslo, Norway; Göteborg and Stockholm, Sweden; and Copenhagen, Denmark.[47] In 1928 Genoa, Naples, and Palermo in Italy and the Czechoslovak city of Prague opened stations. On 1 July 1930, a station opened in Vienna, Austria.[48]

From 1926 to 1930 consular notification rates varied between 9 percent and 13 percent.[49] (PHS officers notified a high of 23.78% of prospective immigrants in Northern Ireland in 1930; they notified a low of 3.98% in Denmark in 1926.) Overall, the PHS issued notifications to 10.24 percent of immigrants applying for visas.[50] The PHS issued the vast majority of medical notifications for Class B conditions. Only in Poland, Italy, and Ireland did Class A notifications for any year exceed 2 percent; they never exceeded 5 percent. Throughout the entire period and at all consulates, officials rejected 100 percent of immigrants notified for Class A conditions. Rejection rates for Class B conditions varied from a low of 34 percent in 1930 to a high of 44 percent in 1927. Of the total examined in all countries, overall rejection rates attributable to medical conditions ranged from a low of 4.4 percent in 1927 to a high of 5.5 percent in 1930. Overall, 4.83 percent of those examined were eventually refused visas for medical reasons. The rejection rate of nearly 5 percent abroad represented a 400 percent increase over the medical exclusion rate of approximately 1 percent that had prevailed in the United States since 1891.

According to the PHS, "the true nature of the medical examination abroad is most clearly revealed by the character of the defect causing the applicant to be refused a visa."[51] Table E-2 shows the different consular rates of notification for each of the conditions for which immigrants were notified. In viewing the changing disease distributions from 1927 to 1930, most striking is the decline of the "more serious disabilities" and the rise of notifications in the category "all other defects," as well as a new interest in reporting notifications for a whole range of mental conditions. Whether the new interest in reporting reflected an increase in notifications for these conditions we cannot know, but it

TABLE E-2. Distribution of Diseases Certified within Country of Inspection, 1927–1930

Condition	Belgium 1927	1928	1929	1930	England 1927	1928	1929	1930	Scotland 1927	1928	1929	1930	North Ireland 1927	1928	1929	1930	Irish Free State 1927	1928	1929	1930	Germany 1927	1928	1929	1930	Holland 1927	1928	1929	1930
Cardiac disease	3.2	2.4	0.0		8.9	4.5			4.4	2.7			6.8	8.6			5.0	5.1			14.6	9.8			3.3	5.4		
Defective vision	31.9	40.8			31.1	29.4			17.2	13.4			5.9	26.2			35.4	23.5			25.9	31.2			38.1	48.0		
Deformities	0.0	7.6			3.6	5.8			0.0	18.6			0.0	2.5			8.3	5.3			6.7	7.9			2.9	5.4		
Dental sepsis	4.9	0.8			0.0	0.8			0.0	0.1			5.0	4.9			18.0	1.3			1.1	0.2			0.0	0.0		
Flat foot		0.4				0.0				0.0				0.9				8.9				0.8				0.4		
Hernia, all forms	8.6	4.8			7.9	6.7			8.5	7.7			6.2	7.4			3.9	2.5			4.3	4.2			0.0	5.8		
Otitis media		0.0				10.6				3.0				0.6				4.0				1.3				0.4		
Epilepsy	0.0	0.0	0.5	0.0	0.1	0.1	0.3	0.1	0.0	0.0	0.6	0.2	0.6	0.6	0.4	0.4	0.1	0.0			0.0	0.1	0.1	0.3	0.0	0.4	0.5	0.1
Favus	0.5	0.0	0.0	0.0	0.0	0.0	0.3	0.0	0.0	0.0	0.0	0.0	0.3	0.0	0.0	0.0	0.0	0.0			0.0	0.3	0.1	0.2	1.0	0.4	0.9	0.0
Loathsome contagious diseases	0.0	0.0	0.0	0.0	0.0	0.0	0.0	0.0	0.0	0.1	0.1	0.1	1.6	0.6	0.4	0.4	4.9	7.1	2.2	1.3	0.4	0.1	0.1	0.3	0.0	0.9	0.0	0.3
Trachoma	13.5	4.4	3.0	1.8	0.3	0.4	0.3	0.8	0.0	0.0	0.0	0.0	0.3	0.0	0.0	0.0	0.5	0.7	0.5	0.5	0.4	0.3	0.1	0.2	0.0	0.0	0.0	0.2
Tuberculosis																												
Pulmonary	1.1	0.8	0.0	0.0	1.0	0.3	0.1	0.8	0.4	0.1	0.1	0.1	0.3	0.9	0.6	0.6	1.0	1.8	1.1	1.5	0.8	1.6	2.7	3.2	1.9	0.4	0.9	0.4
Other forms	1.1	0.8	0.5	0.0	0.1	0.0	0.1	0.0	0.8	0.6	0.3	0.5	0.6	2.5	0.4	0.0	0.4	0.2	0.4	0.6	0.1	0.3	0.3	1.0	0.0	0.0	0.5	0.0
Venereal disease	0.0	0.4	1.0	0.9	0.6	0.5	0.8	0.3	0.0	0.7	0.2	0.0	0.0	0.6	1.0	0.5	0.1	0.6	1.1	0.5	1.0	0.2	1.0	1.2	0.5	0.0	0.0	0.5
Varicose veins	4.9	9.2			4.3	4.5			13.5	10.7			4.0	3.7			8.5	4.7			7.4	5.8			0.0	3.6		
Alcoholism							0.0				0.1				0.0	0.0			0.1	0.1			0.1	0.0			0.0	0.0
Mental conditions																												
Dementia praecox				0.0				0.0				0.0			0.0	0.1				0.0				0.1				0.0
Feeble-mindedness	0.0	0.0	0.0	3.5	0.6	0.1	0.6	0.2	0.1	0.1	0.0	0.2	0.6	0.3	0.3	0.3	0.6	0.5	0.3	0.0	2.7	1.6	1.2	0.3	3.8	3.6	1.4	1.3
Imbecility	0.0	0.4	0.0	0.0	0.3	0.1	0.0	0.0	0.0	0.1	0.1	0.0	0.0	0.0	0.0	0.0	0.0	0.0	0.0	0.0	0.1	0.0	0.1	0.2	0.0	0.0	0.0	0.0
Insanity	0.0	0.4	0.0	0.0	0.1	0.2	0.0	0.1	0.0	0.1	0.0	0.1	0.0	0.0	0.0	0.0	0.1	0.3	0.0	0.0	0.1	0.1	0.2	0.4	0.0	0.4	0.0	0.0
Idiocy								0.0				0.0				0.0				0.0				0.0				0.0
Manic depressive psychosis																												
Mentally defective	0.5	0.4	2.5	0.0	0.5	1.0	0.7	0.5	1.8	4.4	3.8	7.8	0.9	4.0	9.5	8.1	7.0	9.3	10.5	8.2	3.0	5.6	5.8	3.1	0.0	1.3	0.0	3.1
Mentally depressed	0.0	0.0	0.0	0.0	0.0	0.0	0.0	0.0	0.0	0.0	0.0	0.1	0.0	0.0														
Mentally retarded								0.3								0.0			0.0	0.0				1.5				0.0
Nervous instability		0.0	0.0	0.0			0.0	0.0				0.3			0.0	0.0			0.0	0.0			0.2	0.2		0.0	0.0	0.0
Organic nerve disease				0.0				0.0				0.0				0.0				0.0				0.0				0.0
Psychopathic inferiority	0.0	0.0	0.0	1.8	0.5	0.5	0.6	0.1	0.1	1.3	0.3	0.3	0.0	2.2	0.8	0.0	1.2	0.9	1.1	0.6	0.6	1.0	1.5	1.3	0.0	0.9	0.5	1.0
Senile dementia		0.0	0.0	0.0			0.0	0.0			0.0	0.0			0.0	0.0			0.0	0.0			0.0	0.1	0.5	0.0	0.0	0.0
All other	29.7	26.4	92.5	92.1	39.8	34.8	96.6	96.5	53.3	36.3	94.6	91.4	66.8	34.0	87.0	90.1	5.0	23.3	82.6	86.7	30.7	27.9	86.5	86.7	48.6	22.4	95.4	89.6

TABLE E-2. Continued

Condition	Denmark				Norway				Sweden				Poland				Italy				Czechoslovakia			
	1927	1928	1929	1930	1927	1928	1929	1930	1927	1928	1929	1930	1927	1928	1929	1930	1927	1928	1929	1930	1927	1928	1929	1930
Cardiac disease	19.0	6.1			8.3	4.3			6.8	9.6			13.3	4.0				3.7				6.2		
Defective vision	28.5	28.1			37.5	38.7			15.8	21.5			1.6	8.3				9.1				11.6		
Deformities	12.4	5.6			10.2	7.7			0.0	3.3			6.5	3.6				2.7				4.8		
Dental sepsis	4.4	9.2			4.9	5.3			62.2	13.9			34.2	31.2				1.6				2.4		
Flat foot		0.0				2.0				0.2				0.2				0.4				0.0		
Hernia, all forms	9.5	11.2			5.8	8.5			2.9	5.0			10.1	6.3				7.7				8.8		
Otitis media		0.0				3.6				1.6				0.1				1.1				0.2		
Epilepsy	0.0	0.0	0.0	0.0	0.0	0.0	0.0	0.0	0.0	0.0	0.0	0.3	0.1	0.0	0.0	0.0		0.1				0.0		
Favus	0.0	0.0	0.0	0.0	0.0	0.0	0.0	0.0	0.0	0.0	0.0	0.0	0.0	0.0	0.0	0.0		0.0				0.0		
Loathsome contagious diseases	1.5	0.0	0.0	0.0	0.2	0.9	0.2	0.0	0.2	0.9	3.4	0.3	3.8	22.4	5.2	3.5		0.5	1.7	1.0		0.2	0.5	0.7
Trachoma	0.0	0.0	0.0	0.0	0.0	0.0	0.0	0.0	0.2	0.2	0.1	0.0	8.2	4.2	8.2	7.7		11.6	7.8	11.0		2.0	2.3	6.7
Tuberculosis,																								
Pulmonary	0.0	0.5	0.0	0.0	1.3	1.2	0.6	0.8	0.4	2.7	1.2	1.0	1.0	1.0	2.3	2.8		1.2	0.8	0.7		2.2	1.7	2.2
Other forms	0.0	0.5	0.0	0.0	0.9	2.2	1.3	0.2	0.1	0.5	0.5	0.3	0.4	0.2	0.1	0.7		0.7	0.4	0.5		0.7	1.3	1.1
Venereal disease	0.0	0.0	0.3	0.0	2.8	1.4	1.4	0.6	1.2	0.8	1.6	0.5	0.2	0.4	0.2	0.2		0.6	0.4	0.3		1.1	0.5	0.4
Varicose veins	19.0	14.3			4.7	5.3			8.3	8.3			5.6	5.7				2.0				9.0		
Alcoholism		0.0	0.0	0.0		0.0	0.2	0.0		0.0	0.0	0.0		0.0	0.0	0.0		0.0	0.0	0.0		0.0	0.0	0.0
Mental conditions																	0.1				0.0			
Dementia praecox	0.7	0.0		0.0	0.0	0.5	0.3	0.0	0.9	0.2	1.4	0.3	0.0					0.6	1.0	1.6		0.4	1.0	0.9
Feeble-mindedness	0.0	0.0	1.3	0.7	0.0	0.2	0.0	0.0	0.1	0.0	0.2	0.0	5.6	5.1	3.3	1.8		0.2	0.2	0.2		0.0	0.5	0.2
Imbecility	0.0	0.0	0.0	0.0	1.1	0.2	0.0	0.2	0.0	0.1	0.1	0.0	0.2	0.3	0.4	0.3		0.1	0.1	0.0		0.9	0.0	0.0
Insanity	0.7	0.0	0.0	0.7									0.1	0.0	0.0	0.1								
Idiocy				0.0				0.0						0.0	0.0	0.0				0.0				
Manic depressive psychosis	2.9	1.5			1.1	0.8			0.7	1.3			8.7	4.4				0.7				7.0		
Mentally defective			0.0	0.0			0.8	0.8			2.2	1.8			3.1	1.8			1.7	2.4			12.8	9.5
Mentally depressed			0.0	0.0			0.0	0.0			0.0	0.0			0.0	0.0			0.0	0.0			0.0	0.0
Mentally retarded				0.0				0.0			0.0	0.0			0.0	0.0			0.0	0.2			0.0	0.0
Nervous instability				0.0								1.3												
Organic nerve disease			0.0	0.0			0.0	0.0			0.0	0.0			0.0	0.0			0.0	0.0			0.0	0.0
Psychopathic inferiority	1.5	0.5	0.0	0.0	0.0	0.8	0.3	0.0	0.3	1.6	0.6	1.0	0.4	0.0	0.1	0.1		0.1	0.4	0.3		0.0	0.5	0.2
Senile dementia		0.0	0.0	0.0		0.0	0.0	0.0			0.6	1.0		0.0	0.0	0.0		0.0	0.0	0.1		0.0	0.5	0.0
All other	0.0	22.4	98.4	98.6	21.3	16.6	94.9	97.3	0.0	28.4	88.8	93.4	0.0	2.4	77.1	81.1		55.8	84.7	80.0		42.4	78.4	78.1

NOTE: Blank cells may indicate 0 certifications or may indicate that the condition was reported in the "All other" category. For example, it is unlikely that there were no certifications for defective vision after 1928; more likely, it was no longer deemed a condition worth listing separately. Italy and Czechoslovakia did not have consular stations in 1927.

does show what conditions the IS considered significant. Looking closely at the diseases themselves, we find a list of "serious disabilities" that did not differ substantially from previous lists. The traditional immigrant diseases—trachoma, favus, tuberculosis, venereal diseases, insanity—still ranked as "serious," and now so did defective vision, dental sepsis, flat feet, varicose veins, and otitis media. Except in 1927 and 1928, when certifications for "defective vision" constituted more than 30 percent of certifications across most consular offices, none of the "serious defects" ever accounted for more than 14 percent of certifications; typically, the "serious defects" accounted for only 5 to 9 percent of certifications. By 1930, although certifications were no doubt made for the majority of the "serious defects" in most countries, the IS ceased to report data for these conditions.

While this shift in the language of disease did not necessarily point to a deemphasis of industrial citizenship, it did coincide with a new and different philosophy, more in accord with military examinations for selecting only the fittest recruits than with the old PHS and IS industrial philosophy.[52] The nature of disease among immigrants mattered less in the new context of immigration restriction. Disease in the post-1924 period was no longer an indicator of suitability for the industrial labor force. It was, rather, an indication of the environmental and social conditions to which the immigrant had been exposed. Thus, in 1927 the surgeon general attributed differences in rates of medical certification among immigrants to "the economic, sanitary, and social environment" from which the applicants came.[53] In explaining differences in the certification rate for Class A conditions—those that had been so influential in marking the immigrant as undesirable in the years before restriction and inspection abroad—the surgeon general hypothesized that the "predominant basic factor was the intellectual, social, and economic status of the communities from which the individual applicants came."[54] Disease and disability were now associated with the "industrial" centers of Europe.[55] Gone was discussion of "trachoma types" or "sanitary types" of immigrants, in which the presence of disease spoke to fixed qualities of an individual or immigrant group.

Neither the PHS officers nor other immigration officials working at U.S. consulates abroad in the 1920s viewed the immigrant as a potential laborer. They did not imagine the potential immigrant living or working in the United States. Rather officials saw the immigrant primarily within the context of the European environment. Thus, disease became less an indicator of an immigrant's economic potential than a reflection of European conditions: immi-

grant health reflected urban industrial conditions, national commitment to education and sanitation, and national moral and cultural values, all of which had
little bearing on conditions within the United States or on the role the immigrant would play in America's industrial economy. To be sure, all immigrants
with Class A notifications were now denied a visa and prevented from immigrating to the United States. Nevertheless, the subtle shift in the determinants
of health or disease signaled a loss of medical authority that necessarily went
hand in hand with the changing industrial, ideological, and cultural context of
the 1920s.

EXAMINATION ABROAD AND CHANGING
MEDICAL AUTHORITY

On a governmental and international level, the new plan for inspecting immigrants was considered an unqualified success. The *Christian Science Monitor* reported in 1925 that "European governments are clamoring for general
institution of the system, now successfully on trial in the British Islands, for examination of American-bound emigrants before their departure from Europe." The article continued, "Not the slightest danger—physical, moral or
mental—is seen by American immigration officials in having all future citizens
passed on in Europe."[56] The commissioner of immigration at New York was
certain that the system improved the quality of arriving immigrants, concluding, "There is less of European and more of American . . . in their make-up as
we now get them. . . . Most of the credit is due to the consular force. . . . Their
job of sifting at the source has been remarkably done."[57]

On the surface, changes in the immigrant medical inspection abroad
seemed to represent a vastly expanded medical authority. The medical inspection administered to immigrants abroad differed fundamentally from that
carried out at domestic ports, in both form and intent. The medical exam provided at the consular offices was not intended to manage the laboring body—
to inform potential workers of the values and expectations of the industrial
workplace and to exclude them if necessary. Rather, it was envisioned as a
means of "improving the physical and mental types of immigrants coming to
the United States."[58]

Abroad, immigrants did not file past the PHS officer in a line. Instead, after an initial visit to the U.S. consulate, the prospective immigrant was referred
to the PHS and given a private appointment: "The fact that the examinations

of prospective immigrants abroad are made according to schedule and by appointment makes it possible for them to be more thorough and painstaking as compared with the necessarily hurried examinations when large ship loads of immigrants arrive at U.S. ports."[59] Here, the PHS officer reviewed the immigrant's visa application, containing information on the race, age, marital status, birthplace, previous residences, and occupation of each immigrant.[60]

As a procedure that placed the physician face to face with the individual immigrant, the exam—unlike that accorded immigrants at U.S. ports—was regularized along the lines of routine exams performed in the United States for life insurance purposes or in industrial medicine. After 1924 PHS officers adopted a standardized medical examination form, which required the officer to record detailed information on all parts of the body. A checklist now guided the PHS examiner through the medical exam; he recorded detailed information regarding the prospective immigrant's general appearance, facial appearance, condition of head and scalp, condition of the lungs, presence of parasites, and mental condition of the applicant. Thus, the PHS officer created a medical record of a host of conditions and medical observations. He did not merely issue a single medical certificate that stated only the causes for which the immigrant should be rejected. Figure E-1 shows the form used by inspectors in 1925 and the instructions for information to be provided under each aspect of the exam they were to perform. Compare this to figures E-2 and E-3, examples of examination forms used in U.S. industries.

The "routine" or "periodic" medical exam as practiced in the United States in the 1920s typically included a personal and family history of disease, a description of working conditions, and several recommended technological diagnostic procedures, like a measurement of blood pressure or a chest x-ray.[61] Such questions were not part of the standard immigrant inspection except as presented on the visa application. Nevertheless, the immigrant medical inspection took a similar anatomical approach for examination purposes, requiring the physician to take a more "compartmentalized" look at the body. This was an approach scorned by the old masters of the gaze. "There is a tendency," warned an Ellis Island PHS physician as early as 1911, "not to look at the patient at all, but rather to look at his individual and component parts, gazing at these so minutely that the picture becomes so completely resolved into its elementary colors as to render a comprehension of its general effect well nigh impossible."[62]

Thus, the increased rigor of the medical exam also represented a setback.

occupation	abdomen
identification marks	herniae
general appearance	genitalia
facial appearance	extremities
personal hygiene	evidence of intestinal parasites
head and scalp	other abnormalities
vermin	nervous
mouth hygiene	mental
eyes	remarks
skin, nails	passed—yes/no
heart	rejected for:_____
lungs	suspended for:_____

General Appearance: Record should be made as to whether the applicant is of 'poor', 'fair', or 'good physique'. Pay attention to nutrition, musculature, color and condition of mucous membranes, oedema, granular enlargement, rickets, acquired or congenital deformities, gout, rheumatism, ankyloses, "disfiguring" scars or deformities, "premature" senility, etc.

Facial Appearance: Under this heading one or more of the following descriptive terms should be used: 'Pale', 'florid', 'usual', 'oedematous', 'cyanotic', 'jaundiced', 'dissipated', etc.

Head and Scalp: The contour of the head should be noted and look for parasitic infection, favus, etc.

Lungs: Special care should be taken to detect tuberculosis, but until the demonstration of tubercle bacilli in the sputum a definite diagnosis should not be made. In the meantime action should be suspended in the suspected cases until the necessary laboratory and x-ray reports have been furnished by the applicant from a reliable source.

Evidence of Internal Parasites: Discretion should be exercised in respect to examinations for the diseases caused by them. When special laboratory tests are performed record should be made of the fact.

Nervous and Mental: Diseases and defects of these classes are the most important from the standpoint of the immigration laws. Special effort should be made therefore to pass only those persons who are found to be normal.

FIGURE E-1. "Information Collected During and Guide for the Medical Examination Abroad," 1925. (Asst. Surgeon General J. W. Kerr, "Information for Medical Officers Engaged in Examination of Aliens Prior to Granting of Consular Visas at Foreign Ports," 16 July 1925, RG 90, General Subject File, 1924–35, Box 941, File No. 0950-56, Doc. No. H-373)

FORM 100.—*A record of physical examination that requires little clerical work, and provides space for three subsequent examinations. Diagrams are on the reverse side.*

PHYSICAL EXAMINATION.

Name..Check No...................Age..........

Duties recommended for..

Date of examinations.	R.	L.	R.	L.	R.	L.	R.	L.	Date of examinations.	R.	L.	R.	L.	R.	L.	R.	L.
EYES.									**INGUINAL REGION.**								
1. Defective vision....									44. Inguinal hernia....								
2. Old injury.........									45. Inguinal adenitis..								
3. Conjunctivitis.....									46.								
4. Trachoma.........																	
5. Interstitial keratitis									**GENITO-URINARY.**								
6.									47. Chancre...........								
									48. Varicocele........								
EARS.									49. Hydrocele........								
7. Wax in ear........									50. Undescended tes-								
8. Otitis media.......									ticle............								
9. Deafness from other									51. Epididymitis......								
causes............									52.								
10.																	
									EXTREMITIES.								
NOSE.									53. Old fracture.......								
11. Old fracture.......									54. Old mutilation....								
12. Obstruction.......									55. Varicose veins.....								
13.									56. Ankylosed digits...								
									57. Wrist deformities..								
THROAT AND MOUTH.									58. Flat foot..........								
14. Pharyngitis.......									59. Bunion...........								
15. Enlarged tonsils...									60. Ingrowing toenails.								
16. New growths......									61.								
17. Syphilis..........																	
18.									**SKIN.**								
									62. Acne.............								
TEETH.									63. Eczema..........								
19. Defective teeth....									64. Psoriasis.........								
20. Malocclusion......									65. New growth								
21.									66. Syphilis..........								
									67. Other infectious								
TONGUE.									diseases........								
22. New growth.......									68. Scars or identifica-								
23. Syphilis..........									tion marks......								
24.									69.								
									ARTERIES.								
NECK.									70. Arterio sclerosis....								
25. Goitre............									71. Aneurism.........								
26. New growth.......									72.								
27.																	
									BLOOD PRESSURE.								
LUNGS.									73. Systolic..........								
28. Pulmonary tuber-									74. Diastolic..........								
culosis..........									75. Difference or pulse								
29. Pleurisy..........									pressure........								
30. Acute bronchitis...									76. Ratio of P. P. to								
31. Asthma...........									diastolic........								
32. Emphysema.......																	
33.									**GENERAL.**								
									77. Weight in pounds .								
HEART.									78. Height...........								
34. Valvular disease...									79. General appearance								
35. Myocarditis.......									80.								
36.																	
									URINALYSIS.								
ABDOMEN.									81. Sugar............								
37. Enlarged liver.....									82. Albumin..........								
38. Enlarged spleen....									83. Specific gravity....								
39. Chronic appendi-									**MISCELLANEOUS.**								
citis............									84. Stiff joints........								
40. Ventral hernia.....									85.								
41. New growths......									86.								
42. Kidney lesions.....									87.								
43.																	

NOTE.—Mark X to indicate defects that do not require medical measures; XX that require medical measures, but do not disqualify; XXX that disqualify.
Describe XX and XXX conditions on other side, using reference numbers and diagrams where necessary.

PICKANDS, MATHER & CO.

FIGURE E-2. Sample medical examination form used in industry, c. 1910s. (C. D. Selby, "Studies of the Medical and Surgical Care of Industrial Workers," Public Health Bulletin 99 [Washington, D.C.: Government Printing Office, 1919], 100)

PHYSICAL EXAMINATION—MEN.

Name			Clock No.		Dept. No.
Address			Married Single		Age
Nationality	Occupation	Date employed	Date examined		Referred by

What diseases have you had?

Nature		Date	Duration	Complications

What injuries, accidents, or surgical operations have you had?

Nature		Date	Duration	Results

Have you ever had:

Hernia	Rheumatism	Fistula	Venereal disease

Signed.............................

Height	Weight	Temperature	Inspection and palpation of head and neck
Tongue	Teeth	Gums	Throat	Nasal passages

FORM 99.—*Reverse of card shown in figure 98.*

Right Vision:	Left	Color blind	Wear glasses	Right Hearing:	Left
Auscultation Lungs:			Percussion		
Sounds Heart:	Rhythm	Size	Blood pressure: Systolic	Diastolic	Pulse pressure
Character of Pulse:		Condition of arteries	Inguinal or femoral hernia		

Condition of abdominal viscera:

Urinalysis:	Spec. gravity	Albumen	Sugar	Sediment	Microscopic
Pupils	Tremors	Stellwag Groef's Romberg	Spine	Glands	Reflexes

Scars or deformities from operation, injury, or disease...

Evidence of infectious disease...

Accepted	Physically unfit	Rejected

Why?...

(Signed)............................
Examining physician.

FIGURE E-3. Another medical examination form used in industry, c. 1910s. (C. D. Selby, "Studies of the Medical and Surgical Care of Industrial Workers," Public Health Bulletin 99 [Washington, D.C.: Government Printing Office, 1919], 99)

Although diagnostic and laboratory tests were not indiscriminately incorporated into the immigrant medical examination abroad, officers were instructed that "when the facilities for laboratory, X-ray and other technical tests are not available, medical officers should defer reporting until the passenger himself presents reliable reports of such tests made at his expense." Consistent with the now explicit instructions for officers to "satisfy themselves as to the absence among those examined of diseases and defects inadmissible under the immigration laws," medical approval could not be obtained when "objective" tests or procedures were thought to be necessary but were unavailable.[63]

On the one hand, this is a testament to how successfully the PHS had reined in the laboratory: even as diagnostic potential increased, the PHS examiner was not the life insurance examiner who had no occasion to place his own judgment above that of the diagnostic tests he was required to perform. On the other hand, in the context of medical inspection abroad, authority over the lab ceased to have meaning. The onus rested squarely on the immigrant himself or herself to provide "reliable" proof of health when the PHS lacked the necessary means. Health, once read as suitability for taking a place in the industrial work force, was no longer the exclusive domain of the PHS. Rather, the prospective immigrant carried the responsibility to establish the state of his or her own health: the prospective immigrant gained access to the laboratory. Immigrants could now demand, from the network of local physicians and labs established to aid in the medical inspection of immigrants, laboratory tests to help establish that they were disease-free. The relationship between immigrant and doctor, then, was altered: the "burden of proof of eligibility is on the applicant."[64] Consequently, the prospective immigrant paid a "minimum" fee for the PHS examination and was required to pay consultation fees to the network of local specialists and laboratories set up to confirm diagnoses or provide further tests.[65] The medical exam certainly represented a means of screening the potential immigrant, but it also represented a service to which the applicant was *entitled* and for which he or she paid.

In the end, the PHS officers were told to "pass only those persons who are found to be normal."[66] Although at first blush this, too, seems a more rigorous standard of health and one that increased the authority of the PHS officer, in fact, the officer was not asked to exercise much medical judgment, for his recommendation carried no legal weight. Unlike the medical "certificate" PHS officers issued in the United States, the medical "notification" issued abroad had the status of a recommendation only. Consular officers were able to exercise greater discretion in rejecting or approving visa applications than IS officers stationed in the United States had been in excluding immigrants from admission. Nonetheless, the medical notifications could not become the objects of appeal, nor did they mandate exclusion.

Thus, while PHS officers still categorized diseases and defects as Class A, Class B, or Class C, they were no longer required to meet any exacting standards for using these classifications: a consular officer could exclude on the basis of any medical condition, regardless of whether it was a loathsome or dangerous contagious disease, regardless of whether it affected an immigrant's ability to earn a living, and regardless of whether it was serious or minor. Indeed, exactly what constituted "normal" health was an object of concern among those developing the life insurance exams during this period.[67] Stanley Reiser notes that discoveries of the level of disease and disability in the general population were shocking. The rate of infirmities among a sample of 10,000 industrial workers was 100 percent, among 872 immigrants it was 97 percent, and among a 5,000-person sample of the general population the rate was 77 percent. Although most of the defects detected were not serious, assessments of the seriousness of these findings "depended upon defining what the 'normal' was—a problem that many viewed with concern."[68] According to Allan McLaughlin, "The health examination will disclose how healthy a person may be. What is a normal person? Perhaps the term 'normal' is not the best, as it may represent an average and not an ideal. If we take the average we get measurements and figures which do not actually show the ideal, because the average person is somewhat below par. . . . Three out of four persons have been found by actual experience to have physical impairments."[69]

There was no discussion within the PHS about what constituted "normal": "normal" meant, simply, the *absence* of disease or deformity. In this sense, very few could attain the normal when ear infections and defective vision were at issue, making the concept too broad to help differentiate between one immigrant and another, between one class and another, between one group of people and another. Negotiating the meaning of the broad categories of medical exclusion, the meaning of specific diseases, the proper role of the laboratory, the meaning of terms such as *contagion, communicable,* and *normal* as a means of defining and managing the largest industrial work force in the world had once represented the essence of PHS power and authority.

OBSCURING THE GAZE

Back in the United States, immigrant medical inspection on the line was gradually abandoned. By the mid-1920s medical examination at Ellis Island and other immigration stations in the nation was but a shadow of what it had once been. Thousands of arriving immigrants no longer crowded Ellis Island. Immigration at other ports dwindled, and immigrant medical inspection was increasingly accomplished as part of the quarantine exam, for reasons of expediency. Immigrants arriving in the United States from countries where the visa system was not yet in place still received what IS officials referred to as an "intensive line examination," in which they were "stripped to the waist."[70] And although the PHS was still required to examine all second- and third-class passengers carrying approved visas before they were officially landed in the United States, 90 percent of these "confirmatory" exams were conducted aboard ship.[71] The PHS continued to make the medical inspection at U.S. ports because approximately four months would typically pass between inspection of the immigrant abroad and his or her arrival in the United States. "Hence it is most essential for the protection of this country, as well as for that of the alien, that the final medical inspection be made just prior to landing in the United States according to law." Only those immigrants carrying a medical memorandum indicating a potential medical problem issued by consuls abroad were taken to Ellis Island to be reexamined.[72] An immigrant arriving with an "unchecked visa and no accompanying medical memorandum" was cause for suspicion; it was assumed that the immigrant had destroyed his or her memorandum. These immigrants were almost always sent to Ellis Island for reexamination.[73]

In 1925 the chief medical officer at Ellis Island described the significance of the end of the line, voicing a new disdain for the old procedures (see the epigraph). He was quick to dismiss the philosophy of the medical gaze. The only explanation for the existence of former inspection procedures, in his mind, was that "the remedy did not lie within our power. This was the situation: fifteen medical officers, well trained though they were, endeavoring to isolate from an avalanche of 5,000 persons a day all of the persons suffering from one or another of physical or mental conditions specified in the immigration law, and this by the simple process of having these aliens file past them all day long at a distance of about one rod apart. . . . Now, fortunately, this is a thing of the past, although its results will be with us for many years to come."[74]

As the urgency of inspecting immigrants at U.S. borders dissipated and the form and philosophy of immigrant medical inspection was transformed, the PHS altered its perception of the value of the old Ellis Island tradition of medical inspection. Many long-time champions of the gaze endorsed new inspection standards, claiming that inspection abroad raised inspection at U.S. ports to a new level. In 1925, for instance, Assistant Surgeon General Samuel B. Grubbs boasted that "at present a great majority of those passing our immigration stations receive an intensive medical examination that is one compa-

rable to a periodic overhauling by a family physician or a life insurance examination."[75] Yet, despite the rosy gloss that Grubbs gave the improved examination procedures, a new cadre of physicians working with or in the PHS in
the 1920s began to express a certain scorn for medical inspection of immigrants and PHS officers themselves.

Dr. Bernard Notes, who briefly served with the PHS as an Ellis Island in-
tern, declined to take the exam to become a commissioned officer on the
grounds that the organization was becoming too bureaucratic: "I was young
and ambitious and full of vigor, and I wanted to make something of myself—
I didn't want to get a position and draw my check and that was it. Otherwise,
I would have gone into business, or something else."[76] Even by the end of the
war, the prestige of the PHS line officer and his medical gaze had begun to di-
minish. Thus, Dr. T. Bruce H. Anderson, who worked in the Ellis Island hospital, no longer admired the officers who worked the line, speculating that perhaps they had been assigned there as a punishment of sorts: "The officer,
frequently middle-aged, assigned to Line duty where the medical abilities
were not actually used, and where the work physically was difficult, was not
desirable."[77] The new PHS officer wanted to practice scientific medicine, not
master the art of snapshot diagnosis, now no longer consistent with an industrial philosophy of discipline.

Just as physicians associated with the PHS began to express doubts about
their technique and devotion to the once elite corps, the immigrant medical
exam at Ellis Island became the target of often harsh criticism in the period
following World War I. In 1921, for example, Commissioner of Immigration
Frederick A. Wallis was quoted as saying at a University Club meeting, "I believe the medical examinations of to-day are farces. . . . The examination is superficial. Many pass through with governmental permission who are diseased
inwardly."[78] To be sure, the examination of the 1920s was not the same as that
conducted up to the war. Yet the rapidity with which immigrant medical inspection at domestic stations became superfluous was astounding.

Not all memories were so short, however, and the new system of consular
inspections abroad was not greeted with universal praise. The chief medical
officer at Ellis Island, Dr. Lavinder, proved to be a most bitter opponent. As
he complained to the assistant surgeon general in 1930, "something must be
done. Examining officers in Europe are making too many blunders. I do not
know what is the matter with them but it seems to me they need pulling up a
bit. . . . I am very anxious to support to the fullest extent all of the actions of

our medical officers in Europe. It seems to me however that they should display more care and judgement and thereby keep me out of the most embarassing situations, and what is of far more importance, prevent us from discrediting both our work and the work in Europe. That is the important point of the whole thing—the discrediting of our certificates in the minds of laymen."[79] For Lavinder, medical "judgment" in the old tradition of the PHS physician was in decline. Although he was certainly threatened by the increased rate of certification overseas, which seemed to discredit more than two decades of inspection work, it was the supposed indiscriminate certification and imprecise diagnostic classification against which Lavinder railed. In his era the diagnosis had defined the immigrant and his proper place in American industrial society.

The certification of a twenty-five-year-old German woman particularly roused Lavinder's ire. The woman, he alleged, had been certified abroad by Dr. Holt as having "nervous instability—anxiety neurosis." Holt had labeled this as a Class B condition. Lavinder claimed that upon the woman's arrival at Ellis Island, it was clear that she had no mental illness whatsoever. Moreover, had she had a mental condition, it should have been certified as Class A, not Class B: "Perhaps I may be wrong," sniped Lavinder, "but I cannot escape the conclusion that for some reason or reasons unknown to me and operative previous to your arrival in Europe, there has been produced in the minds of all examining officers an impression that the more certificates they render the more commendable their work."[80] Lavinder threatened that "if this business does not stop, we'll have to change our policy and start sending people to Ellis Island again."[81]

Lavinder was not alone in his worries. The commissioner of immigration at Ellis Island was also suspicious of medical inspection abroad and accused the secretary of state of plotting to shut the island down, arguing that the medical inspection system abroad was only experimental and did not justify cursory shipboard inspections at U.S. ports: "It is only at Ellis Island that we can carry out a complete and competent medical examination of those immigrants who are coming to our country for the rest of their lives and for the lives of their children. It is little enough to ask of all immigrants that they come to Ellis Island for two or three hours for such examination, before they come into our country for the rest of their lives. We do not know today how much disease will walk down the gangplank, just because it cannot be detected on the ship or abroad."[82]

The congressional representative from Lima, Ohio, did not disavow inspection abroad but expressed his fear that Ellis Island would be closed based on the success of the new consular inspections (see epigraph). The secretary of labor promptly reassured him that "we will never give up inspection of aliens upon arrival."[83] But while Ellis Island remained open until 1943, the United States had already given up the primary purpose of the exam at the nation's threshold. Inspection at the nation's borders was no longer the first site for disciplining the immigrant work force.[84] This function was lost altogether. Opinion like that of the Ellis Island commissioner of immigration, Frederick Wallis, prevailed. Wallis fully supported inspection abroad from the outset: "'I do not believe that an alien should be permitted to land in this country until he is stripped and thoroughly examined from head to foot, mentally and physically."[85]

In the end, Lavinder accurately assessed the state of medical authority. On the one hand, with his angry yet plaintive protests and empty yet passionate threats, Lavinder seemed to resist progress and begrudge a shift in authority to physicians overseas. On the other, he tried to hold the fort as one of the last true arbiters of industrial citizenship. Despite his best efforts, PHS officers saw their authority and influence limited in the 1920s as the expectations and strategies for managing industrial citizens changed. Once moved abroad, the PHS officer was reduced to the status of a "technical advisor." The medical gaze and the art of snap diagnosis fell into disrepute. The old rules ceased to apply. By 1930, for example, though familiarity with different races was still given lip service, the PHS regulations for the medical inspection of immigrants minimized the diagnostic value of race: "The extent to which the primary examination described in the foregoing may serve to disclose indications of disease will depend somewhat on the medical officer's personal acquaintance with the normal and abnormal characteristics of different types of immigrants, but it is chiefly a matter of the practical utilization of the knowledge of the objective signs of disease to be found described in a comprehensive textbook on physical diagnosis."[86]

There was a new ease of international travel, and improvements had been made in steerage accommodations that altered the way people traveled for both immigration and tourism purposes. Automobiles were increasingly used to cross the border. These changes meant that class also ceased to determine the rigor or form of the immigrant medical exam. The Ellis Island commissioner of immigration explained the breakdown of the correspondence between social class and passage class, noting that the difference between those traveling in cabin and steerage was almost equivalent to the distinction between visitors and immigrants: "The visitors should receive every convenience in landing. Immigrants should receive the most thorough examination, medical and otherwise. The former are our guests, the latter are about to become part of us for life, and the life of our children to come. . . . The present distinction is one of the price of a ticket. This is unfair, un-American and indefensible."[87] Tourist passengers were also beginning to travel in third class—individuals the IS considered "first and second class passengers, from an intellectual and sanitary standpoint."[88] Different inspection standards for different classes were, by the mid 1920s, considered "un-American."[89] As Dr. Eugene Cohn explained to Senator Charles Deneen, "The second cabin on modern lines is as good or better as first cabin used to be twenty years ago. The types of citizens traveling second cabin compare mighty favorably to those traveling first cabin. Am I to presume," he asked with genuine incredulity, "that second class passengers are made of different clay from first cabin travelers? Are they less healthy, less clean, less reliable and less respectable than their wealthier fellows?"[90] The question, of course, implies that third-class passengers were indeed made of a different clay, revealing that Progressive era ideas about immutable differences between populations, even if muted, persisted. Nonetheless, in January 1925 the IS eliminated differential inspections by class at all immigration stations, mandating that "all classes of passengers will receive the same consideration at the hands of the Public Health officials" and would be examined on shipboard.[91]

Historically, however, all classes, all persons, all diseases did not receive the same consideration by the PHS. And herein lay the agency's authority: it had the power to focus the medical gaze as it deemed fit. PHS officers identified not only disease but also those immigrants presenting a broader social and economic threat to the nation. The gaze was, moreover, a means of introducing immigrants to American industrial standards for working and living. The medical gaze—the intensity of which could be varied as needed by region—transformed problems of labor, race, class, and international politics into medical problems resolved either by communicating expectations through the ordeal of inspection or, in more limited instances, through higher rates of exclusion.

APPENDIX

Note on Data Collection, Cleaning, Coding, and Analysis

Limitations of the Data

Most of the data for this book were taken from the Public Health Service (PHS) and Immigration Service (IS) annual reports. I used relational database software (Microsoft Access) to create several databases that could later be linked to bring together data for a particular port or region for a particular year or time period. I transferred linked data to Microsoft Excel for cleaning, coding, analysis, and presentation. The PHS and IS reports contained a vast amount of data, much of it unrelated to immigrant medical inspection. In general, I entered all data directly related to medical inspection and much data more generally related to exclusion. While initially entering the data, I made the variables as specific or discrete as possible.

Although I created several databases, the vast majority of data for this study come from three main databases: the first contains IS data and includes among its 79 variables port name, number of arrivals or admissions, gender of arrivals or admissions, number of debarments, and categories of debarment; the second contains PHS data and includes among its 132 variables port name, number of immigrants examined, and data regarding medical certification and exclusion; the third database contains among its 98 variables IS data on arrivals, certifications, and medical exclusions (both on arrival and after initial admission to the United States) by race or nationality. Other databases contain national rather than port-specific data for each year, data on seamen, and data on medical inspections abroad.

Data given in the annual reports varied over time. Moreover, the two services often reported the same piece of data differently or reported it differently in different years. For example, there were inconsistencies in how overall immigration was reported. The PHS reported the number of immigrants actually examined. It appears that unless specifically broken down into steerage and cabin passengers, the total examined as reported by the PHS includes both steerage and cabin passengers and sometimes nonimmigrants (and in some cases, U.S. citizens). The IS reported total arrived (typically broken into immigrant and nonimmigrant arrivals, as discussed below). In cases in which both pieces of information were available, the number of immigrants examined by the PHS usually equaled the number of immigrants arrived or admitted as reported by the IS; when the numbers were not equal, the PHS figure exceeded the IS figure in about half the cases (293), and the IS figure exceeded the PHS figure in the rest (271). Where possible, I used the PHS figure for number examined as the denominator for estimates of the percentage of immigrants medically certified and then excluded.

The IS was far less consistent in reporting total number of immigrants. First, the IS did not begin to distinguish between "immigrants" and "nonimmigrants" until 1907. Therefore, all figures before 1907 using the IS arrival data as the denominator are probably underestimated. Second, from 1892 to 1903 the IS reported the number of arrivals, while for other years it reported the number of admissions. To calculate percentages, I calculated the total number of arrivals for each year where this information was not reported by summing total admitted and total deported. In addition, to err on the side of conservatism, I used the total arrivals for immigrants and nonimmigrant aliens after 1909 (as opposed to the total arrivals for immigrants only) unless otherwise indicated.

DISEASE CATEGORIES

The disease-specific data in the annual reports of the PHS and IS presented unique problems. The reports contained two different types of disease-related data. First were the broad categories of medical exclusion (Classes A, B, and C), reported for each port by the PHS or the IS. In addition, the PHS reported specific "important" diseases, though it did so inconsistently, sometimes grouping different diseases or conditions and sometimes separating them.

There was a second, detailed set of disease-specific data for selected ports and selected years listing the exact diagnosis for each arriving immigrant certified. For these data I grouped diseases into larger categories. "Mental conditions" includes insanity, imbecility, idiocy, feeblemindedness, and epilepsy (though each year or time period does not include each of these diseases). "Parasitic infections" includes tinea tonsurans (ringworm of the scalp), clonorchiasis (liver fluke), and hookworm. "Tuberculosis" includes—using the lan-

guage of the PHS annual reports—both contagious and noncontagious tuberculosis. Because so few cases of noncontagious tuberculosis were reported,
combining these categories was necessary for data presentation. Similarly,
syphilis, soft chancre, and chancroid are grouped together.

Immigrants were certified relatively rarely as having favus. Since discussion
of favus often went hand in hand with discussion of trachoma and carried many
of the same medical, social, and cultural connotations, I grouped these two
conditions together. Similarly, I combined gonorrhea, syphilis, soft chancre,
and chancroid in the larger category "venereal diseases." Finally, I merged the
category "alcoholism" into "all other diseases and conditions" since, like favus,
it was rarely diagnosed and could not be easily collapsed into another category.

DISEASE CODING

The disease-specific data that were reported in detail for specific ports came
from either the IS or PHS annual reports or, more typically, from Record
Groups 85 and 90 in the National Archives and Records Administration, College Park, Maryland, which preserved many of the original annual reports submitted by individual PHS officers in charge at different ports. I have been able
to report data only for years in which they are available, but, in general, the
data cover a sufficient time-spread to show trends over time. Unlike the data
in the PHS and IS annual reports, these data were quite detailed and specific,
though not comparable to what might be found in a patient chart. I recorded
diseases and conditions certified exactly as they appeared in the original reports.

Although historians have traditionally used modern coding schemes—usually the ninth or tenth revision of the International Classification of Diseases
(ICD)—to code and summarize mortality data, such a methodology would not
work for the immigrant certification data, for language and meaning were
paramount.[1] In general, historians working with mortality data are more in-

---

[1] Kenneth H. Fleiss and Myron P. Gutmann, "Mortality Analysis through Parochial Burial
Registers: The Case of Texas in the Nineteenth Century," *Journal of the History of Medicine
and Allied Sciences* 54 (1999): 296–311; Kenneth H. Fliess, "Mortality Transition among the
Wends of Serbin, Texas, 1854–1884: Investigation of Chances in the Pattern of Death Using
Parochial Records," *Social Biology* 38 (1991): 266–76; A. R. Omran, "The Epidemiological
Transition: A Theory of the Epidemiology of Population Change," *Milbank Memorial Fund
Quarterly* 49 (1971): 509–38; Omran, "Epidemiological Transition in the U.S.: The Health
Factor in Population Change," *Population Bulletin* 32 (1977): 3–42; Robert William Fogel,

terested in "actual" cause of death than in cause of death as recorded. Consequently, symptom-based causes of death, often recorded by priests or physicians in church or parish records, are typically translated into current terminology and then given ICD codes. Imposing this classification system onto early twentieth-century diseases and conditions could distort understandings of different diseases and might result in misinterpretation of patterns of certification.

For example, in ICD-10, "hookworm" is classified under "certain infectious and parasitic diseases." Yet in 1874 it was not recognized, and in 1916 and 1921 hookworm and other parasitic infections were separated from other infectious diseases. Syphilis, like hookworm, is also classified under "certain infectious and parasitic diseases" in ICD-10. In the 1874 PHS "Nomenclature of Diseases," syphilis is classified under "general diseases." In 1916 syphilis was considered "communicable." In 1921 it was classified as a "venereal disease." Finally, trachoma—one of the most rhetorically important of the immigrant diseases—is classified under "certain infectious and parasitic diseases" in ICD-10, yet in the 1874 nomenclature it is unlisted (though, noting this, I coded it under diseases of the eye); in the 1916 nomenclature it is classified under diseases of the eye and annexa, and then in 1921 it is considered a parasitic condition.

Thus, given the importance of using contemporary classifications to capture changes in those classifications and, more important, to present a picture of disease that was consistent with contemporary understandings, rather than using ICD-10, I chose to use the PHS "Nomenclature of Diseases" and coded the data according to the most recent revisions in the classification scheme: up to 1915, I used the nomenclature for 1874; for 1916–20, I used the nomenclature for 1916; and for 1921–30, I used the nomenclature for 1921. Although each classification system provided a code for each disease, I used only the code for the larger category into which each disease or condition was classified.

For each year in which I classified diseases, there were several diseases and conditions that I could not easily classify, because they either did not appear in the PHS nomenclature or did not correspond to the disease or condition listed. If the disease or condition reported by a port was not listed and there was no other disease or condition that seemed a reasonable match, I coded it

"New Sources and New Techniques for the Study of Secular Trends in Nutritional Status, Health, Mortality, and the Process of Aging," *Historical Methods* 26 (winter 1993): 5–43.

as "unable to code" for each time period. For example, in 1919 the PHS officer in Boston certified a case of "very defective speech." Similarly, I was unable to code conditions like "poor physique" or "poor physical development."
In a situation like this, however, the absence of the condition in official nosol-
ogy served as the basis for further discussion (see chapter 5). This category,
then, should not be considered meaningless, for it serves as a measure of how
closely the PHS medical inspectors followed the official nomenclature. Alternatively, it represents the ability of the nomenclature to accommodate the full
range of immigrant diseases.

In cases where the disease was not listed exactly as reported yet seemed to
correspond to another condition or to belong to a particular category based on
the classification of other conditions, I used that category rather than the "un-
known/unlisted" or "miscellaneous" category. For example, "senile imbecility"
was not listed as a disease in the 1874 nomenclature, yet the PHS officer at
Boston reported this condition in 1893. "Senility" was considered a miscellaneous condition, but "imbecility" was considered an intellectual impairment;
because "senility" meant only old age, I coded the condition "senile imbecility" as I would have coded imbecility. Similarly, for 1874 and 1916 "blindness"
was not listed as a specific disease or condition. Nevertheless, I coded it as a
disease or injury of the "eye or annexa."

Once I had coded each disease according to the relevant nomenclature, I
created a master nomenclature based on the categories in the three official
PHS nomenclatures. For the most part, there was a great deal of correspondence between the categories for each nomenclature, and I was able to collapse several categories. When the relationship between categories was not obvious, as in the case of "organs of locomotion" (1874 nomenclature), "diseases
of the joints and bursae" (1916 nomenclature), and "diseases of the muscles
and bones" (1921 nomenclature), going through the process of coding each
disease helped to determine that these categories corresponded closely to
one another. In another example, the 1874 category "diseases of the skin"
contained most of the diseases classified as "parasitic" in the 1916 and 1921
nomenclatures. Therefore, I created a single category, "skin and parasitic diseases." In this manner many, though not all, of the diseases or conditions that
shifted categories over time are captured in the same category. I also collapsed
those categories that were rarely or never used into the "miscellaneous" category. (This category appeared in each of the three official PHS nomenclatures,
though it was not always termed "miscellaneous.")

Because the PHS tended to certify many of the same diseases in immigrants again and again each year, the categories in the different nomenclatures that I decided were equivalent might not have appeared to be equivalent using another data set. Therefore, I would not recommend adopting this coding scheme for another data set without confirming that the same set of diseases is generally included in each of the categories that are linked by the master nomenclature.

Data were available for several ports for several years. Rather than coding the data for each port, I selected the ports to code in two ways. First, I tried to use the major port in a region. Because of the importance of Ellis Island during this period, I also included data from the Ellis Island annual reports.

For the Atlantic coast, Pacific coast, Gulf coast, and Mexican border, several detailed annual reports covering at least several years spread reasonably over the time period were available for Boston, El Paso, and New Orleans. For the Canadian border, a sufficient number of annual reports was not available for a single port. Nevertheless, there were several ports for which an annual report was available for two or three years. Therefore, for the Canadian border, my goal was to represent the entire time period sufficiently, while maintaining regional consistency. Thus, I used data from those ports along the midwestern to eastern portion of the Canadian border only, as this portion of the border showed a pattern slightly different from the western portion of the border. There were not enough data available for the western portion of the Canadian border to present separately. I could find no existing annual reports for stations along the Canadian border from 1906 to 1916, leaving a regrettable gap. There is a similar gap for New Orleans and El Paso.

USE OF STANDARD DEVIATIONS

In chapter 6, I wanted a means of comparing the magnitude of differences in certification and exclusion rates among different races and different U.S. consulates abroad. For all the graphs that make comparisons in rates in terms of standard deviations, I followed these steps:

1. I divided the number certified or excluded by the number arrived for each race for each year to determine each immigrant group's annual rate of certification or annual rate of exclusion.
2. I calculated the mean rate of certification or exclusion for all immigrants for

each year by dividing the total number of all immigrants certified or excluded by the total number of all immigrants arrived that year.

3. I determined the rate of deviation from the mean for each immigrant group's annual certification or exclusion rate by first subtracting the mean annual rate of certification or exclusion (step 2) from each group's individual annual rate of certification or exclusion (step 1) and then dividing this difference by the mean annual rate of certification or exclusion (step 2).

4. I then calculated the standard deviation for each year's set of rates of deviation from the mean.

5. To determine the number of standard deviations of each group's rate from its annual mean, I divided its rate of deviation from the mean (step 3) by its standard deviation (step 4).

Note that I am not claiming to measure statistically significant differences in certification and deportation rates. One determines statistical significance to rule out with great certainty the chance occurrence that a sample from a larger universe shows differences in the variable of interest—in this case, race—where none exist. Finding statistically significant differences between two groups says nothing about the magnitude of the differences.

Typically, significance levels are set at a value of .05, .01, or .001, corresponding to 95%, 99%, and 99.9% certainty. Finding a statistically significant result at the .05 level, then, would allow us to say that we can be 95% certain that the estimated population parameter is significantly different from zero; alternatively, we say that there is a 5% chance that the parameter we estimated is really equivalent to zero.[2] According to J. L. Fleiss, "The proper inference from a statistically significant result is that a nonzero association or difference has been established; it is not necessarily strong, sizable or important, just different from zero. Likewise, the proper conclusion from a statistically nonsignificant result is that the data have failed to establish the reality of the effect under investigation."[3]

My data represent not a sample of arriving immigrants, but rather the entire population of arriving immigrants. Therefore, any differences found are "significant" inasmuch as this means that they are "real." Nevertheless, the

[2] J. Cohen and P. Cohen, *Applied Multiple Regression/Correlation Analysis for the Behavioral Sciences* (Hillsdale, N.J.: Lawrence Erlbaum, 1983).

[3] J. L. Fleiss, "Significance Tests Have a Role in Epidemiologic Research: Reactions to A. M. Walker," *American Journal of Public Health* 76 (1986): 587.

question becomes, Do we care about these differences? Are they important? Expressing difference in terms of the standard deviation does not answer these questions, but it does provide a systematic measure of the magnitude of differences between rates. The standard deviation is the discrepancy of each value or rate in a sample or population of measures from its mean, relative to the variability in all the values or rates.[4] Approximately 67% of the values in a distribution will fall within one standard deviation of the mean. In effect, the standard deviation standardizes data, for it is an example of a linear transformation in which every score is changed by multiplying or dividing by a constant value and adding or subtracting a constant value.

I interpreted all points falling more than one standard deviation away from the mean rate as deviating "substantially," representing a rate either meaningfully higher or lower than expected. In cases where a race falls within one standard deviation of the mean but does so inconsistently, with consistent periods of substantially higher (or lower) rates, I interpreted this as a trend toward meaningful deviation.

[4] Cohen and Cohen, *Applied Multiple Regression*.

NOTES

INTRODUCTION. Immigration by the Numbers

1. Congress created the Public Health Service in 1798 as the U.S. Marine Hospital Service under the jurisdiction of the Treasury Department, where it remained until 1939. Medical officers provided medical care to merchant marines. Congress renamed the service several times. Not until 1912 was it known as the Public Health Service (PHS; the term that I use throughout this book). See Ralph Chester Williams, M.D., *The United States Public Health Service, 1798–1950* (Washington, D.C.: Government Printing Office, 1951).

2. Paul Sigrist, interview with Emma Greiner, 3 March 1991, Ellis Island Oral History Project (hereafter cited as EIOHP).

3. Jastrow described her father's examination at Ellis as interminable: "Hour after hour the examination went on" (Marie Jastrow, *A Time to Remember: Growing Up in New York before the Great War* [New York: W.W. North, 1979], 21, 45). See also Joan Morrison and Charlotte Fox Zabusky, *American Mosaic: The Immigrant Experience in the Words of Those Who Lived It* (New York: E. P. Dutton, 1980), 24.

4. Peter Morton Coan, *Ellis Island Interviews: In Their Own Words* (New York: Facts on File, 1997), 284.

5. Detainees inscribed some 135 legible poems, which historians date to the peak years of Asian immigration before 1930. Eight of the poems make some allusion to the writer's health, the medical examination, or medical treatment of Asian immigrants at Angel Island.

6. Quoted in Him Mark Lai, Genny Lim, and Judy Young, *Island: Poetry and History of Chinese Immigrants on Angel Island 1910–1940* (San Francisco: Chinese Culture Foundation, 1980), 163. While European immigrants produced many narratives of their lives, which often encompassed the journey to America, these poems represent one of the few indicators of the Asian experience. Chinese, Japanese, and Korean exclusion, whether formal or informal, limited the immigration of laborers; those narrative accounts that are available describe the experience of Asian immigrants traveling first-class. See No-Yong Park, *Chinaman's Chance* (Boston: Meador Publishing,

1940); and Etsu Inagaki Sugimoto, *A Daughter of the Samurai* (New York: Doubleday, Page, 1925).

7. Within the category of "dependency," I counted aliens deemed likely to become a public charge, aliens under sixteen years of age and unaccompanied by a parent, assisted aliens (this usually meant aliens who had received financial assistance to travel to the United States), accompanying aliens (immigrants deported because they had to return a dependent or, more typically it seems, a diseased or physically defective relative to the foreign port of origin), and stowaways. For a history of the development of immigrant exclusions as LPC, see Patricia R. Evans, "'Likely to Become a Public Charge': Immigration in the Backwaters of Administrative Law," Ph.D. diss., George Washington University, 1987. Overall, LPC exclusions represented 91% of the "dependency" category. From 1892 to 1916, at least 90% of immigrants I have included under the heading "dependency" were excluded as LPC; from 1917 to 1925, that percentage varied between 69% and 90%; beginning in 1926, LPC consistently represented over 90% of "dependency" exclusions. Nonetheless, whether the subject is disease or the threat of economic dependency stemming from a nonmedical cause, federal immigration statistics convey the overwhelming impression that few immigrants were ever denied entry to the United States for any cause.

8. Dr. Alfred C. Reed, assistant surgeon, U.S. Public Health and Marine Hospital Service, New York City, "The Medical Side of Immigration," *Popular Science Monthly* 80 (1912): 385–87. For another primary-source description of inspection procedures, see Samuel B. Grubbs, *By Order of the Surgeon General* (Greenfield, Ind.: Mitchell, 1943). Secondary-source descriptions include Alan M. Kraut, "Silent Travelers: Germs, Genes, and American Efficiency, 1890–1924," *Social Science History* 12 (winter 1988): 377–94; Elizabeth Yew, "Medical Inspection of Immigrants at Ellis Island, 1891–1924," *Bulletin of the New York Academy of Medicine* 56 (June 1980): 488–510; Fitzhugh Mullan, *Plagues and Politics: The Story of the United States Public Health Service* (New York: Basic Books, 1989); and Rosebud T. Solis-Cohen, "The Exclusion of Aliens from the United States for Physical Defects," *Bulletin of the History of Medicine* 21 (1947): 33–50.

9. Mullan, *Plagues and Politics*.

10. David H. Bennett, *The Party of Fear: From Nativist Movements to the New Right in American History* (New York: Vintage Books, 1990); Richard Hofstadter, *The Age of Reform* (New York: Vintage Books, 1955); John Higham, *Strangers in the Land: Patterns of American Nativism 1850–1925* (New York: Atheneum, 1967); and Robert Wiebe, *The Search for Order, 1877–1920* (New York: Hill and Wang, 1967).

11. Howard Markel, *Quarantine! East European Jewish Immigrants and the New York City Epidemics of 1892* (Baltimore: Johns Hopkins University Press, 1997); and Alan M. Kraut, *Silent Travelers: Germs, Genes, and the Immigrant Menace* (New York: Basic Books, 1994; Baltimore: Johns Hopkins University Press, 1995), 57.

12. Ira Berlin, *Many Thousands Gone: The First Two Centuries of Slavery in America* (Cambridge, Mass.: Belknap Press of Harvard University Press, 1998), 1.

13. Particularly in Britain, where eugenics was also about heredity and class differences in heredity value and fertility, the concept of race was closely intertwined with that of class. See, for example, Donald McKenzie, *Statistics in Britain, 1865–1930: The Social Construction of Scientific Knowledge* (Edinburgh: Edinburgh University Press, 1981); Greta Jones, *Social Darwinism and English Thought: The Interaction between Biological and Social Theory* (Brighton, Sussex: Harvester Press, 1980); G. R. Searle, *Eugenics and Politics in Britain, 1900–1914* (Leiden: Noordhoff International, 1976). In the United States, Nicole Hahn Rafter has shown how the concept of biological unfitness was used to construct the notion of a white southern underclass (*White Trash: The Eugenic Family Studies, 1877–1919* [Boston: Northeastern University Press, 1988]). Nonetheless, the ways in which class overlaps and intertwines with the concepts of race or ethnicity remain largely untheorized.

14. Eric Foner, *The Story of American Freedom* (New York: W.W. Norton, 1998).

15. Matthew Frye Jacobson, *Whiteness of a Different Color: European Immigrants and the Alchemy of Race* (Cambridge, Mass: Harvard University Press, 1998), 6.

16. Foner, *Story of American Freedom*; Ian F. Haney-Lopez, *White by Law: The Legal Construction of Race* (New York: New York University Press, 1996).

17. Rather than specifying that unskilled laborers were barred from entry, the law made exceptions for government officers, ministers, missionaries, lawyers, physicians, chemists, civil engineers, teachers, students, authors, artists, merchants, and travelers, along with their wives and children under age sixteen (5 February 1917, Ch. 29, 39 Stat. 874).

18. Exceptions were made for ministers and professors, along with their wives and unmarried children (24 May 1924, Ch. 182, 43 Stat. 140).

19. Whiteness is a notion whose time has come for challenge and further scrutiny. See, for example, Eric Arnesen, "Whitness and the Historians' Imagination"; David Brody, "Charismatic History: Pros and Cons"; and Barbara J. Fields, "Whitness, Racism, and Identity"; all in *International Labor and Working-Class History* 60 (fall 2001): 3–32, 43–47, and 48–56, respectively. I do not seek to redress the current criticisms of whiteness studies, but this book does open a window onto the thinking of two groups—the PHS and IS—concerned with defining the nation, and the way in which they ultimately ordered the peoples of the world.

20. Foner, *Story of American Freedom*, 112.

21. Jacobson, *Whiteness*, 20.

22. In his groundbreaking study explaining how the Chinese community in San Francisco mobilized to transform itself and the status of its residents in the twentieth century, Nayan Shah relies on a notion of citizen-subject as a way to capture community integration in addition to political or civic participation. In so doing, Shah is able

not only to focus on the Chinese as targets of public health exclusion, but to view them in terms of inclusion and accommodation after the 1920s, as public health practice began to emphasize more welfare-like projects and programs touching on areas of Chinese life such as recreation and housing. Nayan Shah, *Contagious Divides: Epidemics and Race in San Francisco's Chinatown* (Berkeley: University of California Press, 2001).

23. Democratic National Committee, *The Political Reformation of 1884: A Democratic Campaign Handbook* (1884), quoted in Gwendolyn Mink, *Old Labor and New Immigrants in American Political Development: Union, Party, and State, 1875–1920* (Ithaca, N.Y.: Cornell University Press, 1986), 107.

24. Mink, *Old Labor and New Immigrants*, 90–91, 96.

25. Alexander Saxton, *The Indispensable Enemy: Labor and the Anti-Chinese Movement in California* (Berkeley: University of California Press, 1971).

26. In addition to the enduring work of John Higham, see also the fascinating and provocative work of Jacobson, *Whiteness*.

27. Higham, *Strangers in the Land*, 103.

28. Francis A. Walker, "Restriction of Immigration," *Atlantic Monthly*, June 1896, 828. While Walker, in my reading, emphasized questions of civic citizenship, he did not divorce them from questions of industrial citizenship, arguing for "protecting the American rate of wages" and the "American standard of living."

29. Higham, *Strangers in the Land*, 43–44, 48–49, 73, 99–100, 112, 129–30, 202–4, 221.

30. Foner argues that "reconstruction helped to solidify the separation of political and economic spheres" (*Story of American Freedom*, 113).

31. The Page Law of 1875, primarily an anti-Chinese measure, also prohibited the "importation" of prostitutes and persons convicted of crimes of "mortal turpitude" (18 Stat. 477 [1875]). The Immigration Act of 1882, however, is generally regarded as the first federal effort to control immigration. See Higham, *Strangers in the Land*, 356n.19; and Benjamin Klebaner, "State and Local Immigration Regulation in the United States before 1882," *International Review of Social History* 3 (1958): 269–95.

32. Higham, *Strangers in the Land*, 44. See also John Higham, "Origins of Immigration Restriction, 1882–1897: A Social Analysis," *Mississippi Valley Historical Review* 39, no. 1 (June 1952): 79–80.

33. Higham, *Strangers in the Land*, 43–44.

34. Higham, "Origins of Immigration Restriction, 1882–1897."

35. In its 1886 ruling in the *Wabash* case, the Supreme Court recognized corporations as "persons" protected under the Fourteenth Amendment to the Constitution. The ruling also prohibited states from regulating interstate commerce, giving sole regulatory authority to the federal government.

36. Foner, *Story of American Freedom*, 122–23, 142–45.

37. Ford Motor Company, *Helpful Hints and Advice to Employees to Help them Grasp the Opportunities which Are Presented to them by the Ford Profit Sharing Plan* (Detroit, 1915), and *A Brief Account of the Educational Work of the Ford Motor Company* (Detroit, 1916), quoted in Stephen Meyer, "Adapting the Immigrant to the Line: Americanization in the Ford Factory, 1914–1921," *Journal of Social History* (fall 1980): 81.

38. Bureau of Public Health and Marine-Hospital Service, *Book of Instructions for the Medical Inspection of Immigrants* (Washington, D.C.: Government Printing Office, 1903), 5, 10–11.

39. Higham, *Strangers in the Land*, 114–15.

40. Frederick W. Taylor, *Principles of Scientific Management* (Atlanta: Engineering and Managing Press, 1911; rpt., 1998), 48.

41. The most influential study of deviance and stigma has been Erving Goffman, *Stigma: Notes on the Management of a Spoiled Identity* (New York: Simon and Schuster, 1963).

42. Elias Norbert, *The Civilizing Process* (New York: Urizen Books, 1978); Jean-Pierre Goubert, *The Conquest of Water: The Advent of Health in the Industrial Age* (Cambridge: Polity Press, 1986, 1989); Alain Corbin, *The Foul and the Fragrant: Odor and the French Social Imagination* (Cambridge, Mass.: Harvard University Press, 1986).

43. Michel Foucault, *Discipline and Punish: The Birth of the Prison* (New York: Vintage Books, 1979), 25–26. Although a common criticism of Foucault's notion of power is that it leaves no room for individual or group agency, his is an appropriate paradigm for understanding in the context of immigrant medical inspection, where, from the time they left their homes until they arrived in America, immigrants could exercise very little autonomy. I discuss the question of power and agency further at the end of chapter 2. Shah has more directly taken up the question of public health, power, and agency in his *Contagious Divides*.

44. Foucault, *Discipline and Punish,* 138.

45. William D. Haywood and Frank Bohn, *Industrial Socialism* (Chicago, n.d.), 25, quoted in David Montgomery, *The Fall of the House of Labor: The Workplace, the State, and American Labor Activism, 1865–1925* (Cambridge: Cambridge University Press, 1987), 45.

46. D. J. Kevles, *In the Name of Eugenics: Genetics and the Uses of Human Heredity* (Berkeley: University of California Press, 1985); M. H. Haller, *Eugenics: Hereditarian Attitudes in American Thought* (New Brunswick, N.J.: Rutgers University Press, 1963); and K. Ludmerer, *Genetics and American Society: A Historical Appraisal* (Baltimore: Johns Hopkins University Press, 1972).

ONE. Immigrants and the New Industrial Economy

Epigraph: Bureau of Public Health and Marine-Hospital Service, *Book of Instructions for the Medical Inspection of Aliens* (Washington, D.C.: Government Printing Office, 1910). See also letter from A. H. Glennan, acting surgeon general, to the commissioner general of immigration, 7 May 1909. RG 90, Central File, 1897–1923, Box 36, File No. 219, NARA.

1. Emma Lazarus, "The New Colossus" (1883).

2. John Higham, "The Transformation of the Statue of Liberty," in *Send These to Me: Immigrants in Urban America* (Baltimore: Johns Hopkins University Press, 1975, 1984), 73, 75, 76.

3. The Knights of Labor, previously opposed only to contract labor, favored broad restriction of immigration in 1892—the year the Ellis Island immigration station opened. The American Federation of Labor, which still had a large foreign-born membership, followed suit, and by 1893 union endorsement of immigration restriction was relatively widespread. John Higham, *Strangers in the Land: Patterns of American Nativism 1850–1925* (New York: Atheneum, 1967), 71.

4. David Montgomery, *The Fall of the House of Labor: The Workplace, the State, and American Labor Activism, 1865–1925* (Cambridge: Cambridge University Press, 1987), 12. See also David M. Gordon, Richard Edwards, and Michael Reich, *Segmented Work, Divided Workers: The Historical Transformaion of Labor in the United States* (Cambridge: Cambridge University Press, 1982); and David Brody, *Workers in Industrial America: Essays on the Twentieth Century Struggle* (New York: Oxford University Press, 1980).

5. *National Labor Tribune*, 26 August 1882, quoted in Montgomery, *Fall of the House of Labor*, 15. See also Frederick W. Taylor, *Principles of Scientific Management* (Atlanta: Engineering and Managing Press, 1911, 1998).

6. Montgomery, *Fall of the House of Labor*, 24, 25; Roger Daniels, "Two Cheers for Immigration," in *Debating American Immigration, 1882–Present,* ed. Roger Daniels and Otis L. Graham (Lanham, Md.: Rowman and Littlefield, 2001), 5–24; and Kitty Calavita, "U.S. Immigration Policymaking: Contradictions, Myths, and Backlash," in *Regulation of Immigration: International Experiences,* ed. Anita Böcker et al. (Amsterdam: Het Sphinhuis, 1998), 140.

7. Michael McGovern, *Labor Lyrics, and Other Poems* (Youngstown, Ohio, 1899), 27–28, quoted in Montgomery, *Fall of the House of Labor,* 25.

8. Edith Abbott, "The Wages of Unskilled Labor in the United States," *Journal of Political Economy* 13 (June 1905): 324.

9. Terence Vincent Powderly, "With the Board of Review: A Plea for Better Immigration Laws" (n.d.), Terence V. Powderly Papers, Reel 83, Document 9, Part 4, Pamphlets, 1883–1905, Bancroft Library, University of California, Berkeley.

10. Daniel Nelson, *Frederick W. Taylor and the Rise of Scientific Management* (Madison: University of Wisconsin Press, 1980), 10, 11. See also Daniel Nelson, *Managers and Workers: Origins of the New Factory System in the United States, 1880–1920* (Madison: University of Wisconsin Press, 1975).

11. Gordon, Edwards, and Reich refer to this as the "drive system" and categorize it in three ways. First, it involved "reorganization of work, facilitated by both mechanization and job restructuring, which produced increasingly homogeneous employment for production workers." Second, it was characterized by "a rapid increase in plant size . . . , which reinforced the spreading impersonality of wage labor." Finally, it allowed the "continuing expansion of the freeman's role, which added an insistent supervisory impetus to the new system of employer control" (*Segmented Work, Divided Workers,* 128).

12. Taylor, *Principles of Scientific Management,* 89.

13. H. A. Worman, "Recruiting the Workforce. IV—Hiring the Unskilled Workman," *Factory* 1 (February 1908):158, quoted in Montgomery, *Fall of the House of Labor,* 61.

14. Alter Abelson, "The Designer," in *Songs of Labor* (Newburgh, N.Y.: Paebar Co. Publishers, 1947), 67.

15. Taylor, *Principles of Scientific Management,* 30–31, 83.

16. Abelson, "The Song of the Boss," in *Songs of Labor,* 82.

17. Nelson, for example, convincingly argues that Taylor's management system, "both as symbol and substance of the factory revolution, profoundly affected American industry, but its impact in practical terms consisted mostly of changes in machine operations, plant layout, and managerial activities" (*Taylor and the Rise of Scientific Management,* x). He concludes that "the effects of scientific management on the workers were minimal" (151). Although some scholars, like Melvyn Dubofsky, feel that Montgomery failed to respond to the challenge raised by Daniel Nelson that Taylorism failed to change industrial management or alter the nature of work, Nelson and Montgomery treat scientific management very differently. Nelson is more interested in the internal history of scientific management as defined by Taylor and his disciples, while Montgomery treats Taylorism as reflecting broader social and industrial norms. Both agree on the profound change in the nature of factory work that made Taylor such an important cultural and political figure. Melvyn Dubofsky, review of Montgomery, *Journal of American History* 75, no. 1 (June 1988): 215–17.

18. *Taylor and the Rise of Scientific Management,* 20, 168. Nelson argues that *The Principles of Scientific Management* was far more "than a promotional tract" (ibid., 173).

19. Montgomery, *Fall of the House of Labor,* 215, 249, 223.

20. Quoted in Brody, *Workers in Industrial America,* 4.

21. F. Paul Miceli, *Pride of Sicily* (New York: Gaus' Sons, 1950), 60. See also

Lorenzo D. Gillespie, *Songs of Labor* (Salina: Central Kansas Publishing Co., 1904), 52, where the factory is described as "consuming" the worker's good health.

22. Morris Rosenfeld, "The Pale Operator," in *Songs from the Ghetto* (Boston: Copeland and Day, 1898; rpt., Upper Saddle River, N.J.: Literature House/Gregg Press, 1970), 9. In this same volume, Rosenfeld's poem "In the Sweat-Shop" relates the shop floor to a "bloody battlefield," where "[t]he corpses fight for strangers, for strangers! / and they battle, and fall, and disappear into the night" (5).

23. Quoted in Brody, *Workers in Industrial America,* 6–7. Pauline Newman, who emigrated from Lithuania in 1901, wrote of garment work: "What I had to do was not really very difficult. It was just monotonous. . . . [Y]ou did the same thing from seven-thirty in the morning till nine at night." Indeed, "the employers didn't recognize anyone working for them as a human being. You were not allowed to sing. . . . We weren't allowed to talk to each other" (Joan Morrison and Charlotte Fox Zabusky, *American Mosaic: The Immigrant Experience in the Words of Those Who Lived It* [New York: E. P. Dutton, 1980], 10).

24. Rosenfeld, "In the Sweat-Shop," in *Songs from the Ghetto,* 3. See also Charles Denby, *Indignant Heart: A Black Worker's Journal* (Detroit: Wayne State University Press, 1978), 31. Denby explains that in the Detroit foundries the shift to piecework in 1924 meant that "we had to work just like a machine." See also Alma Herbst, *The Negro in the Slaughtering and Meat-Packing Industry in Chicago* (Boston: Houghton Mifflin, 1932), 6, who says that "'robots' became the preferred workmen."

25. Montgomery, *Fall of the House of Labor,* 116.

26. Rose Cohen, *Out of the Shadow* (New York: J. S. Ozer, 1971), 74, 81, 25.

27. See also the narrative of a Lithuanian meat packer, c. 1903, in Rhoda Hoff, *America's Immigrants: Adventures in Eyewitness History* (New York: Henry Z. Walck, 1967), 110, who describes the imperative for speedy work.

28. Alter Abelson, "A Typist Plaint," in *Songs of Labor,* 56.

29. Ibid., 56–57.

30. Rosenfeld, "In the Sweat-Shop," in *Songs from the Ghetto,* 3, 5.

31. Daniels, "Two Cheers for Immigration," 16.

32. Frank P. Sargent, commissioner general of immigration, "The Need of Closer Inspection and Greater Restriction of Immigrants," *Century Magazine* 67 (January 1904): 470–72.

33. Terence Vincent Powderly, quoted in Fitzhugh Mullan, *Plagues and Politics: The Story of the United States Public Health Service* (New York: Basic Books, 1989). The nation as hospital was a popular metaphor: "America must not be made a lazaretto, either physical or moral" ("An Alien Antidumping Bill," *Literary Digest* 69 [7 May 1921]: 13).

34. Herman J. Schulteis, *Report on European Immigration to the United States of America and the Causes Which Incite the Same; with Recommendations for the Further*

Restriction of Undesirable Immigration and the Establishment of a National Quarantine, Submitted January 19, 1892 (Washington, D.C.: Government Printing Office, 1892), 25, quoted in Howard Markel, *Quarantine! East European Jewish Immigrants and the New York City Epidemics of 1892* (Baltimore: Johns Hopkins University Press, 1997), 9.

35. Victor Heiser, *An American Doctor's Odyssey: Adventures in Forty-five Countries* (New York: W. W. Norton, 1936), 15, 37–38.

36. Victor Safford, *Immigration Problems: Personal Experiences of an Official* (New York: Dodd, Mead, 1925), 252.

37. Judith Walzer Leavitt, "'Typhoid Mary' Strikes Back: Bacteriological Theory and Practice in Early Twentieth-Century Public Health," *Isis* 83 (1992): 608–29; Naomi Rogers, *Dirt and Disease: Polio before FDR* (New Brunswick, N.J.: Rutgers University Press, 1992); Nancy Tomes, "The Private Side of Public Health: Sanitary Science, Domestic Hygiene, and the Germ Theory, 1870–1900," *Bulletin of the History of Medicine* 64 (1990): 509–39.

38. Maynard W. Swanson, "The Sanitation Syndrome: Bubonic Plague and Urban Native Policy in the Cape Colony, 1900–1909," *Journal of African History* 18 (1977): 387–410; John W. Cell, "Anglo-Indian Medical Theory and the Origins of Segregation in West Africa," *American Historical Review* 91 (1986): 307–35; Randall M. Packard, *White Plague, Black Labor: Tuberculosis and the Political Economy of Health and Disease in South Africa* (Berkeley: University of California Press, 1989), 52–66.

39. Barbara Bates, *Bargaining for Life: A Social History of Tuberculosis, 1876–1938* (Philadelphia: University of Pennsylvania Press, 1992),16–18; Georgina D. Feldberg, *Disease and Class: Tuberculosis and the Shaping of Modern North American Society* (New Brunswick, N.J.: Rutgers University Press, 1995), 14, 44, 3–5; Sheila M. Rothman, *Living in the Shadow of Death: Tuberculosis and the Social Experience of Illness in American History* (New York: Basic Books, 1994), 13–15.

40. Bureau of Public Health and Marine-Hospital Service, *Book of Instructions for the Medical Inspection of Immigrants* (Washington, D.C.: Government Printing Office, 1903), 5, 10–11 (hereafter cited as PHS, *1903 Book of Instructions*).

41. Bureau of Public Health and Marine-Hospital Service, *Book of Instructions for the Medical Inspection of Aliens* (1910), 5 (hereafter cited as PHS, *1910 Book of Instructions*).

42. PHS, *1903 Book of Instructions*, 10.

43. The only disease the PHS ever explicitly listed as one of the "minor" Class C conditions was pregnancy (PHS, *1910 Book of Instructions*, 20).

44. Ibid., 18.

45. Letter from George Stoner to the surgeon general, 15 July 1907, RG 90, Central File, 1897–1923, Box 36, File No. 219, NARA.

46. Ibid.

47. Safford, *Immigration Problems*, 276–79.

48. Letter from George Stoner to the surgeon general, 15 July 1907, RG 90, Central File, 1897–1923, Box 36, File No. 219, NARA.

49. Letter from George Stoner to the surgeon general, 19 July 1907, RG 90, Central File, 1897–1923, Box 36, File No. 219, NARA.

50. Letter from George Stoner to the surgeon general, 12 July 1907, RG 90, Central File, 1897–1923, Box 36, File No. 219, NARA; Memorandum of Conference Held in Office of Commissioner-General, 23 October 907, RG 85, Box 55, File No. 51758/3, NARA; Memorandum re Changes Which Should be Made in Said Act to Overcome Such Difficulties, 14 January 1908, RG 85, Box 4, File No. 51389/14E, NARA; Memorandum in re Difficulties Encountered during Six Months' Experience with the Administration of the Immigration Act Approved February 20, 1907, 7 January 1908, RG 85, File No. 51538/6; letter from Robe Carl White, second assistant secretary, to the surgeon general, 26 December 1924, RG 85, Box 265, File No. 5461/13, NARA.

51. In his annual report the surgeon general of the PHS rarely reported data regarding immigrant medical certification for more than one or two ports before 1909. By that time, however, careful recording of the disease status of immigrants was paramount.

52. The pattern of certifications according to the "immigrant nomenclature" is very similar for the different regions of the United States. Although Class B conditions typically represented the majority of certifications, along the Pacific and Gulf coasts Class A and B conditions represented approximately the same proportion of certifications. Class B conditions accounted for the largest percentage of certifications along the Canadian border.

53. PHS, *1903 Book of Instructions*, 7. See also George W. Stoner, "Immigration—The Medical Examination of Immigrants and What the Nation Is Doing to Debar Aliens Afflicted with Trachoma," *Medical News* (10 June 1905): 1070. This rationale was also used for the exclusion of immigrants with eczema and other skin diseases. See L. Duncan Bulkley, "On the Exclusion of Immigrants Affected with Diseases of the Skin," *American Academy of Medicine* 14 (1913): 259.

54. Allan J. McLaughlin, *Personal Hygiene: The Rules for Right Living* (New York: Funk and Wagnalls, 1924), 40. See also Martin Cohen, "The Importance of Ophthalmological Examinations in Immigrants," *New York State Journal of Medicine* 13 (November 1913): 603.

55. Heiser, *An American Doctor's Odyssey*, p. 14.

56. Letter from the commissioner of immigration, Ellis Island, to the commissioner general of immigration, 26 October 1897, RG 85, Box 265 (renumbered 228), File No. 54261/12; letter from attorneys for Seropian, Nassau St., New York, to Treasury Department, U.S. Immigration Service, on the matter of the appeal of Nazaret Seropian, from the decision of the Board of Special Inquiry, debarring him from landing, n.d., RG 85, Box 265 (renumbered 228), File No. 54261/12, NARA.

57. Letter from Frank H. Larned, acting commissioner general, to the commis-

sioner of immigration, Ellis Island, 30 October 1897, RG 85, Box 265 (renumbered 228), File No. 54261/12, NARA.

58. Letter from Marcus Braun to Commissioner General Frank Sargent, 26 September 1903, RG 85, Box 27, File No. 51463/A, NARA.

59. Report on trachoma for the Mexican government, 2 March 1907, RG 85, Box 27, File Nos. 51463/A and B, NARA. Mexico passed its first immigration law, modeled on U.S. law, in 1909. Republic of Mexico, Department of the Interior, Division of Immigration, Decree Designating the Frontier Places Authorized for the Entry of Passengers into the Republic and Regulations for the Inspection of Immigrants, 1909, RG 85, Box 439, File No. 55609/551, NARA.

60. CGAR, 1914, 10–11.

61. Letter from the commissioner general to the acting secretary, 9 December 1911. See also the cases of ingratitude described in a letter from William Williams, commissioner of immigration, Ellis Island, to Commissioner General Keefe, 27 July 1911. During 1911 the IS claimed to allow treatment on a more limited basis. Department Circular No. 235, Bureau of Immigration and Naturalization, "Imposition of Fines under Section 9 of the Immigration Act," Department of Commerce and Labor, Office of the Secretary, Washington, D.C., 25 November 1911; letter from Commissioner General Keefe to commissioner of immigration, Angel Island, 23 August 1912; letter from Commissioner General A. Caminetti to commissioner of immigration, San Francisco, 29 September 1913; letter from William Williams to the commissioner general, 2 March 1913. All in RG 85, Box 121, File No. 52516/11A, NARA.

62. "Cases of Loathsome and Dangerous Contagious Diseases Ordered Held by the Department of Commerce and Labor for Treatment Which Illustrates Some of the Difficulties Presented by Cases of This Character," 22 March 1913, RG 85, Box 121, File No. 52516/11B. See also memorandum from Commissioner General Keefe to the acting secretary, 29 July 1909, RG 85, Box 121, File No. 52516/11A, NARA.

63. Draft of a manuscript entitled "Medical Inspection of Aliens," 1929, RG 90, General Subject File, 1924–35, Box 941, File No. 0950-56, NARA. In Boston, for example, PHS inspectors complained of the endless stream of legal cases that immigrants filed upon rejection. Letter from Victor Safford to Surgeon General Rupert Blue, 20 August 1913, RG 90, Central File, 1897–1923, Box 58, File No. 409, NARA.

64. CGAR, 1914, 10–11.

65. The law exacted a head tax of fifty cents per immigrant to cover the expenses of those who defaulted on their hospital bills. By 1917 each immigrant paid a head tax of eight dollars.

66. Memorandum from J. W. Kerr, chief medical officer, to the commissioner of immigration, 28 October 1919, RG 85, Box 262, File No. 54202/17, NARA.

67. Hospital reports for the different immigration stations across the nation can be found in the CGAR and scattered throughout RG 85 and RG 90, NARA.

68. Letter from J. McMullen, chief medical officer, Ellis Island, to the surgeon general, 10 January 1921, RG 90, Central File, 1897–1923, Box 38, File No. 219, NARA.

69. CGAR, 1914, 195.

70. See Charles Rosenberg, *The Cholera Years* (Chicago: University of Chicago Press, 1962, 1987); Francois Delaporte, *Disease and Civilization: The Cholera in Paris, 1932* (Cambridge, Mass.: MIT Press, 1986); S. L. Gilman, *Disease and Representation: Images of Illness from Madness to AIDS* (Ithaca, N.Y.: Cornell University Press, 1988).

71. Andrew McClary, "Germs Are Everywhere: The Germ Threat as Seen in Magazine Articles, 1890–1920," *Journal of American Culture* 3 (1980): 33–46; Suellen Hoy, *Chasing Dirt: The American Pursuit of Cleanliness* (New York: Oxford University Press, 1995); Terra Ziporyn, *Disease in the Popular American Press: The Case of Diphtheria, Typhoid Fever, and Syphilis, 1870 to 1920* (New York: Greenwood Press, 1988), 23, 36, 147. Ziporyn also points out that while both "low-brow" and "sophisticated" magazines often used the term *bacteria* without explanation, the "middle genres" described bacteria as "microbes" and "germs," "often personifying them as evil little people" (147). Regardless, the lay public was inundated with the language of germs; as a collection of letters written to scientists at the Rockefeller Foundation regarding polio demonstrates, people were comfortable discussing germ theory and bacteriology, sometimes in sophisticated terms. See Rogers, *Dirt and Disease,* chap. 4. The foreign-language press also familiarized readers with the language and concepts of germ theory. See, for example, an article in the Polish press that discussed tuberculosis as a "contagious" disease "caused by little organisms called bacilli." "Health and Sanitation," *Dziennik Zwiazkowy,* 7 November 1911, Chicago Foreign Language Press Survey, Immigration History Research Center (IHRC), University of Minnesota, St. Paul. See also the Louisiana State Board of Health, Almanac for 1918, which emphasized the importance of "germs" in a discussion of "infectious" conditions such as malarial, typhoid, tuberculosis, and gum diseases. In its 1919 almanac the Louisiana board boldly stated that "disease germs are the greatest enemies of mankind" (24). Again in 1927, the almanac listed as one of its "health aphorisms" on the inside back cover, "Remember that bacteria are our greatest enemies." The almanac also discussed the role of science in disease, emphasizing Koch's and Pasteur's contributions to bacteriology and Lister's to antiseptic surgery (State Museum, New Orleans).

72. Tomes, "Private Side," 511–12. McClary argues that writers often exaggerated the importance of germs in the spread of disease, and articles increasingly focused on "the many ways in which germs are able to reach us" (McClary, "Germs Are Everywhere," 34).

73. McClary, "Germs Are Everywhere," 34–39; Rogers, *Dirt and Disease,* 42. For an extended list of disease vectors, see Charles Chapin, *How to Avoid Infection* (Cambridge: Harvard University Press, 1917), 44–51; and Charles V. Chapin, *The Sources and Modes of Infection* (New York: John Wiley and Sons, 1910), 126, 146–53, 170–

212. Within the IS, we find evidence that an immigrant inspector in Key West believed that syphilis could be transmitted via cigars. The officer had observed that local cigar makers moistened the tips of the cigar with their mouths to seal the tobacco leaf. Moreover, he believed that syphilis could be diagnosed "by pulling down the lower lid of the eye, or examination of the features." See letter from Joseph Y. Porter, M.D., state health officer, Florida, to Surgeon General Wyman, 27 July 1903; letter from Inspector Eager to Commissioner General Sargent, 29 June 1903; both in RG 90, Central File, 1897–1923, Box 485, File No. 4446, NARA.

74. From Franz Lehar's operetta *Alone at Last,* 1915.

75. William Z. Ripley, "Races in the United States," *Atlantic Monthly* 102 (December 1908): 747.

76. Remsen Crawford, "The Deportation of Undesirable Aliens," *Current History* 30 (1929): 1077.

77. James Davenport Whelpley, "Control of Emigration in Europe," *North American Review* 80 (June 1905): 867.

78. James Davenport Whelpley, "International Control of Immigration: The Startling Facts about the Organized Movement of Undesirable Populations—The Physical and Economic Dangers to the United States Are So Great That Both European and American Regulation Is Necessary," *World's Work* 8 (September 1904): 5255; W. E. Chandler, "Methods of Restricting Immigration," *The Forum* 13 (March 1892): 128.

79. Oswald Ottendorfer, "Are Our Immigrants to Blame," *The Forum* 11 (July 1891): 543; John Weber, commissioner of immigration, and Charles Steward, New York Chamber of Commerce, "Our National Dumping-Ground: A Study of Immigration," *North American Review* 154 (1892): 424–38; Terence Vincent Powderly, "Immigration's Menace to the National Health," *North American Review* 175 (1902): 55–56; Roland P. Falkner, "Some Aspects of the Immigration Problem," *Political Science Quarterly* 19 (March 1904): 32–49; Whelpley, "International Control of Immigration," 5254; U.S. Senate, *Report of the Immigration Commission, Statements and Recommendations Submitted by Societies and Organizations Interested in the Subject of Immigration,* vol. 2 (Washington, D.C.: Government Printing Office, 1911), Statement of the Immigration Restriction League and Statement of the American Federation of Labor; Alfred C. Reed, "The Relation of Ellis Island to the Public Health," *New York Medical Journal* 98 (1913): 173; Charles T. Nesbitt, "The Health Menace of Alien Races," *World's Work* 27 (November 1913): 75; Arthur H. Gleason, "The Yellow Peril," *Harper's Weekly* 58 (2 May 1914): 8; "Social Deterioration of the United States from the Stream of Backward Immigrants," *Current Opinion* 57 (November 1914): 340; "Keep America 'White'!" *Current Opinion* 74 (1923): 399; and "Sifting Immigration," *The Nation* 128 (3 April 1929): 389. Immigrants and those sympathetic to immigrants also invoked such metaphors. See, for example, Abraham Cahan, *Yekl and the Imported Bridegroom and Other Stories of the New York Ghetto* (New York: Dover, 1970), 13.

80. Robert A. Woods, ed., *Americans in Process: A Settlement Study by Residents and Associates of the South End House* (Boston: Houghton, Mifflin, 1903), 40, 44.

81. Weber and Steward, "Our National Dumping-Ground"; Francis A. Walker, "Restriction of Immigration," *Atlantic Monthly* 77 (June 1896): 828; Powderly, "Immigration's Menace to the National Health," 53; Whelpley, "International Control of Immigration," 5255; Rederick Austin, "What an Immigrant Inspector Found in Europe," *World To-Day* 11 (August 1906): 803–7; Ripley, "Races in the United States," 747; U.S. Senate, *Report of the Immigration Commission, Statements and Recommendations Submitted by Societies and Organizations Interested in the Subject of Immigration,* vol. 2, Statement of the American Federation of Labor; H. F. Sherwood, "Those Who Go Back," *Harper's Weekly* 56 (20 July 1912):18; "No Dumping Here," *The Independent* 105 (7 May 1921): 485; "An Alien Antidumping Bill," *Literary Digest* 69 (7 May 1921): 12–13; "Keep America 'White'!" 401; Henry H. Curran, "Smuggling Aliens," *Saturday Evening Post* 197 (31 January 1925): 145; William T. Ellis, "Americans on Guard," *Saturday Evening Post,* 196 (25 August 1923): 80; *New York Times,* 29 August 1892, 1.

82. Chandler, "Methods of Restricting Immigration," 133; Ripley, "Races in the United States," 746; Herman J. Schulteis, *Report on European Immigration to the United States of America and the Causes Which Incite the Same . . . Submitted January 19, 1892* (Washington, D.C.: Government Printing Office, 1892), 25; U.S. Senate, *Report of the Immigration Commission, Statements and Recommendations Submitted by Societies and Organizations Interested in the Subject of Immigration,* vol. 2, Statement of Junior Order of American Mechanics, Statement of the Immigration Restriction League, Statement of the American Federation of Labor, and letter from T. J. Bassett, De Paul University Academy; and "An Alien Antidumping Bill," 13.

83. Walker, "Restriction of Immigration," 828. See also Ripley, "Races in the United States," 747. For examples from testimony before the U.S. Immigration Commission, see the following items in U.S. Senate, *Report of the Immigration Commission, Statements and Recommendations Submitted by Societies and Organizations Interested in the Subject of Immigration,* vol. 2: Statement of the Junior Order of American Mechanics; Statement of the Farmer's Educational and Cooperative Union of America; clipping from the *Farmer's Union News,* 10 February 1909; and Immigration Restriction League, Exhibit 30, letter from Allen G. Braxton, reading, "For years past the Atlantic steamship lines have acted as a siphon to draw off, as it were, the sewage from the cesspools of Europe and discharge it into the choice place of America." See also, in the same volume, Immigration Restriction League exhibit 39, letter from James N. Arnold, historian and genealogist, and exhibit 123, letter from James F. Ailshie, Justice, Supreme Court of Idaho.

84. George Lakoff and Mark Johnson, *Metaphors We Live By* (Chicago: University of Chicago Press, 1980).

85. Timothy Christenfeld, "Wretched Refuse Is Just the Start," *New York Times,* 10 March 1996. In contemporary Germany, for example, the water or "liquid" metaphors are frequently used to describe immigration. Combined with the notion of the nation as a "bottle," these metaphors, read in light of the pressures created by a unified German state, convey the notion of overflow—pushing social and economic resources beyond their capacity. Meredith Anne Green, "Bottles, Buildings, and War: Metaphor and Racism in Contemporary German Discourse," master's thesis, University of Arizona, 1995, 25.

86. For an alternative discussion of disease and metaphor, see Susan Sontag, *Illness as Metaphor and AIDS and Its Metaphors* (New York: Doubleday Anchor Books, 1978). Sontag, for example, identifies the year 1882, when the tuberculosis bacillus was identified, along with 1944 and 1952, when streptomycin and isoniazid were developed, as critical both in augmenting the power of tuberculosis as metaphor and shifting the host of metaphors used to describe the disease. Other diseases, such as insanity and cancer, Sontag argues, simply inherit the metaphorical legacy of tuberculosis; similarly, AIDS inherits the metaphorical history of syphilis. Yet, in this analysis, disease as metaphor and the metaphors used to define disease are transhistorical, waiting to attach themselves to new, little-understood, and untreatable diseases, rather than historically contingent or constructed.

87. Lakoff and Johnson, as part of a larger project on theorizing about the place of metaphors in our conceptual system, argue that "the reason we need two metaphors is because there is no one metaphor that will do the job—there is no one metaphor that will allow us to get a handle simultaneously on both [aspects of the thing being described]." "These two purposes cannot be served at once by a single metaphor." See Lakoff and Johnson, *Metaphors,* 95.

88. Hoy, *Chasing Dirt,* 132.

89. Lys Ann Shore reminded me that this was also the central image of Emile Zola's late nineteenth-century novel *Nana:* the fly on the dungheap of society infecting the higher classes. Emile Zola, *Nana* (Bath, England: Absolute Classics, 1990). Rogers, *Dirt and Disease,* chap. 1; Naomi Rogers, "Dirt, Flies and Immigrants: Explaining the Epidemiology of Poliomyelitis, 1900–1916," *Journal of the History of Medicine* 44 (October 1989): 486–505.

90. Alexander Keyssar, *Out of Work: The First Century of Unemployment in Massachusetts* (Cambridge: Cambridge University Press, 1986), 25.

91. Massachusetts Bureau of Labor Statistics, *Second Annual Report,* 2, quoted in ibid., 36.

92. Woods, ed. *Americans In Process,* 135.

93. Keyssar, *Out of Work,* 47.

94. Joan Morrison and Charlotte Fox Zabusky, *American Mosaic: The Immigrant Experience in the Words of Those Who Lived It* (New York: E. P. Dutton, 1980), 11.

95. Abraham Koosis, *Child of War and Revolution: The Memoirs of Abe Koosis* (Oakland, Calif.: Sea Urchin Press, 1984), 33, 35.

96. Keyssar, *Out of Work,* 50, 74–75.

97. Quoted in ibid., 143.

98. Unemployment Committee of the National Federation of Settlements, *Case Studies of Unemployment* (Philadelphia: University of Pennsylvania Press, 1931), 71.

99. Ibid., 253, 352.

100. Emma O. Lundberg, *Unemployment and Child Welfare: A Study Made in a Middle Western and Eastern City during the Industrial Depression of 1921 and 1922,* U.S. Department of Labor, Children's Bureau Publication 125 (Washington, D.C.: Government Printing Office, 1923), 82.

101. Unemployment Committee of the National Federation of Settlements, *Case Studies in Unemployment,* 307.

102. Lizabeth Cohen, *Making a New Deal: Industrial Workers in Chicago, 1919–1939* (Cambridge: Cambridge University Press, 1990), 57.

103. Keyssar, *Out of Work,* 151–52, 155, 164–65.

104. See, for example, the CGAR for 1905–8, detailing the characteristics of immigrants in penal, reformatory, and charitable institutions.

105. *Reports of the Immigration Commission, Immigrants as Charity Seekers,* vol. 1 (Washington, D.C.: Government Printing Office, 1911), 28–29.

106. Elizabeth Hughes, U.S. Department of Labor, Children's Bureau, *Children of Preschool Age in Gary, Ind. Part I. General Conditions Affecting Child Welfare,* Children's Bureau Publication 122 (Washington, D.C.: Government Printing Office, 1922), 35.

107. CGAR, 1898, 2. See also memorandum abstracting information, in "The Alien as Charity Seeker," Children's Bureau, U.S. Department of Labor, vol. 4, no. 29 (October 1927). The annual report of the Charity Organization Society of New York, 30 September 1930, listed the physical problems of the families served: tuberculosis (261 families, 4.9% of clients), asthma (87 families, 1.6%), cancer (48 families, 0.9%), cardiac trouble (357 families, 6.7%), syphilis (244 families, 4.6%), gonorrhea (30 families, 0.6%), vaginitis (24 families, 0.5%), gynecological trouble (244, 4.6%), maternity (583, 11%), endocrine disturbance (61, 1.1%), malnutrition (371, 7%), disease of the respiratory system (214, 4%), other chronic illness (920, 17.3%), other acute illness (670, 12.6%), need of dental care (500, 9.4%), need of optical care (289, 5.4%), blindness or sight seriously impaired (168, 3.2%), paralyzed or crippled (318, 6%), disability due to old age (460, 8.6%), other physical disability (630, 11.8%), death (179, 3.4%) (American Council for Nationalities Service, Shipment 2, Box 1, General, Aliens and Charity, IHRC). Similarly, the Immigration Commission found that "death or disability of the breadwinner" or another member of the family was responsible for 47.6% of the cases in which public assistance was given. Death or disability was the primary cause of de-

pendency in 49.7% of foreign-born charity seekers as compared to 46.1% of native-born charity seekers (U.S. Senate, *Reports of the Immigration Commission, Abstracts of Reports of the Immigration Commission, Abstract of the Report on Immigrants as Charity Seekers,* vol. 2 [Washington, D.C.: Government Printing Office, 1911], 116).

108. Ford Motor Company, *Helpful Hints and Advice to Employees to Help Them Grasp the Opportunities Which Are Presented to Them by the Ford Profit Sharing Plan* (Detroit, 1915), and *A Brief Account of the Educational Work of the Ford Motor Company* (Detroit, 1916), cited in Stephen Meyer, "Adapting the Immigrant to the Line: Americanization in the Ford Factory, 1914–1921," *Journal of Social History* (Fall 1980): 81, 70.

109. Miriam Blaustein, ed., *Memoirs of David Blaustein* (New York, 1913), 127–37, cited in Daniel Bender, "'A Hero . . . for the Weak': Work, Consumption, and the Enfeebled Jewish Worker, 1881–1924," *International Labor and Working-Class History* 56 (Fall 1999): 9.

110. *Congressional Record* (1902): 5763–64.

111. Keyssar, *Out of Work,* 88–90.

112. Morrison and Zabusky, *American Mosaic,* 42. As an employee Fitzgerald described herself as a "useful girl."

113. Paul Sigrist, interview with Sadie Guttman Kaplan, 2 July 1992, EIOHP.

Two. The Function of Medical Inspection

Epigraph. Foreign Language Information Service report for submission to the Commonwealth Fund, c. 1919, 4, quoting a Yugoslav newspaper, Josephine Aspinwall Roche Papers, IHRC.

1. See, for example, the case of Sammy Goldman, a Russian immigrant awaiting deportation whom the Hearst newspapers championed. The papers led a campaign to convince President Harding to block Goldman's deportation on the grounds that he was feebleminded. "Washington Orders Sam Goldman Back to Russia: Boy's Only Hope Now Rests in Harding: Deportation Order Signed in Washington—Will Be Sent Away on April 1," *Syracuse Telegram,* 25 March 1923; "New Move to Save Sammy: New Examination Is Granted; Lad Ordered Sent Back to Russia," *Syracuse Telegram,* 29 March 1923.

2. See RG 90, Central File, 1897–1923, Box 36, File No. 219, NARA.

3. "In Re Hearing Given Gotlieb Herdenreder, alias Harcheurader, Under Department Warrant of Arrest No. 51890/79, Dated March 14, 1908," 17 March 1908, RG 85, 519806/79, NARA.

4. There are no records available regarding the immigrant medical examination provided by steamship company physicians for any of the steamship lines carrying passengers from Europe, Asia, or Latin America during this period.

5. See also interview with Celia Soloway, in Giles R. Wright, *The Journey from Home* (Trenton: New Jersey Historical Commission, Department of State, 1986), 34, who describes quarantine at Boston in terms of a prison.

6. Weindling notes that Antin underwent a cleansing only and not disinfection for lice. In Germany, notably, all immigrants were screened and disinfected for lice after the connection was made to typhus in 1913. This, Weindling argues, medicalized notions of racial purification and eventually helped to rationalize genocide. See Paul Julian Weindling, *Epidemics and Genocide in Eastern Europe: 1890–1945* (Oxford: Oxford University Press, 2000), 69.

7. Mary Antin, *The Promised Land* (New York: Modern Library, 2001),146, 148–151. Notably, much of Antin's account of the inspection procedures abroad is quoted from a long letter she wrote to an uncle during her first months in America, allowing "Memory . . . a rest" (145).

8. Medical disinfection procedures were established in the regulations governing immigration. See, for example, U.S. Treasury Department, *Immigration Laws and Regulations* (Washington, D.C.: Government Printing Office, 11 March 1893), 4.

9. Antin, *Promised Land,* 148.

10. Quoted in Michael La Sorte, *La Merica: Images of Italian Greenhorn Experience* (Philadelphia: Temple University Press, 1985), 53.

11. Antin, *Promised Land,* 152. See also Dana Gumb, interview with Renee Berkoff, 14 August 1985, EIOHP.

12. Quoted in Peter Morton Coan, *Ellis Island Interviews: In Their Own Words* (New York: Facts on File, 1997), 169.

13. 15 February 1893, 27 Stat. L., 449, 452.

14. Letter from H. E. Conger, Legation of the United States of America, Peking, to Prince Ch'ing, China, 30 September 1903, Doc. No. 557, RG 85, Box 117, File No. 52495/49, NARA; letter from Lloyd C. Griscom, U.S. Legation, Tokyo, to John Hay, secretary of state, 28 September 1903, RG 85, Box 117, File No. 52495/49, NARA. In this file see also letter from John Hay to the secretary of labor and commerce, 22 October 1903, and letter from Dunlop Moore, assistant surgeon, Yokohama, Japan, to the surgeon general, 23 September 1903.

15. Letter from Wyman to PHS officers stationed in Japan and China, 21 August 1903, SGAR, 1904, 196.

16. SGAR, 1900, 523.

17. La Sorte, *La Merica,* 28. Agents granted immigrants transport. They also accompanied immigrants whom they had booked on a ship during the immigration interrogation and answered questions in their presence, as the agent was responsible for the immigrant's return in case of rejection. See U.S. Treasury Department, *Immigration Laws and Regulations,* 6.

18. Evelyn Berkowitz, quoted in Coan, *Ellis Island Interviews,* 315. Evelyn emigrated from Hungary with her family in 1912 at age twelve.

19. Quoted in Coan, *Ellis Island Interviews,* 115.

20. La Sorte, *La Merica,* 29. These figures were based on an Italian government immigration bulletin.

21. Letter from Surgeon General Walter Wyman to secretary of state, 15 December 1909, RG 85, Box 117, File No. 52495/49a, NARA.

22. Letter from commissioner general to secretary of commerce and labor, 14 May 1904, RG 85, Box 348 (to be renumbered 318), File No. 55224/371a, folder 2, NARA. The commissioner general noted that, although a Senate bill that would have provided explicit authority for the medical inspection of immigrants abroad had recently failed to pass, he believed such authority already existed.

23. Fiorello La Guardia, *The Making of an Insurgent: An Autobiography: 1882–1919* (Westport, Conn.: Greenwood Press, 1948), 53.

24. Ibid., 54–55.

25. Ibid., 56. For La Guardia's account of this story, see 53–57. La Guardia described an inspection procedure very similar to that carried out in other foreign ports and, indeed, the United States: the immigrants "all walked up to a platform, stepped up to a doctor, who examined them for physical defects, took their temperature and then passed them on to the next doctor for examination of their eyes for trachoma, which was the most prevalent infectious disease at the time." He continued, "The thousands of emigrants rejected every year at Ellis Island because of trachoma went back to their native lands disappointed, heartbroken, and 'broke,' for they had sold all their property and taken all their savings to pay for their passage to the land of hope" (57).

26. Weindling, *Epidemics and Genocide,* 49, 59, 61, 62, 64. Specifically, Americans pressed officials in Bremen and Hamburg to make inspections more rigid. Weindling notes, "The German medical controls attempted to screen large numbers of passengers at low cost. As rival shipping lines like Cunard, HAPAG (the acronym for Hamburg Amerikanische Paketfahrt-Akteingesellschaft), and the Bremer Lloyd competed, they had to balance the expense of inspection and accommodation with the need to prevent outbreaks of infections on board ship. HAPAG was technologically innovative, pioneering new sanitary procedures. . . . With the rapid rise in migrants, in the cholera year of 1892 HAPAG opened migration halls housing 1,400 people at a time in eight quayside sheds to house, feed, and medically screen the migrants, and to disinfect their baggage. Yet the quayside halls rapidly became overcrowded, and their lack of adequate sanitation and washing facilities meant that they could become a source of infections" (*Epidemics and Genocide,* 62).

27. Henry Diedrich, 1903 report, quoted in Irving Howe, *World of Our Fathers* (New York: Galahad Books, 1976), 37.

28. Staatsarchiv, Bremen, A.4 no. 300a, report by Tjaden and Nocht, 20 July 1906, cited in Weindling, *Epidemics and Genocide,* 65.

29. Officers were sent to foreign ports to monitor and report back on epidemic diseases. They also had the authority to inspect passengers for quarantinable diseases before the departure of ships bound for the United States. See Ralph Chester Williams, *The United States Public Health Service, 1798 to 1950* (Washington, D.C.: Government Printing Office, 1951).

30. Translation of letter from the minister of foreign affairs in Belgium forwarded by Lawrence Townsend, Legation of the United States, to John Hay, secretary of state, 31 March 1904; and letter from John Cassatt, chargé d'affaires ad interim, to John Hay, secretary of state, 29 January 1904; both in RG 85, Box 348 (to be renumbered 318), File No. 55224/371a, folder 2, NARA. The Netherlands explicitly cited conflicts between the American PHS officer and the Rotterdam Committee over the transport of vegetables, assuming that further intervention "will prove detrimental to trade." The Belgian official added that the inspection conducted in Antwerp by the steamship lines was "carried out in the most satisfactory and most conscientious manner."

31. Letter from Charles Wilson to John Hay, secretary of state, 30 September 1903, RG 85, Box 348 (to be renumbered 318), File No. 55224/371a, folder 2, NARA.

32. Letter from Chandler Hale, Embassy of the United States, Vienna, to John Hay, secretary of state, 19 July 1904, RG 85, Box 348 (to be renumbered 318), File No. 55224/371a, folder 2, NARA.

33. Letter from Lord Lansdowne, Great Britain, to Joseph H. Choate, 11 November 1903, RG 85, Box 348 (to be renumbered 318), File No. 55224/371a, folder 2, NARA. See also Weindling, *Epidemics and Genocide,* 61.

34. Memorandum for the acting secretary from Commissioner General Keefe, 11 March 1910, RG 85, Box 117, File No. 52495/49a, NARA.

35. Ibid.; letter from inspector in charge, Honolulu, to commissioner general, 30 November 1909, RG 85, Box 60, File No. 51831/31, NARA.

36. Letter from Commissioner North, San Francisco, to commissioner general, 18 May 1910, RG 85, Box 117, File No. 52495/49b, NARA. The PHS and IS characterized substitution as a serious, persistent problem in both Asian and European ports prior to initiation of the visa system in 1924. The IS believed that, if forced, steamship companies would make more rigid inspections in order to shield themselves from fines. Accordingly, the IS inspector in Hawaii noted a "marked diminution" of diseased immigrants after the PHS exams among Asian immigrants were discontinued. Letter from Richard L. Halsey, acting inspector in charge, Honolulu, to commissioner general, 11 May 1910, RG 85, Box 117, File No. 52495/49b, NARA. See also letter from commissioner general to inspector in charge, Honolulu, 15 December 1909, RG 85, Box 60, File No. 51831/31, NARA.

37. Surgeon general: letter from Surgeon General Wyman to secretary of state, 15

December 1909, RG 85, Box 117, File No. 52495/49a, NARA. Pacific steamship companies: letters from Charles Nagel, secretary, to R. P. Schwerin, Pacific Mail Steamship Co., 17 and 19 May 1911, RG 85, Box 117, File No. 52495/49b, NARA. Japanese government: letter from K. Matsui, Imperial Japanese Embassy, Washington, D.C., to Ransford S. Miller, State Department, 18 January 1910, RG 85, Box 117, File No. 52495/49a, NARA. The Japanese government, apparently not understanding that immigrants were again inspected upon arrival in the United States regardless of any inspections conducted abroad, argued that "it is feared that those emigrants will constantly be subject hereafter to reexamination and suffer no small embarrassment upon their arrival [in the United States]."

38. Letter from Commissioner General A. Caminetti to commissioners of immigration and inspectors in charge at Atlantic and Gulf ports of entry, 15 April 1920; letter from Acting Secretary of State Loeb to secretary of labor, 24 February 1920; both in RG 85, Box 348 (to be renumbered 318), File No. 55224/371A, folder 2, NARA.

39. See Weindling, *Epidemics and Genocide*. Although the disinfection procedures established in Germany would have profound implications for Jews under the Nazi regime, at the turn of the century they failed to win German endorsement—particularly the endorsement of bacteriologists. The medical officer in Bremen, for example, favored a more traditional medical inspection over disinfection (67–68).

40. Don Gussow, *Chaia Sonia: A Family's Odyssey Russian Style* (New York: Cherry and Scammel Books, 1980), 206, 207. See also Weindling, *Epidemics and Genocide*, 69; and D. Dwork and R. van Pelt, *Auschwitz: 1270 to the Present* (New Haven: Yale University Press, 1996), 52–54.

41. Paul Sigrist, interview with Manny Steen, 22 March 1991, EIOHP. Edward Hemmer, a twelve-year-old from northern Italy who arrived at Ellis Island in 1925, shared a similar recollection: "There was one thing that I resented very much as a kid. They asked the boys and the men to form a line and to take out their penises. The doctor went from one man to the other. I don't know . . . I was brought up very conservative. I resented showing somebody my penis" (quoted in Coan, *Ellis Island Interviews*, 70).

42. Janet Levine, interview with Enid Griffiths Jones, 18 April 1993, EIOHP.

43. Eugene L. Fisk, "The Value of Complete Routine Examinations in Supposedly Healthy People," *Boston Medical and Surgical Journal* 195 (14 October 1926): 740.

44. Michel Foucault, *Discipline and Punish: The Birth of the Prison* (New York: Vintage Books, 1979), 34, 111.

45. On the importance of bureaucratic thought to the ordering of society along the workings of the factory and the principles of scientific management, see Robert H. Wiebe, *The Search for Order, 1877–1920* (New York: Hill and Wang, 1967).

46. Gussow, *Chaia Sonia*, 208–9.

47. La Sorte, *La Merica*, 48.

48. Herbert Gutman, "Work, Culture, and Society in Industrializing America, 1815–1915," *American Historical Review* 78 (1973): 543.

49. Stephen Meyer, "Adapting the Immigrant to the Line: Americanization in the Ford Factory, 1914–1921," *Journal of Social History* 14 (fall 1980): 68.

50. Alter Abelson, "The Finisher" and "The Blind Operator," in *Songs of Labor* (Newburgh, N.Y.: Paebar Co., 1947), 75, 77.

51. 5 February 1917, c. 29, 39 Stat. 874; H. D. Smith and H. Guy Herring, *The Bureau of Immigration: Its History, Activities and Organization* (Baltimore: Johns Hopkins Press, 1924), 34–111.

52. Paul Sigrist, interview with Manny Steen, 22 March 1991, EIOHP. Immigration regulations specified that "prior to or at the time of embarkation. . . . a ticket, on which shall be written his name, a number or letter" was to be be issued to each immigrant or head of the family for purposes of identification at the port of arrival. U.S. Treasury Department, *Immigration Laws and Regulations,* 6.

53. La Sorte, *La Merica,* 29–30.

54. Bessie Kriesberg, *Hard Soil, Tough Roots: An Immigrant Woman's Story* (Jericho, N.Y.: Exposition Press, 1973), 138.

55. U.S. Treasury Department, *Immigration Laws and Regulations,* 5–6.

56. Paul Sigrist, interview with Regina Tepper, 21 February 1991, EIOHP.

57. Narrative of Louis Adamic in Rhoda Hoff, *America's Immigrants: Adventures in Eyewitness History* (New York: Henry Z. Walck, 1967), 124, 125. See also Abraham Mitrie Rihbany, *A Far Journey* (Boston: Houghton Mifflin, 1914), 180–81, who describes passage from Syria in 1891.

58. Weindling adds, "For those unfamiliar with modern medical routines, medical inspections could be terrifying, arousing fears of robbery and murder" (*Epidemics and Genocide,* 68). See also J. Wertheimer, *Unwelcome Strangers: East European Jews in Imperial Germany* (New York: Oxford University Press, 1987), 50–51.

59. U.S. Department of Commerce and Labor, Bureau of Immigration and Naturalization, *Immigrant Laws and Regulations of July 1, 1907* (Washington, D.C.: Government Printing Office, 5 October 1908), 66–67. See Ernesto Galarza, *Barrio Boy: The Story of a Boy's Acculturation* (Notre Dame, Ind.: University of Notre Dame Press, 1971), 182.

60. Marie Jastrow, *A Time to Remember: Growing Up in New York before the Great War* (New York: W. W. Norton, 1979), 45.

61. Marge Glasgow, quoted in Coan, *Ellis Island Interviews,* 135. Marge arrived from Scotland in 1922 at age fifteen.

62. Paulina Caramando, quoted in Coan, *Ellis Island Interviews,* 43. See also the interview with Susan Angel in Mimi E. Handlin and Marilyn Smith Layton, *Let Me Hear Your Voice: Portraits of Aging Immigrant Jews* (Seattle: University of Washington Press, 1983), 12.

63. Jack Weinstock, quoted in Coan, *Ellis Island Interviews*, 231. See also the oral history interview of Jake Kreider, ibid., 223. Kreider, who immigrated to the United States in 1911 at age eleven, recalls being "driven in [the Great Hall] like a bunch of cattle."

64. Margaret Wertle, quoted in Coan, *Ellis Island Interviews*, 310.

65. Paul Sigrist, interview with Rachel Shapiro Chenitz, 14 July 1991, EIOHP. On harsh treatment and immigrants as cattle, see also Gerado Ferreri, *Gli Italiani in America: Impressioni di un viaggio agli Stati Uniti* (Rome: Farro, 1907), 24; Bartolomeo Vanzetti, *The Story of a Proletarian Life* (Boston: New Trial League, 1923), 12; and Edward Steiner, *From Alien to Citizen* (New York: Arno Press, 1975), 126.

66. Abelson, "The Designer" and "The Finisher" (75), in *Songs of Labor*, 68, 75. Many other pieces in this volume make reference to the shop as prison or cage. See, for example, "The Cutter" (69), "The Blind Operator" (78), "So Vast the World" (103), and "The Cry of the Crushed" (111). For other references to workers as cattle, see "The Finisher" (75). See also Lorenzo D. Gillespie, *Songs of Labor* (Salina, Kans.: Central Kansas Publishing, 1904), 52.

67. Kriesberg, *Hard Soil, Tough Roots*, 138, 139.

68. The IS explicitly referred to the process as an interrogation. U.S. Treasury Department, *Immigration Laws and Regulations*, 6.

69. Case of Samuel and Ester Nelkin, No. 8, 23 January 1926, RG 85, Box 391 (renumbered 361), File No. 55476/151, NARA. See also the other transcripts in this box.

70. Willa Appel, interview with Elizabeth (Betty) Nimmo, 2 February 1986, EIOHP.

71. J. Andrew Mendelsohn, "'Typhoid Mary' Strikes Again: The Social and the Scientific in the Making of Modern Public Health," *Isis* 83 (1995): 268.

72. John Birge Sawyer Papers, Manuscripts, 81/62c, vol. 2, Bancroft Library, University of California, Berkeley. For a discussion of the significance of personal accounts of health and disease, see Roy Porter, "The Patient's View: Doing Medical History from Below," *Theory and Society* 14 (1985): 175–98; and Roy Porter and Dorothy Porter, *In Sickness and in Health: The British Experience, 1650–1850* (London: Fourth Estate, 1988).

73. This notion that health was linked to other social and economic domains is not to be confused with an older, nineteenth-century understanding of disease as a unique product of the interaction of the individual's heredity, morality, behavior, diet, and lifestyle with environment. See David Rosner's introduction to *Hives of Sickness: Public Health and Epidemics in New York City*, ed. Rosner (New Brunswick, N.J.: Rutgers University Press, 1995), 9; Charles Rosenberg, "The Therapeutic Revolution: Medicine, Meaning, and Social Change in Nineteenth-Century America," *Perspectives in Biology and Medicine* 20 (summer 1977): 485–507.

74. North American Civic League for Immigrants, *Messages for New Comers to the United States,* teacher's edition (Boston, n.d.); YMCA, International Committee Records, Reel 4, Folder 25, IHRC. As for other organizations and bureaus that provided service to immigrants, health was a major concern, and medical services often constituted a large percentage of services provided. See also International Institute of Boston, annual and monthly reports, Box 1, IHRC. As part of its services, for example, the YMCA offered twelve educational classes for working men on "health, hygiene, and first aid to the injured," five classes on tuberculosis, and eight classes on alcoholism. Fred H. Rindge, Industrial Department, International Committee of Young Men's Christian Associations, *Educational Classes and Other Service with Workingmen* (New York: Association Press, n.d.); YMCA, International Committee Records, Reel 4, Folder 25, IHRC.

75. Coan, *Ellis Island Interviews,* 310.

76. Unemployment Committee of the National Federation of Settlements, *Case Studies of Unemployment* (Philadelphia: University of Pennsylvania Press, 1931), 362–63.

77. Foucault, *Discipline and Punish.*

78. *Reports of the Immigration Commission, Immigrants in Industries (in Twenty-Five Parts), Part 23: Summary Report on Immigrants in Manufacturing and Mining,* vol.2: *General Tables* (Washington, D.C.: Government Printing Office, 1911). Settlement workers, too, helped to reinforce behavior through surveillance. Although they did not quantify the cleanliness of different races in the city, Boston's settlement house workers conducted a similar study in 1903, carefully assessing each race in the city's North and West Ends according to personal cleanliness, general sanitation, and predilection for overcrowding. Robert A. Woods, ed., *Americans In Process: A Settlement Study by Residents and Associates of the South End House* (Boston: Houghton, Mifflin, 1903), 40–70.

79. David Montgomery, *The Fall of the House of Labor: The Workplace, the State, and American Labor Activism, 1865–1925* (Cambridge: Cambridge University Press, 1987), 237, 151, 243. See also Bruno Ramirez, *When Workers Fight: The Politics of Industrial Relations in the Progressive Era, 1898–1916* (Westport, Conn., 1978), 151; and Robert Ozanne, *A Century of Labor-Management Relations at McCormick and International Harvester* (Madison: University of Wisconsin Press, 1967), 169.

80. Montgomery, *Fall of the House of Labor,* 235, 236.

81. Ford Motor Company, *Helpful Hints and Advice to Employees to Help Them Grasp the Opportunities Which Are Presented to Them by the Ford Profit Sharing Plan* (Detroit, 1915), and *A Brief Account of the Educational Work of the Ford Motor Company* (Detroit, 1916), quoted in Meyer, "Adapting the Immigrant to the Line," 81, 70, 71.

82. Montgomery notes that shortly after the profit-sharing plan went into effect,

Ford fired some nine hundred workers for celebrating Eastern Orthodox Christmas, stating that those who wished to remain in America should observe American customs and holidays (*Fall of the House of Labor,* 236).

83. Frederick W. Taylor, *Principles of Scientific Management* (Norcross, Ga.: Engineering and Managing Press, 1911, 1998), 35–36.

84. Ibid., 62.

85. *Reports of the Immigration Commission, Abstracts of Reports of the Immigration Commission with Conclusions and Recommendations and Views of the Minority,* vol. 1 (Washington, D.C.: Government Printing Office, 1911), 36–37.

86. "Ten Years of Industrial Self Control: Tenth Annual Report of the Board of Sanitary Control" (New York: Joint Board of Sanitary Control, 1921), 53.

87. "Workers' Health Bulletin" (1915), ILGWU Archives, UHC Papers, Box 10, Folders 10, 12, cited in Daniel Bender, "Inspecting Workers: Medical Examination, Labor Organizing, and the Evidence of Sexual Difference," *Radical History Review* 80 (2001), 75, n. 98.

88. Myron P. Gutmann, W. Parker Frisbie, Peter DeTurk, and K. Stephen Blanchard, "Dating the Origins of the Epidemiological Paradox among Mexican Americans," Texas Population Research Center Papers, 1997–98, Paper No. 97-98-07. Gutmann and his coauthors relate the story of Charles Bellinger, an African American businessman and political power broker who in 1918 created a black voting bloc and used it to win paved streets, sewer lines, water lines, fire services, parks, and electricity for the historically neglected, predominantly black, west side of San Antonio. The result was marked improvement in mortality rates among black residents. See also Stanley K. Schultz and Clay McShane, "To Engineer the Metropolis: Sewers, Sanitation, and City Planning in Late-Nineteenth-Century America," *Journal of American History* 65, no. 2 (September 1978): 389–411.

89. Foreign Language Information Service report for submission to the Commonwealth Fund, c. 1919, 4, quoting a Yugoslav newspaper, Josephine Aspinwall Roche Papers, IHRC. Similarly, the YMCA advised employers to keep their workers healthy and counseled the public to freely give "health suggestions" to "your foreign-born brother." Irving Fisher, "How Can the Employer Help the Worker Satisfy His Fundamental Human Instincts?" *New American,* n.d.; International Committee of Young Men's Christian Associations, *My Foreign-Born Brother* (New York: YMCA, 1917); YMCA, International Committee Records, Reel 3, Folder 23, IHRC.

90. In addition to the articles in the Chicago Foreign Language Press Survey, see Neil M. Cowan and Ruth Schwartz Cowan, *Our Parents' Lives: The Americanization of Eastern European Jews* (New York: Basic Books, 1992), chap. 4; and Suellen Hoy, *Chasing Dirt: The American Pursuit of Cleanliness* (New York: Oxford University Press, 1995), chap. 4.

91. "Health," *Scandia,* 8 January 1927, Chicago Foreign Language Press Survey,

IHRC. The article continued, "Let us be temperate in all things. Keep our thoughts clean, blood red, muscles hard, digestion good, body erect, nerves steady." An editorial in the Swedish press noted that "it is often jokingly said that the present generation suffers from 'health fever,' and many people seem to think we would be better off if we paid less attention to our health and concentrated on improving our lot in other respects. These critics are wrong, for health is the most important thing in life, and the world needs healthy people. . . . Physical and mental hygiene are coming into their own, and medical science is humanity's best friend" ("Health Fever," *Svenska Tribunen-Nyheter,* 24 May 1910, Chicago Foreign Language Press Survey, IHRC).

92. "Backwardness in 'Croatian Unity,'" *Radnicka Straza,* 4 June 1913, Chicago Foreign Language Press Survey, IHRC.

93. J. Leibner, "From the Public Rostrum," *Sunday Jewish Courier,* 28 March 1920, quoted in Lizabeth Cohen, *Making a New Deal: Industrial Workers in Chicago, 1919–1939* (Cambridge: Cambridge University Press, 1990), 57.

94. Cohen, *Making a New Deal,* 60.

95. Him Mark Lai, Genny Lim, and Judy Young, *Island: Poetry and History of Chinese Immigrants on Angel Island 1910–1940* (San Francisco: Chinese Culture Foundation, 1980), 100.

96. Nayan Shah, *Contagious Divides: Epidemics and Race in San Francisco's Chinatown* (Berkeley: University of California Press, 2001), 12, 15.

97. See, for example, Rudolph J. Vecoli, "Contadini in Chicago: A Critique of *The Uprooted,*" *Journal of American History* 51 (December 1964): 404–17; Herbert George Gutman, *Work, Culture, and Society in Industrial America* (New York: Albert A. Knopf, 1976); Virginia Yans-McLaughlin, *Family and Community: Italian Immigrants in Buffalo, 1880–1930* (Ithaca: Cornell University Press, 1982); John E. Bodnar, *Workers' World: Kinship, Community, and Protest in an Industrial Society* (Baltimore: Johns Hopkins University Press, 1982); and Bodnar, *The Transplanted: A History of Immigrants to Urban America* (Bloomington: Indiana University Press, 1985).

98. James R. Barrett, "Americanization from the Bottom Up: Immigration and the Remaking of the Working Class in the United States, 1880–1930," *Journal of American History* 79, no. 3 (December 1992): 998. See also Herbert Gutman, "The Workers' Search for Power: Labor in the Gilded Age," in *Power and Culture: Essays on the American Working Class* (New York: Pantheon Books, 1987), 70–92; Gutman, "Protestantism and the American Labor Movement: The Christian Spirit in the Gilded Age," *American Historical Review* 72 (1966): 74–101; and Gutman, "Work, Culture and Society in Industrializing America," *American Historical Review* 78 (1973): 531–88.

99. David Montgomery, *Workers' Control in America: Studies in the History of*

Work, Technology, and Labor Struggles (New York: Cambridge University Press, 1979), 43.

100. Barrett, "Immigration from the Bottom Up," 1015. Ephraim Morris Wagner, who worked as a presser in New York City for several decades beginning in the late 1880s, describes telling one boss "that I would not be a freight horse any more," which represented the beginning of his union activity (RG 102, Folder 45, American Jewish Autobiography Collection, YIVO, Center for Jewish History, New York). For other self-descriptions of workers as machines, see Alter Abelson, "The Blind Operator" and "My Wrath Is Growing Young and Strong," in *Songs of Labor,* 77, 83.

101. Daniel Bender, "'A Hero . . . for the Weak': Work, Consumption, and the Enfeebled Jewish Worker, 1881–1924," *International Labor and Working-Class History* 56 (fall 1999): 5, 3, 12.

102. Montgomery, *Fall of the House of Labor,* 247.

103. Kingdom of Belgium, Ministère de l'Industrie, du Travail et du Ravitaillement, *Le travail industriel aux Etats-Unis: Rapports de la Mission d'Enquête* (Brussels, 1920), vol. 1, 215, quoted in Montgomery, *Fall of the House of Labor,* 249.

104. Werner Sollors, *Beyond Ethnicity: Consent and Descent in American Culture* (New York, 1986), 5; and Israel Zangwill, *The Melting-Pot* (New York, 1909, 1923), 33.

105. Antin, *Promised Land,* 3.

106. Edward Steiner, *From Alien to Citizen* (New York: Arno Press, 1975), 36. Dhan Gopal Mukerji similarly noted that after seventeen days in steerage en route from India, it all "seemed like the experience of another man the very moment the immigration authorities gave me permission to enter the United States" (*Caste and Outcast* [New York: E. P. Dutton, 1923], 165).

107. See also Rihbany, *Far Journey,* 183, where he asks, on arriving in New York, "Was I still my old self, or had some subtle, unconscious transformation already taken place in me?"

108. Gary Gerstle, *Working-Class Americanism: The Politics of Labor in a Textile City, 1914–1960* (New York: Cambridge University Press, 1989), 115, 153–95. Gerstle, in his essay, "Liberty, Coercion, and the Making of Americans," *Journal of American History* (September 1997), www.jstor.org, identifies other literature in the same vein: Andrew Neather, "Labor Republicanism, Race, and Popular Patriotism in the Era of Empire, 1890–1914," in *Bonds of Affection: Americans Define Their Patriotism,* ed. John Bodnar (Princeton: Princeton University Press, 1996), 82–101; Joseph McCartin, "'An American Feeling': Workers, Managers, and the Struggle over Industrial Democracy in the World War I Era," in *Industrial Democracy: The Ambiguous Promise,* ed. Nelson Lichtenstein and Howell John Harris (New York: Cambridge University Press, 1993), 67–86.

109. Gerstle, "Liberty, Coercion, and the Making of Americans."

110. Montgomery, *Fall of the House of Labor,* 249.

111. Barrett writes that immigrants were creating "identities and embracing values that reflected situations they faced in the workplace" ("Americanization from the Bottom Up," 1004).

112. Foucault, *Discipline and Punish,* 26–27.

113. Erving Goffman, *Asylums* (New York: Doubleday, 1961), 4, 6.

114. Writes Weindling, "Having escaped tsarist repression and terrifying pogroms, migrants could expect prison-like regimentation by German medical personnel, border guards, and officials. Most control centers were guarded and encircled by walls. After disembarking from the packed trains, disinfection involved the separation of male and female passengers and consequent breaking up of families [and] the confiscation of all clothing and possessions" before disinfection (*Epidemics and Genocide,* 67).

THREE. The Medical Gaze

Epigraph. Alfred C. Reed, "Scientific Medical Inspection at Ellis Island," *Medical Review of Reviews* (1912): 541.

1. Six military surgeons were expected to examine no more than fifteen hundred men per six-hour day; typically, they examined between one hundred and nine hundred per day. See Colonel Charles Willcox, "Physical Examination of Large Bodies of Men," *Military Surgeon* (1917): 414. Nugent describes industrial physicians as racing "through the tests, while their counterparts in private practice were admonished to reserve an hour for an adequate history and physical exam." Some physicians claimed to be able to complete the exam in six minutes, while a study in thirty-four industries found that, on average, physicians completed the exam in eleven minutes. See Angela Nugent, "Fit for Work: The Introduction of Physical Examinations in Industry," in *Readings in American Health Care: Current Issues in Socio-Historical Perspective,* ed. William G. Rothstein (Madison: University of Wisconsin Press, 1995), 45.

2. Victor Heiser, *An American Doctor's Odyssey: Adventures in Forty-five Countries* (New York: W. W. Norton, 1936), 13.

3. SGAR, 1905, 128–29.

4. Letter from Dr. A. J. Nute, surgeon, to the surgeon general, 28 January 1924, RG 90, General Subject File, 1924–35, Domestic Stations, Boston Immigration, Box 137, File No. 0950-56, NARA.

5. The PHS officer in San Francisco requested a microscope for use at the quarantine station in 1891, and the officer stationed in New Orleans requested a new microscope in 1893 to replace an obsolete one. Letter from W. P. McIntosh, PA surgeon, Angel Island, to Surgeon P. H. Bailhache, PHS, 15 September 1891, RG 90, Marine Hospital Service, Incoming Correspondence, Letters Received from Marine Hospitals, 1860–61, 1870–75, San Francisco, 1889–91, NARA; and undated letter, c. 1893,

RG 90, General Subject File, 1924–35, Domestic Stations, New Orleans Quarantine, Box 84, NARA. The immigrant inspector stationed in El Paso clearly had a microscope as early as 1904, though it was destroyed in a fire and had not been replaced as of 1906. Letter from Surgeon General Wyman to commissioner general, 13 August 1906; and letter from Luther Steward, inspector in charge, El Paso, 24 July 1906, Doc. No. 224, RG 85, Box 92, File No. 52271/62, NARA.

6. Letter from D. A. Carmichael, senior surgeon, Buffalo, to surgeon general, 15 January 1913, RG 90, Central File, 1897–1923, Box 37, File No. 219, NARA.

7. Letter from Preston H. Bailhache, surgeon in command, Middle Atlantic District, New York, to the surgeon general, 5 June 1893, RG 90, Marine Hospital Service, Letters Received from Marine Hospitals, 1860–61, 1870–75, New York, 1893, NARA. In 1903 the San Francisco quarantine station ordered a microscope in addition to slides and cover glasses. RG 90, Central File, 1897–1923, Box 784, File 16090, NARA.

8. Elizabeth Fee, for example, describes the influence that the science of bacteriology had in shaping the curriculum and research of the Johns Hopkins School of Hygiene and Public Health during the early decades of the twentieth century: Elizabeth Fee, *Disease and Discovery: A History of the Johns Hopkins School of Hygiene and Public Health, 1916–1939* (Baltimore: Johns Hopkins University Press, 1987).

9. Paul Starr, *The Social Transformation of American Medicine: The Rise of a Sovereign Profession and the Making of a Vast Industry* (New York: Basic Books, 1982); and Bruno Latour, *The Pasteurization of France* (Cambridge, Mass.: Harvard University Press, 1988). See also Kenneth M. Ludmerer, *Learning to Heal: The Development of American Medical Education* (New York: Basic Books, 1985); John H. Warner, *The Therapeutic Perspective: Medical Practice, Knowledge, and Identity in America, 1820–1885* (Cambridge. Mass.: Harvard University Press, 1986).

10. Daniel M. Fox, "Social Policy and City Politics: Tuberculosis Reporting in New York, 1889–1900," *Bulletin of the History of Medicine* 49 (1975): 169–95; George Rosen, *A History of Public Health* (New York: MD Publications, 1958); Judith Walzer Leavitt, "'Typhoid Mary' Strikes Back: Bacteriological Theory and Practice in Early Twentieth-Century Public Health," *Isis* 83 (1992): 608–29; Barbara Gutmann Rosenkrantz, *Public Health and the State: Changing Views in Massachusetts, 1842–1936* (Cambridge, Mass.: Harvard University Press, 1972); and Fee, *Disease and Discovery.*

11. For a detailed history of the PHS, highlighting the central role of the Hygienic Laboratory and bacteriological work in general in the careers of PHS officers, see Bess Furman with Ralph Chester Williams, *A Profile of the United States Public Health Service, 1798–1948* (Washington, D.C.: Government Printing Office, 1973).

12. Heiser, *An American Doctor's Odyssey*; Ralph Chester Williams, *The United States Public Health Service, 1798–1950* (Washington, D.C.: Government Printing Office, 1951).

13. After the turn of the century, as the number of immigration stations rapidly expanded, the majority of the immigrant inspection work at all but the nation's largest stations (where, of course, commissioned officers of the PHS processed the vast majority of immigrants) fell to noncommissioned officers, whom we cannot necessarily assume to have maintained the ethos of the PHS. But these civil surgeons, upon accepting their commissions, swore to "support and defend the Constitution of the United States against all enemies, foreign and domestic." Moreover, despite the presence of noncommissioned officers at many smaller stations, the commissioned officers at the nation's major immigration stations inspected the vast majority of immigrants and defined the purpose and intent of the immigrant medical exam. See Form 1902 of the Public Health and Marine Hospital Service. For an example, see the oath of office of Charles William Harrison in taking the position of acting assistant surgeon at Hidalgo, signed 11 March 1908, RG 90, Central File, 1897–1923, Box 123, File No. 1371, NARA.

14. Fitzhugh Mullan, *Plagues and Politics: The Story of the United States Public Health Service* (New York: Basic Books, 1989); Elizabeth Yew, interviews with John Heller, M.D., 14 September 1977, with Dr. T. Bruce H. Anderson, 22 September 1977, with Dr. Bernard Notes, 4 September 1977, all in EIOHP. Drs. Heller and Anderson speculated that the PHS appealed to southerners because of the job stability.

15. Elizabeth Yew, interview with John Heller, M.D., 14 September 1977, EIOHP. John C. Thill, who served as an acting assistant surgeon at Ellis Island in the late 1920s but who was unable to join the PHS as a commissioned officer, remembered PHS officers on duty at Ellis Island as "a very high type of officers, most of them dedicated men." Elizabeth Yew, interview with John C. Thill, M.D., 13 September 1977, EIOHP.

16. Heiser, *An American Doctor's Odyssey*, 12. See also Elizabeth Yew, interview with John Heller, M.D., 14 September 1977, EIOHP.

17. Kitty Calavita, *U.S. Immigration Law and the Control of Labor, 1820–1924* (London: Academic Press, 1984).

18. Letter from Surgeon General Wyman to medical officer in command, San Francisco Quarantine Station, 8 October 1891, RG 90, Letters from the Surgeon General to the Medical Officer in Charge, 1 July 1891–1 July 1918, Box 13, Vol. 1, NARA.

19. Robert H. Wiebe, *The Search for Order, 1877–1920* (New York: Hill and Wang, 1967), 146.

20. Dana Gumb, interview with Dr. Robert Leslie, 14 August 1985, EIOHP. Dr. Leslie attended Johns Hopkins Medical School and served as a medical examiner at Ellis Island from 1912 to 1914.

21. Rose Cohen, *Out of the Shadow* (New York: J. S. Ozer, 1971), 57, 62, 71.

22. Autobiography of Lena Karelitz Rosenman, American Jewish Autobiography Collection, YIVO, Center for Jewish History, New York City, RG 102, Folder 31. Like Cohen, Rosenman describes a long journey by train, foot, and wagon from Warsaw to Rotterdam as the fourteen-year-old girl traveled alone with her young brother.

23. Quoted in Irving Howe, *World of Our Fathers* (New York: Galahad Books, 1976), 41. See also the autobiography of Lena Karelitz Rosenman, American Jewish Autobiography Collection, YIVO, Center for Jewish History, New York City, RG 102, Folder 31, who explains that enduring the vomit from those in upper bunks was something that persisted throughout the voyage. See also autobiography of Benjamin Kopp, American Jewish Autobiography Collection, YIVO, Center for Jewish History, New York City, RG 102, Folder 300; and Daniel Bender, "'A Hero . . . for the Weak': Work, Consumption, and the Enfeebled Jewish Worker, 1881–1924," *International Labor and Working-Class History* 56 (fall 1999): 1, 5.

24. Quoted in Howe, *World of Our Fathers*, 41, 42.

25. Edward Steiner, *From Alien to Citizen* (New York: Arno Press, 1975), 35.

26. PHS regulations presumed that at all stations immigrants underwent an initial quarantine examination. In New York, quarantine inspectors boarded ships as they entered the harbor, miles away from the city's docks. At the port of New York, quarantine inspectors would muster all passengers into groups of about twenty-five into "a good-sized, well-lighted apartment" and take their temperatures. Officers made ready at least fifty thermometers and then brought immigrants down. While the immigrants held the thermometers in their mouths, an officer walked among them "to see that the thermometers are kept in place." Quarantine officers recorded the temperature of each individual. Those whose temperatures exceeded 99.5 degrees Fahrenheit were "immediately conducted to the boat which is waiting to transfer them to Hoffman Island" (A. H. Doty, "The Use of the Clinical Thermometer as an Aid in Quarantine Inspection," *Medical Record* [1 November 1902]: 690). The quarantine officer then visually inspected all arriving passengers for signs of smallpox, plague, or other quarantinable diseases. See A. H. Doty, "Modification of Present Port Inspection," *American Public Health Association Reports* 21 (1906): 260.

27. In addition to the many immigrant biographies, see also Giles R. Wright, *The Journey from Home* (Trenton: New Jersey Historical Commission, Department of State, 1986). Normal passage on the steamships ranged from ten to twenty-one days, depending on weather.

28. Aileen Vroom, quoted in Wright, *Journey from Home*, 22. Vroom left Ireland in 1929 when she was twenty-four.

29. Steiner, *From Alien to Citizen*, 37.

30. Quoted in Howe, *World of Our Fathers*, 43.

31. Samuel Chotzinoff, *A Lost Paradise* (New York: Alfred A. Knopf, 1955), 58.

32. David Montgomery, *The Fall of the House of Labor: The Workplace, the State, and American Labor Activism, 1865–1925* (Cambridge: Cambridge University Press, 1987), 220–21.

33. The anatomo-clinical gaze so eloquently described by Foucault had as its focus not disease, but lesions, pathology. In the late eighteenth century, before the develop-

ment of pathology, the medical gaze was redirected toward reading the visible *symptoms* (which, once properly read, became *signs*) of disease. Pathology focused the gaze to the ultimate seat of disease in the tissue, the lesion. Foucault argues that this anatomo-clinical gaze rejected mediation by the microscope or chemical analysis, which were the fundamental tools of bacteriology. Michel Foucault, *Birth of the Clinic: An Archaeology of Medical Perception* (New York: Vintage Books, 1973).

34. Bureau of Public Health and Marine Hospital Service, *Book of Instructions for the Medical Inspection of Aliens* (Washington, D.C.: Government Printing Office, 1910), 25. These instructions closely mirrored those first promulgated in 1903, but emphasized the importance of visual inspection from head to toe and also alerted the PHS to more diseases. By 1917, though the inspection procedure was fundamentally the same, the medical officer was explicitly instructed to question the immigrant in order to detect mental conditions and was alerted to several other diseases. As always, PHS officers were warned of the methods immigrants might use to hide diseases or defects. See PHS, *1917 Regulations*, 18.

35. A. J. Nute, passed assistant surgeon, Boston, "Medical Inspection of Immigrants at the Port of Boston," *Boston Medical and Surgical Journal* 170 (23 April 1914): 644.

36. Nute, "Medical Inspection," 644. See also J. G. Wilson, M.D., Ellis Island, "Some Remarks Concerning Diagnosis by Inspection," *New York Medical Journal* 94 (1911): 95, from whom Nute apparently took this observation, word for word.

37. Letter from Assistant Surgeon General H. D. Geddings to the surgeon general, 16 November 1923, RG 90, Central File, 1897–1923, Box 36, File No. 219, NARA. The PHS officers at Ellis Island developed an elaborate chalk coding scheme that was apparently used at no other port: "Sc would indicate scalp; P, physical and lungs; F, face; Ft, feet; E, eyes; C, conjunctivitis; CT, trachoma; S, senility; G, goiter; N, neck; B, back. The words hand, skin, temperature, and measles, which are often used, are written out in full" (E. H. Mullan, "The Medical Inspection of Immigrants at Ellis Island," *Medical Record* [27 December 1913]: 1168).

38. Terence Powderly, "Immigration's Menace to the National Health," *North American Review* 75 (1902): 59.

39. Letter from inspector in charge, Douglas, Arizona, to regional supervisor of the Immigration Service in El Paso, 8 December 1923, RG 85, Box 164, File No. 52903/29, NARA.

40. Assistant Surgeon General L. E. Cofer, "The Medical Examination of Arriving Aliens," *Bulletin of the American Academy of Medicine* (1912): 405.

41. Elizabeth Yew, interview with Grover Kempf, M.D., 10 and 11 September 1977, EIOHP.

42. Samuel B. Grubbs, *By Order of the Surgeon General* (Greenfield, Ind.: Mitchell, 1943), 71–72.

43. Victor Safford, *Immigration Problems: Personal Experiences of an Official* (New York: Dodd, Mead, 1925), 245, 246. Safford also argued that more was to be learned about the condition of an immigrant's heart by watching him after having carried baggage up a flight of stairs than by stripping him in an examination room (248).

44. Alfred C. Reed, "Immigration and the Public Health," *Popular Science Monthly* (October 1913): 319. Laborers in other regions of the world also found means to avoid the detection of disease. Randall Packard, *White Plague, Black Labor: Tuberculosis and the Political Economy of Health and Disease in South Africa* (Berkeley: University of California Press, 1989), 177.

45. E. H. Mullan, "The Medical Inspection of Immigrants at Ellis Island," *Medical Record* 84 (1913): 1167; letter from Commissioner General Keefe to commissioners of immigration and inspectors in charge at ports of entry, 15 January 1912, RG 85, Box 167, 52903/69, NARA.

46. SGAR, 1921, 250–51; letter from Commissioner General Keefe to commissioners of immigration and inspectors in charge at ports of entry, 15 January 1912, RG 85, Box 167, 52903/69, NARA. See also letter from A. J. Nute, PHS, Boston, to Surgeon General R. H. Creel, 21 July 1921, RG 90, Central File, 1897–1923, Box 59, File No. 409, NARA.

47. Letter from J. W. Kerr, chief medical officer, Ellis Island, to the surgeon general, 30 June 1921, RG 90, Central File, 1897–1923, Box 38, File No. 219, NARA.

48. Mullan, "Medical Inspection of Immigrants at Ellis Island," 1167–68. Alan Kraut relates a story told by Dr. John C. Thill, who witnessed Dr. W. C. Billings approach a woman on the line, issuing the order, "Nehmen Sie die Perücke ab" ("take the wig off"). "Thill and his colleagues were astounded to see before them 'a totally bald lady who had favus'" (Alan M. Kraut, *Silent Travelers: Germs, Genes, and the "Immigrant Menace"* [New York: Basic Books, 1994], 62–63).

49. See articles from the *San Francisco Chronicle* for 1905, including "Trachoma Used as Source of Revenue" and "Diseased Japanese Deceive Immigration Surgeons: Infected Coolies Are Landed in Hordes," RG 90, Central File, 1897–1923, Box 785, File 16090, NARA.

50. Case 13222/4-4, BSI Transcript, 20 February 1914, RG 95, Box 775, Arrival Investigation Case Files, 1884–1944, 02/02/1914–02/05/1914, NARA, San Bruno, Calif.

51. Howard Markel, *Quarantine! East European Jewish Immigrants and the New York City Epidemics of 1892* (Baltimore: Johns Hopkins University Press, 1997), 80.

52. Central though bacteriology, anatomy, and other sciences were to professional status and authority in an intensely competitive environment, they also created a rift between those who saw moral order in the world and a divine logic to disease causation and those who would recast germs as the sole cause of disease, and between the practice of medicine as an art or as a science. Regina Morantz-Sanchez, for example, explores the philosophical tensions in the medical profession created, in particular, by

the new discoveries in bacteriology: Regina Morantz-Sanchez, *Sympathy and Science: Women Physicians in American Medicine* (New York: Oxford University Press, 1985). By the turn of the century, the majority of physicians were resistant to bacteriology not so much as a theory of disease causation, but rather as a set of technologies that threatened their traditional practices: the laboratory might remove them from the bedside. Russell C. Maulitz, "'Physician versus Bacteriologist': The Ideology of Science in Clinical Medicine," and Gerald L. Geison, "'Divided We Stand': Physiologists and Clinicians in the American Context," in *The Therapeutic Revolution: Essays in the Social History of American Medicine*, ed. Morris J. Vogel and Charles E. Rosenberg (Philadelphia: University of Pennsylvania Press, 1979), 67–90, 91–108; John Harley Warner, "Ideals of Science and Their Discontents in Late Nineteenth-Century American Medicine," *Isis* 82 (1991): 454–78.

53. Audrey B. Davis, "Life Insurance and the Physical Examination: A Chapter in the Rise of American Medical Technology," *Bulletin of the History of Medicine* 55 (1981): 393; Stanley J. Reiser, "The Emergence of the Concept of Screening for Disease," *Milbank Memorial Fund Quarterly* 56, no. 4 (1978): 403–4, 406. By 1924 the lay public supported routine medical exams as conducted by life insurance companies: "The promotion of this ideal of health is the function of the modern physician. The doctor of today is not and should not be merely one who cures disease and rectifies impairments. . . . A few years ago a physician would look with considerable surprise at an apparently healthy individual who wandered into his office and requested such a radical thing as a 'health' examination. It is even not entirely beyond the realms of possibility that he would promptly advise the patient to get out and come back some time when he felt sick" (James A. Tobey, "A Layman's View of Health Examinations," *Boston Medical and Surgical Journal* 191, no. 19 [6 November 1924]: 876, 877).

54. George M. Gould, "A System of Personal Biologic Examinations the Condition of Adequate Medical and Scientific Conduct of Life," *JAMA* 35, no. 3 (1900): 134–37.

55. Historian Angela Nugent reports that a 1914 Industrial Commission of Ohio investigation found that among large industrial firms 10% had examined applicants and 4% had examined employees during the previous year. Studies of plants maintaining welfare departments between 1919 and 1930 found that more than 50% required exams of applicants and about 10% required exams of employees. See Nugent, "Fit for Work," 43. In industry, the examination was used as a means of rejecting unfit workers and of controlling tuberculosis, venereal disease, and other contagious diseases within industrial plants; the industrial medical exam was streamlined, often excluding the medical history of the worker in favor of diagnostic tests (see note 1, above).

56. Conference Board on Safety and Sanitation, "Measuring the Workman's Physical Fitness for His Job," *The Spirit of Caution* 4 (August–September 1916): 3; Nugent, "Fit for Work," 44.

57. For a discussion of the need to examine systematically and rapidly large groups of men, see Willcox, "Physical Examination of Large Bodies of Men."

58. Jas. H. Hamilton, "Physical Examination of Drafted Men," *Vermont Medicine* (December 1917): 285–88.

59. J. E. Tuckerman, "Some Impressions Gained in Examining under the Selective Draft Plan," *Ohio Medical Journal* (1917): 601. Tuckerman continued, "This refers mainly to the matter of flat feet, varicocele, and in less degree to what does and does not constitute hemorrhoids. I have no doubt that a number of individuals will be rejected at the mobilization camps who have been passed by the local board because the local boards have no criterion by which to judge. For instance! How large must a varicocele be to impede locomotion, or how flat must a foot be to be cause for rejection? Obviously the civilian physician has no definite experience upon which to base his judgement." The military surgeon also noted that it was often quite difficult to decide whether or not to reject a recruit. See W. G. Farwell, "Additional Border-Line Cases at the Recruiting Office," *United States Navy Medical Bulletin* (1917): 584.

60. Charles Y. Brownlee, "General Medical Examination of Recruits," *Boston Medical and Surgical Journal* (16 August 1917): 232–33. Brownlee delivered this address before members of the medical examination boards of Brooklyn County and Queens County, New York.

61. In addition, the infantryman "must have well-formed, well arched, elastic and strong feet. He must be well muscled and supple, neither adipose nor lean. He must have excellent heart and lungs in a capacious chest, to supply him in his exertions; a good frame to carry his pack, a good eye to sight his rifle, a good ear to hear the enemy patrol, good teeth to chew and good digestion to assimilate his plain but nourishing ration." Nevertheless, as the "military body" was conceived as "a physiological entity," different physical standards were suggested for different types of troops. The cavalryman did not need to have the frame and musculature of the infantryman, nor be so tall or so heavy, nor did he need such perfect feet and legs. Overall, then, "the military surgeon, when acting as a recruiting officer, must bear in mind the functions of the branch of the service for which the recruit desires to enlist and the fitness of the applicant to perform those functions" (Major Frank T. Woodbury, Medical Corps, U.S. Army, "Recruiting for Military Service," *Military Surgeon* [1917]: 18–19).

62. Ibid., 21–22.

63. J. A. Hofheimer, "An Analysis of the Recent Physical Examinations for a Local Exemption Board," *International Journal of Surgery* (September 1917): 274–78. Hofheimer explains that local exemption boards exercised a great deal of discretion in rejecting recruits for medical causes. See also Harry D. Orr, "Examination of Recruits for the Army and Militia," *AJPH* (1917): 485, who comments that a 75% rejection rate is not uncommon.

64. Albert P. Francine, J. Woods Price, and Francis B. Trudeau, "Cardiovascular Lesions and Tuberculosis: Methods and Results of Examination by the Cardiovascular and Tuberculosis Commission at the Second Plattsburg Training Camp for Reserve Officers," *JAMA* (22 December 1917): 2110–15. As Major Charles Brownlee instructed the medical examiners serving local exemption boards, which examined all recruits before they were examined at military camps, "a Sphygmomanometer and microscope are occasionally needed, as is an ophthalmoscope" (Brownlee, "General Medical Examination of Recruits," 232).

65. A. Farehnold, Medical Inspector, U.S. Navy, "A Plea for Greater Care in the Performance of Duty by Medical Officers at Recruiting Stations," *United States Navy Medical Bulletin* (1917): 320 (emphasis added).

66. Woodbury, "Recruiting," 20–21.

67. The diseases detected among military recruits were of the same order as those found among immigrants. Out of 525 men inspected at Local Exemption Board No. 17 in New York City, for example, 105 (20%) were rejected. The major cause of rejection (40%) was for defects in limbs (flat foot accounting for 52% of this category), followed by optical defects (24%, of which defective vision represented 76% of rejections), and circulatory defects (21%, which included cardiac dilatation, varicocele, and varicose veins). Hofheimer, "Analysis of the Recent Physical Examinations for a Local Exemption Board," 276. See also Lieutenant Calvin H. Goddard, "A Study of about 2,000 Physical Examinations of Officers and Applicants for Commission Made at the Army Medical School," *Military Surgeon* (1917): 578–88. Here, the largest percentage of candidates were rejected as underweight (21%); 14% were rejected for defective vision, and 11.1% for nephritis. Only 1 candidate (0.5%) was rejected for tuberculosis; none was rejected for venereal diseases or any of the Class A immigrant conditions. See also Charles A. Costello, "The Principal Defects Found in Persons Examined for Service in the United States Navy," *American Journal of Public Health* (1917): 489–92; and Hamilton, "Physical Examination," 287–88.

68. Erving Goffman, *Asylums* (New York: Doubleday, 1961).

69. Dr. Victor Safford in Boston coined this phrase. SGAR, 1904, 199. PHS regulations specified, "The examiner should detain any alien or aliens as long as may be necessary to insure a correct diagnosis" (Bureau of Public Health and Marine Hospital Service, *Book of Instructions for the Medical Inspection of Immigrants* [Washington, D.C.: Government Printing Office, 1903], 6).

70. "Personal" letter from George Stoner, Ellis Island, to the surgeon general, 14 August 1909, RG 90, Central File, 1897–1923, Box 36, File No. 219, NARA.

71. In 1915, Wassermann reactions and throat cultures for diphtheria together accounted for the greatest percentage of procedures, each representing 21%; examinations for favus and ringworm represented 11% of procedures.

72. San Francisco began reporting laboratory tests in 1924, later than New York,

though it is clear that laboratory tests for parasites were done long before that date. Letter from W. A. Korn, surgeon, Angel Island Quarantine Station, to surgeon general, 11 November 1916, RG 90, Letters to the Surgeon General from the Medical Officer in Charge, 1 July 1903–1 March 1926, Box 9, Vol. B4, NARA. In 1922 the medical officer in charge stated that "laboratory work plays an important role in connection with the examination of aliens at this station. Five thousand and fifty-seven specimens of feces were examined for ova of intestinal parasites, in addition to the routine examinations of blood, sputum, urine, etc." (SGAR, 1922, 208). Because San Francisco only reported on parasitic conditions, we do not know what percentage of all laboratory tests these made up. The emphasis on fecal analysis indicates that laboratory tests for other conditions were of secondary importance. In 1926, for example, the officer in charge reported that "the examination of feces has long been regarded as a most productive feature of the laboratory work" (SGAR, 1926, 183).

73. Wilson, "Some Remarks," 94.

74. Nute, "Medical Inspection," 644.

75. See Williams, *United States Public Health Service*; and Mullan, *Plagues and Politics.*

76. SGAR, 1926. See also SGAR, 1927, which indicates that the Wassermann tests were actually performed at the William Beaumont General Hospital by army physicians, allowing "a substantial saving of the public funds without in the least decreasing the efficiency of the laboratory service" (213).

77. Brownsville, for example, reported in 1926 that 221 laboratory examinations (representing 16% of all immigrants arrived and examined, and 100% of all immigrants certified) were performed. In 1917 the immigrant inspector in Port Huron reported that the IS had designated a small room for use as a laboratory and had supplied the necessary equipment: "This laboratory has greatly facilitated exact diagnosis in conditions where microscopical confirmation was necessary" (SGAR, 1918, 256).

78. SGAR, 1922, 190. Winnipeg had no laboratory facilities as late as 1918. SGAR, 1918, 259.

79. Two months later he was authorized to get competitive bids for this equipment (except for the examining table and screen, which the PHS thought the IS should purchase). Letter from Erwin Eveleth, AA surgeon, Detroit, to Dr. P. L. Prentis, inspector in charge, 19 January 1921, and letter from surgeon general to Eveleth, 18 March 1921, RG 90, Central Correspondence, 1897–1923, Box 373, File No. 3690, NARA.

80. Dowbiggin's excellent account of the relation of psychiatry to immigration restriction in the United States makes clear that the burden that insane immigrants placed on institutions was a primary motivating force among psychiatrists, who did not uniformly align with eugenicists. See Ian Robert Dowbiggin, *Keeping America Sane: Psychiatry and Eugenics in the United States and Canada, 1880–1940* (Ithaca: Cornell University Press, 1997), 191–231.

81. Some, like Thomas Salmon, had worked as commissioned officers of the PHS on the line at Ellis Island. Salmon left the PHS in 1911 to work as the chief medical examiner of the Board of Alienists in New York State, where he continued to press for better mental examination of aliens on the line at the nation's immigration stations. While in the PHS, Salmon became particularly critical of the mental examination. Thomas Salmon, "Immigration and the Problem of the Alien Insane: Discussion," *American Journal of Psychiatry* 4 (1925): 465–66. Dowbiggin stresses that while psychiatrists may not have uniformly grounded their support for better mental examinations at the nation's borders, they all agreed that such examinations were best left in the hands of trained psychiatrists (*Keeping America Sane,* 219).

82. E. H. Mullan, "Mental Examination of Immigrants: Administration and Line Inspection at Ellis Island," *Public Health Reports* (18 May 1917): 737–38.

83. Alter Abelson, "The Designer," in *Songs of Labor* (Newburgh, N.Y.: Paebar Company Publishers, 1947), 67.

84. Frederick W. Taylor, *Principles of Scientific Management* (Norcross, Ga.: Engineering and Managing Press, 1911, 1998), 48.

85. Mullan, "Mental Examination," 744–45.

86. Dana Gumb, interview with Dr. Robert Leslie, 14 August 1985, EIOHP.

87. Henry H. Goddard, "The Feeble Minded Immigrant," *The Training School* 9 (November–December 1912): 109–13; Goddard, "Mental Tests and the Immigrant," *Journal of Delinquency* 2 (September 1917): 243–77.

88. Howard A. Knox, "The Moron and the Study of Alien Defectives," *JAMA* 60 (1913): 105.

89. Letter from E. H. Mullan to surgeon general, 14 June 1919, RG 90, Central File, 1897–1923, Box 38, File No. 219, NARA.

90. Alfred C. Reed, "The Relation of Ellis Island to the Public Health," *New York Medical Journal* 98 (1913): 172; Mullan, "Mental Examination." The relation of race to the mental and physical examination is also discussed in chapter 6.

91. Mullan, "Medical Inspection," 1168; and Reed, "Relation of Ellis Island to the Public Health," 173.

92. The first indication that an immigrant inspector possessed an x-ray machine comes from Pittsburgh in 1910. Letter from J. A. Nydegget, surgeon, Pittsburgh, to surgeon general, 23 June 1910, RG 90, Central File, 1897–1923, Box 513, File No. 4651, NARA. This corresponds to the timing of x-ray technology's introduction into hospital practice. Joel Howell, *Technology in the Hospital: Transforming Patient Care in the Early Twentieth Century* (Baltimore: Johns Hopkins University Press, 1995).

93. The x-ray was also commonly used for head and scalp treatments. Beginning in the 1920s, the PHS used x-ray to treat cases of favus as well as ringworm. Of course, this technology was also used against the PHS, for immigrants would receive x-ray

treatment for favus before departing for the United States. The PHS held these "bald heads" for observation for favus until their hair grew back and the true condition of their scalps could be assessed. Letter from J. W. Kerr, chief medical officer, Ellis Island, to surgeon general, 30 June 1921, RG 90, Central File, 1897–1923, Box 38, File No. 219, NARA.

94. In 1901 the commissioner general noted, "The officers of the Government are powerless to secure evidence to controvert the claims of the Chinese. Thus, in the case of an alleged minor son of a domiciled merchant, the facts alleged to sustain such claim must all have existed in China, to wit, the marriage of the parents, and the birth of the son, as well as the date of birth, and are therefore proved by Chinese testimony exclusively. The Government has no means of refuting a prima facie case thus established, and must admit an applicant upon such evidence, unless it can find, in minor variations in the testimony of different Chinese witnesses, ground for discrediting it altogether. The simple statement of the case is sufficient to show with what comparative ease the law may be defied by the entrance of Chinamen, without limit practically, who can be made to pass muster as minors" (CGAR, 1901, 46–47).

95. Ronald Takai, *Strangers from a Different Shore: A History of Asian Americans* (New York: Penguin Books, 1989), 234–36. Immigrants with forged birth certificates were known as "paper sons" or "paper daughters" because the relationship claimed existed on paper only. Previously, only the wives and children of merchants or businessmen could legally immigrate to the United States; in an 1883 court ruling, the wives and children of laborers were determined also to have the status of laborers and were barred from entry.

96. "Proved a Minor by an X-Ray Machine," *Boston Sunday Post,* 5 March 1911.

97. Letter from Dr. Victor Safford to commissioner of immigration, Boston, 9 March 1911, RG 90, Central File, 1897–1923, Box 58, File No. 409, NARA.

98. Ibid. This case is of interest because it is the first reported use of the x-ray machine to determine age by examining the development of long bones and joints. In the same location, see also "X-Ray to Tell Age of Youth: Chinaman's Right to Enter Country to be Tested," *Boston Sunday Post,* 19 February 1911.

99. Notably, the opinion of the PHS physicians in the case of determining age was just that, a medical opinion that lacked the legal status of the medical certificate. But Safford insisted that the "conclusions" provided by outside physicians also amounted to no more than medical opinion, regardless of the procedure used. It was up to the IS, where suspicion of an immigrant's stated age always originated, to make a final determination. By 1930 the commissioner general conclusively ruled that PHS age determinations did not carry the authority of a medical certificate. Letter from Harry E. Hull, commissioner general, to surgeon general H. S. Cumming, 23 April 1930, RG 90, General Subject File, 1924–35, Box 941, File No. 0950-56, NARA; letter from Sur-

geon General Cumming to chief medical officer, San Francisco, 18 July 1930, RG 90, General Subject File, 1924–35, Domestic Stations, San Francisco Immigration, Box 28, File No. 0950-56, NARA.

100. Ibid.

101. Disease-causing organisms identified between 1880 and 1898 included the following: leprosy (1873), malaria (1880), tuberculosis (1882), glanders (1882), cholera (1883), streptococcus/erysipelas (1883), diphtheria (1884), typhoid (1884), staphylococcus (1884), streptococcus (1884), tetanus (1884), coli (1885), pneumococcus (1886), Malta fever (1887), soft chancre (1887), gas gangrene (1892), plague (1894), botulism (1894), dysentery bacillus (1898). See Rosen, *History of Public Health,* 314.

102. These proportions remained relatively stable in 1917, though tests for favus and ringworm (11%) surpassed those for tuberculosis (8%). By 1920 urethral smears for gonococcus infection led the procedures performed (36%), followed by Wassermann reactions (27%), and then cultures for diphtheria (13%). Urine tests accounted for 10% of procedures performed, and sputum tests for tuberculosis (2.91%) edged out exams for favus and ringworm (2.74%).

103. Frederick W. Taylor, "Principles of Scientific Management," in *Scientific Management* (Westport, Conn.: Greenwood Press, 1972, 1911), 114–15.

104. The SGAR published data for each port aggregated into larger categories: Class A, Class B, and Class C. The surgeon general, however, received quite detailed data from officers at each immigration station. Figures 3-5 through 3-10 take the original data that were available for key immigration stations around the nation and group together separate but related diseases and conditions using the official PHS nomenclature. I created a composite nomenclature, described in detail in the appendix, based on the three official PHS nomenclatures in use from 1891 to 1930. I coded each original immigrant diagnosis according to the broad disease categories of the official PHS nomenclature.

105. Unemployment Committee of the National Federation of Settlements, *Case Studies of Unemployment* (Philadelphia: University of Pennsylvania Press, 1931), 357.

106. *Songs of Labor,* 67.

107. Only ports along the Atlantic Coast experienced a wartime depression in immigration. Consequently, New York and Boston experimented with developing a system of intensive primary examination of aliens.

108. The officers estimated that under the old system, each immigrant at Ellis Island was examined for about forty seconds, with anywhere from 15% to 20% being turned aside for more intensive examination.

109. "These observations bring up the question as to what the result would be if there were a sufficient number of medical officers at all immigration stations to give all arriving aliens a careful physical examination. It is thought that an increase might be expected in the class of diseases known as 'loathsome, contagious diseases.' It is also

thought that venereal diseases would be disclosed in a much larger proportion of cases than is possible at the present time. Then, too, it is thought that conditions could be found in a number of arriving aliens which, while not diseases or defects directly affecting the public health, might prove to be disabling disorders which would not only severely handicap them in the struggle for existence in this country but would sooner or later make them public charges upon the States or municipalities." The chief medical officer concluded that the need to conduct intensive examinations after the war would only increase, as countries would want to "dispose of . . . those persons who will be least useful at home" (SGAR, 1915, 189).

110. Annual Report for Ellis Island, 1918, RG 90, Central File, 1897–1923, Box 38, File No. 219, NARA. On 16 March 1920 medical inspections at Ellis Island resumed (SGAR, 1920, 187).

111. In 1921 Dr. Kerr, the chief medical officer at Ellis Island, recommended putting a permanent intensive examination in place that would provide for examination only of eyes and mental conditions on the line. All aliens would then be taken to special secondary examination rooms where, examining them without their baggage, physicians could get more than a superficial look at their bodies, could see their natural gait, and so on. He argued that some conditions, like heart, lung, and genitourinary conditions, could be found only through close physical examination. Enactment of the quota law of 1921, Kerr reasoned, would make this plan more feasible. Letter from Dr. Kerr, Ellis Island, to surgeon general, 24 April 1921, RG 90, Central File, 1897–1923, Box 38, File No. 219, NARA. Regardless of the surgeon general's stated opposition in the SGAR for 1921, by 1922 he reported that due to the new immigration law of 1921, reduced immigration had allowed "a more critical examination. Especially at Boston and New York, the ports which received practically all of the European immigrants, examinations were made in a more leisurely manner than formerly, with the result that a greater percentage of disease and physical defects was found" (SGAR, 1922, 188). Nonetheless, as I argue in the epilogue, the advent of more intensive examination procedures in the 1920s heralded the end of traditional medical authority in the immigration arena and of the disciplinary function of the medical exam.

112. SGAR, 1921, 234. See also Antonio Stella, "The Immigrant and the Health of the Nation," *Medical Times* (May 1922): 140.

113. Wiebe, *Search for Order,* 147.

114. Taylor, *Principles of Scientific Management,* 103, 18–19, 112.

115. Assistant Surgeon General L. E. Cofer, "The Medical Examination of Arriving Aliens," *Bulletin of the American Academy of Medicine* (1912): 405.

116. David Rosner and Gerald Markowitz, "The Early Movement for Occupational Safety and Health, 1900–1917," in *Sickness and Health in America: Readings in the History of Medicine and Public Health,* ed. Judith Walzer Leavitt and Ronald L. Numbers (Madison: University of Wisconsin Press, 1985), 508.

117. Immigrant health on arrival compared favorably to that of populations within the United States. In a 1923 Life Extension Institute study of 100 postal workers, none was without some form of physical defect: 31% were deemed to have "moderate defects," 57% to have "advanced physical impairment requiring systematic medical or surgical attention," and 12% to have "serious physical defects requiring immediate surgical or medical attention" (Eugene Lyman Fisk, "Extending the Health Span and Life Span after Forty," *Southern Medical Journal* 16 [June 1923]: 447–58). A comparable study among physicians, conducted by the Kings County Medical Society in 1924, produced similar results. Mitchell H. Charap, "The Periodic Health Examination: Genesis of a Myth," *Annals of Internal Medicine* 95 (1981): 734. Similarly, among 2,000 "average" native-born working men applying for positions at the Norton Company in Worcester, Massachusetts, in 1917, 2.3% were rejected for employment on the basis of their physical conditions. The diseases diagnosed were of the same order as those typically diagnosed among immigrants, with the exception of the absence of contagious or parasitic diseases. Of the workers examined, 9.4% were diagnosed with some form of defective vision, 1.3% with diseases or defects of the lungs (pulmonary tuberculosis representing 37% of these conditions), 18.6% with defects of the extremities (including flat feet, varicose veins, ankylosis, and amputations), 16.3% with hernia, 9.0% with defects of the genito-urinary region (varicocele accounting for 86% of these conditions), 1.2% with defects of the spine, and 3.4% with skin lesions (74% accounted for by acne). At least 18.56% of applicants had some defect. Raymond W. Cutler, "Physical Examination of Factory Employees: Two Thousand Consecutive Cases and the Defects Found," *Boston Medical and Surgical Journal* 177 (1917): 627–31.

The physical condition of the native working population was at least comparable to, if not worse than, the condition of immigrants arriving in most regions for most years. For foreign-born groups living in the United States, however, adult and infant mortality was much higher than for the native-born. Contributing to the mortality differential were social and economic factors and concentration of immigrants in urban areas. Samuel H. Preston, Douglas Ewbank, and Mark Hereward, "Child Mortality Differences by Ethnicity and Race in the United States: 1900–1910," in *After Ellis Island: Newcomers and Natives in the 1910 Census*, ed. Susan Cotts Watkins (New York: Russell Sage Foundation, 1994).

118. Gould, "System of Personal Biologic Examinations," 134–36.

119. Conference Board on Safety and Sanitation, "Measuring the Workman's Physical Fitness for His Job," *Spirit of Caution* 4 (August–September 1916): 4; W. Irving Clark, "Physical Examination and Medical Supervision of Factory Employees," *Boston Medical and Surgical Journal* 176, no. 7(15 February 1917): 239.

120. Clark, "Physical Examination and Medical Supervision of Factory Employees," 239.

121. James A. Tobey, "A Layman's View of Health Examinations," *Boston Medical and Surgical Journal* 191, no. 19 (6 November 1924): 876–77.

122. Industrial demands shaped the form of medical examinations in other national contexts, too. In South Africa, for example, Randall Packard finds that inspection of mine workers closely resembled the immigrant medical inspection in the United States, but he notes that the rationale for the examination was to reduce worker mortality in order to prevent the government from cutting off a critical supply of tropical laborers. Packard, *White Plague, Black Labor,* 159–93.

FOUR. The Shape of the Line

Epigraph. Elizabeth Yew, interview with Dr. T. Bruce H. Anderson, 22 September 1977, EIOHP. See also Samuel B. Grubbs, *By Order of the Surgeon General* (Greenfield, Ind.: Mitchell, 1943); and Victor Heiser, *An American Doctor's Odyssey: Adventures in Forty-five Countries* (New York: W. W. Norton, 1936).

1. Although the details are discussed in the appendix, I would like to stress, with regard to all figures in this chapter, that for any year in which it appears that certification or exclusion was zero, this zero represents missing data. Missing data posed a particular problem for Canadian border stations and ports in the island possessions of the United States (including Puerto Rico, the Philippines, and Hawaii), which provided detailed data only for a limited number of years. I provided an estimate for missing data if no more than two consecutive years were missing, but I did not attempt to estimate data where they were missing for more than two consecutive years.

2. Georges Perre, *Ellis Island* (New York: New Press, 1995); Wilton S. Tifft, *Ellis Island* (Chicago: Contemporary Books, 1990); Barbara Benton, *Ellis Island: A Pictorial History* (New York: New Facts on File, 1985); U.S. Public Health Service, *Ellis Island: America's Immigration Cornerstone* (Washington, D.C.: Department of Health and Human Services, 1993); Ivan Chermayeff, *Ellis Island: An Illustrated History of the Immigrant Experience* (New York: Macmillan, 1991); Norman Kotker, *Ellis Island: Echoes from a Nation's Past,* ed. Susan Jones (New York: Aperture Foundation, 1989); Pamela Reeves, *Ellis Island: Gateway to the American Dream* (New York: Dorset Press, 1991); New York Landmarks Preservation Commission, *Ellis Island Historic District* (New York: Landmarks Preservation Commission, 1993); Edward Corsi, *In the Shadow of Liberty* (New York: Arno Press, 1969); David M. Brownstone, Irene M. Frank, and Douglass L. Brownstone, *Island of Hope, Island of Tears* (New York: Rawson, Wade Publishers, 1979); Thomas M. Pitkin, *Keepers of the Gate: A History of Ellis Island* (New York: New York University Press, 1975); Willard Allison Heaps, *The Story of Ellis Island* (New York: Seabury Press, 1976); Ann Novotny, *Strangers at the Door: Ellis Island, Castle Garden, and the Great Migration to America* (Riverside,

Conn.: Chatham Press, 1971); James B. Bell, *In Search of Liberty: The Story of the Statue of Liberty and Ellis Island* (Garden City, N.Y.: Doubleday, 1984); and Eleni Mylonas, *Journey through Ellis Island* (New York: Sammass Productions, 1984).

3. Alan M. Kraut, *Silent Travelers: Germs, Genes, and the "Immigrant Menace"* (New York: Basic Books, 1994); Elizabeth Yew, "Medical Inspection of Immigrants at Ellis Island, 1891–1924," *Bulletin of the New York Academy of Medicine* 56 (June 1980): 488–510; and Fitzhugh Mullan, *Plagues and Politics: The Story of the United States Public Health Service* (New York: Basic Books, 1989). Although his focus is much different, Bernard Marinbach discusses the medical inspection of Jewish immigrants at Galveston, Texas, in *Galveston: Ellis Island of the West* (Albany: State University of New York Press, 1983).

4. Paul Sigrist, interview with Sadie Guttman Kaplan, 2 July 1992, EIOHP. Kaplan immigrated to the United States from Russia in approximately 1905 at around twelve years of age. She later worked at the Hebrew Immigrant Aid Society on Ellis Island in 1920.

5. L. E. Cofer, "The Medical Examination of Arriving Aliens," *Bulletin of the American Academy of Medicine* (1912): 400.

6. See the appendix for a discussion of differences between the number of immigrants inspected and the number of immigrants arrived, and the use of these two figures in the denominator of different calculations.

7. Regarding the role of the steamship companies in obscuring class lines, Williams noted: "On some lines, particularly those running on southern routes, the difference in price between the steerage and second cabin ticket may be as low as $10, resulting in many persons of the immigrant type traveling in the second cabin. Again some steamship officials (fortunately few) attempt evasion of the immigration law by sending obviously ineligible immigrants in the cabin, and sometimes families are separated in the expectation that the ineligible one traveling in the cabin may pass unnoticed" (William Williams, "Ellis Island: Its Organization and Some of its Work," c. August 1911, RG 85, Box 120, File No. 52516/1A, NARA). Steamship companies would also accuse other lines of this practice. The view that cabin passengers were less likely to be diseased held until at least 1921, when the surgeon general wrote in his annual report, "Because they have lived under a better hygienic condition, are better nourished, and of a sufficiently higher intelligence to exercise a certain amount of self-care, cabin passengers as a rule may be considered as suffering from a smaller percentage of defects or diseases than is found amongst steerage" (SGAR, 1921, 282). See also "Organization of the US Immigrant Station at Ellis Island, New York, Together with a Brief Description of the Work Done in Each of Its Divisions," October 1903, RG 90, Box 120, File No. 52516/1, NARA.

8. SGAR, 1921, 241.

9. A less rigid examination on shipboard was accorded to first- and second-cabin

passengers at all ports, though shipboard examinations of steerage, which prevailed at many of the smaller ports, tended to make the examination less rigid for steerage passengers because of the limits of space, light, facilities, and, above all else, time—especially when a cargo ship waiting to dock had to be held in quarantine until all passengers were examined. In Boston, Philadelphia, and Tampa, for instance, all immigrants were examined on shipboard. Letter from J. A. Nydeggen, surgeon, Baltimore, to J. B. Densmore, solicitor, Department of Labor, 18 February 1916, RG 85, Box 208, File No. 53438/15D, NARA; letter from E. E. Greenawalt, commissioner, Philadelphia Immigration Station, to Densmore, 9 February 1916, Doc. No. 1785, RG 90, Box 1, 311398; Letters to the Surgeon General from the Medical Officer in Charge, 1 July 1903–1 March 1926, vol. 21, NARA. Also letter from Dr. Cumming to surgeon general, 28 October 1903; letter from George B. Schumaker, acting inspector in charge, Tampa, to commissioner general, 7 January 1909; letter from Schumaker to commissioner general, 4 January 1909; all in RG 85, Box 60, File No. 51831/28, NARA.

10. Letter from Assistant Surgeon General H. D. Geddings to surgeon general, 16 November 1906, RG 90, Central File, 1897–1923, Box 36, File No. 219, NARA (emphasis added). The Texas Department of Health maintained a similar attitude. The quarantine law relating to venereal diseases specified that "persons not known to the QUARANTINE Officers, but having every appearance as coming from an elevated social level, may, at the discretion of the Quarantine Officer be examined or voluntarily sign [an affidavit stating that they are not] suffering from any venereal or other contagious disease" (memorandum from Bureau of Venereal Diseases, "Enforcement of 'Law Relating to Venereal Diseases,'" Austin, Texas, to the Texas State Quarantine Service, 15 September 1918, RG 85, Box 292 (renumbered 262), File No. 54549/381, NARA.

11. Letter from William H. Howitt to the British consul general in New York, 8 February 1906, RG 90,Central File, 1897–1923, Box 36, File No. 219, NARA.

12. Letter from Dr. William Ward to surgeon general, 7 March 1906, RG 90, Central File, 1897–1923, Box 36, File 219, NARA. Upon further investigation, Surgeon General Walter Wyman concluded that Dr. Ward had acted properly and that—as was apparent upon reading Howitt's full account of the incident—he was clearly an opponent of immigrant medical inspection, calling it "obnoxious and dangerous." Letter from Walter Wyman to secretary of the treasury, 7 March 1906, RG 90, Central File, 1897–1923, Box 36, File No. 219, NARA.

13. Joan Morrison and Charlotte Fox Zabusky, *American Mosaic: The Immigrant Experience in the Words of Those Who Lived It* (New York: E. P. Dutton, 1980), 70.

14. Richard Hofstadter, *The Age of Reform* (New York: Vintage Books, 1955); George Mowry, *The California Progressives* (Berkeley: University of California Press, 1951); Gabriel Kolko, *The Triumph of Conservatism: A Re-Interpretation of American History* (New York: Free Press, 1963); Robert Wiebe, *The Search for Order, 1877–*

1920 (New York: Hill and Wang, 1967); Samuel P. Hays, *The Response to Industrialism, 1885–1914* (Chicago: University of Chicago Press, 1957); and John D. Buenker, *Urban Liberalism and Progressive Reform* (New York: Scribner's, 1973). Hofstadter and Mowry both saw social and cultural elites at the heart of the Progressive movement; Kolko and Hays saw businessmen and industrialists; Wiebe saw a new "middle class" and new professionals; and Buenker saw working-class immigrants.

15. Timothy J. Meagher, "Immigration through Boston: A Comment," in *Forgotten Doors: The Other Ports of Entry to the United States,* ed. M. Mark Stolarik (Philadelphia: Balch Institute Press, 1988), 32–33. See also Lawrence H. Fuchs, "Immigration through the Port of Boston," in *Forgotten Doors,* 17–25.

16. SGAR 1911, 201. Safford attributed the preference of the diseased for second or first cabin to a variety of conditions: "The less efficient precautions taken abroad to prevent diseased or defective persons from embarking as cabin passengers, the comparatively small difference in price between a steerage and second-class passage, certain privileges which it is customary to extend on arrival here to passengers who may be designate cabin passengers, and conference agreements penalizing a company for exceeding its prescribed quota of steerage passengers, are all tending constantly to increase the proportion of so-called second-cabin passengers aboard ships in the Atlantic trade" (SGAR, 1913, 148).

17. Although New York, San Francisco, and Philadelphia reported certification rates by class on one or more occasions, this was not a consistent part of reporting, as it was in Boston. Moreover, second-cabin passengers were never distinguished from first-cabin passengers, and the certification rates for cabin passengers never exceeded those for steerage. SGAR, 1900, 1902; letter from J. C. Perry, medical director, District No. 5, San Francisco, to surgeon general, 4 November 1927, RG 90, General Subject File, 1924–35, Domestic Stations, San Francisco Immigration, Box 28, File No. 0950-56, NARA. In some years, Boston only reported certifications by class.

18. The certification rates for stowaways were usually high, though it is significant that in 1913 and 1917 a greater percentage of second-cabin passengers were certified. The PHS and IS viewed stowaways with extreme suspicion, for they were typically individuals attempting to enter the United States illegally, trying to avoid inspection entirely.

19. Transcript of committee to discuss the problem of the examination of second-cabin passengers, 3–4 March 1916, RG 85, Box 208, File No. 53438/15E, NARA (emphasis added).

20. The United States, for example, carefully monitored the activities of Eng Hok Fong and his shipping company during his 1908 trip to Mexico. U.S. officials understood that Eng Hok Fong traveled to Mexico in an attempt to convince the Mexican government not to enforce medical inspection regulations. Letter from Alvey A. Adee,

acting secretary, Department of State, to secretary of commerce and labor, 26 September 1909, RG 85, Box 81, File No. 52082/43, NARA; letter from Oscare S. Straus to Elihu Root, secretary of state, 22 September 1908, RG 85, Box 81, File No. 52082/44, NARA.

21. "[In the fall of 1913] there was considerable protest on the part of the steamship companies, railroads and civic organizations and other associations in New York and its suburbs, and the Secretary gave these people a hearing in Washington [in 1915]; and after that hearing he suspended the operation of the order and appointed a committee to investigate the matter and see if there were some arrangements that could be made with the steamship officials whereby facilities could be provided on board the ships or on the piers for this examination" (transcript of committee to discuss the problem of the examination of second-cabin passengers, 3–4 March 1916, RG 85, Box 208, File No. 53438/15E, NARA. Although the New York shipping lines agreed to adhere to any regulations regarding light or space on either the ships or the piers, the Boston steamship companies, though willing to allow pier examinations, were unwilling to provide proper lighting and facilities on the ships. The major concern of the PHS officers regarding shipboard medical examinations was related to proper lighting. While the cabin was considered suitable for the first-class passengers, who were expected to be free from disease, the artificial lighting in the ship's cabin was considered insufficient to properly illuminate the second-cabin passengers. Thus, inspection on the upper deck of a ship was preferred for second-class passengers. The possibility of overcast days made even this option inferior to inspection in proper examination facilities at the pier or in an immigration station.

22. Again in 1924, the PHS and IS expressed "the growing feeling that unfit immigrants will try to travel in second-cabin and that more intensive examination than shipboard inspection is necessary to catch the types of conditions they present" (letter from W. C. Billings, chief medical officer, Ellis Island, to surgeon general, 16 April 1924, RG 90, General Subject file, 1924–35, Domestic Stations, New York, Box 163, File No. 950-56, NARA). That same year, the IS planned to conduct the inspection of second-cabin aliens at Ellis Island and, for the first time, suggested that this would even be ideal procedure for first-cabin passengers. Yet by 1924 the philosophy of the immigrant medical exam was undergoing fundamental changes, as discussed in the epilogue. Letter from Henry H. Curran, commissioner, Ellis Island, to Sidney E. Morse, Esq., secretary, Trans-Atlantic Passenger Conference, 29 February 1924, RG 90, General Subject file, 1924–35, Domestic Stations, New York, Box 163, File No. 950-56, NARA.

23. "With the cooperation of the Commissioner the practice of inspecting alien second class passengers on the dock has been continued wherever possible. It has been a success from a public service point of view, although it has never had the approval of the transportation companies" (letter from Dr. A. J. Nute, medical officer in charge, to

surgeon general, 10 July 1922, and Boston's annual report for 1922, RG 90, General Subject File, 1924–35, Domestic Stations, Boston Immigration, Box 137, File No. 1850-15, NARA.

24. In 1921 the medical examiner at Boston claimed that examination of second-cabin passengers on the docks worked well, "although it was difficult at times to overcome the tradition in the minds of some officials that having paid a cabin fare such passengers were entitled to special privileges" (SGAR, 1921, 241).

25. Report of M. V. Stafford, Boston, in SGAR, 1909, 182; letter from W. C. Billings, chief medical officer, Ellis Island, to surgeon general, 16 April 1924, RG 90, General Subject File, 1924–35, Domestic Stations, New York, Box 163, File No. 950-56, NARA.

26. Letter to Surgeon R. M. Woodward from Dr. Safford, 21 January 1905, RG 90, Central File, 1897–1923, Box 58, File No. 409, NARA.

27. Weindling observes a similar pattern for Germany: "Despite the anxiety concerning epidemics, in 1905 the vast majority (5,272 out of c.112,000) of migrants turned back at the border control stations and at Ruhleben had eye diseases; a further 98 were refused passage because of the condition of their hair, and 146 were condemned as too decrepit, but only 40 persons were denied passage for harbouring infectious diseases like cholera, plague, and typhus" (Staatsarchiv, Bremen, A.4 no. 300a, report by Tjaden and Nocht, 20 July 1906, cited in Paul Julian Weindling, *Epidemics and Genocide in Eastern Europe: 1980–1945* [Oxford: Oxford University Press, 2000], 65).

28. Charles Wollenberg, "Immigration through the Port of San Francisco," in *Forgotten Doors,* 144–47.

29. San Francisco was the largest port of entry for Chinese immigrants; more Japanese entered the port of Los Angeles. Ibid., 150.

30. Ibid., 146–47. See also Ira B. Cross, *A History of the Labor Movement in California* (Berkeley: University of California Press, 1935).

31. Alexander Saxton, *The Indispensable Enemy: Labor and the Anti-Chinese Movement in California* (Berkeley: University of California Press, 1971); Elmer C. Sandmeyer, *The Anti-Chinese Movement in California* (Urbana: University of Illinois Press, 1939); Neil L. Shumsky, "San Francisco Workingmen Respond to the Modern City," *California Historical Quarterly* 55, no. 1 (1976): 46–57; and Alan M. Kraut, *Silent Travelers: Germs, Genes, and the "Immigrant Menace"* (New York: Basic Books, 1994), 78–104.

32. Letter from Hugh Cumming, Quarantine Station, to Treasury Department, 27 April 1903, and letter from Hugh Cumming to surgeon general, 22 April 1903, RG 90, Central File, 1897–1923, Box 784, File 16090, NARA.

33. SGAR, 1903, 207.

34. This process was very similar to that described in 1903. Letter from Dr. Trotter to surgeon general, 18 July 1907, RG 90, Central File, 1897–1923, Box 785, File 16090, NARA.

35. Ibid.

36. SGAR, 1904, 210.

37. Annual Report for San Francisco, 1910, RG 90, Central File, 1897–1923, Box 785, File No. 16090, NARA. See also SGAR, 1910, 172.

38. The officer in charge reported that of the immigrants inspected on shipboard, "over 7,810 were brought to Angel Island for further and more extended examination. . . . Previous to the year just closed some few second and third cabin passengers were released directly from the vessel, but during the year 1916 all second and third cabin passengers were brought to Angel Island" (SGAR, 1916, 226).

39. SGAR, 1922, 207.

40. W. C. Billings, "Oriental Immigration," *Journal of Heredity* 6 (1915): 467. Billings continued, "The slightest suspicion of any untoward condition, such as a lung or heart involvement, a skin eruption, evidence of temperature, etc., etc., is followed by a sufficiently extensive examination appropriate to the condition to determine the diagnosis; but in the absence of any suspicious signs this concludes the physical examination," except for the final microscopic examination of the feces.

41. Ibid.

42. Enclosure explaining desired protocol in letter from B. J. Lloyd, PA surgeon, Seattle, to surgeon general, 14 November 1911, RG 90, Central File, 1897–1923, Box 783, File No. 16058, NARA.

43. Letter from assistant commissioner general to commissioner of immigration, San Francisco, 27 June 1923, Doc. No. 5421/13, RG 85, Box 265, File No. 5461/13, NARA.

44. W. Crewdson, "Japanese Emigrants," *Nineteenth Century* 56 (November 1904): 814.

45. SGAR, 1911, 142.

46. SGAR, 1914, 220–21.

47. Letter from Dr. Trotter, Angel Island Quarantine, to surgeon general, 22 August 1911, RG 90, Central File, 1897–1923, Box 785, File No. 16090, NARA.

48. Stool samples were to be collected from all Asian immigrants with the exception of returning Chinese merchants and laborers. Enclosure explaining desired protocol in letter from B. J. Lloyd, PA surgeon, Seattle, to surgeon general, 14 November 1911, RG 90, Central File, 1897–1923, Box 783, File No. 16058, NARA.

49. Ibid.

50. Letter from Secretary Franklin MacVeagh to secretary of commerce and labor, 20 December 1912, RG 90, Central File, 1897–1923, Box 677, File No. 5738, NARA.

51. Letter from surgeon general to commissioned medical officers and acting assistant surgeons engaged in medical examination of arriving aliens, December 1912, RG 90, Central File, 1897–1923, Box 677, File No. 5738, NARA.

52. Aside from Boston and New York (which only broached the issue for Chinese in transit through that port), the only port ever to mention routine inspection of the Chinese for parasites was New Orleans, which in 1922 ordered the elimination of "bacteriological examinations of Chinese passing in transit through this port." Letter from J. T. Scott, acting assistant surgeon, forwarded to R. E. Ebersol, surgeon in charge, New Orleans, to surgeon general, 18 April 1922, RG 90, Central File, 1897–1923, Box 52, File No. 377, NARA. Of course, this was part of a larger decision on the part of the IS to end the requirement that passengers merely traveling through the United States be subjected to examinations.

53. Billings, "Oriental Immigration," 462–67. Other ports that admitted Chinese immigrants included San Diego; Portland, Oregon; New York; New Orleans; Port Townsend, Washington; Seattle; Tampa; Honolulu; and San Juan, in addition to stations along the Mexican border, such as El Paso.

54. Here the PHS was discussing parasitic infection among African natives of the Cape Verde Islands. SGAR, 1917, 159.

55. Letter from Dr. A. J. Nute to surgeon general, 25 February 1922, RG 90, Central File, 1897–1923, Box 59, File No. 409, NARA.

56. The IS provided the necessary laboratory equipment and "as a result, 91 Chinese were found to be infected with various types of intestinal parasites. There were 12 cases of clonorchiasis and 25 of uncinariasis. In some cases the infections were so mixed that aliens were veritable walking zoological gardens" (SGAR, 1922, 199).

57. Annual Report for Boston, 1930, 3, RG 90, General Subject File, 1924–35, Domestic Stations, Boston Immigration, Box 137, File No. 1850-15, NARA; SGAR, 1930, 201.

58. Letter from J. C. Perry, surgeon, medical director, District No. 5, to surgeon general, 29 October 1923, RG 90, General Subject File, 1924–35, Domestic Stations, San Francisco Immigration, Box 28, File No. 0950-56, NARA. As of 31 December 1924 Asian steerage passengers with "return certificates"—indicating that they had been living in the United States and stating that they had been medically inspected before departure back to Asia—were no longer required to go to Angel Island for medical examination.

59. SGAR, 1925, 176; 1926, 183; 1928, 216.

60. SGAR, 1927, 223.

61. In 1926 the officer in charge reported, "The laboratory facilities, though somewhat limited, have proved invaluable in connection with the medical examination of aliens. . . . A larger proportion of the incoming aliens originate in the southern Chinese Province of Kwantung, whose indigenes present an intestinal fauna of perhaps unsur-

passed richness and variety, so that the examination of feces has long been regarded as a most productive feature of the laboratory work" (SGAR, 1926, 183).

62. In 1905 and 1908 scabies accounted for 100% of the parasitic and skin conditions. In 1910 scabies still represented more than 65% of parasitic skin conditions, though the "new" parasitic diseases like ringworm and hookworm began to prevail. By 1917 parasitic diseases consisted almost entirely of nematode worms, flukes, and roundworms.

63. SGAR, 1923, 182.

64. Apparently no one knew how the practice of making night inspections began at this port. The writer said that the steamship companies claimed that an exception had been made for them because their role as mail carriers from Florida to Havana made any extra delay out of the question. The postmaster at Key West, however, felt he could only benefit from a delay in the arrival of mail. Letter from immigrant inspector in charge, Tampa, to commissioner general, in reply to a request to explain why inspections were being conducted at night despite explicit prohibitions, 9 February 1907, RG 90, Central File, 1897–1923, Box 485, File No. 4446, NARA.

65. SGAR, 1916, 224; 1917, 183.

66. Correspondence between the PHS and IS, January and February 1918. In particular see letter from Surgeon General Blue to commissioner general, 19 December 1918, RG 90, Box 52, File No. 377, Central File, 1897–1923, NARA.

67. Annual Report for New Orleans, 1 July 1924, RG 90, General Subject File, 1924–35, Domestic Stations, New Orleans Quarantine, Box 115, File No. 1850-15, NARA; Report of the Medical Inspection of Immigrants, during the year ending 30 June 1917, and Annual Report for New Orleans, 1917, both in RG 90, Box 52, File No. 377, Central File, 1897–1923, NARA; SGAR, 1917, 178. As described by the physician in charge, "every person on board is examined in a screened room adjacent to the wharf and their temperatures are taken invariably as a routine procedure throughout the year as part of the quarantine inspection." The physician in charge estimated that the combined exam took approximately five times as long as the quarantine exam alone. Letter from Dr. Fauntleroy to surgeon general, 8 May 1919, RG 90, Box 52, File No. 377, Central File, 1897–1923, NARA. The quarantine service took almost complete charge of immigrant and alien crew inspections by 1924, yet according to Dr. C. L. Williams, the surgeon in charge at the New Orleans Quarantine Station, "in the majority of cases only preliminary inspection is performed." See Annual Report for New Orleans, 1925, RG 90, General File, 1924–35, Domestic Stations, New Orleans Quarantine, Box 115, File No. 1850-15, NARA. See also the annual reports for 1924–30 in the same location.

68. Letter from Dr. Scott to G. M. Corput, PHS, 2 February 1918, RG 90, Box 52, File No. 377, Central File, 1897–1923, NARA.

69. Letter from Commissioner S. E. Redfern to Dr. Scott, 17 December 1910, RG 90, Box 52, File No. 377, Central File, 1897–1923, NARA.

70. Letter from Surgeon General Walter Wyman to Dr. Fairbanks, medical officer in command, New Orleans Quarantine, 29 December 1910, RG 90, Box 52, File No. 377, Central File, 1897–1923, NARA.

71. "New Orleans Certain to Have Immigration Station," *New Orleans Item,* 20 January 1907; "First Immigrants at New Station Here," *Daily Picayune,* 3 May 1913; "Where the Immigrant Finds a Welcome: Art Has Helped Nature Make New Orleans Station Antithesis of Ellis Island," *Times Picayune,* 19 June 1927, Magazine Section. Notably, New Orleans newspapers provided very little coverage of immigration. Of the four articles concerning immigration appearing in New Orleans newspapers from 1891 to 1930, three covered the building, opening, or general operation of the immigration station at Algiers (serving New Orleans).

72. John Higham, *Strangers in the Land: Patterns of American Nativism 1850– 1925* (New York: Atheneum, 1967), 113–14. See also Joseph Logsdon, "Immigration through New Orleans," in *Forgotten Doors,* ed. Stolarik, 110.

73. John Smith Kendall, *Histories of New Orleans,* vol. 3 (Chicago: Lewis Publishing, 1922); see also letter from Justin Denechaud to Mr. N. Behar, managing director, National Liberal Immigration League, New York City, 10 March 1913, Box 5, RG 33, Justin Francis Denechaud Papers, Louisiana State Archives (LSA), New Orleans.

74. Oscar Dowling, president of Louisiana State Board of Health, "Health Conditions in Louisiana" (1911), 6. See also the biennial reports of the Louisiana State Board of Health, 1900–1923.

75. Circular letter (addressee unknown) from Justin Denechaud, January 1912, RG 33, Justin Francis Denechaud Papers, Box 1, LSA. The Denechaud papers date from 1908 to 1914. After the completion of the Panama Canal, Louisiana hoped to attract U.S. citizens who had grown accustomed to a tropical climate to come to Louisiana. In response to an editorial in a local newspaper, Denechaud flatly denied that he was engaged in efforts to bring undesirable immigrants to the state: "It is not the purpose of our Zone office to bring to Louisiana an undesirable class of immigrants such as some would consider the class of laborers now at work building the Canal, but what we hope to interest is the American citizen and I am sure that both he and his family, as well as his money, will be looked upon as desirable acquisitions to our State, and no fault should be found with the Agricultural and Immigration Department for endeavoring to draw such to Louisiana" (letter from Denechaud to editor of the *Lake Charles Times,* 11 July 1913).

76. Letter from Denechaud to Dr. W. M. Perkins, secretary, Louisiana State Board of Health, 16 October 1913, RG 33, Justin Francis Denechaud Papers, Box 1, LSA.

77. Letter from Mr. Olaf Huseby, Empire Land Company, Moorhead, Minnesota, to Mr. W. R. Dodson, secretary of agriculture, 2 August 1913, RG 33, Justin Francis Denechaud Papers, Box 1, LSA.

78. Charles Caldwell, "Thoughts on the Probable Destiny of New Orleans, in Re-

lation to Health, Population and Commerce," *Philadelphia Journal of the Medical and Physical Sciences* 6 (1823): 1.

79. Letter from Justin Denechaud, secretary, Immigration Division, Louisiana State Board of Agriculture and Immigration, 23 July 1913, to president and members of the police jury (circular letter sent to all parishes), RG 33, Justin Francis Denechaud Papers, Box 1, LSA.

80. Letter from Justin Denechaud to Mr. A. Long Hargrave, Cleveland, 14 September 1912, RG 33, Justin Francis Denechaud Papers, Box 3, LSA. Similarly, in 1911 Oscar Dowling, president of the Louisiana State Board of Health, wrote, "Climate in the first is the primary element [of health]. The idea is almost universal among the people of the North, East and West, that we have a non-salubrious climate, whereas in reality a comparison of weather conditions here with those in states considered more desirable, will show points in favor of Louisiana" (Dowling, "Health Conditions in Louisiana," 1).

81. "State Board of Health Will Hold a Special Session Early in January to Perfect Plans to Keep Out Disease by Way of Texas and Mexico," *Daily Picayune,* 17 November 1903; "City Offers a Site for Quarantine Station in Limits," *Daily Picayune,* 10 March 1916.

82. Letter from Surgeon General Walter Wyman to S. D. McEnery, U.S. Senate, 13 November 1909, RG 90, Central File, 1897–1923, Box 122, File No. 1339, NARA.

83. Letter from C. H. Ellis, manager, United Fruit Company, to Surgeon General Walter Wyman, 23 October 1909, RG 90, Central File, 1897–1923, Box 112, File No. 1339, NARA; see also, in the same location, letter from Mr. Ellis, New Orleans Board of Trade, Ltd., to Surgeon General Wyman, 20 October 20, 1909.

84. Edwin Adams Davis, *The Story of Louisiana, Volume I* (New Orleans: J. F. Hyer Publishing, 1960), 294–305.

85. Gerald Markowitz and David Rosner, *Deceit and Denial* (Berkeley: University of California Press, 2002).

86. James C. Young, "Breaking into the United States," *World's Work* 49 (1924): 55; "Canadian Inspection of Aliens," *San Francisco Call,* 14 November 1901; "Records from Canada and Mexico," *San Francisco Call,* 16 December 1920; "Canadian Route in Evasion of U.S. Laws," *New York Times,* 10 August 1900; "Why Canada Gets 'Cream' and U.S. Gets 'Skim Milk,'" *New York Times,* 26 September 1890.

87. CGAR, 1902, 40.

88. Heiser, *An American Doctor's Odyssey*, 31.

89. Ibid.

90. See the extensive correspondence in RG 85, Box 74, File Nos. 51931/14a, 51931/14b, and 51931/21; Box 38, File Nos. 51564/4(-1), 51564/4(-2), 51564/4(-4), and 51564/4(-5); Box 412(382), File No. 55599/39; Box 63, File No.51831/157, NARA. For a copy of the Canadian Agreement, see Bureau of Immigration, "Relating to the

Inspection of Aliens Landed at the Ports of the Dominion of Canada," Department Circular No. 97, 1901, RG 85, Box 38, File No. 551564/4(-2), NARA.

91. *Annual Report of the Department of the Interior for the Year 1903 to 1904* (Ottawa: S. E. Dawson, Printer to the King's Most Excellent Majesty, 1905), 148–49. Although Canadian inspectors were instructed to look for a narrower range of diseases, the Canadian classification of immigrant diseases and the instructions for the immigrant medical inspection were modeled on the United States, although with the added imperative to "further carefully examine the marks or stigmata of degeneracy" (150). Canadian regulations, moreover, did not emphasize the medical gaze or the notion that the examining physician could detect disease at a glance.

92. Report of the chief medical officer, in ibid.,147.

93. Cited in A. Sears, "Immigration Controls as Social Policy: The Case of Canadian Medical Inspection 1900–1920," *Studies in Political Economy* 33 (autumn 1990): 97. Using the specific example of trachoma, Sears points out that Canadian officials were not particularly concerned with its contagiousness, but with the low social condition and habits of life of trachoma sufferers.

94. Dominion of Canada, Ministry of the Interior, "Immigration Facts and Figures" (Ottawa: Printer to the King's Most Excellent Majesty, 1911).

95. *Annual Report of Department of the Interior for the Fiscal Year Ending March 31, 1916,* Sessional Paper No. 25 (Ottawa: Printer to the King's Most Excellent Majesty, 1916).

96. Letter from John E. Brooks, acting assistant surgeon, Eastport, Maine, to surgeon general, 1 July 1915, RG 90, Central File, 1897–1923, Box 533, File No. 4716, NARA.

97. Although in many respects Montreal was very much a Canadian border port, it differed from most other stations in that it was a Canadian port in which a U.S. PHS officer was stationed. Here, the line inspection resembled that of New York, though it also included routine examination of the groin and chest. The United States offered a similar privilege to Canadian medical inspectors who wished to inspect immigrants bound for Canada who arrived at Ellis Island. As of 1911 the inspection at Montreal was still done on shipboard. But, as explained by C. A. Bailey of the PHS in Montreal, "the Canadian immigration department is contemplating the erection of a modern administration building, of about 700 feet in length by 80 feet in width, in which a space of 140 feet by 80 feet is to be set aside for the United States Immigration Service, when the medical officer will probably be given better facilities for the conduct of the medical examination" (SGAR, 1911, 207).

98. In 1913, for example, Earl Coe, the inspector in charge of immigration at Saulte Ste. Marie, Michigan, was coping with what he perceived to be a deteriorating situation. He complained to his superiors that Dr. Fred Townsend cared more for his private practice than his immigration duties. Letter from Earl F. Coe, inspector in charge,

Saulte Ste. Marie, to commissioner of immigration, Montreal, 6 December 1913, Doc. No. 252 JHJ, RG 85, Box 181, File No. 53033/9, NARA. Mr. Coe insisted that he harbored no ill will toward Dr. Townsend: "What I desire . . . is simply to emphasize the fact and belief that the medical examination of aliens at this port will never be satisfactory so long as it is conducted by private practitioners." The previous September the commissioner general had made a similar assessment of Dr. Townsend, whom he described as "so taken up with his private practice that our Service is greatly hindered in the prompt examination of aliens. . . . The service of the medical officer provided at the Soo has not been other than unsatisfactory for years, owing to such officer being absent so much in connection with private practice" (letter from commissioner general to John Clark, commissioner at Montreal, 27 September 1913, Doc. No. 10900/332, RG 85, Box 181, File No. 53033/9, NARA).

Coe filed similar charges against Dr. Winslow, who had been hired to replace Townsend for three months while Townsend looked into transferring his private practice to Mt. Clemens. Coe complained to Dr. Winslow that his private practice was seriously interfering with the medical examination of aliens, primarily because the IS had great difficulty reaching Winslow when he was needed. Letter from Coe to R. C. Winslow, acting assistant surgeon, 30 November 1913, Doc. No. 169 JHJ, RG 85, Box 181, File No. 53033/9, NARA. Winslow, who also served as the city coroner, agreed, responding, "I also recognize that the frequent calls to the Federal Building, together with calls to trains, for the purpose of examining aliens, are entirely incompatible with a private practice, and that it is beyond reason to hope to secure satisfactory service in examining immigrants from a physician and surgeon engaged in active practice." Nevertheless, he added, "I sincerely believe that you are now receiving services fully commensurate with the salary of one hundred dollars a month . . . when we consider the volume of services required" (letter from R. C. Winslow to Coe, 6 December 1913, RG 85, Box 181, File No. 53033/9, NARA).

In Detroit Dr. Wollenberg, who did not maintain a private practice but instead was stationed at the Marine hospital and served as Detroit's medical inspector as needed, complained that the IS had unreasonable expectations, arguing that he was "criticized by the Immigration Service if he comes even half an hour after being called," even if he were in the middle of an operation at the time (letter from Fairfax Irwin, surgeon, Marine Hospital, Detroit, to surgeon general, 10 December 1906, RG 90, Central Correspondence, 1897–1923, Box 373, File No. 3690, NARA).

Dr. Winslow's perspective stands in stark contrast to commissioned officers within the PHS. He felt no inclination to serve beyond the limits of his salary. His professional duty was to his patients; he would inspect immigrants only when called and only when it was convenient for him to do so. The surgeon general, upon hearing of the situation at "the Soo," was not eager to replace Dr. Fred Townsend. The expectations for noncommissioned officers or acting assistant surgeons were much lower than for commissioned officers. As one PHS officer explained, there was, "certainly, a difference in philosophy. The regular officer entered the Service as a career, the Acting Assistant as a job" (Elizabeth Yew, interview with Dr. T. Bruce H. Anderson, 22 September 1977, EIOHP). As Walter Wyman saw it, however, there was a "great advantage of having a well appointed physician, such as Dr. Townsend," at Saulte Ste. Marie. While the PHS might hire another full-time officer, the possession of Townsend "is much more important than to have a man who, while he might be on hand in the office constantly, might in the end fail to reflect the credit on the work which Dr. Townsend is capable of doing; indeed, from all reports is doing" (letter from Surgeon General Wyman to commissioner general, 31 March 1911, RG 85, Box 181, File No. 53033/9, NARA).

The surgeon general, then, expressed different expectations for Townsend, a civil servant. Yet at the same time he also sought to protect the reputation of the PHS at "the Soo," which had suffered humiliation and even a loss of credibility following controversy (related in chapter 5) over a supposed misdiagnosis made by Dr. Wesley Townsend (Dr. Fred Townsend's brother) while allegedly intoxicated. Fred Townsend, in the end, was only a part-time employee working at a station with light immigration. Thus, Wyman was willing to tolerate the demands and whims of private physicians who reflected positively upon the PHS and performed with competence if not devotion.

99. Overwork was common to many ports, but in places like Boston and Ellis Island it was typically shared by a staff of physicians, though the staff was small considering the volume of immigration. The Canadian border was noteworthy for the demand it placed on individuals working alone. I found only one other such case, at a station not located on the Canadian border. Dr. W. P. Woodall, stationed in Hidalgo, Texas, complained, "The work has become so strenuous that I have to devoted my whole time and attention thereto; in fact I am so overworked and underpaid, and with the forthcoming cold weather and incident pediculis vestimenti, things do not appear to be getting better fast. My facilities and my help are wholly inadequate to do the divers duties that are devolved upon me. . . . I would like to have urged upon the surgeon general authority to employ a reliable assistant to assist in examinations and in vaccinations. Also an appropriation for an adequate building to clean up and disinfect and vaccinate arriving aliens" (letter from Dr. W. P. Woodall, Hidalgo, Texas, to Dr. J. B. Lloyds, passed assistant surgeon, El Paso, 5 October 1916, RG 90, Central File, 1897–1923, Box 123, File No. 1371, NARA). Woodall got his raise (one hundred dollars a month) and permission to hire assistants (at seventy-five dollars per month). Letter from surgeon general to Senior Surgeon C. C. Pierce, Eagle Pass, 26 October 1916, RG 90, Central File, 1897–1923, Box 123, File No. 1371, NARA. Certainly, it was possible for PHS inspectors along the Canadian and other borders to be underworked, though such complaints or reports were extremely rare. Dr. Markley, stationed in Bellingham, Washington, reportedly examined immigrants from only two to three ships each year.

Letter from L. L. Williams in Vancouver to surgeon general, RG 90, Central File, 1897–1923, Box 141, File No. 1582, NARA.

100. In the annual report of the PHS, the surgeon general described the changing role of the PHS along the Canadian border: "Up to 1901 the immigration work at Buffalo, N.Y., was carried on in a perfunctory manner, and its offices were located in the Federal Building in the city of Buffalo. There was little or no attention paid to immigrants coming to the United States, and only occasionally was the Public Health and Marine-Hospital Service called upon to examine aliens. Such as were examined were detained by the customs officers and the immigration service notified. The chief work of the [Immigration] service at the port of Buffalo prior to 1901 was the looking up of violations of the laws governing alien contract labor. . . . During the year of the Pan-American Exposition at Buffalo the work assumed a new and more extensive form, and the immigrants coming to the United States were subjected to a rigid examination, both physically and mentally. At this time an extra inspector was placed at Black Rock, and the office removed from the Federal building to its present location on the banks of the Niagara River, and an officer of the Public Health and Marine-Hospital Service was placed in charge of the medical examination of all immigrants seeking admission to the United States. . . . At first the work was light. It was not long, however, before the increase in work necessitated the assignment of another immigrant inspector to duty there. The flow of immigration to this country at Black Rock from and through Canada has increased to such an extent during the past eight years that the immigration office force now numbers 14 persons. Notwithstanding the increase in immigration work there has been no increase in the number of Public Health and Marine-Hospital Service officers detailed to examining the incoming immigrants, the one on duty being obliged to hold himself at readiness at all times, both day and night, to report at the immigration office for the examination of aliens, independent of the regular daily office hours" (SGAR, 1909, 184–85).

101. SGAR, 1912, 134. In 1914 the inspector in charge of the Canadian ports in District No. 1 complained of congestion and overwork, and criticized the federal government: "It seems needless to dwell upon the inefficiency of an inspection system whereby the Government expends no inconsiderable amount of money to enforce the immigration laws at one point and at the same time maintains a wide open door but a short distance away" (CGAR, 1914, 171).

102. Letter from J. H. Galliger, U.S. Senate, to surgeon general, 24 March 1906, RG 90, Central File, 1897–1923, Box 60, File No. 460, NARA.

103. In 1913 Dr. Safford alerted the surgeon general that he and Dr. Nute bore a crushing workload. Letter from Dr. Victor Safford to surgeon general, 19 August 1913, RG 90, Central File, 1897–1923, Box 58, File No. 409, NARA.

104. Letter from J. J. Siffer, surgeon, Port Huron, to surgeon general, 21 Novem-

ber 1906, and letter from Acting Commissioner General Larned to surgeon general, 28 November 1906, RG 90, Central File, 1897–1923, Box 60, File No. 460, NARA. In Siffer's case, the IS was willing to increase his pay but lacked the funds to do so.

105. Letter from Dr. B. L. Schuster, acting assistant surgeon, Port Huron, to surgeon general, 27 October 1912, RG 90, Central File, 1897–1923, Box 60, File No. 460, NARA.

106. Letter from surgeon general to Schuster, 8 November 1912, RG 90, Central File, 1897–1923, Box 60, File No. 460, NARA. This was the surgeon general's second warning to Dr. Schuster, the first having been delivered on 24 October 1912, in response to a charge by the IS that Schuster worked only four hours a day.

107. CGAR, 1906, 69.

108. Letter from American Consulate General, Mexico City, to assistant secretary of state, 27 November 1907, RG 85, Box 27, File 51463/B, Folder 2, NARA; letter sent from Frank Sargent to National Railroad of Mexico, Mexico City, 3 February 1903, RG 85, Box 27, File 51463/A, NARA.

109. *Transactions of the Second International Convention of the American Republics,* held at the New Willard Hotel, Washington, D.C., 9–14 October 1905, under the auspices of the governing board of the International Union of the American Republics (Washington, D.C.: Government Printing Office, 1906), 114.

110. *Transactions of the Third International Sanitary Conference of the American Republics,* held at the National Palace, City of Mexico, 2–7 December 1907 (Washington, D.C.: Press of Byron S. Adams, for the International Bureau of the American Republics, 1907).

111. The United States was heavily involved in negotiations with the Mexican railroads and Mexican government from 1903 to 1906. The Mexican government consistently put off the question of forging a "Mexican agreement" by insisting that it was developing its own immigration legislation that would remedy the problem of diseased immigrants seeking entry to the United States through Mexico. See the correspondence in RG 85, Box 27, File Nos. 51463/A and 51463/B; Box 439, File No. 55609/551; Box 28, File Nos. 51463/c and 51423/1a; and Box 27, File No. 51463/b, NARA.

112. *Transactions of the Third International Sanitary Conference,* 1907, 20.

113. Ibid., 219.

114. PHS translation (1909) of Republic of Mexico, Department of the Interior, Division of Immigration, Decree Designating the Frontier Places Authorized for the Entry of Passengers into the Republic and Regulations for the Inspection of Immigrants, RG 85, Box 439, File No. 55609/551, NARA.

115. *Transactions of the Fourth International Sanitary Conference of the American Republics,* held at San Jose, Costa Rica, 25 December 1909–3 January 1910 (Washington, D.C.: Press of Byron S. Adams, for the Pan-American Union, 1910), 167.

116. *Transactions of the Third International Sanitary Conference*, 1907, 65.

117. *Proceedings of the Immigration Convention of Texas, Convened in Dallas, Texas*, 20–21 December 1887 (Dallas: A. D. Aldridge and Co., 1888); *Proceedings of the State Immigration Committee of Texas*, convened in Dallas, 29 December 1887, Texas State Archives, Austin. Mr. J. S. Daugherty, chairman, noted that Texas needed and welcomed hardworking, law-abiding immigrants and claimed that Texas offered the immigrant "good health and educational and social advantages for himself and family" (12).

118. See the *Annual Report on Quarantine* (Austin: State Printer) for 1882, 1883–84, 1885–86, 1887–88; *Report of the State Health Officer for the State of Texas* (Austin: State Printer) for 1891–92, 1893–94, 1895–96, and 1897–98; *Biennial Report of the State Health Officer to the Governor* (Austin: State Printer) for 1 December 1898 to 1 November 1900, from 1 November 1900 to 31 August 1902, from 1 September 1902 to 31 August 1904, from 1 September 1904 to 31 August 1906, from 1 September 1906 to 31 August 1908, from 1 September 1912 to 31 August 1914, from 1 September 1916 to 31 August 1918, and for the fiscal years 1927–28, all in Texas State Archives, Austin. In the 1904–6 biennial reports, the state health officer began to comment regularly on good public health relations with Mexico: "The sanitary conditions under the direction of the distinguished president of the Superior Board of Health of Mexico, Dr. E. Licéaga, has undergone wonderful improvement, and yellow fever in that country is rapidly disappearing. Our relations with the Mexican officials are most pleasant" (27).

119. This quarantine procedure was similar to what immigrants had to undergo overseas before sailing for the United States, in both Europe and Asia. At Chinese and Japanese ports, PHS quarantine officers inspected immigrants, as they did at selected European ports, to give them final quarantine clearance before departure; this practice predated immigrant inspection in the United States. Disinfection and bathing was a routine part of this clearance. Alternatively, steamship companies bathed and disinfected immigrants before departure for the United States. An immigrant's luggage was typically disinfected, too. In Japan "steerage passengers are required to present themselves at the disinfecting station twenty-four hours before sailing, when they are bathed with hot water and soap, followed by a mild disinfecting bath, their clothing and all other effects disinfected by steam as far as possible or by formaldehyde or sulphur gas, and detained in the barracks until the hour of sailing of their steamer" (SGAR, 1907, 154). See also H. Phelps Whitmarsh, "The Steerage of To-day: A Personal Experience," *Century* 55 (February 1898): 533.

120. Letter from Dr. Tappan to surgeon general, 12 July 1924, RG 90, General Subject File, 1924–35, Domestic Stations, Texas, El Paso, Box 248, File No. 950-56, NARA.

121. See, for example, John W. Tappan, "The Medical Inspection of Immigrants

with Special Reference to the Texas-Mexican Border," *Texas State Journal of Medicine* (July 1919): 120–24; SGAR, 1916, 147; John W. Tappan, "Medical Inspection of Alien Immigrants at El Paso," *Southwestern Medicine* 2 (1918–19): 8–9.

122. Weindling, *Epidemics and Genocide*, 65.

123. Weindling is careful to clarify that "any concern with lice was merely inciden-tal until 1913, when the link with typhus began to be recognized" (ibid.). He further notes that procedures could vary drastically from one station to another (ibid., 67).

124. Rules, 19 February 1916, RG 90, Central File, 1897–1923, Box 770, File 159260, NARA. Officers stationed at the ferry were to observe the same rules.

125. Ibid. (emphasis in original). Note that a civilian was making this judgment rather than a PHS officer, so presumably admissibility was made on the base of class assessments, into which race most likely figured, rather than disease assessments.

126. Letter from Will E. Soult, inspector in charge, El Paso, to supervisor, Immi-gration Service, El Paso, 13 December 1923, Doc. No. 3007/1, RG 85, Box 164, File No. 52903/29, NARA.

127. SGAR, 1911, 205.; letter from assistant surgeon general to Dr. Scott, 31 Jan-uary 1911, RG 90, Box 52, File No. 377, Central File, 1897–1923, NARA.

128. Letter from Will E. Soult, inspector in charge, El Paso, to supervisor, Immi-gration Service, El Paso, 13 December 1923, Doc. No. 3007/1, RG 85, Box 164, File No. 52903/29, NARA.

129. Letter from Irving McNeil, acting assistant surgeon at El Paso, to J. W. Tap-pan, medical officer in charge at El Paso, 22 December 1923, RG 85, Box 164, File No. 52903/29, NARA. See also letter from Dr. Tappan to surgeon general, 12 July 1924, RG 90, General Subject File, 1924–35, Domestic Stations, Texas, El Paso, Box 248, File No. 950-56, NARA. The procedure of disinfection, bathing, and examination of entirely disrobed immigrants continued along the Mexican border through 1930. See annual report for Eagle Pass, 15 July 1930, RG 90, General Subject File, 1924–35, Do-mestic Stations, Texas, Eagle Pass, Box 245, File No. 1850-15; annual report for El Paso, 1930, RG 90, Records Relating to Mexican Immigration, 1922–33, Correspon-dence between Border Stations and New Orleans Headquarters, Box 4, NARA.

130. Tzvetan Todorov, *The Conquest of America: The Question of the Other*, trans. Richard Howard (New York: Harper and Row, 1984), 185.

131. Charles H. Hufford, "The Social and Economic Effects of the Mexican Mi-gration into Texas" (Howard Payne College, 1925; reprinted, San Francisco: R. and E. Research Associates, 1971), 3, Barker Library, Center for American History, Univer-sity of Texas, Austin.

132. C. A. Hawley, *Life along the Border: A Personal Narrative of Events and Ex-periences between 1905 and 1914* (Spokane, Washington: Shaw and Borden Co., Print-ers, 1955), 60, Barker Library, Center for American History, University of Texas, Austin.

133. The incident that stirred ill will was one in which Woodall, having asked a Mexican woman to show him her vaccination mark, allegedly "tore the sleeve of her shirt down to the wrist. . . . The Consul of Mexico at Brownsville reports to me that the gentleman is of an excitable and neurasthenic disposition and at times is rude to the Mexicans who are passing through his town" (translation of letter from Y. Bonillas, ambassador of Mexico, Embassy of Mexico, Washington, D.C., to secretary of state, 26 October 1917, RG 90, Central File, 1897–1923, Box 123, File No. 1371, NARA).

134. Woodall insisted he had never been discourteous to any immigrant or Mexican citizen; moreover, in the past two years not a single case of contagion could be traced to Hidalgo. See letter from Woodall to surgeon general, 8 November 1917, RG 90, Central File, 1897–1923, Box 123, File No. 1371, NARA.

135. Letter from assistant secretary of state to secretary of state, 22 November 1917, RG 90, Central File, 1897–1923, Box 123, File No. 1371, NARA.

136. Letter from Fairbanks to surgeon general, 28 October 1919, and letter from Assistant Surgeon General Creel to Fairbanks, Brownsville, 7 November 1919, RG 90, Central File, 1897–1923, Box 770, File 159260, NARA.

137. James Davis to Albert Johnson, 14 February 1929, HR71A-F16.1, Records of the House of Representatives, RG 233, quoted in Mae Ngai, "The Architecture of Race in American Immigration Law: A Reexamination of the Immigration Act of 1924," *Journal of American History* 86, no. 1(June 1999): para. 51.

138. *In re Rodriguez*, 81 Fed. 337–338 (W.D. Texas, 1897).

139. Mae Ngai, "Architecture of Race in American Immigration Law," *Journal of American History* 86, no. 1 (June 1999): paras. 52–59, 60.

140. Joseph Hill, "Composition of the American Population by Race and Country of Origin," *Annals of the American Academy of Political and Social Sciences* 188 (November 1936): 177–84.

141. George J. Sanches, *Becoming Mexican American: Ethnicity, Culture and Identity in Chicano Los Angeles, 1900–1945* (New York: Oxford University Press, 1993), 30.

142. A. A. Seraphic, "Report on Conditions in Mexico and on the Mexican Border to the Commissioner General," 8 January 1908, RG 85, Box 28, File No. 51423/1, NARA.

143. Marcus Braun, U.S. immigrant inspector, New York, report to Commissioner General Sargent, 12 February 1907, RG 85, Box 95, File No. 52320/1, Folder 2, NARA.

144. Letter from Commissioner General Frank Sargent to National Railroad of Mexico, Mexico City, 3 February 1903, RG 85, Box 27, File No. 51463/A, NARA.

145. Letter from Sargent, commissioner general, to A. A. Robinson, Esq., president, Mexican Central Railroad Co., City of Mexico, 25 March 1903, RG 85, Box 27, File No. 41563/A, NARA.

146. Marcus Braun, report to Commissioner General Sargent, 10 June 1907; mem-

orandum, Department of Commerce and Labor, 13 June 1907; letter from D. E. Thompson to Elihu Root, secretary of state, 19 June 1906; Immigration Service, memorandum re Mexican ports, 11 November 1907; letter from D. E. Thompson, American Embassy, to Mr. Ignacio Mariscal, Minister for Foreign Affairs, 3 January 1906, all in RG 85, Box 27, File No. 51463/A, NARA.

147. Letter from Marcus Braun, U.S. immigration inspector, New York, to Commissioner General Sargent, 12 February 1907, RG 85, Box No. 52320/1, Folder 2, NARA.

148. A. A. Seraphic, "Report on Conditions in Mexico and on the Mexican Border to the Commissioner General," 8 January 1908, RG 85, Box 28, File No. 51423/1, NARA.

149. Unfortunately, no data are available to indicate changes in the patterns of certification during the period from 1911 to 1920.

FIVE. At the Borders of Science

Epigraph. RG 90, Central File, 1897–1923, Box 770, File 159260, NARA.

1. Unemployment Committee of the National Federation of Settlements, *Case Studies of Unemployment* (Philadelphia: University of Pennsylvania Press, 1931), 88–89.

2. The law specified that all "persons afflicted with tuberculosis or with a loathsome or dangerous contagious disease" were "excluded from admission to the United States" (20 February 1907, c. 1134, 34 Stat. 898:899, 898). See Barbara Bates, *Bargaining for Life: A Social History of Tuberculosis, 1876–1938* (Philadelphia: University of Pennsylvania Press, 1992), 16–18; Georgina D. Feldberg, *Disease and Class: Tuberculosis and the Shaping of Modern North American Society* (New Brunswick, N.J.: Rutgers University Press, 1995), 14, 44, 3–5; Sheila M. Rothman, *Living in the Shadow of Death: Tuberculosis and the Social Experience of Illness in American History* (New York: Basic Books, 1994),13–15.

3. Tuberculosis mortality among immigrants living in the United States was tremendously high. In 1890, for example, the New York City death rate from tuberculosis was 390 per 100,000 population. In affluent neighborhoods, the rate was 49 per 100,000. In immigrant neighborhoods the mortality rate was a shocking 776 per 100,000. In 1900 immigrant tuberculosis mortality was 500 per 100,000. Rothman, *Shadow of Death,* 184.

4. Victor Safford, *Immigration Problems: Personal Experiences of an Official* (New York: Dodd, Mead, 1925), 256–57.

5. CGAR, 1901, 33–34.

6. Letter from Francis Tracy Tobin, attorney for Thomas P. Boden, to Theodore Roosevelt, 4 December 1901, RG 90, Central File, 1897–1923, Box 38, File No. 219, NARA.

7. "The Health Board and Compulsory Reports," *Medical Record* 51 (23 January 1897): 126.

8. "Compulsory Reporting of Cases of Pulmonary Tuberculosis," *Medical Record* (27 March 1897): 459.

9. Hermann Biggs, "Sanitary Measures for the Prevention of Tuberculosis in New York City and Their Results," *JAMA* 34 (1902): 1635.

10. Letter from H. W. Austin, surgeon, chairman; L. L. Williams, surgeon; R. M. Woodward, surgeon; members of the board to consider the case of "immigrant Boden," to surgeon general, 7 December 1901, RG 90, Central File, 1897–1923, Box 38, File No. 219, NARA.

11. Memorandum from surgeon general to secretary of the Treasury, 21 December 1901, reprinted in SGAR, 1902, 395–96 (emphasis added).

12. Ibid.

13. Letter from surgeon general to Dr. Stoner, 20 November 1901, and telegram from Stoner to surgeon general, 20 November 1901, RG 90, Central File, 1897–1923, Box 38, File No, NARA.

14. Memorandum in re Difficulties Encountered during Six Months' Experience with the Administration of the Immigration Act Approved February 20, 1907, 7 January 1908, RG 85, Box 35, File No. 51538/6, NARA; Minutes from Commissioner General's Conference, 23 October 1907, RG 85, Box 55, File No. 51758/3, NARA.

15. Public Health Service Memorandum, 10 July 1909, RG 85, File No. 128 52600/ 30 142, NARA.

16. Bureau of Public Health and Marine Hospital Service, *Book of Instructions for the Medical Inspection of Immigrants* (Washington, D.C.: Government Printing Office, 1903), 12.

17. Letter from Commissioner General Larned to Surgeon General Walter Wyman, 22 December 1906, RG 85, Box 30, File No. 51490/19, NARA.

18. CGAR, 1906, 61–62.

19. Letter from Acting Surgeon General Taliaferro Clark to Carl Ramus, Boston, 5 May 1930, RG 90, General Subject File, 1924–35, Box 944, File No. 0950–121, NARA. Nevertheless, even the PHS could blur the scientific boundaries between bacteriology and eugenics, as it did in the definition of "poor physique."

20. Letter from Schereschewsky to commissioner of immigration, Baltimore, 28 March 1905, and letter from Commissioner General Larned to Surgeon General Walter Wyman, 22 December 1906, RG 85, Box 30, File No. 51490/19, NARA.

21. E. A. Ross, *The Old World in the New: The Significance of Past and Present Immigration to the American People* (New York: Century, 1914), 289; and Madison Grant, *The Passing of the Great Race* (New York: Scribner's, 1916), 16.

22. Kraut, *Silent Travelers,* 138.

23. Maurice Fishberg, *Tuberculosis among the Jews* (New York, 1908).

24. Letter from Commissioner General Sargent to commissioners of immigration at Montreal, Boston, Ellis Island, Philadelphia, Baltimore, San Francisco, and inspectors in charge at New Orleans, Seattle, Norfolk, Eagle Pass, Laredo, El Paso, and Tucson, 17 April 1905, RG 85, Box 30, File No. 51490/19, NARA.

25. Letter from Commissioner General Larned to Surgeon General Walter Wyman, 22 December 1906, RG 85, Box 30, File No. 51490/19, NARA.

26. Minutes of Medical Conference to Discuss the Medical Examination of Immigrants, 8 February 1907, RG 85, Box 30, File No. 51490/19, NARA.

27. CGAR, 1910, 5. See also CGAR, 1914, 5.

28. Quoted in Robert DeC. Ward, "Higher Mental and Physical Standards for Immigrants," *Scientific Monthly* 19 (1924): 533–47.

29. Francis Galton, a cousin of Charles Darwin, coined the term *eugenics* in 1883. As historian Nancy Leys Stepan argues, for some eugenicists better breeding involved not only improving the genetic quality of the human race by encouraging the reproduction of the "fit" and discouraging that of the "unfit," but also preserving the genetic purity of "superior" races. Nancy Leys Stepan, *In the Hour of Eugenics: Race, Gender, and Nation in Latin America* (Ithaca: Cornell University Press, 1991).

30. Letter from Acting Surgeon General Taliaferro Clark to Carl Ramus, Boston, 5 May 1930, RG 90, General Subject File, 1924–35, Box 944, File No. 0950–121, NARA.

31. Handwritten note from Acting Surgeon General Taliaferro Clark to Dr. Carmelia, attached to letter from Dr. Carl Ramus, Boston, to surgeon general, 7 April 1930, RG 90, General Subject File, 1924–35, Box 944, File No. 0950–121, NARA.

32. While officers in the field did abandon the term "poor physique" and diagnosed suspected pulmonary tuberculosis as "physical signs of tuberculosis," in its place sprang up a number of terms that received no official sanction, censure, or comment: "undersized," "lack of physical development," "constitutional inferiority." Letter from Alfred Hampton, assistant commissioner general to supervising inspector, Immigration Service, El Paso, 5 May 1916, RG 85, Box 118, File No. 52495/65, NARA; letter from L. L. Williams to surgeon general, 25 August 1913, RG 90, Central File, 1897–1923, NARA; SGAR, 1917, 175.

33. In El Paso, while PHS officers used the diagnosis consistently in the late 1920s, it represented less than 1% of certifications.

34. Letter from Commissioner General Keefe to surgeon general, 27 September 1910, RG 85, Box 181, File No. 53033/9, NARA.

35. Letter from Coe, inspector in charge, Sault Ste. Marie, 29 June 1910; letter from Commissioner Clark, Montreal, to commissioner general, 6 July 1910; letter from Clark to commissioner general, 4 October 1910; all in RG 85, Box 167, File No. 52903/69, NARA.

36. Certificate of A. S. McCraig, 15 August 1910, RG 85, Box 167, File No. 52903/69, NARA.

37. Certificate of A. S. McCraig, 20 August 1910, RG 85, Box 167, File No. 52903/69, NARA (emphasis added).

38. Certificate of A. S. McCraig, 14 September 1910, and letter from Coe to Commissioner Clark, Montreal, 14 September 1910, RG 85, Box 181, File No. 53033/9, NARA.

39. The citizenship status of Native Americans born after 1924 remained in question until 1940. For an explication of the legal history of Native Americans, whiteness, and eligibility for citizenship, see Ian F. Haney-Lopez, *White by Law: The Legal Construction of Race* (New York: New York University Press, 1996), 40–41.

40. X-ray examination was never contemplated in the PHS regulations governing the diagnosis of tuberculosis, and x-ray machines appear not to have been widely available to PHS officers inspecting immigrants. X-ray diagnosis of tuberculosis was complicated. Barron Lerner notes that physicians were generally skeptical of the value of x-rays in the diagnosis of tuberculosis even in the second decade of the twentieth century: x-rays "might appear normal when a patient's sputum showed no bacilli; alternatively, they might appear normal when the sputum was positive" (Barron Lerner, "The Perils of 'X-Ray Vision': How Radiographic Images Have Historically Influenced Perception," *Perspectives in Biology and Medicine* 35, no. 3 [spring 1992]: 389). David Rosner and Gerald Markowitz note the same general diagnostic limitations, focusing specifically on the problem of differentiating tuberculosis from silicosis. David Rosner and Gerald Markowitz, *Deadly Dust: Silicosis and the Politics of Occupational Disease in Twentieth-Century America* (Princeton: Princeton University Press, 1991), 32–33. See also "A Discussion on the Use of X-Rays in the Diagnosis of Pulmonary Tuberculosis," *Archives of the Roentgen Ray* 17 (1913): 477–85; N. Bridge, *Tuberculosis* (Philadelphia: W. B. Saunders, 1903); S. F. Achsner, "Pulmonary Tuberculosis: Contributions of Radiology in Diagnosis and Treatment," *Southern Medical Journal* 79 (1986): 1416–24; H. Sewall and S. B. Childs, "A Comparison of Physical Signs and X-Ray Pictures of the Chest in Early Stages of Tuberculosis," *Archives of Internal Medicine* 10 (1912): 85–89; and C. Reviere, *The Early Diagnosis of Tubercle* (London: Henry Frowde, 1914).

41. William Osler, *The Principles and Practice of Medicine*, 8th ed. (New York: D. Appleton, 1916), 207.

42. The diagnostic standard for diagnosing tuberculosis in an immigrant was less exacting than those recommended by the National Tuberculosis Association, which listed five diagnostic criteria. The PHS rejected these criteria. The medical chief of Ellis Island's Hospital Division noted that "practically all writers on the subject mention the following five cardinal diagnostic points: Presence of tubercle bacilli. Moderately

coarse rales. Parenchymatous X-ray lesion in the upper quadrant. Hemorrhage from the lung (5 c.c. or more). Pleurisy with effusion." He noted that while the presence of the bacillus was sufficient for exclusion in the immigrant, he recommended that two of the other criteria be met (letter from chief, Hospital Division, Ellis Island, to Dr. Long, 5 May 1925, RG 90, General Subject file, 1924–33, Box 942, File No. 0950–56, NARA).

43. Letter from Victor Safford, medical officer in charge, Boston, to commissioner of immigration, 23 October 1911, RG 85, Box 167, File No. 52903/69, NARA.

44. Alfred C. Reed, "Immigration and the Public Health," *Popular Science Monthly* (October 1913): 320. Although the requirement for the bacteriological confirmation of tuberculosis in the sputum was first established in the PHS regulations for medical inspection in 1910, in 1902 the surgeon general informed the officer stationed at the Portland quarantine station that certificates for tuberculosis should not be issued "until the diagnosis is confirmed by bacteriological examination" (letter from surgeon general to Dr. P. C. Kalloch, 2 February 1902, RG 90, Central File, 1897–1923, Box 32, File No. 182, NARA).

45. "Such cases are held under observation of a sufficient length of time to enable the certifying physician to arrive at a satisfactory conclusion" (SGAR, 1917, 175).

46. U.S. Public Health Service, *Regulations Governing Medical Inspection of Aliens* (Washington: Government Printing Office, 1917), 31.

47. Allan M. Brandt, *No Magic Bullet: A Social History of Venereal Disease in the United States since 1880* (New York: Oxford University Press, 1987), 20–21.

48. Letter from Dr. C. W. Peckham, PHS surgeon in charge, to surgeon general, 29 July 1903, RG 90, Central File, 1897–1923, Box 38, File No. 219, NARA.

49. Letter from Stoner to surgeon general, 15 August 1903, RG 90, Central File, 1897–1923, Box 38, File No. 219, NARA.

50. Letter from Surgeon General Walter Wyman to Stoner, 5 September 1903, RG 90, Central File, 1897–1923, Box 38, File No. 219, NARA.

51. Brandt explains that uncertain diagnostics resulted in wildly different estimates of venereal infection in the American population before about 1910. The development of the Wassermann test for syphilis (1906), advances in microscopic techniques to detect gonorrhea, and the development of effective treatments for syphilis (1909) were important antecedents to the wartime effort to control venereal diseases among the military and to the introduction of intensive venereal exams for immigrants.

52. Before the war the IS had been interested in flushing out prostitutes to prevent moral decay in the nation and among U.S. troops.

53. Letter from J. W. Kerr, Ellis Island, to S. G. Creel, 11 March 1921, RG 90, Central File, 1897–1923, Box 38, File No. 219, NARA.

54. Letter from Safford to surgeon general, 17 November 1916, RG 90, Central File, 1897–1923, Box 59, File No. 409, NARA.

55. Kerr continued, "I have had two or three newspaper men inquiring into venereal disease problems one of them from the New York Tribune, and I suspect he is going to make a considerable story. He told Dr. Corput last night that the Commissioner had stated to him that our present examinations for venereal disease are a farce" (letter from J. W. Kerr, Ellis Island, to Surgeon General Creel, 11 March 1921, RG 90, Central File, 1897–1923, Box 38, File No. 219, NARA.

56. Annual Report from Brownsville, 1 July 1919, RG 90, Central File, 1897–1923, Box 770, File 159260, NARA.

57. Annual Report for Montreal, 1924, RG 90, General Subject File, 1924–35, Foreign Stations, Immigration, Montreal, Box 786, File No. 1850-15, NARA.

58. Letter from Dr. Kerr to surgeon general, 28 July 1919, RG 90, Central File, 1897–1923, Box 38, File No. 219, NARA.

59. The regulations in 1910 merely stipulated that "cases of syphilis in an active communicable stage, in which there can be no question as to the diagnosis, should be certified immediately" (Bureau of Public Health and Marine Hospital Service, *Book of Instructions for the Medical Inspection of Aliens* [Washington, D.C.: Government Printing Office, 1910], 16).

60. Letter from Safford to surgeon general, 1 February 1915, RG 90, Central File, 1897–1923, Box 59, File No. 409, NARA.

61. Ludwik Fleck discusses the complexities underlying the theory behind and interpretation of the Wassermann reaction. See Ludwik Fleck, *The Genesis and Development of a Scientific Fact* (Chicago: University of Chicago Press, 1979; originally published in 1935), 52–81. In 1935 Fleck noted that the meaning of the Wassermann was still in question: "Only a continuous, regular, and well-organized execution of the procedure for the reaction, always with many blood samples, several taken from each series for comparison with the next, will yield results of the necessary reliability. A clinical control of these results must of course also be carried out, involving a comparison of the laboratory results with the clinical results and an appropriate adjustment of the mode of procedure. Despite every safeguard and mechanization, however, new and unexpected findings continually emerge" (52). As Osler explained, "There is seldom any doubt concerning the recognition of syphilitic lesions; but the number of persons, without any evident sign of the disease, in whom a positive Wassermann reaction is found proves that a negative diagnosis cannot be based on the absence of history and clinical manifestations." He recommended the use of the Wassermann reaction only when it was performed "in good hands" (Osler, *Principles and Practice of Medicine*, 276–77).

62. Letter from Safford to surgeon general, 11 March 1915, RG 90, Central File, 1897–1923, Box 59, File No. 409, NARA.

63. Letter from Lavinder to Williams, Ellis Island, 6 February 1915, RG 90, Central File, 1897–1923, Box 59, File No. 409, NARA.

64. Statement of Dr. Williams, 6 February 1915, and letter from Williams to surgeon general, 24 February 1915, RG 90, Central File, 1897–1923, Box 59, File No. 409, NARA.

65. Statement of Dr. Safford, 15 February 1915, RG 90, Central File, 1897–1923, Box 59, File No. 409, NARA.

66. Letter from Safford to surgeon general, 11 March 1915, RG 90, Central File, 1897–1923, Box 59, File No. 409, NARA.

67. Letter from Dr. John Anderson, director of Hygienic Laboratory, to surgeon general, 25 March 1915, RG 90, Central File, 1897–1923, Box 59, File No. 409, NARA.

68. Letter from Board of Commissioned Officers convened to consider the Wassermann reaction as related to the issuance of medical certificates (L. L. Williams, M. V. Safford, and C. H. Lavinder) to surgeon general, 29 April 1915, RG 90, Central File, 1897–1923, Box 59, File No. 409, NARA.

69. PHS, *1917 Regulations,* 37.

70. See, for example, the annual reports of the chief medical officer at Ellis Island to the surgeon general for 1925 and 1930, which noted that some immigrants with a positive Wassermann (as opposed to syphilis) are certified as Class A and some as Class B. RG 90, General Subject File, 1924–35, Domestic Stations, Ellis Island Immigration, Box 165, File 1850-15, NARA.

71. Letter from Assistant Surgeon General Creel to Dr. Fairbanks, 15 January 1917, RG 90, Central File, 1897–1923, Box 770, File 159260, NARA.

72. Letter from Mrs. Marron to Surgeon General Rupert Blue, 10 July 1918, RG 90, Central File, 1897–1923, Box 770, File 159260, NARA. See the epigraph to this chapter. There seem to have been few complaints of arbitrary medical certification. In isolated districts, a PHS physician or IS official could conceivably have acted out of malice, but there is only one such case in the archives of either agency. In 1905 H. G. Dubose, the immigrant inspector at Eagle Pass, Texas, accused his superior of arbitrarily deporting aliens and abusing fellow officers, immigrants, and residents of Eagle Pass. See letter from H. G. Dubose, immigrant inspector, Eagle Pass, to C.O.C. Cowley, special immigrant inspector, Eagle Pass, 1 May 1905, and letter from Walter L. O'Neil, immigrant inspector, Eagle Pass, to C.O.C. Cowley, 1 May 1905, RG 85, Box 50, File No. 51701/3b, NARA.

73. Letter from Dr. Fairbanks to Surgeon General Blue, 22 July 1918, RG 90, Central File, 1897–1923, Box 770, File 159260, NARA.

74. Letter from Surgeon General Rupert Blue to Fairbanks, 29 July 1918, RG 90, Central File, 1897–1923, Box 770, File 159260, NARA.

75. Letter from Fairbanks to Blue, 8 August 1918, RG 90, Central File, 1897–1923, Box 770, File 159260, NARA.

76. Letter from Assistant Surgeon General Creel to Fairbanks, 14 August 1918, RG 90, Central File, 1897–1923, Box 770, File 159260, NARA.

77. Letter from Mrs. Victoria Fernandez Marron to Surgeon General Blue, 9 October 1918, RG 90, Central File, 1897–1923, Box 770, File 159260, NARA.

78. Letter from Surgeon General Blue to Mrs. Marron, 18 October 1918, RG 90, Central File, 1897–1923, Box 770, File 159260, NARA.

79. Letter from Dr. Billings to surgeon general, 24 June 1922, RG 90, Central File, 1897–1923, Box 39, File No. 219, NARA.

80. Letter from Sanitary Board to surgeon general, 3 July 1922, and letter from C. C. Pierce, assistant surgeon general, to surgeon general, 28 June 1922, RG 90, Central File, 1897–1923, Box 39, File No. 219, NARA.

81. Letter from Surgeon General Cumming to Dr. Billings, Ellis Island, 1 July 1922, RG 90, Central File, 1897–1923, Box 39, File No. 219, NARA.

82. Letter from C. H. Lavinder, member of the Board for Revising Inspection Instructions, Stapleton, New York, to Dr. Perry, Chief Medical Officer at Ellis Island, 30 May 1917, RG90, Central File, 1897 to 1923, Box 677, File No. 5738, NARA.

83. Letter from Dr. Victor Safford, Boston, to J. S. Perry, senior surgeon, chairman of the Board for Revising Inspection Instructions, Ellis Island, 31 May 1917, RG 90, Central File, 1897–1923, Box 677, File No. 5738, NARA.

84. For a discussion of the treatment of trachoma, see the correspondence in RG 85, Box 79, File No. 52065, Folder No. 1, and RG 85, Box 265 (renumbered 228), File No. 54261/12, NARA. For documents related to general policy on granting treatment, see the correspondence in RG 85, Box 121, File Nos. 52516/11, 52516/11A, 52516/11B, 52516/11C, and in Box 167, 52903/69, NARA.

85. See John Ettling, *The Germ of Laziness: Rockefeller Philanthropy and Public Health in the New South* (Cambridge, Mass.: Harvard University Press, 1981), 2, 5, 228n.12. PHS officers treated hookworm with various combinations of thymol, a chloroform–castor oil mixture, and sodium bicarbonate.

86. "Japs Bring Frightful Disease," *San Francisco Chronicle*, 2 April 1905; "Diseased Japanse Deceive Immigration Surgeons: Infected Coolies Are Landed in Hordes," *San Francisco Chronicle*, 31 August 1915; letter from Dr. M. W. Glover to commissioner of immigration, San Francisco, 13 February 1911, RG 90, Central File, 1897–1923, Box 785, File No. 16090, NARA.

87. For related discussions of "coolie" labor and the American standard of living, see Nayan Shah, *Contagious Divides: Epidemics and Race in San Francisco's Chinatown* (Berkeley: University of California Press, 2001); and Alexander Saxton, *The Indispensable Enemy: Labor and the Anti-Chinese Movement in California* (Berkeley: University of California Press, 1971).

88. Letter from E. C. Berry, Vallejo Trade and Labor Council, Vallejo, California, to Commissioner General A. Caminetti, 13 September 1913, RG 85, Box 168, File No. 52903/110a, NARA.

89. The league, in 1908, discussed "the immodest and filthy habits of the Hindoos";

in *Collier's Weekly* in 1910 it referred to them as "inferior workmen." Ronald Takai, *Strangers from a Different Shore: A History of Asian Americans* (New York: Penguin Books, 1989), 296–97.

90. The Asiatic Barred Zone included all the territory from Afghanistan through the Pacific, with the exception of Japan (where restriction was "self-imposed") and the Philippines (a U.S. territory).

91. *Congressional Record* 13, part 2, 47th Cong., 1st sess., 1482–85, 2211.

92. Letter from George Broadman, vice president, San Francisco Chamber of Commerce, to Surgeon General Rupert Blue, 14 March 1919, RG 90, Central File, 1897–1923, NARA.

93. Letter from Surgeon General Rupert Blue to San Francisco Chamber of Commerce, 22 March 1919, RG 90, Central File, 1897–1923, Box 677, File No. 5738, NARA; SGAR, 1922, 190–91.

94. While the quota system formally included non-Western countries like China and Japan, these countries received only the minimum quota of 100 immigrants per year. Significantly, however, that quota could not be filled with natives of Asian countries, for they were legally ineligible for U.S. citizenship.

95. In 1930 it was removed from the list of Class A conditions. Memorandum from J. D. Long, assistant surgeon general, to surgeon general, 30 July 1924, RG 90, General Subject File, 1924–35, Box 944, File No. 0950–121, NARA.

96. Letter from White to IS officers, 7 March 1919, RG 85, Box 271 (renumbered 241), File No. 54261/184, NARA.

97. According to White, the "inaccurate and misleading press notices" concerning his handling of the hookworm incident "are very humiliating and embarrassing to me." White fretted over the possibility that Dr. Billings was leaking information to the press and discrediting him in the eyes of the public. Letter from Edward White, commissioner, Angel Island, to commissioner general, 9 May 1919, RG 85, Box 271(241), File No. 54261/184, NARA.

98. Letter from Edward White, commissioner, San Francisco, to W. C. Billings, Angel Island, 11 April 1919, RG 85, Box 271 (renumbered 241), File No. 54261/184, NARA.

99. Letter from W. C. Billings, PHS, Angel Island, to Commissioner of Immigration White, 17 April 1919, RG 85, Box 271 (renumbered 241), File No. 54261/184, NARA.

100. Telegram from Billings to Surgeon General Blue, 12 March 1919, RG 85, Box 271 (renumbered 241), File No. 54261/184, NARA.

101. Letter from Commissioner General A. Caminetti to White, San Francisco, 10 May 1919, RG 85, Box 271 (renumbered 241), File No. 54261/184, NARA.

102. Letter from commissioner general to commissioners of immigration and in-

spectors in charge at all seaports, 18 March 1919, RG 85, Box 271 (renumbered 241), File No. 54261/184, NARA.

103. Letter from White to commissioner general, 19 April 1919, RG 85, Box 271 (renumbered 241), File No. 54261/184, NARA.

104. Letter from Edward White, commissioner, Angel Island, to commissioner general, 9 May 1919, RG 85, Box 271(241), File No. 54261/184, NARA.

105. Letter from Commissioner General A. Caminetti to Commissioner White, San Francisco, 10 May 1919, RG 85, Box 271 (renumbered 241), File No. 54261/184, NARA.

106. "Hookworm Crawls into Official Row: Commissioner White Held to Blame If Undesirables Got In over Billings' Protest; Public Health Service Bureau Issues Statement Concerning Friction at Immigration Station," *San Francisco Examiner,* 9 May 1919, RG 85, Box 271 (renumbered 241), File No. 54261/184, NARA.

107. Barbara Fields emphasizes the "bigotry, discrimination, and exploitation to which Americans of non-African, non-European origin have been subjected," but argues that "the situations of Americans of non-African, non-European origin can no more be equated with each other than with the situation of Afro-Americans ("Whiteness, Racism, and Identity," *International Labor and Working-Class History* 60 (fall 2001): 55n.9.

108. Rogers M. Smith, *Civic Ideals: Conflicting Visions of Citizenship in U.S. History* (New Haven: Yale University Press, 1997), 372.

109. Osofsky observes, "Historians, impressed by the enormity of changes that occurred at the time of the 'Great War,' have tended to overlook or underestimate the significance of the pre–World War I migration of Negroes to northern cities." Although "a few discerning analysts were aware of this new shift in Negro migration in the 1890s . . . by the first decade of the twentieth century the migration was well recognized" (Gilbert Osofsky, *Harlem: The Making of a Ghetto,* 2d ed. [Chicago: Elephant Paperbacks, Ivan R. Dee, Publisher, 1996], 18, 19, 20, 17). See also Alma Herbst, *The Negro in the Slaughtering and Meat-Packing Industry in Chicago* (Buffalo: Houghton Mifflin, 1932).

110. Eric Foner, *The Story of American Freedom* (New York: W.W. Norton, 1998), 173–74.

111. Gwendolyn Mink, *Old Labor and New Immigrants in American Political Development: Union, Party, and State, 1875–1920* (Ithaca: Cornell University Press, 1986), 89–90.

112. Samuel Gompers, "Talks on Labor," *American Federationist* (September 1905): 636–37. Although Gompers also implied that the new southern and eastern European immigrants were also equivalent, he refers to these groups only obliquely as "any others," demonstrating that he held a conception of race that fell along popularly

understood color lines rather than lines of nationality. See also Thomas Lane's review of Mink's *Old Labor and New Immigrants, American Historical Review* 94, no. 3 (June 1989): 881.

113. Foner, *Story of American Freedom,* 174.

114. Henry Yu, *Thinking Orientals: Migration, Contact, and Exoticism in Modern America* (Oxford: Oxford University Press, 2001), ix.

115. Eric Arnesen, "'Like Banquo's Ghost, It Will Not Down': The Race Question and the American Railroad Brotherhoods, 1880–1920," *American Historical Review* 99, no. 5 (December 1994): 1604.

116. Daniel Letwin, "Interracial Unionism, Gender, and 'Social Equality' in the Alabama Coalfields, 1878–1908," *Journal of Southern History* 61, no. 3 (August 1995): 522–23. See also Herbert G. Gutman, "The Negro and the United Mine Workers of America: The Career and Letters of Richard L. Davis and Something of Their Meaning: 1890–1900," in *Work, Culture, and Society in Industrializing America: Essays in American Working-Class and Social History* (New York: Alfred A Knopf, 1976), 123. Gutman notes that by 1900 African Americans represented up to 15% of the nation's coal miners.

117. Thomas N. Maloney and Warren C. Whatley, "Making the Effort: The Contours of Racial Discrimination in Detroit's Labor Markets, 1920–1940," *Journal of Economic History* 55, no. 3 (September 1995): 465–93. See also Charles Denby, *Indignant Heart: A Black Worker's Journal* (Detroit: Wayne State University Press, 1989).

118. Arnesen, "Like Banquo's Ghost," 1604.

119. Gutman, "The Negro and the United Mine Workers of America," 121–23. See also Herbst, *The Negro in the Slaughtering and Meat-Packing Industry in Chicago;* and Osofsky, *Harlem,* 23. African Americans did not uniformly play the role of strikebreaker, nor did any other immigrant group. See, for example, Eric Arnesen, *Waterfront Workers of New Orleans: Race, Class, and Politics, 1863–1923* (New York: Oxford University Press, 1991).

120. Lizabeth Cohen, *Making a New Deal: Industrial Workers in Chicago, 1919–1939* (Cambridge: Cambridge University Press, 1990), 42–45,165–67.

121. Arnesen, "Like Banquo's Ghost," 1606.

122. Memo on "Negro Exclusion from Canada" from Consul General, Winnipeg, Manitoba, to U.S. secretary of state, 22 April 1911, RG 85, Box 63, File No.51831/157, NARA.

123. African or black immigrants are mentioned in only three instances in the materials that I uncovered. Roman Dobler, an immigration inspector sent to Puerto Rico, reported to T. V. Powderly, commissioner general, on 30 June 1900, "I was informed by the customs officials and others at San Juan that the said Tortolo and St. Thomas as well as the St. Kitts Island negroes are a lawless and intemperate class, and do not live peaceably with the natives" (CGAR, 1892, 44). Victor Safford reported to the surgeon

general on 12 June 1915 that "the Cape Verde Island immigrants are nearly all wholly or partly of negro blood. The majority go to southeastern New England as agricultural laborers." He noted that New Bedford had no facilities for detecting hookworm (RG 90, Central File, 1897–1923, Box 59, File No. 409, NARA). The surgeon general's annual report for 1915 noted that among African immigrants from Cape Verde, hookworm infection was common and intestinal parasites universal (159).

124. Susan Craddock, *City of Plagues: Disease, Poverty, and Deviance in San Francisco* (Minneapolis: University of Minnesota Press, 2000), 9.

125. See also Yu, *Thinking Orientals*.

SIX. Drawing the Color Line

1. Bob Barde reports on the case of one Chinese immigrant detained at Angel Island from September 1916 to April 1918. "An Alleged Wife: A Tale of Angel Island," *Passages: The Quarterly Newsletter of the Angel Island Immigration Station Foundation* 3, no. 4 (summer 2000), located at http://www.haas.berkeley.edu/~barde/immigration/Passages%20article.pdf. Between 1908 and 1909, Barde estimates that, on average, Chinese immigrants not immediately approved for entry spent an average of 23 days in detention. This figure, however, he believes overestimates the length of detention. Barde finds that the vast majority of immigrants to San Francisco were kept for fewer than 3 days and 5% were detained a month or more. Among the Chinese, nearly 20% of those sent to Angel Island were kept for over two weeks. All but 10% of Japanese immigrants were sent to Angel Island, but those immigrants were kept only 2–3 days on average. Bob Barde, "Detained at Angel Island: Some Data," unpublished report, April 2000.

2. Waltraud Ernst, "Historical and Contemporary Perspectives on Race, Science and Medicine," in *Race, Science and Medicine, 1700–1960,* ed. Waltraud Ernst and Bernard Harris (London: Routledge, 1999), 1–28.

3. See text of *Ah Yup* and *Ozawa* decisions in Ian F. Haney Lopez, *White by Law: The Legal Construction of Race* (New York: New York University Press, 1996), app. A. See also Mae Ngai, "The Architecture of Race in American Immigration Law: A Reexamination of the Immigration Act of 1924," *Journal of American History* 86, no. 1 (June 1999): para. 9. Available at: http://www.historycooperative.org/journals/jah/86.1//ngai.html

4. Lopez, *White by Law,* 6, 9.

5. Most of this literature focuses on Great Britain. See, for example, Bernard Harris, "Pro-alienism, Anti-alienism and the Medical Profession in Late-Victorian and Edwardian Britain," 189–217; Michael Worboys, "Tuberculosis and Race in Britain and its Empire, 1900–1950," 144–66; and Mathew Thomson, "'Savage Civilisation': Race, Culture and Mind in Britain, 1898–1939," 235–58, in *Race, Science and Medicine,*

1700–1960, ed. Waltraud Ernst and Bernard Harris (London: Routledge, 1999). See also Elazar Barkan, *The Retreat of Scientific Racism: Changing Concepts of Race in Britain and the United States between the World Wars* (Cambridge: Cambridge University Press, 1991); Henrika Kuklik, *The Savage Within: The Social History of British Anthropology, 1885–1945* (Cambridge: Cambridge University Press, 1991); Paul B. Rich, *Prospero's Return? Historical Essays on Race, Culture and British Society* (London: Hansib, 1994); Graham Richards, *'Race', Racism and Psychology: Towards a Reflexive History* (London: Routledge, 1997); and George Stocking, *Race, Culture and Evolution: Essays in the History of Anthropology* (New York: Free Press, 1969).

6. *In re Ah Yup,* 1 F.Cas. 223 (C.C.D.Cal. 1878) and *United States v. Thind,* 261 U.S. 204 (1923), quoted in Lopez, *White by Law,* 210, 222.

7. In 1923, for example, the IS employed 1 commissioner general (salary $6,500), 8 assistant commissioners general ($4,500), 10 assistant commissioners ($1,980–$4,800), 2 supervisors ($4,200 and $4,000), 24 inspectors in charge ($1,980–$3,240), 2 special immigrant inspectors ($3,000 and $3,500), 25 special agents ($3,000–$3,600), 768 immigrant inspectors ($100–$4,800), 306 clerks ($960–$2,220), 84 interpreters ($300–$1,860), 243 guards ($30–$1,500), 106 laborers ($900–$1,187), 41 matrons ($300–$1,200), 23 janitors ($30–$1,080), 23 charwomen ($144–$900), 4 miscellaneous employees ($1,980–$3,500), and 122 in the general station force ($960–$2,400). See Darrel H. Smith and H. Guy Herring, *The Bureau of Immigration* (Baltimore, 1924), 131–63.

8. Marian Smith, INS historian, personal communication, August 2001. Unfortunately, in contrast to the PHS, little is known about the background of IS employees.

9. Janet Levine, interview with Jacob Auerbach, 14 October 1992, EIOHP.

10. There were also limits to what Sawyer was willing to sacrifice for the sake of employment: "Arnold and Baker [colleagues in the Consular Service] have largely sacrificed their health by life in trying climates. Anderson and Dr. Wilder have suffered in health" (John Birge Sawyer Papers, Manuscripts, 81/62c, vol. 2, Bancroft Library, University of California, Berkeley). See also Sawyer Papers, vols. 3–5.

11. La Guardia served in Congress as a representative from New York from 1917 to 1932. He was elected New York City mayor in 1934.

12. The Ellis Island Oral History Project, which has collected 1,553 oral histories, including interviews with six individuals who were PHS officers in the pre-1930 period, has only one such interview with an IS officer.

13. Fiorello La Guardia, *The Making of an Insurgent: An Autobiography: 1882–1919* (Westport, Conn.: Greenwood Press, 1948), 64.

14. Clifford Alan Perkins, *Border Patrol: With the U.S. Immigration Service on the Mexican Boundary, 1910–54* (El Paso: University of Texas at El Paso, Texas Western Press, 1978).

15. In 1903, for example, the *San Francisco Post* praised North and his staff for fol-

lowing the letter of the law in denying entrance to the insane son of a Salvador millionaire. "Forbidden to Land in City: Insane Son of a Salvador Millionaire Coming Here for Medical Treatment Is Ordered Deported," *San Francisco Post,* 2 December 1903. See also "North, San Francisco, Refuses 306 Japanese," *New York Times,* 23 June 1907; "Sicilian Passengers Rejected by Boston Commissioner Also Excluded at Ellis Island," *New York Times,* 16 January 1895; "Tuberculosis Cases Ordered Excluded," *New York Times,* 2 March 1892.

16. "Disease Restriction Inadequate," *New York Times,* 12 January 1907; "Congress Probes for How Undesirables Evade Law," *New York Times,* 2 March 1909; "Need for U.S. to Enforce Laws," *New York Times,* 4 April 1909; "White Star Passengers Escape Inspection," *New York Times,* 26 September 1890; "Chinese with Clonorchis Permitted to Land," *San Francisco Examiner,* 19, 20, and 23 December 1921; "Clonorchis Releases Investigated," *San Francisco Examiner,* 4 January 1922; "Japanese Cases of Hookworm in Los Angeles," *San Francisco Chronicle,* 3 April 1910.

17. "Commissioner Williams on Abuses," *New York Times,* 1 October 1902; "Commissioner Powderly Removed," *San Francisco Call,* 16 March 1902; "Cabin Passengers Questioned Offensively," *New York Times,* 30 and 31 January 1899; "Ellis Island Indignities," *San Francisco Call,* 23 October 1921; "Angel Island Expose," *San Francisco Examiner,* 7 October 1917; "Angel Island Methods to Be Investigated," *San Francisco Examiner,* 15 December 1928; "Angel Island Expose," *San Francisco Chronicle,* 6 July 1917; "Brazilian Ambassador Suffers Indignities," *San Francisco Call,* 16 November 1905.

18. U.S. Dept. of Labor, *Bureau of Immigration, Problems of the Immigration Service: Papers Presented at a Conference of Commissioners and District Directors of Immigration, Washington D.C., January 1929* (Washington, D.C.: Government Printing Office, 1929), 1–2.

19. Edward Steiner, *From Alien to Citizen* (New York: Arno Press, 1975), 135.

20. Joshua Freeman, "Racial and Ethnic Categorization by New York City Unions after World War II," paper presented at the American Public Health Association Annual Meeting, Boston, 15 November 2000.

21. Carl Degler, *In Search of Human Nature: The Decline and Revival of Darwinism in American Social Thought* (New York: Oxford University Press, 1991), 20–22. See also Stocking, *Race, Culture, and Evolution,* 234–69.

22. CGAR, 1899, 5. This passage is also quoted in a letter from Secretary of Labor Doak to Edward F. Corsi, commissioner of immigration, Ellis Island, 18 May 1932, RG 85, Box 143, File No. 52729/9, NARA.

23. Committee report from Edward F. McSweeney, George S. Rodgers, Richard K. Campbell, and M. Victor Safford, Office of the Commissioner of Immigration, to T. V. Powderly, commissioner general of immigration, 26 June 1898, RG 85, Box 143, File No. 52729/9, NARA.

24. Case of Henry Ryan Kenny Burke, KS-31, 28 January 1926, RG 85, Box 391 (renumbered 361), File No. 55476/151, NARA. See also the other transcripts in this box. See also the case of Samuel and Ester Nelkin, No. 8, 23 January 1926, and other files in this location.

25. Committee report from Edward F. McSweeney, George S. Rodgers, Richard K. Campbell, and M. Victor Safford, Office of the Commissioner of Immigration, to T. V. Powderly, commissioner general of immigration, 26 June 1898, RG 85, Box 143, File No. 52729/9, NARA.

26. Unsigned memorandum, c. 1898, RG 85, Box 143, File No. 52729/9, p. II, NARA (emphasis added).

27. Z. F. McSweeny, "The Character of Our Immigration, Past and Present," *National Geographic Magazine* 16 (January 1905): 7.

28. See, for example, the writings of Edward Alworth Ross, "The Lesser Immigrant Groups in America," *Century Magazine* 88 (October 1914): 934–40; "American and Immigrant Blood: A Study of the Social Effects of Immigration," *Century Magazine* 87 (December 1913): 225–32; "Origins of the American People," *Century Magazine* 87 (March 1914): 712–18; and "Racial Consequences of Immigration," *Century Magazine* 87 (February 1914): 614–22.

29. Alexander E. Cance, James A. Field, Robert DeC. Ward, Prescott F. Hall, "First Report of the Committee on Immigration of the Eugenics Society," *American Breeder's Magazine* 3 (1912): 249–55. See also Albert Allemann, "Immigration and the Future American Race," *Popular Science Monthly* (1909): 586–96; Roger Mitchell, "Recent Jewish Immigration to the United States," *Popular Science Monthly* (1902–3): 334–43.

30. U.S. Congress, House of Representatives, *Hearings before the Special Committee of the House of Representatives to Investigate the Taylor and Other Systems of Shop Management* (Washington, D.C.: Government Printing Office, 1912), 1397.

31. Frederick W. Taylor, *Principles of Scientific Management* (Norcross, Ga.: Engineering and Managing Press, 1911, 1998), 48.

32. Robert A. Woods, ed., *Americans in Process: A Settlement Study by Residents and Associates of the South End House* (Boston: Houghton, Mifflin, 1903), 58, 104. The Irish immigrant, for example, pursued politics "as an end in itself," as a sort of "game which has a fascination all its own," whereas the Jew is "gravely deficient in a civic sense. Business is his great concern, and politics wholly subsidiary to it" (63).

33. Taylor, *Principles of Scientific Management,* viii–xi, 50.

34. David Montgomery, *The Fall of the House of Labor: The Workplace, the State, and American Labor Activism, 1865–1925* (Cambridge: Cambridge University Press, 1987), 252.

35. R. C. Clothier, "The Function of the Employment Department," in *Employment Management: Selected Articles on Employment Management,* ed. D. Bloomfield

(New York: W. H. Wilson, 1922), 159, quoted in Montgomery, *Fall of the House of Labor,* 242.

36. H. A. Worman, "Recruiting the Working Force. II—The Personal Interview of Hiring Men," *Factory* 1 (January 1908): iii, quoted in Montgomery, *Fall of the House of Labor,* 243.

37. Editorial, *Iron Age* 96 (8 July 1915), quoted in Montgomery, *Fall of the House of Labor,* 243.

38. "Racial Adaptability to Various Types of Plant Work," reproduced in Ira DeA. Reid, "The Negro in the Major Industries and Building Trades of Pittsburgh," M.A. thesis, University of Pittsburgh, 1925, app. 3, p. 54, quoted in Montgomery, *Fall of the House of Labor,* 243.

39. David Brody, *Workers in Industrial America: Essays on the Twentieth Century Struggle* (New York: Oxford University Press, 1980), 17.

40. Frank Thistlewaite, "Migration from Europe Overseas in the Nineteenth and Twentieth Centuries," in *A Century of European Migrations, 1830–1930,* ed. Rudolph J. Vecoli and Suzanne M. Sinke (Urbana: University of Illinois Press, 1960), 25.

41. Smith and Herring, *The Bureau of Immigration;* Darrell H. Smith, *The Bureau of Naturalization* (Baltimore, 1926).

42. Department of Commerce and Labor, *Annual Report* (Washington, D.C.: Government Printing Office, 1908), 25.

43. Smith, *Bureau of Naturalization,* 21.

44. Degler, *In Search of Human Nature,* 20–22; Stocking, *Race, Culture, and Evolution,* 234–69.

45. Letter from William Williams, commissioner of immigration, Ellis Island, to Theodore Roosevelt, 5 July 1911, RG 85, Box 103, File No. 52363/25A, NARA.

46. Madison Grant, *The Passing of the Great Race* (New York: Scribner's, 1916), 104.

47. Reports of the U.S. Immigration Commission, *Dictionary of Races and Peoples* (Washington, D.C.: Government Printing Office, 1911), 4.

48. Ibid., 2, 3, 1.

49. In describing the physical aspects of race, the Immigration Commission's dictionary discussed color but displayed a keen concern with the bodily form—particularly the head and brain sizes—of different groups. Notably, the court system explicitly rejected language as a basis for racial classification. See *United States v. Thind,* 261 U.S. 204 (1923), in Lopez, *White by Law,* 223.

50. Immigration Commission, *Dictionary of Races and Peoples,* 15, 28, 75, 84.

51. See, for example, the statements of R. R. Lutz, manager, Washington office, National Industrial Conference Board, Washington, D.C.; Cairoli Gigliotti, editor of the *New Comer,* Chicago; H. S. Jennings, professor of zoology, Johns Hopkins University, Baltimore; C. M. Panunzio, Washington, D.C., instructor in social science at

Willamette University, Salem, Oregon, and fellow in political economy at the Residence Foundation of Washington University, Washington, D.C.; Hon. Thomas W.
Phillips, Jr., representative from Pennsylvania; and Mr. Francis H. Kinnicutt, Immigration Restriction League. Representatives of ethnic groups, while refuting the biological basis for race, nonetheless tended to couch their arguments in the language of
science in response to committee considerations of the racial "inferiority" of various
groups. See, for example, statement of Gedalia Bublick, editor of the *Jewish Daily
News,* New York City, and Rep. Lewis Marshall, New York City, in *Hearings before the
Committee on Immigration and Naturalization, House of Representatives,* 68th Cong.,
1st sess. (Washington, D.C.: Government Printing Office, 1924).

52. 17 April 1920, "Biological Aspects of Immigration,"; 21 November 1922,
"Analysis of America's Modern Melting Pot," in *Hearings before the Committee on Immigration and Naturalization, House of Representatives,* 67th Cong., 3d sess. (Washington, D.C.: Government Printing Office, 1923). Also 8 March 1924, "Europe as an
Immigrant Exporting Continent and the United States as an Immigrant Receiving Na-
tion," in *Hearings before the Committee on Immigration and Naturalization, House of
Representatives,* 68th Cong., 2d sess. (Washington, D.C.: Government Printing Office,
1925); 21 February 1928, "Eugenical Aspects of Deportation," in *Hearings before the
Committee on Immigration and Naturalization of the House of Representatives,* 70th
Cong., 1st sess. (Washington, D.C.: Government Printing Office, 1928).

53. A. J. Nute, "Medical Inspection of Immigrants at the Port of Boston," *Boston
Medical and Surgical Journal* 170 (23 April 1914): 645. See also J. G. Wilson, M.D.,
Ellis Island, "Some Remarks Concerning Diagnosis by Inspection," *New York Medical
Journal* 94 (1911): 94–96.

54. As with all of the PHS's musings on the medical gaze, Nute's remarks on the
importance of race were extensive. He elaborated, "The face of a muscular able-
bodied Italian peasant often is so devoid of fat and muscle tissue that on first sight one
would think that the whole body was thin and undeveloped. The complexion of the
Slavish peasant woman would be suspicious of chlorosis if possessed by a Scandinavian
or English woman of the same class. On the other hand the red cheek of the Scandinavian would arouse thought of a hectic flush if seen in the Polish woman."

"The excitably *[sic]* of the southern Italian and the Hebrew are well known. It is
easy to excite in them almost maniacal action. The stolidity and indifference of the Slav
would suggest melancholia if presented by the Hebrew. The sanity of an Englishman
would be questioned if on slight provocation he evinced the external manifestation of
emotion that would occur in the Sicilian. The German girl takes her examination seriously and her sanity would at once be suspected if she saw the same reason for light
remark and laughter as the girl from Ireland. Some races are extremely emotional, others slow; and unless the normal is known it is impossible to pick the abnormal" (Nute,
"Medical Inspection," 645). See also Wilson, "Some Remarks," 94–96; Alfred C. Reed,

"The Relation of Ellis Island to the Public Health," *New York Medical Journal* 98 (1913): 172–74; and George W. Stoner, "Problems of Immigration from a Hygienic Standpoint," *Transactions of the International Congress on Hygiene and Demography* 5, pt.1 (1912): 261–85.

55. U.S. Public Health Service, *Regulations Governing Medical Inspection of Aliens* (Washington, D.C.: Government Printing Office, 1917), 19.

56. Elizabeth Yew, interview with John C. Thill, M.D., 13 September 1977, EIOHP.

57. SGAR 1905, 134; SGAR 1917, 175, 246; SGAR 1920, 187; draft of a manuscript entitled "Medical Inspection of Aliens," 1929, RG 90, General Subject File, 1924–35, Box 941, File No. 0950–56, NARA. See also William Williams, "Ellis Island. Its Organization and Some of Its Work," c. August 1911, RG 85, Box 120, File No. 52516/1A, NARA.

58. Report of M. V. Safford, Boston, in SGAR, 1909, 182; letter from W. C. Billings, chief medical officer, Ellis Island, to surgeon general, 16 April 1924, RG 90, General Subject File, 1924–35, Domestic Stations, New York, Box 163, File No. 950–56, NARA.

59. J. M. Eager, in SGAR, 1902, 394.

60. SGAR, 1921, 246.

61. Memorandum from J. D. Long, assistant surgeon general, to surgeon general, 30 July 1924, RG 90, General Subject File, 1924–35, Box 944, File No. 0950-121, NARA. In 1930 it was removed from the list of Class A conditions.

62. Letter from Dr. Safford to Surgeon R. M. Woodward, 21 January 1905, RG 90, Central File, 1897–1923, Box 58, File No. 409, NARA.

63. Report of P.A. Surgeon J. M. Eager, in SGAR, 1901, 466.

64. A. J. McLaughlin, "The American's Distrust of the Immigrant," *Popular Science Monthly* (1903): 230–36. McLaughlin constructed these tables for the following races: Lithuanian, Slav, Magyar, Finn, Italian, Syrian, and Hebrew.

65. SGAR, 1911, 207. See also letter from Arthur J. Somers, Portal, North Dakota, to surgeon general, 12 July 1916, RG 90, Central File, 1897–1923, Box 143, File No. 1610, NARA.

66. Annual report, 1928, Surgeon J. G. Wilson in charge from 1 July to 12 September 1928, Surgeon J. R. Hurley in charge from 12 September to 20 June 1929, RG 90, General Subject File, 1924–35, Domestic Stations, Texas, El Paso, Box 248, File No. 1850-15, NARA. See also SGAR, 1911, 208, for a similar description of Chinese and Indian immigrants.

67. Robert DeC. Ward, "Higher Mental and Physical Standards for Immigrants," *Scientific Monthly* 19 (1924): 533–47; Ward, "Our Immigration Laws from the Viewpoint of National Eugenics," *National Geographic Magazine* 23 (1912): 38–41.

68. Statement of Harry Hamilton Laughlin, "Biological Aspects of Immigration," 16–17 April 1920, *Hearings before the House Committee on Immigration and Natu-*

ralization, 66th Cong., 2d sess. (Washington, D.C.: Government Printing Office, 1921), 1–26.

69. Statement of H. S. Jennings, professor of zoology at Johns Hopkins University, Baltimore, in *Hearings before the Committee on Immigration and Naturalization, House of Representatives,* 67th Cong., 3d sess. (Washington, D.C.: Government Printing Office, 1923), 511.

70. Henry Laughlin, "The Relative Numbers of European-Born Defectives from the Chief Sources of European Immigration and the Effect of a Change in the Basis of Admission, from the Census of 1910 to That of 1890," in ibid., 512.

71. Lutz statement, in *Hearings before the House Committee on Immigration and Naturalization,* 68th Cong., 1st sess. (Washington, D.C.: Government Printing Office, 1924), 262.

72. A standard deviation is a standardized unit that measures average discrepancy or variation of each of a set of observations (values) from the mean of these observations or values.

73. Comparing IS data to morbidity data for foreign countries holds little promise. Morbidity data are quite limited for the United States before 1900. See Daniel M. Fox, *Power and Illness: The Failure and Future of American Health Policy* (Berkeley: University of California Press, 1993); and Fox, "Health Policy and Changing Epidemiology in the United States: Chronic Disease in the Twentieth Century," in *Unnatural Causes: The Three Leading Killer Diseases in America,* ed. R. Maulitz (New Brunswick, N.J.: Rutgers University Press, 1989). Organizations that collected such data on an international basis focused almost exclusively on mortality statistics rather than morbidity statistics; morbidity statistics are not readily available until 1938. World Health Organization, *Annual Epidemiological and Vital Statistics, 1939–1946. Part I. Vital Statistics and Causes of Death* (Geneva: WHO, 1951). There is, however, no reason to believe that morbidity in immigrants, who were generally poor, would match that of the population of their country of origin.

74. The PHS consistently noted the degree of severity only in Class B and C certifications for defective vision. Certificates typically read "very defective vision," "slightly defective vision," or "blind" in one or both eyes.

75. With regard to the category of "African Blacks": Barbara Fields, in a critique of whiteness studies, has noted that few "ask how African and Afro-Caribbean immigrants became black" ("Whiteness, Racism, and Identity," *International Labor and Working-Class History* 60 [fall 2001]: 51). See also Mary C. Waters, *Black Identities: West Indian Immigrant Dreams and American Realities* (Cambridge, Mass.: Harvard University Press, 1999); Milton Vickerman, *Crosscurrents: West Indian Immigrants and Race* (New York: Oxford University Press, 1999); and Winston James, "Explaining Afro-Caribbean Social Mobility in the United States: Beyond the Sowell Thesis," *Comparative Studies in Society and History* 44, no. 2 (April 2002): 218–62.

76. In contrast to the instance of Asians and trachoma, the PHS in certifying mental illness among the so-called Nordic races was not acting in an exclusionary capacity. Rather, this represents an instance in which the PHS acted in its most unremarkable capacity—to screen out disease. In the case of mental conditions, the number of certifications made was quite small, while the number of European immigrants arriving was large. In the case of trachoma and hookworm, the number of certifications made was large, while the number of Asians arriving was small, allowing the medical exam to serve the ends of racial exclusion.

77. 26 May 1924, c. 190, 43 Stat. 153. The Immigration Act of 1921 restricted immigration to 3% of the number of each nationality recorded in the U.S. Census of 1910. 19 May 1921, c. 8, 42 Stat. 5. Immigration restrictionists argued that the quotas established originally in 1921, based on the 1910 census, did not go far enough in reducing southern and eastern European immigration, allowing such immigrants to fill fully 45% of the quota. The 1924 act, by relying on the distribution of immigrants in the 1890 census, reduced the proportion of southern and eastern Europeans to 16%. See D. J. Kevles, *In the Name of Eugenics: Genetics and the Uses of Human Heredity* (Berkeley: University of California Press, 1985); M. H. Haller, *Eugenics: Hereditarian Attitudes in American Thought* (New Brunswick, N.J.: Rutgers University Press, 1963); K. Ludmerer, *Genetics and American Society: A Historical Appraisal* (Baltimore: Johns Hopkins University Press, 1972); and Robert A. Divine, *American Immigration Policy, 1924–1952* (New York: Da Capo Press, 1972, 1957).

78. John Higham, *Strangers in the Land: Patterns of American Nativism 1850–1925* (New York: Atheneum, 1967), 300. See his discussion of the decline of nativism in the period following the passage of the 1924 restriction act (324–30). See also Gary Gerstle, *Working-Class Americanism: The Politics of Labor in a Textile City, 1914–1960* (Cambridge: Cambridge University Press, 1989), 3.

79. Rudolph J. Vecoli, "Ethnicity: A Neglected Dimension of American History," in *The State of American History*, ed. Herbert J. Bass (Chicago: Quadrangle Books, 1970), 73.

80. Divine, *American Immigration Policy.*

81. Ibid.

82. Ibid.

83. Quoted in ibid., 36. Herbert Hoover also publicly denounced the validity of national origins in the presidential campaign of 1928 and attempted to eliminate the quota system upon taking office in 1929. With certainty that the national origins system would be a thing of the past, lobbying on the part of ethnic minorities virtually ceased, and the U.S. Chamber of Commerce and the AFL, without repudiating quotas based on the 1890 census, declined to give their support to the new national origins system. Ibid., 38–42.

84. Mae M. Ngai, "Illegal Aliens and Alien Citizens: United States Immigration Policy and Racial Formation, 1924–1945," Ph.D. diss., Columbia University, 1998.

85. Act of 26 May 1924, 43 Stat. 153, Sec. 11 (d).

86. Ngai, "Illegal Aliens and Alien Citizens," 36.

87. Ibid., 44–46.

88. Matthew Frye Jacobson, *Whiteness of a Different Color: European Immigrants and the Alchemy of Race* (Cambridge, Mass: Harvard University Press), 3.

89. Eric Arnesen, referring to the work of Ignatiev, Roediger, and others, notes that it was the question of how the Irish became white that launched whiteness studies. Eric Arnesen, "Whiteness and the Historians' Imagination," *International Labor and Working-Class History* 60 (fall 2001): 3–32. See, for example, David R. Roediger, *The Wages of Whiteness: Race and the Making of the American Working Class* (New York: Verso, 1991);and Noel Ignatiev, *How the Irish Became White* (New York: Routledge, 1995).

90. Canadian and Mexican officials also used the term to set Europeans as a whole apart from "black" and "yellow" peoples. See, for example, George Lockwood, U.S. Immigration Service, Naco, Arizona, translation of Department of the Interior, Division of Immigration, Decree Designating the Frontier Places Authorized for the Entry of Passengers into the Republic and Regulations for the Inspection of Immigrants, Mexico, 1909, RG 85, Box 439, File No. 55609/551, NARA; letter from consul general, Winnipeg, Manitoba, to U.S. secretary of state, 22 April 1911, RG 85, Box 63, File No.51831/157, NARA.

91. Letter from Dr. Victor Safford to commissioner of immigration, Boston, 9 March 1911, RG 90, Central File, 1897–1923, Box 58, File No. 409, NARA; report of Dr. M. J. White, China, in SGAR, 1904, 220–21; report of Dr. M. W. Glover, San Francisco, in SGAR, 1911, 209; and SGAR, 1913, 163. Whiteness was also used as a term in opposition to American "Indians" (SGAR, 1913, 27).

92. Letter from Robert Watchhorn, Ellis Island, to Oscar S. Straus, secretary of commerce and labor, 17 December 1907, RG 85, Box 143, File No. 52729/2, NARA.

EPILOGUE. The End of the Line

Epigraph. SGAR, 1925, 172; letter from John L. Cable, Representative, U.S. Congress, Lima, Ohio, to secretary of labor, 24 April 1925, RG 85, Box 348 (to be renumbered 318), File No. 55224/371A, Folder 2, NARA. Earlier, in 1923, the U.S. representative from Texas, John Box, argued against immigrant medical inspection abroad, claiming that it would give foreign countries "a voice in the making of our immigration regulations." In addition, he felt that foreign governments would never give their consent to having prospective immigrants rejected in their ports. Box, however, was one of the few dissenting voices. John C. Box, "Shall U.S. Adopt Policy of Selecting Immigrants Abroad? Con," *Congressional Digest* 2 (July 1923): 312.

1. Kitty Calavita, *U.S. Immigration Law and the Control of Labor, 1820–1924* (London: Academic Press, 1984).

2. Henry Yu, *Thinking Orientals: Migration, Contact, and Exoticism in Modern America* (Oxford: Oxford University Press, 2001), 6–7.

3. David M. Gordon, Richard Edwards, and Michael Reich, *Segmented Work, Divided Workers: The Historical Transformation of Labor in the United States* (Cambridge: Cambridge University Press, 1982), 171.

4. Ibid., 3.

5. Irving Bernstein, *The Lean Years: A History of the American Worker, 1920–1933* (Boston: Houghton Mifflin, 1960), 62.

6. David Brody, *Workers in Industrial America: Essays on the Twentieth Century Struggle* (New York: Oxford University Press, 1980), 12, 13.

7. Lizabeth Cohen, *Making a New Deal: Industrial Workers in Chicago, 1919–1939* (Cambridge: Cambridge University Press, 1990), 162.

8. Brody notes that while only 6.5% of small companies established industrial relations departments, 30% of companies employing between 500 and 2,000, and 50% of those employing over 2,000, established such departments. The National Industrial Conference Board found that welfare capitalism was more widespread, with 90% of companies surveyed offering safety programs; 60%, mutual aid associations; and 70%, group insurance; few offered pensions, stock options, or savings plans. Brody, *Workers in Industrial America,* 59–60.

9. Quoted in Bernstein, *Lean Years,* 51.

10. Ibid., 52.

11. Sumner Slichter, "The Current Labor Policies of American Industries," *Quarterly Journal of Economics* 42 (May 1929): 432, quoted in Brody, *Workers in Industrial America,* 57 (emphasis added).

12. Gerard Swope, Columbia Oral History Collection, quoted in Brody, *Workers in Industrial America,* 58–59 (emphasis added).

13. Gary Gerstle, *Working-Class Americanism: The Politics of Labor in a Textile City, 1914–1960* (Cambridge: Cambridge University Press, 1989), 2. Cohen, while acknowledging that efforts of industrial managers were intended to quash labor uprisings and collective representation, argues that employers saw "welfare capitalism" as "the end of an industrial system that pitted proletarian against owner." "Managers felt that welfare programs were their best defense against appeals to their workers' class consciousness. Harvester and other employers decided to grant hourly employees paid vacations, for example, 'to remove the argument of class distinction which now prevails'" (Cohen, *Making a New Deal,* 175–76).

14. Gordon, Ewards, and Reich, *Segmented Work, Divided Workers,* 172–73; Bernstein, *Lean Years,* 55.

15. Calavita, *U.S. Immigration Law*.

16. Cohen, *Making a New Deal*, 40–41. This, however, was not always the case. Without denying that the deep ethnic and labor divisions in Chicago's steel industry resulted in fragmentation and disunity, James R. Barrett, in his study of Chicago's meatpacking industry in the twentieth century, argues that ethnic residential and occupational clustering was evident but that segregation was not absolute. Workers came into daily contact with one another. This combined with a union structure to maximize ethnic solidarity and facilitate unity in this industry until the early 1920s. James R. Barrett, "Unity and Fragmentation: Class, Race, and Ethnicity on Chicago's South Side, 1900–1922," in *The Work Experience: Labor, Class, and Immigrant Enterprise*, ed. George E. Pozetta (New York: Garland Publishing, 1991), 1–20.

17. Gordon, Edwards, and Reich, *Segmented Work, Divided Workers*, 174; Stephen Meyer, "Adapting the Immigrant to the Line: Americanization in the Ford Factory, 1914–1921," *Journal of Social History* 14, no. 1 (fall 1980): 78–79.

18. Cohen, *Making a New Deal*, 32.

19. Raymond Edward Nelson, "A Study of Isolated Industrial Community: Based on Personal Documents Secured by the Participant Observer Method," M.A. thesis, University of Chicago, 1929, 130; quoted in Cohen, *Making a New Deal*, 165.

20. Yu, *Thinking Orientals*, 6.

21. In Woonsocket, Rhode Island, for example, while "Americanization campaigns and a troubled economy freed the city's two most important working-class groups, traditionalist French Canadians and progressive Franco-Belgians, from the insular ethnic worlds in which they had been confined" and allowed them to make "political action based on a shared class experience a real possibility," this did not amount to the demise of ethnic culture. The French Canadian workers of Woonsocket were able to reshape ethnic values to help negotiate entrance into American society (Gerstle, *Working-Class Americanism*, 4). For other studies exploring the depth of ethnic residential and industrial segregation based on demographic analysis of the 1910 Census, see Michael J. White, Robert F. Dymowski, and Shilian Wang, "Ethnic Neighborhoods and Ethnic Myths: An Examination of Residential Segregation in 1910," and Ann R. Miller, "The Industrial Affiliation of Workers: Differences by Nativity and Country of Origin," both in *After Ellis Island: Newcomers and Natives in the 1910 Census*, ed. Susan Cotts Watkins (New York: Russell Sage Foundation, 1994). White and his coauthors demonstrate that the "older," northern European immigrants were more residentially integrated than southern and eastern Europeans; they attribute this primarily to cultural rather than economic differences between groups (206). Miller shows that in 1910, "native" "white" Americans dominated agriculture, while southern and eastern European immigrant groups dominated other industries. Moreover, she concludes that the transformation of the labor force was dependent on foreign immigration because of the failure of native whites to break into other occupations (300).

22. Mae Ngai, "The Architecture of Race in American Immigration Law: A Reexamination of the Immigration Act of 1924," *Journal of American History* 86, no. 1 (June 1999): para. 9. Available at http://www.historycooperative.org/journals/jah/86.1//ngai.html

23. Joshua Freeman, "Racial and Ethnic Categorization by New York City Unions after World War II," paper presented at the American Public Health Association Annual Meeting, Boston, 15 November 2000.

24. *Monthly Labor Review* 30 (1930): 11–54; Brody, *Workers in Industrial America,* 63–64.

25. Robert H. Wiebe, *The Search for Order, 1877–1920* (New York: Hill and Wang, 1967), 158.

26. Gorge Soule, *Prosperity Decade: From War to Depression: 1917–1929,* vol. 3 of *The Economic History of the United States* (New York: Rinehart, 1947).

27. Bernstein, *Lean Years,* 59.

28. Cohen, *Making a New Deal,* 184.

29. Bernstein, *Lean Years,* 52, 55, 60.

30. Brody, *Workers in Industrial America,* 6, 59.

31. Bernstein, *Lean Years,* 62.

32. Unemployment Committee of the National Federation of Settlements, *Case Studies of Unemployment* (Philadelphia: University of Pennsylvania Press, 1931), 1.

33. Elizabeth Hughes, U.S. Department of Labor, Children's Bureau, *Children of Preschool Age in Gary, Ind. Part I. General Conditions Affecting Child Welfare,* Bureau Publication No. 122 (Washington, D.C.: Government Printing Office, 1922), 35.

34. Emma Octavia Lundberg, U.S. Department of Labor, Children's Bureau, *Unemployment and Child Welfare: A Study Made in a Middle-Western and an Eastern City during the Industrial Depression of 1921 and 1922,* Bureau Publication No. 125 (Washington, D.C.: Government Printing Office, 1923), 85.

35. Reuben Fink, "Visas, Immigration, and Official Anti-Semitism," *The Nation* 112 (22 June 1921): 870.

36. 24 May 1924, Ch. 182, 43 Stat. 140.

37. Letter from W. W. Husband, second assistant secretary, to Surgeon General H. S. Cumming, 22 February 1927, RG85, Box 348 (to be renumbered 318), File No. 55224/371a, Folder 2, NARA.

38. The *New York Times* reported that the plan to inspect immigrants abroad was popular even before the passage of the 1924 Immigration Act, having "met with the instant approval of European Governments and American diplomatic and consular officers overseas" ("Europe for Testing Immigrants There: Caminetti Tells Senate Committee That Project Meets General Favor Abroad," *New York Times,* 26 January 1921).

39. As early as 1920 the commissioner general of immigration advocated inspections abroad for humanitarian reasons: "The tragedies of disappointed men without

homes ought and can be avoided by proper administration" ("Caminetti Advocates Emigration Centres: Reaches Rotterdam on His Trip for Study of Conditions Abroad," *New York Times,* 8 December 1920). The rationale stated in 1928 was quite similar to that set forth in other years.

40. SGAR, 1928, 192.

41. Report of the Conference of Officers of the State, Treasury, and Labor Departments for the Consideration of Matters Relating to the Immigration Act of 1924, 23 April 1925, RG 90, 0950-56, NARA.

42. Darrell H. Smith and H. Guy Herring, *The Bureau of Immigration: Its History, Activities, and Organization* (Baltimore: Johns Hopkins Press, 1924), 14.

43. PHS officers were stationed at Cobh and Dublin in the Irish Free State, Belfast in Northern Ireland, Glasgow in Scotland, and Liverpool, London, and Southampton in England. SGAR, 1926, 186.

44. RG 90, General Subject File, 1924–33, Box 943, File No. 0950-56, NARA. The plan received strong support from the press. "Examine Britons Before Coming Here: New Plan Put in Effect to Relieve Rejected Ones of Useless Trip to America," *New York Times,* 2 August 1925.

45. Letter from secretary of the Treasury to secretary of state, 5 December 1925, and Report of the Conference of Officers of the State, Treasury and Labor Departments for the Consideration of Matters Relating to the Immigration Act of 1924, 23 April 1925, RG 90, 0950-56, NARA. The United States also considered a plan whereby local physicians, rather than PHS officers, would examine prospective immigrants. The PHS did not like this plan because it felt local physicians either would not do the exams or would feel hard-pressed to appropriately certify patients whom they had been treating for a number of years. Indeed, the British Medical Association strongly advised physicians to refuse absolutely to have anything to do with the American certification plan because it required technical knowledge of some rare diseases, placed a great deal of responsibility on the physician in terms of an immigrant's medical status and the liability of shipping companies, and was generally "too much to expect of the general practitioner with no tropical experience" (Report of the Medical Examination of Aliens in the British Isles Conducted from August 1, 1925, to October 31, 1925, by the United States Public Health Service in Cooperation with the Consular and Immigration Services, RG 90, General Subject File, 1924–33, Box 943, File No. 0950-56, NARA).

46. SGAR, 1926, 187.

47. SGAR, 1927.

48. SGAR 1930, 187.

49. For all years I have presented data for "quota" and "nonquota" immigrants combined. "Nonquota" immigrants represented a special category of immigrants in-

cluding only wives and children of U.S. citizens, returning aliens previously legally admitted, students, college professors, and ministers and their families.

50. All these figures, and all subsequent figures discussed, represent baseline estimates of immigrants notified. Many immigrants who began the visa application process and were then found to be diseased or defective by the PHS physicians did not complete the visa application. Those notified who received treatment and were later granted a visa were not counted among those rejected. SGAR, 1928, 184.

51. SGAR, 1928, 193. This notion was repeated almost verbatim in subsequent years.

52. Out of 525 men inspected at Local Exemption Board No. 17 in New York City, for example, 105 (20%) were rejected. The major cause of rejection (40%) was for defects in limbs (flat foot accounting for 52% of this category), followed by optical defects (24%, of which defective vision represented 76% of rejections), and circulatory defects (21%, which included cardiac dilatation, variocele, and varicose veins). J. A. Hofheimer, "An Analysis of the Recent Physical Examinations for a Local Exemption Board," *International Journal of Surgery* (September 1917): 276. See also Lieut. Calvin H. Goddard, "A Study of About 2,000 Physical Examinations of Officers and Applicants for Commission Made at the Army Medical School," *Military Surgeon* (1917): 578–88. Here, the largest percentage of candidates were rejected as underweight (21%); 14% were rejected for defective vision and 11.1% for nephritis. Only 1 candidate (0.5%) was rejected for tuberculosis; none was rejected for venereal diseases or any of the Class A immigrant conditions. See also Charles A. Costello, "The Principal Defects Found in Persons Examined for Service in the United States Navy," *American Journal of Public Health* (1917): 489–92; and Jas. H. Hamilton, "Physical Examination of Drafted Men," *Vermont Medicine* (December 1917): 287–88.

53. SGAR, 1927, 193.

54. SGAR, 1928, 188.

55. Draft of Report on the Medical Examination of Applicants for Immigration Visas at Certain Foreign Ports during the Fiscal Year 1926, RG 90, File 0950-56, NARA.

56. Frederick William Wile, "Immigration Plan Succeeds: New Use May Be Found for Ellis Island Station," *Christian Science Monitor,* 22 December 1925, clipping in RG 90, General Subject File, 1924–35, Box 944, File No. 0950-56, NARA. The PHS reported that "cooperation between the services engaged in immigration work abroad, namely the Department of State, the Department of Labor, and the Public Health Service, has been characterized by the greatest cordiality on the part of the officials at all stations. The attitude of the public toward the examinations as reported from stations is tolerant and favorable in most countries" (SGAR, 1930, 187–88). Beginning in the early 1920s, the secretary of labor began receiving citizens' requests for inspection

abroad. One letter requested screening abroad because the diseased would "simply not believe what is told them and insist on coming nevertheless. . . . The fact is that those [orphans and diseased and inefficient] are the very classes who insist on coming because they find it impossible to maintain themselves in Europe. The able-bodied and efficient are not coming because they find that they can get along fairly well where they are; and there is our problem" (letter from Mrs. Laura B. Collier, chairman, Home Missionary Division, Brooklyn, to Assistant Secretary E. J. Henning, 16 November 1922, RG 85, Box 348 [renumbered 318], File No. 55224/371A, Folder 2, NARA). See also letter from Thos. H. Sherman, Gorham, Maine, to Robe Carl White, second assistant secretary, 1 October 1923, Doc. No. 55224/371, and letter from Walter Davidson, Central Division, American Red Cross, to George W. Powell, chairman, Fulton Red Cross Chapter, 5 April 1922, RG 85, Box 348 (to be renumbered 318), File No. 55224/371A, Folder 2, NARA.

57. Associated Press, "Second Year of Quota Law Brings Higher Type of Immigrant: Younger Men, More Easily Assimilated, Replace Old Type, Curran Says on Anniversary of Policy," *The Evening Star,* 1 July 1925.

58. Report of the Conference of Officers of the State, Treasury, and Labor Departments for the Consideration of Matters Relating to the Immigration Act of 1924, 23 April 1925, RG 90, 0950-56, NARA.

59. Draft of an undated manuscript entitled "Medical Inspection of Aliens," RG 90, General Subject File, 1924–35, Box 941, File No. 0950-56, NARA.

60. 26 May 1924, c. 190, 43 Stat. 153.

61. See, for example, Eugene L. Fisk, "The Value of Complete Routine Examinations in Supposedly Healthy People," *Boston Medical and Surgical Journal* 195 (14 October 1926): 741, where he describes the "Vital Protective" examination, which included as standard procedures a hemoglobin blood test, urinalysis, a chest and dental x-ray, electrocardiogram, tests for levels of blood metabolites, a tonsil culture, a blood smear, and examination of the feces.

62. J. G. Wilson, M.D., Ellis Island, "Some Remarks Concerning Diagnosis by Inspection," *New York Medical Journal* 94 (1911): 94.

63. Information for Medical Officers Engaged in Examination of Aliens Prior to Granting of Consular Visas at Foreign Ports, 16 July 1925, Doc No. H-373, RG 90, General Subject File, 1924–35, Box 941, File No. 0950-56, NARA.

64. SGAR, 1927, 207.

65. The PHS reported that in contrast to the situation in the United States, in which immigrants would often hire local physicians to contradict the diagnosis of the PHS physician and would attempt to use this to reverse decisions, the European consultants usually confirmed the diagnosis of the PHS officer attached to the consulate. SGAR, 1927, 207.

66. Information for Medical Officers Engaged in Examination of Aliens Prior to Granting of Consular Visas at Foreign Ports, 16 July 1925, RG 90, 0950-56, NARA.

67. Audrey B. Davis, "Life Insurance and the Physical Examination: A Chapter in the Rise of American Medical Technology," *Bulletin of the History of Medicine* 55 (1981): 396.

68. Stanley J. Reiser, "The Emergence of the Concept of Screening for Disease," *Milbank Memorial Fund Quarterly* 56, no. 4 (1978): 406. Charap, however, argues that such findings "reflected an absurd standard of health set by the screening agencies" (Mitchell H. Charap, "The Periodic Health Examination: Genesis of a Myth," *Annals of Internal Medicine* 95 [1981]:734).

69. Allan J. McLaughlin, *Personal Hygiene: The Rules for Right Living* (New York: Funk and Wagnalls, 1924).

70. Memorandum from Dr. Lavinder, chief medical officer, to Benjamin M. Day, commissioner of immigration, 18 November 1929, RG 90, General Subject File, 1924–35, Domestic Stations, New York, Box 163, File No. 0950-56, NARA.

71. SGAR, 1928.

72. Memorandum from Dr. Lavinder, chief medical officer, to Benjamin M. Day, commissioner of immigration, 18 November 1929, RG 90, General Subject File, 1924–35, Domestic Stations, New York, Box 163, File No. 0950-56, NARA.

73. Letter from Lavinder to Assistant Surgeon General F. A. Carmelia, 1 August 1929, and letter from Dr. Carmelia to Senior Surgeon Taliaferro Clark, 6 August 1929, RG 90, General Subject File, 1924–35, Domestic Stations, New York, Box 163, File No. 0950-56, NARA.

74. SGAR, 1925, 172.

75. Grubbs explained that the 1924 quota law "has allowed more thorough examinations to be made even than the previous quota laws and infinitely better ones than were possible when the rush was timed only by the capacity of the steamers, because it has insured a steady flow that replaced the wild rush at the beginning of each of the first five months of the fiscal year. These human avalanches engulfed the resources of immigration stations and beat down the physical powers of medical officers in a way that made regular and careful work exceedingly difficult" (letter to Mr. Cook, Department of Labor, 16 October 1925, RG 90, General Subject File, 1924–33, Box 942, File No. 0950-56, NARA).

76. Elizabeth Yew, interview with Dr. Bernard Notes, 4 September 1977, EIOHP.

77. Elizabeth Yew, interview with Dr. T. Bruce H. Anderson, 22 September 1977, EIOHP.

78. "Says the Foreigner Is Not Appreciated: Immigration Not a Menace to Nation, Says Commissioner Wallis. Talks at University Club: Thinks US Agents Should Select Immigrants Abroad," *Brooklyn Standard Union*, 23 March 1921. The examination at

Ellis Island for venereal diseases, in particular, was criticized as a "farce." Letter from J. W. Kerr, Ellis Island, to Surgeon General Creel, 11 March 1921, Central File, 1897–1923, Box 38, File No. 219, NARA. See also discussion of the report of Dr. Spencer Dawes, which cited that, on average, eight diseased or defective aliens were passed by PHS officers at Ellis Island every minute, in *Hearings before the Committee on Immigration and Naturalization, House of Representatives*, 68th Cong., 1st sess. (Washington, D.C.: Government Printing Office, 1925), 149; "Lodge Wants More Doctors to Examine Immigrants," *New York Times*, 18 June 1907; Robert DeC. Ward, "Higher Mental and Physical Standards for Immigrants," *Scientific Monthly* 19 (1924): 535; Alexander E. Cance et al., "First Report of the Committee on Immigration of the Eugenics Section," *American Breeder's Magazine* 3 (1912): 250; Robert DeC. Ward, "Our Immigration Laws from the Viewpoint of National Eugenics," *National Geographic Magazine* 23 (1912): 40; "Insufficient Government Inspection at Ellis Island Is Admitting Large Numbers Each Year That Should Be Deported, and Is Saddling a Heavy Burden on the State," *New York Times,* 7 April 1912. As early as 1914, the New York Academy of Medicine, upon learning of the intensive exams being conducted at Ellis Island, encouraged this endeavor, stating, "The need of more careful work than has hitherto been possible would, therefore, seem to be demonstrated" (letter from E. H. Lewinski-Corwin, executive secretary, New York Academy of Medicine, to Surgeon General Rupert Blue, 9 December 1914, RG 90, Central File, 1897–1923, Box 37, File No. 219, NARA.

79. Memorandum from C. H. Lavinder, chief medical officer, Ellis Island, to Dr. Carmelia, 8 December 1930, RG 90, General Subject File, 1924–35, Box 941, File No. 0950-56, NARA.

80. In another letter to the surgeon general, Lavinder elaborated, characterizing the certifications abroad: "First, by too much zeal in making them out at all. And second, by too great particularity in the language employed." In addition, the diseases and conditions noted in the memoranda that accompanied the immigrants were often "trivial" and not categorized correctly. Letter of 10 December 1930, RG 90, General Subject File, 1924–35, Domestic Stations, New York, Box 163, File No. 0950-56, NARA. Those cases where PHS officers in the United States found disease among immigrants who had been cleared by consuls abroad provoked some criticism of consular inspection. For example, an unusually high number of Ellis Island certifications among Irish immigrants for heart disorders provoked the National Catholic Welfare Conference to demand that consular exams be brought into conformity with those conducted in the United States. "For Irish Immigrants: Director Urges Uniform Examinations Abroad and Here," *New York Times,* 29 March 1925; "Curran Would Test Immigrants Abroad: Also Suggests Handling of Appeals on Ellis Island Instead of at Washington: To Expedite Inspections: Lays Epidemic of Heart Disease among Irish Newcomers to Stiffer Examinations," *New York Times,* 30 March 1925.

81. Letter from Dr. Lavinder to John McMullen, medical director, American Embassy, Paris, 8 December 1930, RG 90, General Subject File, 1924–35, Box 941, File No. 0950-56, NARA.

82. "Curran Sees a Plot to End Ellis Island: Charges Secretary Husband Is One of Officials Planning to Abolish It: Washington Explains: Says Landing of British after Examination Abroad Is Purely Experimental," *New York Times,* 12 August 1925. Curran received support from the medical examiner of the New York State Hospital Commission, who witnessed the inspection abroad and agreed that the exam should be done at Ellis Island. "Dr. Dawes to Study Immigrant Tests: State Hospital Medical Examiner Sails on *Cedric* for Survey in England: Upholds Curran's View: Declares Davis's Regulations Are Impossible and Improper," *New York Times,* 23 August 1925.

83. Letter from secretary of labor to John L. Cable, member of Congress, 27 April 1925, RG 85, Box 348 (to be renumbered 318), File No. 55224/371A, Folder 2, NARA.

84. Immigrant medical inspection was conducted as necessary on Ellis Island until it closed. Bess Furman with Ralph Chester Williams, *A Profile of the United States Public Health Service, 1798–1948* (Washington, D.C.: Government Printing Office, 1973).

85. "Wallis Deplores Methods: Commissioner Says Health Inspection Should Begin Overseas," *New York Times,* 15 February 1921.

86. Public Health Service, *Regulations Governing the Medical Examination of Aliens* (Washington, D.C.: Government Printing Office, 1930), 43.

87. Letter from Henry Curran, commissioner, Ellis Island, to commissioner general, 1 July 1925, Doc. No. 98524/248, RG 85, Box 436, File No. 55607/578, NARA.

88. Memorandum regarding the Preliminary Statement of Henry H. Curran, U.S. Commissioner of Immigration at the Port of New York, c. 1924, RG 90, General Subject File, 1924–35, Domestic Stations, New York, Box 163, File No. 0950-56, NARA. See also letter from Charles D. Hurrey, Committee on Friendly Relations among Foreign Students, to S. Wirt Wiley, 13 June 1925, YMCA, International Committee Records, Reel 1, Folder 6, IHRC.

89. Memorandum from Mr. Hull to G. E. Iman, assistant commissioner general, 1 August 1925, RG 85, Box 436, File No. 55607/578, NARA.

90. Letter from Dr. Eugene Cohn to Sen. Chas. S. Deneen, 22 October 1925, RG 90, General Subject File, 1924–33, Box 942, File No. 0950-56, NARA.

91. Letter from Acting Commissioner Haff, San Francisco, to Commissioner General Husband, 2 January 1925, Doc. No. 12030/15-08 Personal, RG 85, Box 436, File No. 55607/578, NARA. Officials at Ellis Island objected to General Order 39, which established uniform shipboard exams, consistent with their general suspicion of the medical examinations conducted abroad. The commissioner of immigration at Ellis Island, Henry H. Curran, complained to the commissioner general, "We all well know the medical examination aboard ship is a farce." As an example, he related the story of

a man passed medically on the ship who, when taken to Ellis Island for another reason, was discovered to have tuberculosis. Nevertheless, Curran did not advocate that class be maintained as a determinant of the rigor of inspection: "I see no reason why an additional class of arriving aliens should be allowed to escape a thorough medical examination at Ellis Island. If we are to make any change as to this, it should be in the direction of taking more classes of aliens to Ellis Island, instead of allowing them to sift through the negligible medical examination to take place aboard ship. This comment goes directly to the point of protecting our country against incoming contagious diseases and merits most careful attention" (21 November 1924, Doc. No. 98524/167E, RG 85, Box 436, File No. 55607/578, NARA).

Index

circulatory diseases, 131
citizenship, 9–14, 155–56; industrial, 13–
14, 160–61, 173, 183, 200–203, 213–
14, 256, 275
classification of immigrants. *See* medical
certification categories; "typing" of immigrants
class issue, 123–31, 135, 165, 328nn. 7&9,
330n. 16, 344n. 125, 375n. 91; and race,
8–9, 137–39, 150, 154–59, 161, 170,
176–79, 181, 287n. 13. *See also* convergence of race, class, and disease
clonorchiasis (liver fluke), 181–82
Coe, Earl, 338n. 98
Cofer, L. E. (asst. surgeon general), 91, 120
Cohen, Lizabeth, 79, 254, 256, 367n. 13
Cohen, Rose, 86–87
Cohn, Dr. Eugene, 276
Conference Board on Safety and Sanitation, 96
convergence of race, class, and disease,
184–89
Craddock, Susan, 185, 188–89
culture, and performance on mental tests,
103–4
Cumming, Dr. Hugh (surgeon general),
133, 178–79
Cunard Line, 60–61
Curran, Henry H., 375n. 91
curvature of the spine, 108–9

Day, Benjamin, 197
Degler, Carl, 198
Denby, Charles, 292n. 24
Denechaud, Justin, 141–42
Deneen, Sen. Charles, 276
dependency, 32–37, 40, 50–52, 73, 109,
162, 207, 257–58. *See also* unemployment
deportation, 4–5, 40–41, 120, 209–15
deportation cases, 55, 352n. 72; Abdallah,
Josef, 40; Boden, Thomas P., 162–65;
Burke, Henry Ryan Kenny, 198–99;
Goldman, Sammy, 301n. 1; Herdenreder, Gotlieb, 55; Kinoshita, Komume, 190–93; Marron, Manuel, 176–
79; Nelkin, Samuel, 73; Saropian,

Nazaret, 37–39; Zatarian, Marian, 53–
55
detention, 53–55, 99, 357n. 1. *See also*
quarantine
diagnosis: of syphilis, 173–79; of tuberculosis, 162–72. *See also* laboratory techniques, use of; medical gaze
diagnostic technology, 100–101, 104–6,
171–72, 174–80, 182, 269–70, 349n.
40, 350nn. 44&51, 351n. 61
Diedrich, Henry, 61
digestive diseases, 130
disciplining of immigrants, 15–16, 24,
26–30, 51–52, 67–69, 85–86
disease, 40, 49–52, 90–91, 151, 161–65,
205–7, 264–65, 300n. 107; immigrants'
concealment of, 92–94, 96–98; metaphors of, 299n. 86. *See also* convergence of race, class, and disease; medical certification; *names of diseases*
disinfection procedures, 153–54, 302n. 8,
305n. 39, 312n. 114, 343n. 119, 344n.
129
Divine, Robert, 216–17
Dobler, Roman (PHS), 356n. 123
drive system, industrial, 255, 291n. 11
Dubofsky, Melvyn, 291n. 17
Dubose, H. G., 352n. 72
Dymowski, Robert F., 368n. 21

Educational Alliance, New York City, 51
Ellis Island, 1, 41–42, 68, 102–4, 114,
119–20, 130–31, 194–95, 338n. 97;
immigrant medical inspection at, 86–
88, 98–100, 111–12, 271–72, 274–75,
324n. 108, 325n. 111, 331n. 22
Ellis Island Oral History Project, 358n.
12. *See also* immigrant literature
environment, and disease, 264–65
ethnicity, 187, 256, 368nn. 16&21
ethnic solidarity, 255–56
Ettling, John, 180
eugenics, 162, 166–71, 199–201, 204,
206–7, 215–16, 348n. 29
Eugenics Society, 200
Europe, in racial categorization schemes,
218–20. *See also under* immigrants